Principles of Modern Chemistry:
The Molecular Science

John W. Moore

Conrad L. Stanitski

Peter C. Jurs

Prepared by

Judy L. Ozment
The Pennsylvania State University

BROOKS/COLE
CENGAGE Learning

Australia • Brazil • Japan • Korea • Mexico • Singapore • Spain • United Kingdom • United States

For product information and technology assistance, contact us at **Cengage Learning Customer & Sales Support, 1-800-354-9706**

For permission to use material from this text or product, submit all requests online at **www.cengage.com/permissions** Further permissions questions can be emailed to **permissionrequest@cengage.com**

ISBN-13: 978-0-495-39158-6
ISBN-10: 0-495-39158-1

Brooks/Cole
10 Davis Drive
Belmont, CA 94002-3098
USA

Cengage Learning is a leading provider of customized learning solutions with office locations around the globe, including Singapore, the United Kingdom, Australia, Mexico, Brazil, and Japan. Locate your local office at: **www.cengage.com/international**

Cengage Learning products are represented in Canada by Nelson Education, Ltd.

To learn more about Brooks/Cole, visit **www.cengage.com/brookscoles**

Purchase any of our products at your local college store or at our preferred online store **www.ichapters.com**

Printed in the United States of America
1 2 3 4 5 6 7 12 11 10 09

Table of Contents

Solutions to Blue-Numbered Questions

Introduction

This solutions manual was written specifically to accompany the first edition of the textbook *Principles of Chemistry – The Molecular Science* by John Moore, Conrad Stanitski, and Peter Jurs. It presents detailed solutions for some Questions for Review and Thought at the end of each chapter. If a question's number is blue, then its solution will be in this book.

Using this Book

Many of these solutions are presented using the same format described in Chapter 1 and Appendix A. The *Strategy and Explanation* section usually uses three of the four-stage process: define the problem, develop a plan, and execute the plan, then the *Reasonable Answer Check* shows a check of the answer. Following these stages should help to make the methods you learn more readily applicable to similar problems, or the same kind problem in a different context. Some of the solutions are patterned after the methods shown in Problem-Solving Examples throughout the text.

It is important to try to answer any question for yourself, before looking at this solutions book. When you find it necessary to use this book to get hints or directions, try first reading the first part of the *Strategy and Explanation* section to see if that information clarifies how you read the problem. If you find that your interpretation or evaluation of the problem is routinely incorrect or incomplete, then you might seek help from your instructor, a teaching assistant, a learning center staff person, or a tutor in reading word problems for comprehension. Sometimes, the best help for these kinds of problems can be gained from math tutors, since they often have experience helping students specifically with reading word problems.

If you find that you usually defined the problem in a similar fashion as described here, but still need help understanding how to solve the problem, read further to see how the plan is developed. This will give general step-by-step instructions for how the problem is solved. It is at this point where you should try to estimate what answer you anticipate. Do you expect it to be a large or small number? What units and sign do you expect it to have? What significant figures will it have? The more you are able to frame an expectation for the answer, the less likely you are to make mistakes along the way.

Step-by-step solutions are shown after the plan is described. If you can answer the questions and solve the problems before actually looking at these solutions, you will start gaining confidence in your ability to learn how to answer the questions on your own.

Once you have the answer to a question, take a moment to think about whether it makes sense. Often, simple reflection will help you confirm the correctness of an answer, or expose its flaws. For example, if you just determined that an atom of gold weighs four times more than the mass of the entire earth, the *Reasonable Answer Check* stage of the problem-solving method could help you see that you made an error. It is wise to go back to the question and read it again, then ask yourself: Does this result answer this question? Is the result the right size and sign? Is it what you expected? Check the units and the significant figures at this point, also.

A true sign that you are learning how to do chemistry problems is when you find yourself relying on this solutions book less and less. Set goals for yourself whenever you use this book to limit how often you consult it and how extensively. It is easy to use such books as a crutch, and crutches keep you from walking on your own.

There are many ways to solve problems. Often times, the right answer can be derived in several ways that are equally valid. Your instructors may have completely different ways of describing how to work some of the problems solved here. All good problem-solving methods have three things in common: (1) They always give the right answer. (2) They demonstrate how the answer was achieved. (3) And, they make sense. If any one of these three is missing, the method is flawed.

Cautions

Resist the temptation to just read the solutions in this book without trying any work on your own. There are two very obvious reasons for this: (1) Recognition is easier than recall. It is far easier to look over a solution done correctly and believe that you understand it, than it is to look at the same question followed by a blank space (as on a test, for example) and recall how to do it. (2) Most teachers will not let you use this book when they are testing you; hence, these solutions will not be there when you need to know how to do things.

All teachers using the textbook *Principles of Chemistry – The Molecular Science* also have a copy of the solutions in this book. You must resist the temptation to copy work from this book into your assignments. It is a violation of academic integrity to copy these solutions, then call it your work. Besides being unethical, that practice defeats the purpose of an assignment designed to have you practice problem-solving. The instructor is asking you to do your own work, and show what you have learned about the subject.

Learn How to Learn

Finally, keep in mind that you are learning basic chemistry and introductory physical science as building blocks for other things. Those things are much more complicated. Always try to learn a subject with maximum flexibility. Whenever possible, think about how a solution can be generalized. Relying on rote memorization and narrowly defined systems that designed to solve very specific types of problems may work to get you through this course, but it can be detrimental to your learning of science.

> "Learning without thought is labor lost. Thought without learning is perilous."
>
> — Confucius

Acknowledgements

I want to especially thank Karen Pesis, from the American River College in California, for her expert help in checking the extensive references in this book. I also want to thank Leslie Kinsland again, since much of what she helped me learn on the very first solutions manual I wrote is still a part of these newer ventures.

Lastly, I gratefully acknowledge the support of my family, especially Karen, Lynda Webb, Susan Thompson, and Loretta Ozment, and good friends, especially Sara Shriner, Paul Bomboy, Lori Campbell, Jared Lipton, and Carolyn Weber.

Chapter 1: The Nature of Chemistry

Introduction

Principles of Chemistry – The Molecular Science encourages you to think about chemistry at three different levels: the macroscopic, the particulate or nanoscale, and the symbolic.

Section 1.3 discusses how science is done, i.e., how observations lead to a hypothesis, how the hypothesis leads to more observations or experimentation, which in turn can lead to a law or theory. Science is not a step-by-step procedure—it's actually a process that people can use everyday to solve problems and understand the world around them.

Learning chemistry feels a lot like learning a foreign language. You'll see an extensive use of new terminology. You should try to associate terms and concepts with the root or origin of the terms. For example, in this chapter the terms nanoscale and microscale are introduced. Metric prefixes "micro-" and "nano-" are used in these terms for a reason. Learning the connections help you learn the patterns of the language, and reduce the quantity of rote memorization you do.

Whether cooperative learning activities are done during class time or not, it is important to get acquainted with the other students in the class. Exchange personal information, e.g., name, hometown, majors. Share your feelings about taking a college chemistry course. You'll find that others feel the same way that you do about taking chemistry.

Solutions to Blue-Numbered Questions for Review and Thought in Chapter 1

Topical Questions

How Science is Done

8. *Answer:* **(a) quantitative (b) qualitative (c) qualitative (d) quantitative and qualitative (e) qualitative**

   Strategy and Explanation:

   (a) The temperature at which an element melts (29.8 °C) is **quantitative** information. Information about what element it is (gallium) is **qualitative**.

   (b) The color of a compound (blue) and information about specific elements it contains (cobalt and chlorine) are both **qualitative**.

   (c) The fact that a metal conducts electricity and information about what element it is (aluminum) are **qualitative**.

   (d) The temperature at which a compound boils (79 °C) is **quantitative** information. Information about what compound it is (ethanol) is **qualitative**.

   (e) The details of the appearance of the crystals of a compound (shiny, plate-like, and yellow) and information about specific elements it contains (lead and sulfur) are both **qualitative**.

Identifying Matter: Physical Properties

10. *Answer/Explanation:* Bromine is a dark, reddish-brown liquid. Sulfur is a pale yellow powdery solid. Both the melting point and the boiling point of sulfur must be above room temperature. The boiling point, though not the melting point of bromine must be above room temperature. Both substances are colored. They appear to have no other properties in common. The physical phase, shape, color, and appearance are different.

12. *Answer:* **The liquid will boil because your body temperature of 37°C is above that boiling point of 20°C.**

Strategy and Explanation: Many Americans only remember the human body temperature in the Fahrenheit scale. That is 98.6 °F. If that is the case, we can quickly apply the °F to °C conversion equation, so we can compare it to the boiling point.

$$°C = \frac{5}{9} \times \left(°F - 32\right) = \frac{5}{9} \times \left(98.6 - 32\right) = 37.0°C$$

If the liquid boils at a temperature of 20 °C and your hand is 37 °C, the liquid will boil when exposed to the heat energy emitted by your hand when you hold the sample.

14. *Answer:* **copper**

Strategy and Explanation: We have the mass of the metal and some volume information. We need to determine the density. Use the initial and final volumes to find the volume of the metal piece, then use the mass and the volume to get the density. The metal piece displaces the water when it sinks, making the volume level in the graduated cylinder rise. The difference between the starting volume and the final volume, must be the volume of the metal piece:

$$V_{metal} = V_{final} - V_{initial} = (37.2 \text{ mL}) - (25.4 \text{ mL}) = 11.8 \text{ mL}$$

$$d = \frac{m}{V} = \frac{105.5 \text{ g}}{11.8 \text{ mL}} = 8.94 \ \frac{g}{mL}$$

According to Table 1.1, this is very close to the density of copper (d = 8.93 g/mL).

✓ *Reasonable Answer Check:* The metal piece sinks, so the density of the metal piece needed to be higher than water. (Table 1.1 gives water density as 0.998 g/mL.)

16. *Answer:* **aluminum**

Strategy and Explanation: We have the three linear dimensions of a regularly shaped piece of metal.

We also have its mass. We have a table of densities (Table 1.1). We need to determine the identity of the metal.

Use the three linear dimensions to find the volume of the metal piece. Use the volume and the mass to find the density. Use the table of densities to find the identity of the metal.

$$V = (\text{thickness}) \times (\text{width}) \times (\text{length}) = (1.0 \text{ cm}) \times (2.0 \text{ cm}) \times (10.0 \text{ cm}) = 20. \text{ cm}^3$$

Using dimensional analysis, find the volume in mL:

$$20. \text{ cm}^3 \times \frac{1 \text{ mL}}{1 \text{ cm}^3} = 20. \text{ mL}$$

Find the density:

$$d = \frac{m}{V} = \frac{54.0 \text{ g}}{20. \text{ mL}} = 2.7 \ \frac{g}{mL}$$

According to Table 1.1, the density that most closely matches this one is **aluminum** (d = 2.70 g/mL).

✓ *Reasonable Answer Check:* An object whose mass is larger than its volume will have a density larger than one. The mass of this object is between two and three times larger than the volume.

Chemical Changes and Chemical Properties

18. *Answer:* **(a) physical (b) chemical (c) chemical (d) physical**

Strategy and Explanation:

(a) The normal color of bromine is a **physical** property. Determining the color of a substance does not change its chemical form.

(b) The fact that iron can be transformed into rust is a **chemical** property. Iron, originally in the elemental metallic state, is incorporated into a compound, rust, when it is observed to undergo this transformation.

(c) The fact that dynamite can explode is a **chemical** property. The dynamite is chemically changed when it is observed to explode.

(d) Observing the shininess of aluminum does not change it, so this is a **physical** property. Melting aluminum does not change it to a different substance, though it does change its physical state. It is still aluminum, so melting at 387 °C is a **physical** property of aluminum.

20. *Answer:* **(a) chemical (b) chemical (c) physical**

Strategy and Explanation:

(a) Bleaching clothes from purple to pink is a **chemical** change. The purple substance in the clothing reacts with the bleach to make a pink substance. The purple color cannot be brought back nor can the bleach.

(b) The burning of fuel in the space shuttle (hydrogen and oxygen) to form water and energy is a **chemical** change. The two elements react to form a compound.

(c) The ice cube melting in the lemonade is a **physical** change. The H_2O molecules do not change to a different form in the physical state change.

Classifying Matter: Substances and Mixtures

22. *Answer/Explanation:* It is clear by visual inspection that the mixture is non-uniform (**heterogeneous**) at the macroscopic level. Iron could be separated from sand **using a magnet**, since iron is attracted to magnets and the sand is not.

24. *Answer:* **(a) homogenous (b) heterogeneous (c) heterogeneous (d) heterogeneous**

Strategy and Explanation: The terminology "heterogeneous" and "homogeneous" is somewhat subjective. In Section 1.6, the terms are described. A heterogeneous mixture is described as one whose uneven texture can be seen without magnification or with a microscope. A homogeneous mixture is defined as one that is completely uniform, wherein no amount of optical magnification will reveal variations in the mixture's properties. Notice the "gap" between these two terms. The question: "how close do I look?" comes to mind. The bottom line is this: These terms were designed to HELP us classify things, not to create trick questions. If you explain your answer with a valid defense, the answer ought to be right; however, don't go out of your way to imagine unusual circumstances that make the substance more difficult to classify. Think about what you would SEE. Identify whether what you see has variations, then make a case for the proper term.

(a) Vodka is classified as a **homogenous** mixture. It is a clear, colorless solution of alcohol, water, and probably some other minor ingredients.

(b) Blood appears smooth in texture and to our eyes most likely appears to be homogeneous. Upon closer examination it is found to have various particles within the liquid, and might, for that reason, be called **heterogeneous**.

(c) Cowhide is **heterogeneous**. Even folks who have never seen a cowhide might be able to imagine that brown cows, white cows, black cows and spotted cows probably have different coloration to their hide. There are probably pores where the hair grows out, making it rough in texture. Unless presented with a sample that visually changed our perception of what constitutes a cowhide, it is safe to call it heterogeneous.

(d) Bread is **heterogeneous**. The crust is a different color. Some parts of the bread have bigger bubbles than other parts. Some breads have whole grains in them. Some breads are composed of different colors (like the rye/white swirl breads). In general, most samples of bread have identifiable regions that are different from other regions.

Classifying Matter: Elements and Compounds

26. *Answer/Explanation:*

(a) A blue powder turns white and loses mass. The loss of mass is most likely due to the creation of a gaseous product. That suggests that the original material was a **compound that decomposed** into the white substance (a compound or an element) and a gas (a compound or an element).

(b) If three different gases were formed, that suggests that the original material was a **compound that decomposed** into three compounds or elements.

28. *Answer:* **(a) heterogeneous mixture (b) pure compound (c) heterogeneous mixture (d) homogeneous mixture**

Strategy and Explanation:

(a) A piece of newspaper is a **heterogeneous mixture**. Paper and ink are distributed in a non-uniform fashion at the macroscopic level.

(b) Solid granulated sugar is a **pure compound**. Sugar is made of two or more elements.

(c) Fresh squeezed orange juice is a **heterogeneous mixture**. The presence of unfiltered pulp makes part of the mixture solid and part of the mixture liquid.

(d) Gold jewelry is a **homogeneous mixture**, most of the time. Most jewelry is less than 24 carat – pure gold. If it is 18 carat, 14 carat, 10 carat gold, that means that other metals are mixed in with the gold (usually to make it more durable and cheaper). The combining of metals in a fashion that prevents us from seeing variations in the texture or properties of the metal qualifies it as a homogeneous mixture.

30. *Answer:* **(a) No (b) Maybe**

Strategy and Explanation:

(a) The black substance was both the source of the element that contributes to the red-orange substance and the source of the oxygen in the water.

(b) The red-orange substance may be a combination of two or more elements including possibly hydrogen or oxygen, or it maybe an elemental substance, since the water produced could account for the fate of the hydrogen and the oxygen.

Nanoscale Theories and Models

32. *Answer:* **The macroscopic world; a parallelpiped shape; the atom crystal arrangement is parallelpiped shape.**

Strategy and Explanation: The crystal of halite pictured is in the macroscopic world. Its shape is cubic or parallelepiped. It could be expected that the arrangement of particles (atoms and/or ions) in the nanoscale world are also arranged in a cubic or parallelepiped fashion.

34. *Answer/Explanation:* When we open a can of a soft drink, the carbon dioxide gas expands rapidly as it rushes out of the can. At the nanoscale, this can be explained as large number of **carbon dioxide molecules crowded into the unopened can**. When the can is opened, the molecules that were about to hit the surface where the hole was made continue forward through the hole. A large number of the carbon dioxide particles that were contained within the can very quickly **escape through the** same **hole**.

36. *Answer/Explanation:* The atoms in the solid sucrose molecules start off at a relatively low energy and compose a rather complex molecule. A significant amount of heat energy must be added to increase the motion of these atoms so that they are able to break free of the bonds that hold them together in the sugar molecule and to interact with each other to make the "caramelization" products.

38. *Answer:* **(a) 3.275×10^4 m (b) 3.42×10^4 nm (c) 1.21×10^{-3} μm**

Strategy and Explanation:

(a)
$$32.75 \text{ km} \times \frac{1000 \text{ m}}{1 \text{ km}} = 3.275 \times 10^4 \text{ meters}$$

(b)
$$0.0342 \text{ mm} \times \frac{10^{-3} \text{ m}}{1 \text{ mm}} \times \frac{1 \text{ nm}}{10^{-9} \text{ m}} = 3.42 \times 10^4 \text{ nanometers}$$

(c)
$$1.21 \times 10^{-12} \text{ km} \times \frac{10^3 \text{ m}}{1 \text{ km}} \times \frac{1 \text{ μm}}{10^{-6} \text{ m}} = 1.21 \times 10^{-3} \text{ micrometers}$$

The Atomic Theory

40. *Answer/Explanation:* Conservation of mass is easy to see from the point of view of atomic theory. A chemical change is described as the rearrangement of atoms. Because the atoms in the starting materials must all be accounted for in the substances produced, and because the mass of each atom does not change, there would be no change in the mass.

42. *Answer/Explanation:* The law of multiple proportions can be explained using atomic theory. Consider two compounds that both contain the same two elements. The proportion of the two elements in these two compounds must be different. Compounds are composed of a specific number and type of atoms. If you pick samples of each of these compounds such that they contain the same number of atoms of the first element, then you'll find that the ratio of atoms on the other element is a small integer. Consider a concrete example. Fe_2O_3

and Fe_3O_4. Compare a sample containing three Fe_2O_3 to a sample containing two Fe_3O_4. These samples both have six Fe atoms. The first sample has nine oxygen atoms and the second sample has 8 oxygen atoms. Since all oxygen atoms weigh the same, the ratio of mass will also be the ratio of atoms; in this example, the ratio of mass and the ratio of atoms is $\frac{9}{8}$.

The Chemical Elements

44. *Answer/Explanation:* Many responses are equally valid here. Below are a few common examples. These lists are not comprehensive; many other answers are also right. The periodic table on the inside cover of your text is color coded to indicate metals, non-metals and metalloids.

    (a) Common metallic elements: iron, Fe; gold, Au; lead, Pb; copper, Cu; aluminum, Al

    (b) Common non-metallic elements: carbon, C; hydrogen, H; oxygen, O; nitrogen, N

    (c) Metalloids: boron, B; silicon, Si; germanium, Ge; arsenic, As; antimony, Sb; tellurium, Te

    (d) Elements that are diatomic molecules: nitrogen, N_2; oxygen, O_2; hydrogen, H_2; fluorine, F_2; chlorine, Cl_2; bromine, Br_2; iodine, I_2

Communicating Chemistry: Symbolism

46. *Answer/Explanation:* Formula for each substance and nanoscale picture:

    (a) Water H_2O (b) Nitrogen N_2

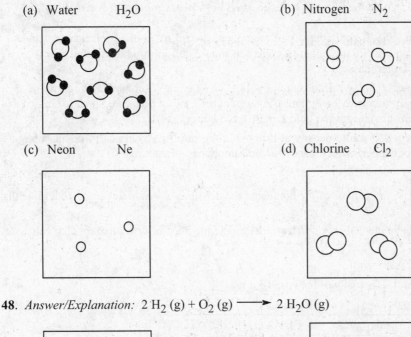

    (c) Neon   Ne   (d) Chlorine Cl_2

48. *Answer/Explanation:* $2\,H_2\,(g) + O_2\,(g) \longrightarrow 2\,H_2O\,(g)$

Hydrogen and oxygen gas   water vapor

General Questions

50. *Answer/Explanation:*

    (a) The mass of the compound (1.456 grams) is quantitative and relates to a physical property. The color (white), the fact that it reacts with a dye, and the color change in the dye (red to colorless) are all qualitative. The colors are related to physical properties. The reaction with the dye is related to a chemical property.

(b)   The mass of the metal (0.6 grams) is quantitative and related to a physical property. The identity of the metal (lithium) and the identities of the chemicals it reacts with and produces (water, lithium hydroxide, and hydrogen) are all qualitative information. The fact that a chemical reaction occurs when the metal is added to water is qualitative information and related to a chemical property.

52.   *Answer/Explanation:* If the density of solid calcium is almost twice that of solid potassium, but their masses are approximately the same size, then the volume must account for the difference. This suggests that the atoms of calcium are smaller than the atoms of potassium:

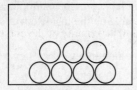   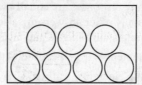

solid calcium	solid potassium
smaller atoms	larger atoms
closer packed	less closely packed
smaller volume	larger volume

Applying Concepts

54.   *Answer:* **(a) bromobenzene sample (b) gold sample (c) lead sample**

Strategy and Explanation:

(a)   (Table 1.1) density of butane = 0.579 g/mL; density of bromobenzene = 1.49 g/mL 1 mL butane weighs less than 1 mL of bromobenzene so, 20 mL butane weighs less than 20 mL of bromobenzene. The bromobenzene sample has a larger mass.

(b)   (Table 1.1) density of benzene = 0.880 g/mL; density of gold = 19.32 g/mL There are 0.880 grams of benzene in 1 mL of benzene, so there are 8.80 grams of benzene in 10 mL of benzene. Since 1.0 mL of gold has a mass of 19.32 grams that means the gold sample has a larger mass.

(c)   (Table 1.1) density of copper = 8.93 g/mL; density of lead = 11.34 g/mL Any volume of lead has a larger mass than the same volume of copper. That means the lead sample has a larger mass.

56.   (a)   *Answer:* **2.7×10^2 mL ice**

Strategy and Explanation: Use the volume of the bottle and the densities of water and ice to determine the volume of ice formed from a fixed amount of water.

Use the volume of the bottle and the density of water to determine the mass of water frozen, then calculate the volume of the ice.

At 25 °C, density of water is 0.997 g/mL.

At 0 °C, density of ice = 0.917 g/mL

$$250 \text{ mL water} \times \frac{0.997 \text{ g H}_2\text{O}(\ell)}{1 \text{ mL water}} \times \frac{1 \text{ g H}_2\text{O}(s)}{1 \text{ g H}_2\text{O}(\ell)} \times \frac{1 \text{ mL ice}}{0.917 \text{ g H}_2\text{O}(s)} = 2.7 \times 10^2 \text{ mL ice}$$

✓ *Reasonable Answer Check:* The density of water is larger than the density of ice. It makes sense that the volume of the ice produced is larger than the volume of water.

(b)   If the bottle is made of flexible plastic, it might be deformed and bulging if not cracked and leaking ice. If the bottle is made of glass and the top came off, there would be ice (approximately 20 mL of it) oozing out of the top. Worst case scenario: if the bottle was glass and the top did not come off, it would be broken.

58.   *Answer/Explanation:*

(a)   (Table 1.1) density of water = 0.998 g/mL, density of bromobenzene = 1.49 g/mL. Since water does not dissolve in bromobenzene, the lower-density water will be the top layer of the immiscible layers.

(b)   If poured slowly and carefully, the ethanol will float on top of the water and slowly dissolve in the water. Both ethanol and water will float on the bromobenzene.

(c)   Stirring will speed up the ethanol dissolving with the water to make one phase. Assuming the new mixture has the average density of the original liquids, the water/ethanol layer (average density is 0.894 g/mL) will sit on top of the bromobenzene layer. (density is 1.49 g/mL)

60. *Answer:* **Drawing (b)**

Strategy and Explanation: The 90 °C mercury atoms will be a little bit further apart and moving somewhat more than the 10 °C mercury, though they still would be the same size atoms. The individual atoms in (c) are bigger – that doesn't happen. The individual atoms in (d) are smaller – that doesn't happen, either.

More Challenging Questions

62. *Answer:* 6.02×10^{-29} **m^3**

Strategy and Explanation: We have the length of the edge of a cube. Find the volume of the cube in m^3:

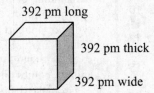

392 pm long

392 pm thick

392 pm wide

Use the linear dimensions to find the volume, then convert the volume to m^3 using metric conversions.

$$V = (\text{thickness}) \times (\text{width}) \times (\text{length}) = (392 \text{ pm}) \times (392 \text{ pm}) \times (392 \text{ pm}) = 6.02 \times 10^7 \text{ pm}^3$$

Using dimensional analysis, find the volume in m^3:

$$6.02 \times 10^7 \text{ pm}^3 \times \left(\frac{10^{-12} \text{ m}}{1 \text{ pm}} \right)^3 = 6.02 \times 10^{-29} \text{ m}^3$$

✓ *Reasonable Answer Check:* The cube is from the nanoscale, so it makes sense that it would be a very small volume using a macroscale unit of measure.

64. *Answer/Explanation:* Look at the periodic table.

(a) Metals are in the **gray and blue areas**.

(b) Nonmetals are in the **lavender area**.

(c) Metalloids are in the **orange area**.

66. *Answer/Explanation:* A substance that can be broken down is not an element. A series of tests will result in a confirmation with one positive test. To prove that something is an element requires a battery of tests that all have negative results. A hypothesis that the substance is an element and cannot be broken down is more difficult to prove.

68. *Answer:* **(a) nickel, lead and magnesium (b) titanium**

Strategy and Explanation:

(a) According to the table of densities, a metal will float if the density is lower. That means that nickel, lead and magnesium will float on liquid mercury.

(b) The more different the densities, the smaller the fraction of the floating element will be below the surface. That means that titanium will float highest on mercury.

70. *Answer/Explanation:* To do the experiment, obtain four or more lemons. Do nothing to one of them, to serve as the "control" case. Perform the designated tasks to the remaining lemons then juice all three lemons recording the results: juice volume and/or noticeable ease of squeeze or ease of juice delivery. Repeat several times to achieve better reliability. Hypothesis: disrupting "juice sacks" inside the pulp of the lemon will release juice.

Chapter 2: Atoms and Elements

Introduction

New terminology continues to be introduced in Chapter 2, so pay careful attention to distinguishing between one term and another.

Many students have simplistic and superficial understanding of significant figures and uncertainty. Proper attention to uncertainty and significant figures is a very important part of understanding the limitations of the data being used. Rough estimates cannot result in answers with huge precision. Just because your calculator comes up with answers that have many digits does not make them significant digits.

The mole concept (Section 2.7) is one of the most important concepts in a first course of introductory chemistry. Pay close attention and ask questions if you find it confusing you. A "mole" is just word representing a fixed numerical quantity. What makes the "mole" difficult to learn is that Avogadro's number is so huge that it's hard to imagine such a large number of things.

Solutions to Blue-Numbered Questions
for Review and Thought in Chapter 2

Topical Questions

Units and Unit Conversions

5. *Answer:* **40,000 cm**

    Strategy and Explanation: If the nucleus is scaled to a diameter of 4 cm, determine the diameter of the atom.

    Find the accepted relationship between the size of the nucleus and the size of the atom. Use size relationships to get the diameter of the "artificially large" atom.

    The atom is about 10,000 times bigger than the nucleus.

$$10,000 \times 4 \text{ cm} = 40,000 \text{ cm}$$

    ✓ *Reasonable Answer Check:* A much larger nucleus means a much larger atom with a large atomic diameter.

8. *Answer:* $\mathbf{1.97 \times 10^3 \text{ cm}^3, 1.97 \text{ L}}$

    Strategy and Explanation: Given displacement volume of 120. in^3, determine the volume in cm^3 and liters.

    Use conversion factor between inches and centimeters to change the units from cubic inches to cubic centimeters. Then use conversion factors between cubic centimeters and milliliters, then milliliters and liters to change the units from cubic centimeters to liters.

    NOTICE: Each of the "inch" units in "cubic inches" must be converted to centimeters. That is why the length conversion factor must cubed to obtain the volume conversion factor:

$$120. \text{ in}^3 \times \frac{2.54 \text{ cm}}{1 \text{ in}} \times \frac{2.54 \text{ cm}}{1 \text{ in}} \times \frac{2.54 \text{ cm}}{1 \text{ in}} = 120. \text{ in}^3 \times \left(\frac{2.54 \text{ cm}}{1 \text{ in}}\right)^3 = 1.97 \times 10^3 \text{ cm}^3$$

$$1.97 \times 10^3 \text{ cm}^3 \times \frac{1 \text{ mL}}{1 \text{ cm}^3} \times \frac{1 \text{ L}}{1000 \text{ mL}} = 1.97 \text{ L}$$

    Notice: The metric definition of "milli-" applied to liters is: $10^{-3} L = 1 mL$. *The 1:1000 ratio used above is a very common variant that appropriately indicates a larger number (1000) of small things (mL) equal to a smaller number (1) of large things (L). Of course, it is perfectly acceptable, but not as common in practice, to use the metric definition directly as the conversion factor to get the same answer:*

$$1.97 \times 10^3 \ cm^3 \times \frac{1 \ mL}{1 \ cm^3} \times \frac{10^{-3} \ L}{1 \ mL} = 1.97 \ L$$

✓ *Reasonable Answer Check:* Centimeters are smaller than inches, so a cubic centimeter is much smaller than a cubic inch, so the number of cubic centimeters should be larger than the number of cubic inches. A liter is larger than a cubic centimeter, so the number of liters should be smaller than the number of cubic inches.

9. *Answer:* **1550 in^2**

Strategy and Explanation: Given one square meter, determine the number of square inches.

Use metric conversion between meters and centimeters, then the relationship between centimeters and inches.

$$1 \ m^2 \times \left(\frac{100 \ cm}{1 \ m} \right)^2 \times \left(\frac{1 \ in}{2.54 \ cm} \right)^2 = 1550 \ in^2$$

Notice: The significant figures are somewhat ambiguous, since the word "one" might be interpreted as 1, which has one significant figure. If that is the case, the answer would be 2000 in^2.

✓ *Reasonable Answer Check:* A meter is larger than an inch, so the number of square inches should be larger than the number of square meters.

11. *Answer:* **2.8 × 10^9 m^3**

Strategy and Explanation: Given a volume in cubic miles, determine the number of cubic meters.

Use the relationship between miles and meters.

$$0.67 \ miles^3 \times \left(\frac{1000 \ m}{0.62137 \ mile} \right)^3 = 2.8 \times 10^9 \ m^3$$

✓ *Reasonable Answer Check:* A meter is smaller than a mile, so the number of cubic meters should be larger than the number of cubic miles.

Significant Figures

13. *Answer:* **(a) four (b) three (c) four (d) four (e) three**

Strategy and Explanation: Given several measured quantities, determine the number of significant figures.

Use rules given in Section 2.4, summarized here: All non-zeros are significant. Zeros that precede (sit to the left of) non-zeros are never significant (e.g., 0.003). Zeros trapped between non-zeros are always significant (e.g., 3.003). Zeros that follow (sit to the right of non-zeros are (a) significant if a decimal point is explicitly given (e.g., 3300) OR (b) not significant, if a decimal point is not specified (e.g., 3300.).

(a) 1374 kg has **four** significant figures (The 1, 3, 7, and 4 digits are each significant.)

(b) 0.00348 s has **three** significant figures (The 3, 4, and 8 digits are each significant. The zeros are all before the first non-zero-digit 3 and therefore they are not significant.)

(c) 5.619 mm has **four** significant figures (The 5, 6, 1, and 9 digits are each significant.)

(d) 2.475 × 10^{-3} cm has **four** significant figures (The 2, 4, 7, and 5 digits are each significant.)

(e) 33.1 mL has **three** significant figures (The 3, 3, and 1 digits are each significant.)

✓ *Reasonable Answer Check:* Only one answer had zeros in it, and those (in (b)) were to the left of the first non-zero digit, so none of the zeros here were significant.

15. *Answer:* **(a) 4.33 × 10^{-4} (b) 4.47 × 10^1 (c) 2.25 × 10^1 (d) 8.84 × 10^{-3}**

Strategy and Explanation: Given several measured quantities, round them to three significant figures and write them in scientific notation. Use rules for rounding given in Section 2.4. If the last digit is below 5, then rounding does not change the digit before it. If the last digit is above a five, the digit before it is made one larger. If the last digit is exactly five, round the digit before it to an even number, up if odd and down if even. After rounding, adjust the appearance of the number to scientific notation (i.e., to a number between 1 and 9.999... that is multiplied by ten to a whole-number power.)

(a) 0.0004332 has four significant figures. To round it to three significant figures, we need to remove the fourth significant figure, the 2. Since 2 is below 5, the result is 0.000433. Putting this value in scientific notation, we get **4.33 × 10^{-4}**.

(b) 44.7337 has six significant figures. To round it to three significant figures, we need to remove the fourth, fifth and sixth significant figures. The removal of the last digit 7 rounds the 3 next to it up to a 4, but the removal of 4 doesn't change the 3 next to it, and the removal of 3 doesn't change the 7 next to it, the result is 44.7. Putting this value in scientific notation, we get **4.47 × 10¹**.

(c) 22.4555 has six significant figures. To round it to three significant figures, we need to remove the fourth, fifth and sixth significant figures. Since the removal of the last 5 rounds the 5 next to it up to a 6, and the removal of 6 rounds the 5 next to it up to a 6, and the removal of 6 rounds the 4 next to it to a 5, the result is 22.5. Putting this value in scientific notation, we get **2.25 × 10¹**.

(d) 0.0088418 has five significant figures. To round it to three significant figures, we need to remove the fourth and fifth significant figures. Since the removal of the last 8 rounds the 1 next to it up to a 2, and the removal of 2 doesn't change the 4 next to it, the result is 0.00884. Putting this value in scientific notation, we get **8.84 × 10⁻³**.

✓ *Reasonable Answer Check:* Small numbers are still small, large numbers are still large, what remains after the rounding is the larger part of the value. Each answer has three significant figures and each number is represented in proper scientific notation.

17. *Answer:* **(a) 1.9 g/mL (b) 218.4 cm³ (c) 0.0217 (d) 5.21 × 10⁻⁵**

Strategy and Explanation: Given some numbers combined using calculations, determine the result with proper significant figures. Perform the mathematical steps according to order of operations, applying the proper significant figures (addition and subtraction retains the least number of decimal places in the result; multiplication and division retain the least number of significant figures in the result). *Notice: if operations are combined that use different rules, it is important to stop and determine the intermediate result any time the rule switches.*

(a)
$$\frac{4.850 \text{ g} - 2.34 \text{ g}}{1.3 \text{ mL}}$$

The numerator uses the subtraction rule. The first number has three decimal places (the 8, the 5, and the 0 are all decimal places – digits that follow the decimal point to the right) and the second number has two decimal places (the numerals 3 and the 4 are both decimal places), so the result of the subtraction has two decimal places.

$$\frac{2.51 \text{ g}}{1.3 \text{ mL}}$$

The ratio uses the division rule. The numerator has three significant figures and the denominator has two significant figures, so the answer will have two significant figures. Therefore, we get **1.9 g/mL**.

(b) $$V = \pi r^3 = (3.1415926) \times (4.112 \text{ cm})^3$$

This whole calculation uses the multiplication rule, with four significant figures, limited by the measurement of r. The value of π should be carried to more than four significant figures, such as 3.14159… The answer comes out **218.4 cm³**.

(c) $$(4.66 \times 10^{-3}) \times 343.2$$

This calculation uses the multiplication rule. The first number, 4.66×10^{-3}, has three significant figures and the second number, 343.2, has four significant figures, so the answer has three significant figures **0.0217**.

(d)
$$\frac{0.003400}{65.2}$$

This calculation uses the division rule. The numerator has four significant figures and the denominator has three significant figures, so the answer has three significant figures 0.0000521 or **5.21 × 10⁻⁵**.

✓ *Reasonable Answer Check:* The proper significant figures rules were used. The size and units of the answers are appropriate.

Percent

19. *Answer:* **80.1% silver and 19.9% copper**

Strategy and Explanation: Given a 17.6-gram bracelet that contains 14.1 grams of silver and the rest copper, determine the percentage silver and the percentage copper.

Determine the percentage of silver by dividing the mass of silver by the total mass and multiplying by 100%. Since the metal is made up of only silver and copper, determine the percentage of copper by subtracting the percentage of silver from 100%.

$$\frac{14.1 \text{ g silver}}{17.6 \text{ g bracelet}} \times 100\% = 80.1\% \text{ silver}$$

$$100\% \text{ total} - 80.1\% \text{ silver} = 19.9\% \text{ copper}$$

Notice: When you are using grams of different substances, be careful to carry enough information in the units so you don't confuse one mass with another.

✓ *Reasonable Answer Check:* A significant majority of the metal in the bracelet is silver, so it makes sense that the percentage of silver is larger than the percentage of copper.

21. *Answer:* **245 g sulfuric acid**

Strategy and Explanation: Given the volume of a battery acid sample, the density and the percentage of sulfuric acid in the battery acid by mass, determine the mass of acid in the battery.

Always start with the sample. Use the density of the solution to create a conversion factor between milliliters and grams, so you can determine the mass of the battery acid in grams. Then use the mass percentage of sulfuric acid in the battery acid as a conversion factor to determine the mass of sulfuric acid in the sample.

Every 1.000 mL of battery acid solution weighs 1.285 grams. Every 100.00 g of battery acid solution contains 38.08 grams of sulfuric acid.

$$500. \text{ mL solution} \times \frac{1.285 \text{ g solution}}{1.000 \text{ mL solution}} \times \frac{38.08 \text{ g sulfuric acid}}{100.00 \text{ g solution}} = 245 \text{ g sulfuric acid}$$

Notice: When you are using grams of different substances, be careful to carry enough information in the units so you don't confuse one mass with another.

✓ *Reasonable Answer Check:* Only about a third of the sample is sulfuric acid, so the mass of sulfuric acid should be smaller than the volume of the solution.

Isotopes

25. *Answer:* **27 protons, 27 electrons, and 33 neutrons**

Strategy and Explanation: Given the identity of an element (cobalt) and the atom's mass number (60), find the number of electrons, protons, and neutrons in the atom.

Look up the symbol for cobalt and find that symbol on the periodic table. The periodic table gives the atomic number. The atomic number is the number of protons. The number of electrons is equal to the number of protons since the atom has no charge. The number of neutrons is the difference between the mass number and the atomic number.

The element technetium has the symbol Co. On the periodic table, we find it listed with the atomic number 27. So, the atom has 27 protons, 27 electrons and (60 – 27 =) 33 neutrons.

✓ *Reasonable Answer Check:* The number protons and electrons must be the same (27=27). The sum of the protons and neutrons is the mass number (27 + 33 = 60). This is correct.

27. *Answer:* **78.92 amu/atom**

Strategy and Explanation: Given the average atomic weight of an element and the percentage abundance of one isotope, determine the atomic weight of the only other isotope.

Using the fact that the sum of the percents must be 100%, determine the percent abundance of the second isotope. Knowing that the weighted average of the isotope masses must be equal to the reported atomic weight, set up a relationship between the known atomic mass and the various isotope masses using a variable to describing the second isotope's atomic weight.

We are told that natural bromine is 49.31% ^{81}Br and that there are only two isotopes. To calculate the percentage abundance of the other isotope, subtract from 100%:

$$100.0\% - 49.31\% = 50.69\%$$

These percentages tell us that every 10000 atoms of bromine contains 4931 atoms of the ^{81}Br isotope and 5069 atoms of the other bromine isotope *(limited to 4 sig figs)*. The atomic weight for Br is given as 79.904 amu/atom. Table 2.2 in Section 2.6 gives the isotopic mass of ^{81}Br isotope as 80.916289 amu/atom. Let X be the atomic mass of the other isotope of bromine.

$$\frac{4931 \text{ atoms } ^{81}\text{Br}}{10000 \text{ Br atoms}} \times \left(\frac{80.916289 \text{ amu}}{1 \text{ atom } ^{81}\text{Br}}\right) + \frac{5069 \text{ atoms other isotope}}{10000 \text{ Br atoms}} \times \left(X \frac{\text{amu}}{\text{atom}}\right) = 79.904 \frac{\text{amu}}{\text{Br atom}}$$

Solve for X

$$39.90 + 0.5069 \, X = 79.904$$

$$X = 78.92 \text{ amu/atom} \quad \textit{(limited to 4 sig figs)}$$

✓ *Reasonable Answer Check:* Table 2.2 in Section 2.6 gives the atomic weight of ^{79}Br to be 78.918336, which is the same as that given in Table 2.2 for ^{79}Br, within the permissible significant figures.

29. *Answer:* **(a) 20 e⁻, 20 p⁺, 20 n⁰   (b) 50 e⁻, 50 p⁺, 69 n⁰   (c) 94 e⁻, 94 p⁺, 150 n⁰**

Strategy and Explanation: Given the atomic symbol $^{A}_{Z}X$ of the isotope, determine the number of electrons, protons, and neutrons. The atomic number (Z) represents the number of protons. In neutral atoms, the number of electrons is equal to the number of protons. To get the number of neutrons, subtract the number of protons from the mass number (A).

(a) The isotope given is $^{40}_{20}$Ca . That means A = 40 and Z = 20. So, the number of protons is 20, the number of electrons is 20, and the number of neutrons is (40 – 20 =) 20.

(b) The isotope given is $^{119}_{50}$Sn. That means A = 119 and Z = 50. So, the number of protons is 50, the number of electrons is 50, and the number of neutrons is (119 – 50 =) 69.

(c) The isotope given is $^{244}_{94}$Pu. That means A = 244 and Z = 94. So, the number of protons is 94, the number of electrons is 94, and the number of neutrons is (244 – 94 =) 150.

✓ *Reasonable Answer Check:* The number of protons and electrons must be equal in neutral atoms. The mass number must be the sum of the protons and neutrons.

31. *Answer:*

Z	A	Number of Neutrons	Element
35	81	**46**	**Br**
46	**108**	62	Pd
77	**192**	115	**Ir**
63	151	**88**	Eu

Strategy and Explanation: Fill in an incomplete table with Z, A, number of neutrons and element identity.

The atomic number (Z) represents the number of protons. The mass number (A) is the number of neutrons and protons. The element's identity can be determined using the periodic table by looking up the atomic number and getting the symbol.

Z	A	Number of Neutrons	Element
35	81	(a)	(b)
(c)	(d)	62	Pd
77	(e)	115	(f)
(g)	151	(h)	Eu

(a) Number of neutrons = A – Z = 81 – 35 = 46

(b) Z = 35. Look up element #35 on periodic table: Element = Br

(c) Look up Pd on the periodic table: Z = 46

(d) A = Z + number of neutrons = 46 + 62 = 108

(e) A = Z + number of neutrons = 77 + 115 = 192

(f) Z = 77. Look up element #77 on periodic table: Element = Ir

(g) Look up Eu on the periodic table: Z = 63

(h) Number of neutrons = A – Z = 151 – 63 = 88

✓ *Reasonable Answer Check:* The atomic number and the symbol must match what is shown on the periodic table. The mass number must be the sum of the atomic number and the number of neutrons.

33. *Answer:* $^{18}_{9}$X, $^{20}_{9}$X, and $^{15}_{9}$X

Strategy and Explanation: The atomic symbol general form is $^{A}_{Z}X$, where A is the mass number and Z is the atomic number. Isotopes of the same element will have the same Z, but A will be different. So, $^{18}_{9}$X, $^{20}_{9}$X, and $^{15}_{9}$X are all isotopes of the same element with an atomic number of 9.

Atomic Weight

35. *Answer:* **(0.07500 × 6.015121 amu/atom ^{6}Li) + (0.9250 × 7.016003 amu/atom ^{7}Li) = 6.941 amu/atom Li**

    Strategy and Explanation: Using the exact mass and the percent abundance of several isotopes of an element, determine the atomic weight.

    Calculate the weighted average of the isotope masses.

    Every 1000 atoms of lithium contains 75.00 atoms of the ^{6}Li isotope and 925.0 atoms of the ^{7}Li isotope.

    $$\frac{75.00 \text{ atoms } ^6\text{Li}}{1000 \text{ Li atoms}} \times \left(\frac{6.015121 \text{ amu}}{1 \text{ atom } ^6\text{Li}}\right) + \frac{925.0 \text{ atoms } ^7\text{Li}}{1000 \text{ Li atoms}} \times \left(\frac{7.016003 \text{ amu}}{1 \text{ atom } ^7\text{Li}}\right) = 6.941 \text{ amu/Li atom}$$

    ✓ *Reasonable Answer Check:* The periodic table value for atomic weight is the same as calculated here.

37. *Answer:* **60.12% ^{69}Ga, 39.87% ^{71}Ga**

    Strategy and Explanation: Using the exact mass of several isotopes and the atomic weight, determine the abundance of the isotopes.

    Establish variables describing the isotope percentages. Set up two relationships between these variables. The sum of the percents must be 100%, and the weighted average of the isotope masses must be the reported atomic mass. X% ^{69}Ga and Y% ^{71}Ga. This means: Every 100 atoms of gallium contains X atoms of the ^{69}Ga isotope and Y atoms of the ^{71}Ga isotope.

    $$\frac{\text{X atoms } ^{69}\text{Ga}}{100 \text{ Ga atoms}} \times \left(\frac{68.9257 \text{ amu}}{1 \text{ atom } ^{69}\text{Ga}}\right) + \frac{\text{Y atoms } ^{71}\text{Ga}}{100 \text{ Ga atoms}} \times \left(\frac{70.9249 \text{ amu}}{1 \text{ atom } ^{71}\text{Ga}}\right) = 69.723 \frac{\text{amu}}{\text{Ga atom}}$$

    And, X + Y = 100%. We now have two equations and two unknowns, so we can solve for X and Y algebraically. Solve the first equation for Y: Y = 100 – X. Plug that in for Y in the second equation. Then solve for X:

    $$\frac{\text{X}}{100} \times \left(68.9257\right) + \frac{100 - \text{X}}{100} \times \left(70.9249\right) = 69.723$$

    $$0.689257\text{X} + 70.9249 - 0.709249\text{X} = 69.723$$

    $$70.9249 - 69.723 = 0.709249\text{X} - 0.689257\text{X} = \left(0.709249 - 0.689257\right)\text{X}$$

    $$1.202 = \left(0.019992\right)\text{X}$$

    $$\text{X} = 60.12, \text{ so there is } 60.12\% \text{ } ^{69}\text{Ga}$$

    Now, plug the value of X in the first equation to get Y.

    $$\text{Y} = 100 - \text{X} = 100 - 60.12 = 39.88, \text{ so there is } 39.88\% \text{ } ^{69}\text{Ga}$$

    Therefore the abundances for these isotopes are: 60.12% ^{69}Ga and 39.88% ^{71}Ga

    ✓ *Reasonable Answer Check:* The periodic table value for the atomic weight is closer to 68.9257 than it is to 70.9249, so it makes sense that the percentage of ^{69}Ga is larger than ^{71}Ga. The sum of the two percentages is 100.00%.

39. *Answer:* **^{7}Li**

    Strategy and Explanation: The two isotopes of lithium are ^{6}Li and ^{7}Li. The mass of ^{6}Li is close to 6 amu and the mass of ^{7}Li is close to 7 amu. Because lithium's atomic weight (6.941 amu) is much closer to 7 amu than to 6 amu, the isotopic **^{7}Li is more abundant** than the isotope ^{6}Li.

The Mole

41. *Answer:* **(a) 27 g B (b) 0.48 g O$_2$ (c) 6.98 × 10^{-2} g Fe (d) 2.61 × 10^3 g H**

    Strategy and Explanation: Determine mass in grams from given quantity in moles.

    Look up the elements on the periodic table to get the atomic weight (with at least four significant figures). If necessary, calculate the molecular weight. Use that number for the molar mass (with units of grams per mole) as a conversion factor between moles and grams.

    Notice: Whenever you use physical constants that you look up, it is important to carry <u>more</u> significant figures *than the rest of the measured numbers, to prevent causing inappropriate round-off errors.*

(a) Boron (B) has atomic number 5 on the periodic table. Its atomic weight is 10.811, so the molar mass is 10.811 g/mol.

$$2.5 \text{ mol B} \times \frac{10.811 \text{ g B}}{1 \text{ mol B}} = 27 \text{ g B}$$

(b) O_2 (diatomic molecular oxygen) is made with two atoms of element with atomic number 8 on the periodic table. Its atomic weight is 15.9994; therefore, the molecular weight of O_2 is $2 \times 15.9994 = 31.9988$, and the molar mass is 31.9988 g/mol.

$$0.015 \text{ mol } O_2 \times \frac{31.9988 \text{ g } O_2}{1 \text{ mol } O_2} = 0.48 \text{ g } O_2$$

(c) Iron (Fe) has atomic number 26 on the periodic table. Its atomic weight is 55.845, so the molar mass is 55.845 g/mol.

$$1.25 \times 10^{-3} \text{ mol Fe} \times \frac{55.845 \text{ g Fe}}{1 \text{ mol Fe}} = 6.98 \times 10^{-2} \text{ g Fe}$$

(d) Helium (He) has atomic number 2 on the periodic table. Its atomic weight is 4.0026, so the molar mass is 4.0026 g/mol.

$$653 \text{ mol He} \times \frac{4.0026 \text{ g He}}{1 \text{ mol He}} = 2.61 \times 10^3 \text{ g He}$$

✓ *Reasonable Answer Check:* The moles units cancel when the factor is multiplied, leaving grams.

43. *Answer:* **(a) 1.9998 mol Cu   (b) 0.499 mol Ca   (c) 0.6208 mol Al   (d) 3.1 × 10⁻⁴ mol K**

(e) 2.1 × 10⁻⁵ mol Am

Strategy and Explanation: Determine the quantity in moles from given mass in grams.

Look up the elements on the periodic table to get the atomic weight. Use that number for the molar mass (with units of grams per mole) as a conversion factor between grams and moles.

Notice: Whenever you use physical constants that you look up, it is important to carry <u>more</u> significant figures than the rest of the measured numbers, to prevent causing inappropriate round-off errors.

(a) Copper (Cu) has atomic number 29 on the periodic table. Its atomic weight is 63.546, so the molar mass is 63.546 g/mol.

$$127.08 \text{ g Cu} \times \frac{1 \text{ mol Cu}}{63.546 \text{ g Cu}} = 1.9998 \text{ mol Cu}$$

(b) Calcium (Ca) has atomic number 20 on the periodic table. Its atomic weight is 40.078, so the molar mass is 40.078 g/mol.

$$20.0 \text{ g Ca} \times \frac{1 \text{ mol Ca}}{40.078 \text{ g Ca}} = 0.499 \text{ mol Ca}$$

(c) Aluminum (Al) has atomic number 13 on the periodic table. Its atomic weight is 26.9815, so the molar mass is 26.9815 g/mol.

$$16.75 \text{ g Al} \times \frac{1 \text{ mol Al}}{26.9815 \text{ g Al}} = 0.6208 \text{ mol Al}$$

(d) Potassium (K) has atomic number 19 on the periodic table. Its atomic weight is 39.0983, so the molar mass is 39.0983 g/mol.

$$0.012 \text{ g K} \times \frac{1 \text{ mol K}}{39.0983 \text{ g K}} = 3.1 \times 10^{-4} \text{ mol K}$$

(e) Radioactive americium (Am) has atomic number 95 on the periodic table. The atomic weight given on the periodic table is the weight of its most stable isotope 243, so the molar mass is 243 g/mol.

Convert milligrams into grams, first.

$$5.0 \text{ mg Am} \times \frac{1 \text{ g Am}}{1000 \text{ mg Am}} \times \frac{1 \text{ mol Am}}{243 \text{ g Am}} = 2.1 \times 10^{-5} \text{ mol Am}$$

✓ *Reasonable Answer Check:* Notice that grams units cancel when the factor is multiplied, leaving moles.

45. *Answer:* **4.131 × 10²³ Cr atoms**

   Strategy and Explanation: Given a chromium sample with known mass, determine the number of atoms in it.

   Start with the mass. Use the molar mass of chromium as a conversion factor between grams and moles. Then use Avogadro's number as a conversion factor between moles of chromium atoms and the actual number of chromium atoms.

$$35.67 \text{ g Cr} \times \frac{1 \text{ mol Cr atoms}}{51.996 \text{ g Cr}} \times \frac{6.0221 \times 10^{23} \text{ Cr atoms}}{1 \text{ mol Cr atoms}} = 4.131 \times 10^{23} \text{ Cr atoms}$$

   ✓ *Reasonable Answer Check:* A sample of chromium that a person can see and hold is macroscopic. It will contain a very large number of atoms.

47. *Answer:* **1.055 × 10⁻²² g Cu**

   Strategy and Explanation: Given the number of copper atoms, determine the mass.

   Start with the quantity. Use Avogadro's number as a conversion factor between the number of copper atoms and moles of copper atoms. Then use the molar mass of copper as a conversion factor between moles and grams.

$$1 \text{ Cu atom} \times \frac{1 \text{ mol Cu atoms}}{6.022 \times 10^{23} \text{ Cu atoms}} \times \frac{63.546 \text{ g Cu}}{1 \text{ mol Cu atoms}} = 1.055 \times 10^{-22} \text{ g Cu}$$

   ✓ *Reasonable Answer Check:* An atom of copper is NOT very large. It will have a very small mass.

The Periodic Table

51. *Answer/Explanation:* There are five elements in Group 4A of the periodic table. They are non-metal: carbon (C), metalloids: silicon (Si) and germanium (Ge), and metals: tin (Sn) and lead (Pb).

53. *Answer:* **(a) I (b) In (c) Ir (d) Fe**

   Strategy and Explanation: Look up the four elements on Figure 2.5 in Section 2.9.

   (a) I, iodine, is the halogen (because it is in Group 7A)

   (b) In, indium, is a main group metal (because it is a metal found in an A group – Group 3A)

   (c) Ir, iridium, is a transition metal (colored blue on Figure 2.5) in period 6. The period number 6 is given to the far left of the row in the periodic table next to the element Cs.

   (d) Fe, iron, is a transition metal (colored blue on Figure 2.5) in period 4. The period number 4 is given to the far left of the row in the periodic table next to the element K.

55. *Answer:* **(a) Mg (b) Na (c) C (d) S (e) I (f) Mg (g) Kr (h) O (i) Ge** *Notice: There are multiple answers to (a), (b) and (i) in this Question. The ones given here are only examples.*

   Strategy and Explanation: Use the periodic table and information given in Section 2.9.

   (a) An element in Group 2A is magnesium (Mg).

   (b) An element in the third period is sodium (Na).

   (c) The element in the second period of Group 4A is carbon (C).

   (d) The element in the third period in Group 6A is sulfur (S).

   (e) The halogen in the fifth period is iodine (I).

   (f) The alkaline earth element in the third period is magnesium (Mg).

   (g) The noble gas element in the fourth period is krypton (Kr).

   (h) The non-metal in Group 6A and the second period is oxygen (O).

   (i) A metalloid in the fourth period is germanium (Ge).

General Questions

58. *Answer:* **(a) 0.197 nm (b) 197 pm**

   Strategy and Explanation: A distance is given in angstroms (Å), which are defined. Determine the distance in nanometers and picometers.

   Use the given relationship between angstroms and meters as a conversion factor to get from angstroms to meters. Then use the metric relationships between meters and the other two units to find the distance in nanometers and picometers.

$$1.97 \text{ Å} \times \frac{1 \times 10^{-10} \text{ m}}{1 \text{ Å}} \times \frac{1 \text{ nm}}{1 \times 10^{-9} \text{ m}} = 0.197 \text{ nm}$$

$$1.97 \text{ Å} \times \frac{1 \times 10^{-10} \text{ m}}{1 \text{ Å}} \times \frac{1 \text{ pm}}{1 \times 10^{-12} \text{ m}} = 197 \text{ pm}$$

✓ *Reasonable Answer Check:* The unit nanometer is larger than an angstrom, so the distance in nanometers should be a smaller number. The unit picometer is smaller than an angstrom, so the distance in picometers should be a larger number.

59. *Answer:* **(a) 0.178 nm^3 (b) 1.78 × 10^{-22} cm^3**

Strategy and Explanation: The edge length of a cube is given in nanometers. Determine the volume of the cube in cubic nanometers and in cubic centimeters.

Cube the edge length in nanometers to get the volume of the cube in cubic nanometers. Use metric relationships to convert nanometers into meters, then meters into centimeters. Cube the edge length in centimeters to get the volume of the cube in cubic centimeters.

$$V = (\text{edge length})^3 = (0.563 \text{ nm})^3 = 0.178 \text{ nm}^3$$

$$0.563 \text{ nm} \times \frac{1 \times 10^{-9} \text{ m}}{1 \text{ nm}} \times \frac{100 \text{ cm}}{1 \text{ m}} = 5.63 \times 10^{-8} \text{ cm}$$

$$V = (\text{edge length})^3 = (5.63 \times 10^{-8} \text{ cm})^3 = 1.78 \times 10^{-22} \text{ cm}^3$$

✓ *Reasonable Answer Check:* Cubing fractional quantities makes the number smaller. The unit centimeter is larger than a nanometer, so the volume in cubic centimeter should be a very small number.

61. *Answer:* **89 tons/yr**

Strategy and Explanation: Start with the sample. Given the number of people in the city, use the volume of water each person needs per day, then calculate everyone's water needs for the day. Using the number of days in a year calculate the total volume of water used per year. Then using the mass of one gallon of water as a conversion factor, determine the grams of water. Convert the grams to tons. Then, using the fluoride concentration as a conversion factor, determine the number of tons of fluoride, and using the mass percentage of fluoride in sodium fluoride to determine the number of tons.

$$150{,}000 \text{ people} \times \frac{175 \text{ gal water / person}}{1 \text{ day}} \times \frac{365 \text{ days}}{1 \text{ year}} \times \frac{8.34 \text{ lb water}}{1 \text{ gal water}} \times \frac{1 \text{ ton water}}{2000 \text{ lb water}}$$

$$\times \frac{1 \text{ ton fluoride}}{1{,}000{,}000 \text{ tons water}} \times \frac{100 \text{ tons sodium fluoride}}{45.0 \text{ tons flouride}} = 89 \frac{\text{tons sodium fluoride}}{\text{year}}$$

✓ *Reasonable Answer Check:* The significant figures are limited to two by the 150,000 figure. The mass units are appropriately labeled. The units cancel appropriately to give tons per year. This is a large number of people using a large amount of water so the large quantity of sodium fluoride makes sense.

62. *Answer/Explanation:* Potassium's atomic weight is 39.0983. The isotopes that contribute most to this mass are ^{39}K and ^{41}K, since the question tells us that ^{40}K has a very low abundance. Since the atomic mass is closer to 39 than 41, that confirms that the **^{39}K isotope** is more abundant.

64. *Answer:* **0.038 mol**

Strategy and Explanation: Given the carat mass of a diamond and the relationship between carat and milligrams, determine how many moles of carbon are in the diamond.

Always start with the sample. Diamond is an allotropic form of pure carbon. Given the carats of the diamond, use the relationship between carats and milligrams as a conversion factor to determine milligrams of carbon. Then using metric relationships to determine grams of carbon, and the molar mass of carbon to determine the moles of carbon.

$$2.3 \text{ carats C} \times \frac{200 \text{ mg C}}{1 \text{ carat C}} \times \frac{1 \text{ g C}}{1000 \text{ mg C}} \times \frac{1 \text{ mol C}}{12.01 \text{ g C}} = 0.038 \text{ mol C}$$

✓ *Reasonable Answer Check:* The significant figures are limited to two by the 2.3 figure. The units cancel appropriately to give moles of carbon.

65. *Answer:* **$5,070**

Strategy and Explanation: Given the moles of gold you want to buy, the relationship between troy ounces and grams, and the price of a troy ounce of gold, determine the amount of money you must spend.

The sample is the 1.00 mole of gold. Use the molar mass of gold to determine the grams of gold and the given relationship between grams and troy ounces to determine the troy ounces of gold, then use the price per troy ounce to determine how much that would cost.

$$1.00 \text{ mol Au} \times \frac{196.9666 \text{ g Au}}{1 \text{ mol Au}} \times \frac{1 \text{ troy ounce Au}}{31.1 \text{ g Au}} \times \frac{\$800.00}{1 \text{ troy ounce Au}} = \$5,070$$

✓ *Reasonable Answer Check:* The number of moles has three significant figures, so the answer must be reported with three significant figures. The units cancel properly. A mole of gold is pretty heavy, so this seems like a reasonable amount of money to spend.

66. *Answer:* **3.4 mol, 2.0×10^{24} atoms**

Strategy and Explanation: Given the dimensions of a piece of copper wire and the density of copper, determine the moles of copper and the number of atoms of copper in the wire.

Use the metric and English length relationships to convert the wire's dimensions to centimeters. Then use those dimensions to find its volume. Then use the density of copper to determine the mass of the wire. Then use the molar mass of copper to determine the moles of copper and Avogadro's number to determine the actual number of copper atoms.

$$\text{wire length in centimeters} = L = 25 \text{ ft long} \times \frac{12 \text{ in}}{1 \text{ ft}} \times \frac{2.54 \text{ cm}}{1 \text{ in}} = 7.6 \times 10^2 \text{ cm long}$$

$$\text{wire diameter in centimeters} = d = 2.0 \text{ mm diameter} \times \frac{1 \text{ m}}{1000 \text{ mm}} \times \frac{100 \text{ cm}}{1 \text{ m}} = 0.20 \text{ cm}$$

$$\text{wire radius in centimeters} = r = \frac{d}{2} = \frac{0.20 \text{ cm}}{2} = 0.10 \text{ cm}$$

$$\text{cylindrical wire's volume} = V = A \times L = (\pi r^2) \times L$$

$$V = (3.14159) \times (0.10 \text{ cm})^2 \times (7.6 \times 10^2 \text{ cm}) = 24 \text{ cm}^3$$

$$24 \text{ cm}^3 \text{ Cu} \times \frac{8.92 \text{ g Cu}}{1 \text{ cm}^3 \text{ Cu}} \times \frac{1 \text{ mol Cu atoms}}{63.546 \text{ g Cu}} = 3.4 \text{ mol Cu atoms}$$

$$3.4 \text{ mol Cu atoms} \times \frac{6.022 \times 10^{23} \text{ Cu atoms}}{1 \text{ mol Cu atoms}} = 2.0 \times 10^{24} \text{ Cu atoms}$$

✓ *Reasonable Answer Check:* The length and diameter of the wire both have two significant figures, so the answer must be reported with two significant figures. The units cancel properly. The number of atoms in a wire you can see and hold is conveniently represented in the quantity unit moles. The actual number of atoms is very large.

Applying Concepts

67. *Answer:* **(a) not possible (b) possible (c) not possible (d) not possible (e) possible (f) not possible**

Strategy and Explanation: Check to see if the following statements are true: The atomic number must be the same as the number of protons and the mass number must be the sum of the number of protons and neutrons.

(a) These values are **not possible**, since the atomic number (42) is not the same as the number of protons (19), and the sum of protons and neutrons (19+23=42) is not the same as the mass number (19).

(b) These values are **possible**. The atomic number (92) is the same as the number of protons (92), and the sum of protons and neutrons (92+143=235) is the same as the mass number (235).

(c) These values are **not possible**. The sum of protons and neutrons (131+79=210) is not the same as the mass number (53).

(d) These values are **not possible**. The sum of protons and neutrons (15+15=30) is not the same as the mass number (32).

(e) These values are **possible**. The atomic number (7) is the same as the number of protons (7), and the sum of protons and neutrons (7+7=14) is not the same as the mass number (14).

(f) These values are **not possible**. The sum of protons and neutrons (18+40=58) is not the same as the mass number (40).

69. *Answer:* **(a) same (b) second (c) same (d) same (e) same (f) second (g) same (h) first (i) second (j) first**

Strategy and Explanation: A particle is an atom or a molecule. The unit mole is a convenient way of describing a large quantity of particles. It is also important to keep in mind that 1 mol of particles contains 6.022×10^{23} particles and the molar mass describes the mass of 1 mol of particles.

(a) A sample containing 1 mol of Cl has the **same** number of particles as a sample containing 1 mol Cl_2, since each sample contains 1 mole of particles. (The masses of the samples are different, the numbers of atoms contained in the samples are different; however, those statements do not answer this question.)

(b) The **second** sample containing 1 mol of O_2 contains 6.022×10^{23} molecules of O_2. The 1-mol sample has more particles than the first sample containing just 1 molecule of O_2.

(c) Each sample contains only one particle. These samples have the **same** number of particles.

(d) A sample containing 6.022×10^{23} molecules of F_2 contains 1 mol of F_2 molecules. This sample has the **same** number of particles as 6.022×10^{23} molecules of F_2.

(e) The molar mass of Ne is 20.18 g/mol, so a sample of 20.2 grams of neon contains 1 mol of neon. Notice that, with the degree of certainty limited to one decimal place, the slightly smaller molar mass is indistinguishable from the sample mass. This 20.2-gram sample has the **same** number of particles as the sample with 1 mol of neon.

(f) The molar mass of bromine (Br_2) is (79.9 g/mol Br)×(2 mol Br) = 159.8 g/mol, so the **second** sample of 159.8 grams of bromine contains 1 mol bromine which equals 6.022×10^{23} molecules. This 159.8-gram sample has more particles than a sample containing just 1 molecule of Br_2.

(g) The molar mass of Ag is 107.9 g/mol, so a sample of 107.9 grams of Ag contains 1 mol of Ag. The molar mass of Li is 6.9 g/mol, so a sample of 6.9 grams of Li contains 1 mol of Li. These samples have the **same** number of particles, since each sample contains 1 mole of particles.

(h) The molar mass of Co is 58.9 g/mol, so the first sample of 58.9 grams of Co contains 1 mol of Co. The molar mass of Cu is 63.55 g/mol, so the second sample of 58.9 grams of Cu contains less than 1 mol of Cu. The 58.9-g **first** sample of Co has more particles than the second 58.9-g sample of Cu.

(i) The second sample containing 6.022×10^{23} atoms of Ca contains 1 mol of Ca atoms. The molar mass of Ca is 40.1 g/mol, so the first sample of 1 gram of Co contains less than 1 mol of Co. The 6.022×10^{23}-atoms **second** sample has more particles than the second 1-gram sample.

(j) A chlorine molecule contains two chlorine atoms. Since chlorine atoms all weigh the same, a chlorine molecule weighs twice as much as a chlorine atom. If the two samples of chlorine both weigh the same, the **first** sample of atomic chlorine (Cl) sample will have more particles in it than in the second sample of molecular sample (Cl_2).

More Challenging Questions

71. (a) *Answer:* **(a) 270 mL (b) No**

Strategy and Explanation: Given the volume of a bottle containing water, the density of water and ice (at different temperatures), determine the volume of ice.

Always start with the sample. Convert the volume of water to grams using the density of water. The entire mass of water freezes at the new temperature. Use the density of water as a conversion factor to determine the volume of ice.

$$250 \text{ mL water} \times \frac{0.997 \text{ g water}}{1 \text{ mL water}} \times \frac{1 \text{ g ice}}{1 \text{ g water}} \times \frac{1 \text{ mL ice}}{0.917 \text{ g ice}} = 272 \text{ mL ice} \approx 270 \text{ mL ice}$$

✓ *Reasonable Answer Check:* The density of ice is lower than the density of water, so the volume should be greater once the water freezes.

(b) The ice could not be contained in the bottle. The volume of ice exceeds the capacity of the bottle.

75. *Answer:* **(a) Se (b) ^{39}K (c) ^{79}Br (d) ^{20}Ne**

Strategy and Explanations:

(a) O is a nonmetal in Group 6A. The atomic number is the same as the number of protons. The number of protons is the same as the number of electrons in an uncharged atom. The element in Group 6A that has 34 electrons is Se.

(b) The alkali metal group is Group 1A. To get the number of neutrons, subtract the atomic number (Z) from the mass number (A). The element in Group 1A that has A – Z = 39 – 19 = 20 is ^{39}K.

(c) A halogen is an element in Group 7A. The atomic number is the number of protons, and the sum of the protons and the neutrons is the mass number. The element in Group 7A with Z = 35 and A = (35 + 44 =) 79 is ^{79}Br.

(d) A noble gas is an element in Group 8A. The atomic number is the number of protons, and the sum of the protons and the neutrons is the mass number. The element in Group 8A with Z = 10 and A = (10 + 10 =) 20 is ^{20}Ne.

Chapter 3: Chemical Compounds

Chapter Contents:

Introduction

It is important to properly interpret the subscripts in chemical formulas to prevent mistakes when calculating molar masses, writing balanced chemical equations, and doing stoichiometric calculations.

While the text and the course encourages you to think conceptually and not rely on algorithms for problem solving, some algorithms, such as the one for naming compounds, are worthwhile.

When an ionic compound dissolves in water, only common ions will form, so it's important to know what those are. It will help you avoid making up incorrect formulas for aqueous ions, such as these: Cl_2^-, Na_2^+, $(OH)_2^{2-}$.

Pay attention to including the charges on formulas of ions—there are big differences between SO_3 and SO_3^{2-} or NO_2 and NO_2^-. An ion's formula is incomplete unless the proper charge is given. A substance is a compound if there is no charge specified.

Solutions to Blue-Numbered Questions
for Review and Thought in Chapter 3

Topical Questions

Molecular and Structural Formulas

5. *Answer/Explanation:*

    (a) Bromine trifluoride is **BrF_3**.

    (b) Xenon difluoride **XeF_2**.

    (c) Diphosphorus tetrafluoride is **P_2F_4**.

    (d) Pentadecane is **$C_{15}H_{32}$**.

    (e) Hydrazine is **N_2H_4**.

7. *Answer/Explanation:*

    (a) Benzene   Molecular Formula: **C_6H_6**

    (b) Vitamin C   Molecular Formula: **$C_6H_8O_6$**

9. *Answer:* **(a) 1 Ca, 2 C, 4 O (b) 8 C, 8 H (c) 2 N, 8 H, 1 S, 4 O (d) 1 Pt, 2 N, 6 H, 2 Cl (e) 4 K, 1 Fe, 6 C, 6 N**

   Strategy and Explanation: Keep in mind that atoms found inside parentheses that are followed by a subscript get multiplied by that subscript.

    (a) CaC_2O_4 contains one atom of calcium, two atoms of carbon, and four atoms of oxygen.

    (b) $C_6H_5CHCH_2$ contains eight atoms of carbon and eight atoms of hydrogen.

    (c) $(NH_4)_2SO_4$ contains two (1×2) atoms of nitrogen, eight (4×2) atoms of hydrogen, one atom of sulfur, and four atoms of oxygen.

    (d) $Pt(NH_3)_2Cl_2$ contains one atom of platinum, two (1×2) atoms of nitrogen, six (3×2) atoms of hydrogen, and two atoms of chlorine.

    (e) $K_4Fe(CN)_6$ contains four atoms of potassium, one atom of iron, six (1×6) atoms of carbon, and six (1×6) atoms of nitrogen.

Constitutional Isomers

11. *Answer/Explanation:*

   (a) Two molecules that are constitutional isomers have the same formula (i.e., on the molecular level, these molecules have **the same number of atoms of each kind**).

   (b) Two molecules that are constitutional isomers of each other have their atoms in **different bonding arrangements**.

Predicting Ion Charges

13. *Answer:* **(a) +2 (b) +2 (c) +2 or +3 (d) +3**

   Strategy and Explanation: A general rule for the charge on a metal cation: the group number represents the number of electrons lost. Hence, the group number will be the cation's positive charge. Transition metals often have a +2 charge. Some have +3 and +1 charged ions, as well.

   (a) magnesium (Group 2A) has a +2 charge. Mg^{2+}

   (b) zinc (Group 8B) has a +2 charge. Zn^{2+}

   (c) iron (a transition metal) has a +2 or +3 charge. Fe^{2+} or Fe^{3+}

   (d) gallium (Group 3A) has a +3 charge. Ga^{3+}

15. *Answer:* **(c) and (d) are correct formulas. (a) $AlCl_3$ (b) NaF**

   Strategy and Explanation:

   (a) Aluminum ion (from Group 3A) is Al^{3+}. Chloride ion (from Group 7A) is Cl^-.

   AlCl is **not** a neutral combination of these two ions. The correct formula would be $AlCl_3$.

   [net charge = $+3 + 3 \times (-1) = 0$]

   (b) Sodium ion (Group 1A) is Na^+. Fluoride ion (from Group 7A) is F^-.

   NaF_2 is **not** a neutral combination of these two ions. The correct formula would be NaF.

   [net charge = $+1 + (-1) = 0$]

   (c) Gallium ion (from Group 3A) is Ga^{3+}. Oxide ion (from Group 6A) is O^{2-}.

   Ga_2O_3 **is** the correct neutral combination of these two ions.

   [net charge = $2 \times (+3) + 3 \times (-2) = 0$]

   (d) Magnesium ion (from Group 2A) is Mg^{2+}. Sulfide ion (from Group 6A) is S^{2-}.

   MgS **is** the correct neutral combination of these two ions.

   [net charge = $+2 + (-2) = 0$]

Polyatomic Ions

17. *Answer:* **(a) Pb^{2+} and NO_3^-; 1 and 2, respectively (b) Ni^{2+} and CO_3^{2-}; 1 and 1, respectively (c) NH_4^+ and PO_4^{3-}; 3 and 1, respectively (d) K^+ and SO_4^{2-}; 2 and 1, respectively**

   Strategy and Explanation:

   (a) $Pb(NO_3)_2$ has one ion of lead(II) (Pb^{2+}) and two ions of nitrate (NO_3^-).

   (b) $NiCO_3$ has one ion of nickel(II) (Ni^{2+}) and one ion of carbonate (CO_3^{2-}).

   (c) $(NH_4)_3PO_4$ has three ions of ammonium (NH_4^+) and one ion of phosphate (PO_4^{3-}).

   (d) K_2SO_4 has two ions of potassium (K^+) and one ion of sulfate (SO_4^{2-}).

19. *Answer:* **(a) $Ni(NO_3)_2$ (b) $NaHCO_3$ (c) LiClO (d) $Mg(ClO_3)_2$ (e) $CaSO_3$**

   Strategy and Explanation:

   (a) Nickel(II) ion is Ni^{2+}. Nitrate ion is NO_3^-. Use one Ni^{2+} and two NO_3^- to make neutral $Ni(NO_3)_2$.

(b) Sodium ion is Na^+. Bicarbonate ion is HCO_3^-. Use one Na^+ and one HCO_3^- to make neutral $NaHCO_3$.

(c) Lithium ion is Li^+. Hypochlorite ion is ClO^-. Use one Li^+ and one ClO^- to make neutral $LiClO$.

(d) Magnesium ion is Mg^{2+}. Chlorate ion is ClO_3^-. Use one Mg^{2+} and two ClO_3^- to make neutral $Mg(ClO_3)_2$.

(e) Calcium ion is Mg^{2+}. Sulfite ion is SO_3^{2-}. Use one Mg^{2+} and one SO_3^{2-} to make neutral $CaSO_3$.

Ionic Compounds

21. *Answer:* **(b), (c), and (e) are ionic.**

Strategy and Explanation: To tell if a compound is ionic or not, look for metals and nonmetals together, or common cations and anions. If a compound contains only nonmetals or metalloids and nonmetals, it is probably not ionic.

(a) CF_4 contains only nonmetals. Not ionic.

(b) $SrBr_2$ has a metal and nonmetal together. Ionic.

(c) $Co(NO_3)_3$ has a metal and nonmetals together. Ionic.

(d) SiO_2 contains a metalloid and a nonmetal. Not ionic.

(e) KCN has a metal and a common diatomic ion (CN^-) together. Ionic.

(f) SCl_2 contains only nonmetals. Not ionic.

23. *Answer:* **(a) $(NH_4)_2CO_3$ (b) CaI_2 (c) $CuBr_2$ (d) $AlPO_4$**

Strategy and Explanation: Make neutral combinations with the common ions involved.

(a) Ammonium (NH_4^+) and carbonate (CO_3^{2-}) must be combined 2:1, to make $(NH_4)_2CO_3$.

(b) Calcium (Ca^{2+}) and iodide (I^-) must be combined 1:2, to make CaI_2.

(c) Copper(II) (Cu^{2+}) and bromide (Br^-) must be combined 1:2, to make $CuBr_2$.

(d) Aluminum (Al^{3+}) and phosphate (PO_4^{3-}) must be combined 1:1, to make $AlPO_4$.

25. *Answer:* **(a) potassium sulfide (b) nickel(II) sulfate (c) ammonium phosphate (d) aluminum hydroxide (e) cobalt(III) sulfate**

Strategy and Explanation: Give the name of the cation then the name of the anion.

(a) K_2S contains cation K^+ called potassium and anion S^{2-} called sulfide, so it is potassium sulfide.

(b) $NiSO_4$ contains cation Ni^{2+} called nickel(II) and anion SO_4^{2-} called sulfate, so it is nickel(II) sulfate.

(c) $(NH_4)_3PO_4$ contains cation NH_4^+ called ammonium and anion PO_4^{3-} called phosphate, so it is ammonium phosphate.

(d) $Al(OH)_3$ contains cation Al^{3+} called aluminum and anion OH^- called hydroxide, so it is aluminum hydroxide.

(e) $Co_2(SO_4)_3$ contains cation Co^{3+} called cobalt(III) and anion SO_4^{2-} called sulfate, so it is cobalt(III) sulfate.

Electrolytes

27. *Answer:* **(a) K^+ and OH^- (b) K^+ and SO_4^{2-} (c) Na^+ and NO_3^- (d) NH_4^+ and Cl^-**

Strategy and Explanation: Identify the common ions present in the compounds.

(a) The ions present in a solution of KOH are K^+ and OH^-.

(b) The ions present in a solution of K_2SO_4 are K^+ and SO_4^{2-}.

(c) The ions present in a solution of $NaNO_3$ are Na^+ and NO_3^-.

(d) The ions present in a solution of NH_4Cl are NH_4^+ and Cl^-.

29. *Answer:* **(a) and (d)**

Strategy and Explanation: Determine if the compound is an electrolyte, by determining if it is ionic.

(a) NaCl is an ionic compound and will ionize to form Na^+ and Cl^-. When NaCl is dissolved in water, the resulting solution will conduct electricity.

(b) $CH_3CH_2CH_3$ is an organic hydrocarbon compound and will not ionize. When $CH_3CH_2CH_3$ is dissolved in water, the resulting solution will not conduct electricity.

(c) CH_3OH is an organic compound and will not ionize. When CH_3OH is dissolved in water, the resulting solution will not conduct electricity.

(d) $Ca(NO_3)_2$ is an ionic compound and will ionize to form Ca^{2+} and NO_3^-. When $Ca(NO_3)_2$ is dissolved in water, the resulting solution will conduct electricity.

Moles of Compounds

31. *Answer:*

	CH_3OH	Carbon	Hydrogen	Oxygen
No. of moles	1 mol	1 mol	4 mol	1 mol
No. of molecules or atoms	6.022×10^{23} molecules	6.022×10^{23} atoms	2.409×10^{24} atoms	6.022×10^{23} atoms
Molar mass (grams per mol methanol)	32.0417	12.0107	4.0316	15.9994

Strategy and Explanation: Consider a sample of 1 mol of methanol. The formula gives the mole ratio of each atom in one mole of methanol. Each mole is 6.022×10^{23} molecules. The molar masses can be determined by looking up each element on the period table, finding the atomic weight (which is also the grams/mol), and multiplying by the number of moles (see the solution to Question 33 for more details).

33. *Answer:* **(a) 159.688 g/mol (b) 67.806 g/mol (c) 44.0128 g/mol (d) 197.905 g/mol (e) 176.1238 g/mol**

Strategy and Explanation: To calculate the molar mass of a compound, we perform a series of steps: First, look up the atomic mass of each element on the periodic table or in a table of atomic masses to identify the molar mass of the element. Second, determine the number of moles of atoms in one mole of the compound by examining the formula. Third, multiply the number of moles of the element by the molar mass of the element. Last, add all individual masses.

(a) 1 mol of Fe_2O_3 contains 2 mol of Fe and 3 mol of O.

$$\left(\frac{2 \text{ mol Fe}}{1 \text{ mol } Fe_2O_3} \times \frac{55.845 \text{ g}}{1 \text{ mol Fe}} \right) + \left(\frac{3 \text{ mol O}}{1 \text{ mol } Fe_2O_3} \times \frac{15.9994 \text{ g}}{1 \text{ mol O}} \right) = 159.688 \frac{\text{g}}{\text{mol } Fe_2O_3}$$

Once we get used to what the numbers represent, this calculation may be abbreviated as follows:

$$2 \times (55.845 \text{ g Fe}) + 3 \times (15.9994 \text{ g O}) = 159.688 \text{ g/mol } Fe_2O_3$$

(b) 1 mol of BF_3 contains 1 mol of B and 3 mol of F.

$$\left(\frac{1 \text{ mol B}}{1 \text{ mol } BF_3} \times \frac{10.811 \text{ g}}{1 \text{ mol B}} \right) + \left(\frac{3 \text{ mol F}}{1 \text{ mol } BF_3} \times \frac{18.9984 \text{ g}}{1 \text{ mol F}} \right) = 67.806 \frac{\text{g}}{\text{mol } BF_3}$$

This calculation may be abbreviated as follows:

$$10.811 \text{ g B} + 3 \times (18.9984 \text{ g F}) = 67.806 \text{ g/mol } BF_3$$

(c) 1 mol of N_2O contains 2 mol of N and 1 mol of O. The abbreviated calculation for the molar mass looks like this:

$$2 \times (14.0067 \text{ g N}) + 15.9994 \text{ g O} = 44.0128 \text{ g/mol } N_2O$$

(d) 1 mol of $MnCl_2 \cdot 4 H_2O$ compound has 1 mol of $MnCl_2$ with 4 mol of H_2O molecules. So, it contains 1 mol

Mn, 2 mol Cl, $(4 \times 2 =)$ 8 mol H and $(4 \times 1 =)$ 4 mol O. The abbreviated calculation for the molar mass looks like this:

$$54.938 \text{ g Mn} + 2 \times (35.453 \text{ g Cl}) + 8 \times (1.0079 \text{ g H}) + 4 \times (15.9994 \text{ g O}) = 197.905 \text{ g/mol MnCl}_2 \cdot 4 \text{ H}_2\text{O}$$

(e) 1 mol of $C_6H_8O_6$ compound contains 6 mol of C, 8 mol of H, and 6 mol of O. The abbreviated calculation for the molar mass looks like this:

$$6 \times (12.0107 \text{ g C}) + 8 \times (1.0079 \text{ g H}) + 6 \times (15.9994 \text{ g O}) = 176.1238 \text{ g/mol } C_6H_8O_6$$

35. *Answer:* **(a) 0.0312 mol (b) 0.0101 mol (c) 0.0125 mol (d) 0.00406 mol (e) 0.00599 mol**

Strategy and Explanation: Determine the number of moles in a given mass of a compound.

Use the formula and the periodic table to calculate the molar mass for the compound, then use the molar mass as a conversion factor between grams and moles.

(a) Molar mass $CH_3OH = (12.0107 \text{ g}) + 4 \times (1.0079 \text{ g}) + (15.9994 \text{ g}) = 32.0417$ g/mol

$$1.00 \text{ g } CH_3OH \times \frac{1 \text{ mol } CH_3OH}{32.0417 \text{ g } CH_3OH} = 0.0312 \text{ mol } CH_3OH$$

(b) Molar mass $Cl_2CO = 2 \times (35.453 \text{ g}) + (12.0107 \text{ g}) + (15.9994 \text{ g}) = 98.916$ g/mol

$$1.00 \text{ g } Cl_2CO \times \frac{1 \text{ mol } Cl_2CO}{98.916 \text{ g } Cl_2CO} = 0.0101 \text{ mol } Cl_2CO$$

(c) Molar mass $NH_4NO_3 = 2 \times (14.0067 \text{ g}) + 4 \times (1.0079 \text{ g}) + 3 \times (15.9994 \text{ g}) = 80.043$ g/mol

$$1.00 \text{ g } NH_4NO_3 \times \frac{1 \text{ mol } NH_4NO_3}{80.043 \text{ g } NH_4NO_3} = 0.0125 \text{ mol } NH_4NO_3$$

(d) Molar mass $MgSO_4 \cdot 7 \text{ H}_2\text{O}$

$$= (24.305 \text{ g}) + (32.065 \text{ g}) + 11 \times (15.9994 \text{ g}) + 14 \times (1.0079 \text{ g}) = 246.474 \text{ g/mol}$$

$$1.00 \text{ g } MgSO_4 \cdot 7H_2O \times \frac{1 \text{ mol } MgSO_4 \cdot 7H_2O}{246.474 \text{ g } MgSO_4 \cdot 7H_2O} = 0.00406 \text{ mol } MgSO_4 \cdot 7H_2O$$

(e) Molar mass $AgCH_3CO_2$

$$= (107.8682 \text{ g}) + 2 \times (12.0107 \text{ g}) + 3 \times (1.0079 \text{ g}) + 2 \times (15.9994 \text{ g}) = 166.9121 \text{ g/mol}$$

$$1.00 \text{ g } AgCH_3CO_2 \times \frac{1 \text{ mol } AgCH_3CO_2}{166.9121 \text{ g } AgCH_3CO_2} = 0.00599 \text{ mol } AgCH_3CO_2$$

✓ *Reasonable Answer Check:* The quantity in moles is always going to be smaller than the mass in grams.

37. *Answer:* **(a) 151.1622 g/mol (b) 0.0352 mol (c) 25.1 g**

Strategy and Explanation: Determine the molar mass of a compound and then determine the mass of a given number of moles and the number of moles in a given mass.

Use the formula and the periodic table to calculate the molar mass for the compound, then use the molar mass as a conversion factor between grams and moles.

(a) Molar mass $C_8H_9O_2N = 8 \times (12.0107 \text{ g}) + 9 \times (1.0079 \text{ g}) + 2 \times (15.9994 \text{ g})$

$$+ (14.0067 \text{ g}) = 151.1622 \text{ g/mol}$$

(b) $$5.32 \text{ g } C_8H_9O_2N \times \frac{1 \text{ mol } C_8H_9O_2N}{151.1622 \text{ g } C_8H_9O_2N} = 0.0352 \text{ mol } C_8H_9O_2N$$

(c) $$0.166 \text{ mol } C_8H_9O_2N \times \frac{151.1622 \text{ g } C_8H_9O_2N}{1 \text{ mol } C_8H_9O_2N} = 25.1 \text{ g } C_8H_9O_2N$$

✓ *Reasonable Answer Check:* The quantity in moles is always going to be smaller than the mass in grams.

39. *Answer:* **1.2×10^{24} molecules**

Strategy and Explanation: Given the mass of a compound, determine the number of molecules. Use the formula and the periodic table to calculate the molar mass for the compound. Use the molar mass as a conversion factor between grams and moles, then use Avogadro's number to determine the actual number of molecules.

Molar mass $CCl_2F_2 = (12.0107 \text{ g}) + 2 \times (35.453 \text{ g}) + 2 \times (18.9984 \text{ g}) = 120.914 \text{ g/mol}$

$$250 \text{ g } CCl_2F_2 \times \frac{1 \text{ mol } CCl_2F_2}{120.914 \text{ g } CCl_2F_2} \times \frac{6.022 \times 10^{23} \text{ molecules } CCl_2F_2}{1 \text{ mol } CCl_2F_2} = 1.2 \times 10^{24} \text{ molecules } CCl_2F_2$$

✓ *Reasonable Answer Check:* The number of molecules for a macroscopic sample will be huge.

40. *Answer:* **(a) 0.250 mol CF_3CH_2F (b) 6.02×10^{23} F atoms**

Strategy and Explanation: Given the mass of a compound, determine the number of moles of that compound, and the number of atoms of one of the elements.

Use the formula and the periodic table to calculate the molar mass for the compound. Use the molar mass as a conversion factor between grams and moles, then use Avogadro's number to determine number of molecules, then use the formula stoichiometry to determine number of atoms of each type.

Molar mass $CF_3CH_2F = 2 \times (12.0107 \text{ g}) + 4 \times (18.9984 \text{ g}) + 2 \times (1.0079 \text{ g}) = 102.0308 \text{ g/mol}$

(a)
$$25.5 \text{ g } CF_3CH_2F \times \frac{1 \text{ mol } CF_3CH_2F}{102.0308 \text{ g } CF_3CH_2F} = 0.250 \text{ mol } CF_3CH_2F$$

(b) Stoichiometry of the chemical formula: Each mol of CF_3CH_2F contains 4 mol of F atoms.

$$0.250 \text{ mol } CF_3CH_2F \times \frac{4 \text{ mol F}}{1 \text{ mol } CF_3CH_2F} \times \frac{6.022 \times 10^{23} \text{ F atoms}}{1 \text{ mol F}} = 6.02 \times 10^{23} \text{ F atoms}$$

✓ *Reasonable Answer Check:* The number of atoms in a macroscopic sample will be huge. The mass of a substance will always be larger than the number of moles.

Percent Composition

41. *Answer:* **(a) 239.3 g/mol PbS, 86.60% Pb, 13.40% S (b) 30.0688 g/mol C_2H_6, 79.8881% C, 20.1119% H (c) 60.0518 g/mol CH_3CO_2H, 40.0011% C, 6.7135% H, 53.2854% O (d) 80.0432 g/mol NH_4NO_3, 34.9979% C, 5.0368% H, 59.9654% O**

Strategy and Explanation: Given the formula of a compound, determine the molar mass, and the mass percent of each element.

Calculate the mass of each element in one mole of compound, while calculating the molar mass of the compound. Divide the calculated mass of the element by the molar mass of the compound and multiply by 100% to get mass percent. To get the last element's mass percent, subtract the other percentages from 100%.

(a)
$$\text{Mass of Pb per mole of PbS} = 207.2 \text{ g Pb}$$
$$\text{Mass of S per mole of PbS} = 32.065 \text{ g S}$$
$$\text{Molar mass PbS} = (207.2 \text{ g}) + (32.065 \text{ g}) = 239.3 \text{ g/mol PbS}$$

$$\% \text{ Pb} = \frac{\text{mass of Pb per mol PbS}}{\text{mass of PbS per mol PbS}} \times 100\% = \frac{207.2 \text{ g Pb}}{239.3 \text{ g PbS}} \times 100\% = 86.60\% \text{ Pb in PbS}$$

$$\% \text{ S} = 100\% - 86.60\% \text{ Pb} = 13.40\% \text{ S in PbS}$$

(b)
$$\text{Mass of C per mole of } C_2H_6 = 2 \times (12.0107 \text{ g}) = 24.0214 \text{ g C}$$
$$\text{Mass of H per mole of } C_2H_6 = 6 \times (1.0079 \text{ g}) = 6.0474 \text{ g H}$$
$$\text{Molar mass } C_2H_6 = (24.0214 \text{ g}) + (6.0474 \text{ g}) = 30.0688 \text{ g/mol } C_2H_6$$

$$\% \text{ C} = \frac{\text{mass of C / mol } C_2H_6}{\text{mass of } C_2H_6 / \text{mol } C_2H_6} \times 100\% = \frac{24.0214 \text{ g C}}{30.0688 \text{ g } C_2H_6} \times 100\% = 79.8881\% \text{ C in } C_2H_6$$

$$\% \text{ H} = 100\% - 79.8881\% \text{ C} = 20.1119\% \text{ H in } C_2H_6$$

(c)
$$\text{Mass of C per mole of } CH_3CO_2H = 2 \times (12.0107 \text{ g}) = 24.0214 \text{ g C}$$
$$\text{Mass of H per mole of } CH_3CO_2H = 4 \times (1.0079 \text{ g}) = 4.0316 \text{ g H}$$

$$\text{Mass of O per mole of } CH_3CO_2H = 2 \times (15.9994 \text{ g}) = 31.9988 \text{ g O}$$

$$\text{Molar mass } CH_3CO_2H = (24.0214 \text{ g}) + (4.0316 \text{ g}) + (31.9988 \text{ g}) = 60.0518 \text{ g/mol } CH_3CO_2H$$

$$\%\,C = \frac{\text{mass of C / mol } CH_3CO_2H}{\text{mass of } CH_3CO_2H \text{ / mol } CH_3CO_2H} \times 100\%$$

$$= \frac{24.0214 \text{ g C}}{60.0518 \text{ g } CH_3CO_2H} \times 100\% = 40.0011\%\ C \text{ in } CH_3CO_2H$$

$$\%\,H = \frac{\text{mass of H / mol } CH_3CO_2H}{\text{mass of } CH_3CO_2H \text{ / mol } CH_3CO_2H} \times 100\%$$

$$= \frac{4.0316 \text{ g H}}{60.0518 \text{ g } CH_3CO_2H} \times 100\% = 6.7135\%\ H \text{ in } CH_3CO_2H$$

$$\%\,O = 100\% - 40.0011\%\ C - 6.7135\%\ H = 53.2854\%\ O \text{ in } CH_3CO_2H$$

(d)
$$\text{Mass of N per mole of } NH_4NO_3 = 2 \times (14.0067 \text{ g}) = 28.0134 \text{ g N}$$

$$\text{Mass of H per mole of } NH_4NO_3 = 4 \times (1.0079 \text{ g}) = 4.0316 \text{ g H}$$

$$\text{Mass of O per mole of } NH_4NO_3 = 3 \times (15.9994 \text{ g}) = 47.9982 \text{ g O}$$

$$\text{Molar mass } NH_4NO_3 = (28.0134 \text{ g}) + (4.0316 \text{ g}) + (47.9982 \text{ g}) = 80.0432 \text{ g/mol } NH_4NO_3$$

$$\%\,N = \frac{\text{mass of N / mol } NH_4NO_3}{\text{mass of } NH_4NO_3 \text{ / mol } NH_4NO_3} \times 100\% = \frac{28.0134 \text{ g N}}{80.0432 \text{ g } NH_4NO_3} \times 100\%$$

$$= 34.9979\%\ N \text{ in } NH_4NO_3$$

$$\%\,H = \frac{\text{mass of H / mol } NH_4NO_3}{\text{mass of } NH_4NO_3 \text{ / mol } NH_4NO_3} \times 100\% = \frac{4.0316 \text{ g H}}{80.0432 \text{ g } NH_4NO_3} \times 100\%$$

$$= 5.0368\%\ H \text{ in } NH_4NO_3$$

$$\%\,O = 100\% - 34.9979\%\ C - 5.0368\%\ H = 59.9654\%\ O \text{ in } NH_4NO_3$$

Notice: When masses of different things are used in the same question, make sure your units clearly specify what each mass refers to.

✓ *Reasonable Answer Check:* Calculating the last element's mass percent using the formula gives the same answer as subtracting the other percentages from 100%.

43. *Answer:* **245.745 g/mol, 25.858% Cu, 22.7992% N, 5.74197% H, 13.048% S, 32.5528% O**

Strategy and Explanation: Given the formula of a compound, determine the molar mass, and the mass percent of each element

Calculate the mass of each element in one mole of compound, while calculating the molar mass of the compound (see full method on Question 33). Divide the calculated mass of the element by the molar mass of the compound and multiply by 100% to get mass percent.

$$\text{Mass of Cu per mole of } Cu(NH_3)_4SO_4 \cdot H_2O = 63.546 \text{ g Cu}$$

$$\text{Mass of N per mole of } Cu(NH_3)_4SO_4 \cdot H_2O = 4 \times (14.0067) = 56.0268 \text{ g N}$$

$$\text{Mass of H per mole of } Cu(NH_3)_4SO_4 \cdot H_2O = (3 \times 4 + 2) \times (1.0079 \text{ g}) = 14 \times (1.0079 \text{ g}) = 14.1106 \text{ g H}$$

$$\text{Mass of S per mole of } Cu(NH_3)_4SO_4 \cdot H_2O = 32.065 \text{ g S}$$

$$\text{Mass of O per mole of } Cu(NH_3)_4SO_4 \cdot H_2O = (4 + 1) \times (15.9994 \text{ g}) = 79.9970 \text{ g O}$$

$$\text{Molar mass } Cu(NH_3)_4SO_4 \cdot H_2O = (63.546 \text{ g}) + (56.0268 \text{ g}) + (14.1106 \text{ g})$$
$$+ (32.065 \text{ g}) + (79.9970 \text{ g}) = 245.745 \text{ g/mol } Cu(NH_3)_4SO_4 \cdot H_2O$$

$$\%\,Cu = \frac{\text{mass of Cu / mol } Co(NH_3)_4SO_4 \cdot H_2O}{\text{mass of } Co(NH_3)_4SO_4 \cdot H_2O \text{ / mol } Co(NH_3)_4SO_4 \cdot H_2O} \times 100\%$$

$$= \frac{63.546 \text{ g Cu}}{245.745 \text{ g } Co(NH_3)_4SO_4 \cdot H_2O} \times 100\% = 25.858\%\ Cu \text{ in } Co(NH_3)_4SO_4 \cdot H_2O$$

$$\% \ N = \frac{\text{mass of N}/\text{mol Co(NH}_3)_4\text{SO}_4 \cdot \text{H}_2\text{O}}{\text{mass of Co(NH}_3)_4\text{SO}_4 \cdot \text{H}_2\text{O}/\text{mol Co(NH}_3)_4\text{SO}_4 \cdot \text{H}_2\text{O}} \times 100\%$$

$$= \frac{56.0268 \text{ g N}}{245.745 \text{ g Co(NH}_3)_4\text{SO}_4 \cdot \text{H}_2\text{O}} \times 100\% = 22.7992\% \text{ N in Co(NH}_3)_4\text{SO}_4 \cdot \text{H}_2\text{O}$$

$$\% \ H = \frac{\text{mass of H}/\text{mol Co(NH}_3)_4\text{SO}_4 \cdot \text{H}_2\text{O}}{\text{mass of Co(NH}_3)_4\text{SO}_4 \cdot \text{H}_2\text{O}/\text{mol Co(NH}_3)_4\text{SO}_4 \cdot \text{H}_2\text{O}} \times 100\%$$

$$= \frac{14.1106 \text{ g H}}{245.745 \text{ g Co(NH}_3)_4\text{SO}_4 \cdot \text{H}_2\text{O}} \times 100\% = 5.74197\% \text{ H in Co(NH}_3)_4\text{SO}_4 \cdot \text{H}_2\text{O}$$

$$\% \ S = \frac{\text{mass of S}/\text{mol Co(NH}_3)_4\text{SO}_4 \cdot \text{H}_2\text{O}}{\text{mass of Co(NH}_3)_4\text{SO}_4 \cdot \text{H}_2\text{O}/\text{mol Co(NH}_3)_4\text{SO}_4 \cdot \text{H}_2\text{O}} \times 100\%$$

$$= \frac{32.065 \text{ g S}}{245.745 \text{ g Co(NH}_3)_4\text{SO}_4 \cdot \text{H}_2\text{O}} \times 100\% = 13.048\% \text{ S in Co(NH}_3)_4\text{SO}_4 \cdot \text{H}_2\text{O}$$

$$\% \ O = \frac{\text{mass of O}/\text{mol Co(NH}_3)_4\text{SO}_4 \cdot \text{H}_2\text{O}}{\text{mass of Co(NH}_3)_4\text{SO}_4 \cdot \text{H}_2\text{O}/\text{mol Co(NH}_3)_4\text{SO}_4 \cdot \text{H}_2\text{O}} \times 100\%$$

$$= \frac{79.9970 \text{ g O}}{245.745 \text{ g Co(NH}_3)_4\text{SO}_4 \cdot \text{H}_2\text{O}} \times 100\% = 32.5528\% \text{ O in Co(NH}_3)_4\text{SO}_4 \cdot \text{H}_2\text{O}$$

✓ *Reasonable Answer Check:* The sum of the percentages add up to 100%.

45. *Answer:* **One**

Strategy and Explanation: Given the molar mass of a compound with an unknown formula and the mass percent of an element in that compound, determine the number of atoms of that element in the compound.

Using a convenient sample size of compound, determine the mass of the element in that compound. Convert both masses to moles, then determine the mole ratio.

In 100.000 grams of the compound carbonic anhydrase (c.a.) there are 0.218 g Zn.

$$100.000 \text{ g c.a.} \times \frac{1 \text{ mole c.a.}}{3.00 \times 10^4 \text{ g c.a.}} = 0.00333 \text{ mole c.a.}$$

$$0.218 \text{ g Zn} \times \frac{1 \text{ mole Zn}}{65.409 \text{ g Zn}} = 0.00333 \text{ mole Zn}$$

$$0.00333 \text{ mole Zn} : 0.00333 \text{ mole c.a.}$$

$$1 \text{ mole Zn} : 1 \text{ mole c.a.}$$

One Zn atom in every molecule of c.a.

✓ *Reasonable Answer Check:* The contribution of the zinc mass to the mass of the molecule is very small, so it makes sense that the number of atoms of zinc is very small in this large molecule.

Empirical and Molecular Formulas

49. *Answer:* $C_4H_4O_4$

Strategy and Explanation: Given the empirical formula of a compound and the molar mass, determine the molecular formula.

Find the mass of 1 mol of the empirical formula. Divide the molar mass of the compound by the empirical mass to get a whole number. Multiply all the subscripts in the empirical formula by the whole number.

The empirical formula is CHO, the molecular formula is $(CHO)_n$.

Mass of 1 mol of CHO = 12.0107 g + 1.0079 g + 15.9994 g = 29.0180 g/mol CHO

$$n = \frac{\text{mass of 1 mol of molecule}}{\text{mass of 1 mol of CHO}} = \frac{116.4 \text{ g}}{29.0180 \text{ g}} = 4.011 \approx 4$$

Molecular Formula is $(CHO)_4 = C_4H_4O_4$

✓ *Reasonable Answer Check:* The molar mass is about 4 times larger than the mass of one mole of the empirical formula, so the molecular formula $C_4H_4O_4$ makes sense.

50. *Answer:* $C_2H_3OF_3$

Strategy and Explanation: Given the molar mass of a compound and the percents by mass of elements in a compound, find the molecular formula.

Choose a convenient sample of $C_xH_yO_zF_w$, such as 100.00 g. Using the percent by mass, determine the number of grams of each element in the sample. Use the molar mass of each element to get the moles of that element. Set up a mole ratio to determine the values of x, y, z, and w. Use those integers as subscripts in the empirical formula. Next, calculate the molar mass of the empirical formula, and divide the given molar mass of the compound into this value to determine the number of empirical formulas in the molecular formula and use that to write the formula of the molecule.

Exactly 100.0 g of $C_xH_yO_zF_w$ contains 24.0 grams C, 3.0 grams H, 16.0 grams O, and 57.0 grams F.

$$24.0 \text{ g C} \times \frac{1 \text{ mol C}}{12.0107 \text{ g C}} = 2.00 \text{ mol C} \qquad 3.0 \text{ g H} \times \frac{1 \text{ mol H}}{1.0079 \text{ g H}} = 3.0 \text{ mol H}$$

$$16.0 \text{ g O} \times \frac{1 \text{ mol O}}{15.9994 \text{ g O}} = 1.00 \text{ mol O} \qquad 57.0 \text{ g F} \times \frac{1 \text{ mol F}}{18.9984 \text{ g F}} = 3.00 \text{ mol F}$$

Set up ratio and simplify: 2.00 mol C : 3.0 mol H : 1.00 mol O : 3.00 mol F

simplify and write the empirical formula: 2 C : 3 H : 1 O : 3 F $C_2H_3OF_3$

Mass of 1 mol of $C_2H_3OF_3$ = 2 × (12.0107 g) + 3 × (1.0079 g) + 15.9994 g + 3 × (18.9984 g)

$$= 100.0397 \text{ g / mol } C_2H_3OF_3$$

$$n = \frac{\text{mass of 1 mol of molecule}}{\text{mass of 1 mol of } C_2H_3OF_3} = \frac{100.0 \text{ g}}{100.0397 \text{ g}} = 1.000 \approx 1$$

Molecular Formula is $C_2H_3OF_3$

✓ *Reasonable Answer Check:* The empirical mass is very close to the given molecular mass.

52. *Answer:* KNO_3

Strategy and Explanation: Given the percents by mass of elements in a compound, find the empirical formula.

Choose a convenient sample of $K_xN_yO_z$, such as 100.00 g. Using the percent by mass, determine the number of grams of each element in the sample. Use the molar mass of each element to get the moles of that element. Set up a mole ratio to determine the values of x, y, z, and w. Use those integers as subscripts in the empirical formula.

Exactly 100.0 g of $K_xN_yO_z$ contains 38.67 grams K, 13.85 grams N, and 47.47 grams O.

$$38.67 \text{ g K} \times \frac{1 \text{ mol K}}{39.0983 \text{ g K}} = 0.9890 \text{ mol K} \qquad 13.85 \text{ g N} \times \frac{1 \text{ mol N}}{14.0067 \text{ g N}} = 0.9888 \text{ mol N}$$

$$47.47 \text{ g O} \times \frac{1 \text{ mol O}}{15.9994 \text{ g O}} = 2.967 \text{ mol O}$$

Set up ratio and simplify: 0.9890 mol K : 0.9888 mol N : 2.967 mol O

simplify the ratio 1 K : 1 N : 3 O and write the empirical formula: KNO_3

✓ *Reasonable Answer Check:* The formula matches a common ionic compound, potassium nitrate.

55. *Answer:* $C_5H_{14}N_2$

Strategy and Explanation: Given the percent by mass of all the elements in a compound and the molar mass, determine the molecular formula.

Choose a convenient sample of $C_xH_yN_z$, such as 100.00 g. Using the percent by mass, determine the number of grams of C, H, and N in the sample. Use the molar masses of C, H, and N as conversion factors to get the moles of C, H and N. Set up a mole ratio to determine the whole number x:y:z ratio for the empirical formula. Find the mass of 1 mol of the empirical formula. Divide the molar mass of the compound by the calculated empirical mass to get a whole number. Multiply all the subscripts in the empirical formula by this whole number.

The compound is 58.77% C, 13.81% H, and 27.42% N by mass. This means that 100.00 g of $C_xH_yN_z$ contains 58.77 grams C, 13.81 grams H, and 27.42 grams N.

Find moles of C, H, and N in the sample:

$$58.77 \text{ g C} \times \frac{1 \text{ mol C}}{12.0107 \text{ g C}} = 4.893 \text{ mol C} \qquad 13.81 \text{ g H} \times \frac{1 \text{ mol H}}{1.0079 \text{ g H}} = 13.70 \text{ mol H}$$

$$27.42 \text{ g N} \times \frac{1 \text{ mol N}}{14.0067 \text{ g N}} = 1.958 \text{ mol N}$$

Mole ratio: 4.893 mol C : 13.70 mol H : 1.958 mol N

Divide each term by the smallest number of moles, 1.958 mol, to get atom ratio: 2.5 C : 7 H : 1 N

Multiply each term by 2 to get whole number atom ratio: 5 C : 14 H : 2 N

The empirical formula is $C_5H_{14}N_2$, so the molecular formula is $(C_5H_{14}N_2)_n$.

Mass of 1 mol $C_5H_{14}N_2$ = 5 × (12.0107 g) + 14 × (1.0079 g) + 2 × (14.0067 g) = 102.1775 g / mol

$$n = \frac{\text{mass of 1 mol of molecule}}{\text{mass of 1 mol of } C_5H_{14}N_2} = \frac{102.2 \text{ g}}{102.1775 \text{ g}} = 1.000 \approx 1$$

Molecular Formula is $(C_5H_{14}N_2)_1 = C_5H_{14}N_2$

✓ *Reasonable Answer Check:* The mole ratios of C, H, and N in the sample are close to integer values, so the empirical formula makes sense. The molar mass is almost the same as the mass of one mole of the empirical formula, so the molecular formula $C_5H_{14}N_2$ makes sense.

56. *Answer:* **x = 7**

Strategy and Explanation: The mass of a sample of a hydrate compound is given. The formula of the hydrate is known except for the amount of water in it. All the water is dried out of the sample using high temperature, leaving a mass of completely dehydrated salt. Determine out the number of water molecules in the formula of the hydrate compound.

Use the molar mass of the dehydrated compound as a conversion factor to convert the mass of the dehydrated compound into moles. Since the only thing lost was water, a mole relationship can be established between the dehydrated and hydrated compound. In addition, the difference between the mass of the hydrated compound and the mass of the dehydrated compound gives the amount of water lost by the sample. Convert that water into moles, using the molar mass of water. Divide the moles of water by the moles of hydrate, to determine how many moles of water are in one mole of hydrate compound.

1.687 g of hydrated compound, $MgSO_4{\cdot}xH_2O$, is dehydrated to make 0.824 g of $MgSO_4$.

Molar Mass $MgSO_4$ = 24.305 g + 32.065 g + 4 × (15.9994 g) = 120.368 g/mol

Molar Mass H_2O = 2 × (1.0079 g) + 2 × (15.9994 g) = 18.0152 g/mol

Find moles of $MgSO_4{\cdot}xH_2O$ in the sample:

$$0.824 \text{ g } MgSO_4 \times \frac{1 \text{ mol } MgSO_4}{120.368 \text{ g } MgSO_4} \times \frac{1 \text{ mol } MgSO_4 \cdot xH_2O}{1 \text{ mol } MgSO_4} = 6.85 \times 10^{-3} \text{ mol } MgSO_4{\cdot}xH_2O$$

Find mass of water lost by the sample: (1.687 g $MgSO_4{\cdot}xH_2O$) – (0.824 g $MgSO_4$) = 0.863 g H_2O

Find moles of H_2O lost from sample: $0.863 \text{ g } H_2O \times \dfrac{1 \text{ mol } H_2O}{18.0152 \text{ g } H_2O} = 0.0479 \text{ mol } H_2O$

$$\text{Mole ratio} = \frac{\text{mol } H_2O \text{ from sample}}{\text{mol hydrate is sample}} = \frac{0.0479 \text{ mol } H_2O}{6.85 \times 10^{-3} \text{ mol hydrate}} = 7.00 \approx 7$$

The proper formula for the hydrate is $MgSO_4{\cdot}7H_2O$ and x = 7.

✓ *Reasonable Answer Check:* The formula for epsom salt is given in Table 3.10 with seven water molecules.

General Questions

61. *Answer:* **(a) (i) chlorine tribromide (ii) nitrogen trichloride (iii) calcium sulfate (iv) heptane (v) xenon tetrafluoride (vi) oxygen difluoride (vii) sodium iodide (viii) aluminum sulfide (ix) phosphorus pentachloride (x) potassium phosphate (b) (iii), (vii), (viii), and (x)**

Strategy and Explanation:

(i)   binary molecular compound (both nonmetals; not ionic): chlorine tribromide

(ii)   binary molecular compound (both nonmetals; not ionic): nitrogen trichloride

(iii)   ionic compound (common cation, Ca^{2+}, and anion, SO_4^{2-}): calcium sulfate

(iv)   organic compound (alkane): heptane

(v)   binary molecular compound (both nonmetals; not ionic): xenon tetrafluoride

(vi)   binary molecular compound (both nonmetals; not ionic): oxygen difluoride

(vii)   ionic compound (common cation, Na^+, and anion, I^-): sodium iodide

(viii) ionic compound (common cation, Al^{3+}, and anion, S^{2-}): aluminum sulfide

(ix)   binary molecular compound (both nonmetals; not ionic): phosphorus pentachloride

(x)   ionic compound (common cation, K^+, and anion, PO_4^{3-}): potassium phosphate

63. *Answer:* **(a) CO_2F_2 (b) CO_2F_2**

Strategy and Explanation: Given the percent by mass of all the elements in a compound and the molar mass, determine the empirical formula and the molecular formula.

Choose a convenient sample of $C_xO_yF_z$, such as 100.0 g. Using the percent by mass, determine the number of grams of C, O, and F in the sample. Use the molar masses of C, O, and F as conversion factors to get the moles of C, O and F. Set up a mole ratio to determine the whole number x:y:z ratio for the empirical formula. Find the mass of 1 mol of the empirical formula. Divide the molar mass of the compound by the calculated empirical mass to get a whole number. Multiply all the subscripts in the empirical formula by this whole number.

(a) The compound is 14.6% C, 39.0% O, and 46.3% F by mass. This means that 100.00 g of $C_xO_yF_z$ contains 14.6 grams C, 39.0 grams O, and 46.3 grams F.

Find moles of C, O, and F in the sample:

$$14.6 \text{ g C} \times \frac{1 \text{ mol C}}{12.0107 \text{ g C}} = 1.22 \text{ mol C} \qquad\qquad 39.0 \text{ g O} \times \frac{1 \text{ mol O}}{15.9994 \text{ g O}} = 2.44 \text{ mol O}$$

$$46.3 \text{ g F} \times \frac{1 \text{ mol F}}{18.9984 \text{ g F}} = 2.44 \text{ mol F}$$

Mole ratio: 1.22 mol C : 2.44 mol O : 2.44 mol F

Divide each term by the smallest number of moles, 1.22 mol

Atom ratio: 1 C : 2 O : 2 F

The empirical formula is CO_2F_2

(b) The molecular formula is $(CO_2F_2)_n$.

Mass of 1 mol CO_2F_2 = 12.0107 g + 2 × (15.9994 g) + 2 × (18.9984 g) = 82.0063 g/mol

$$n = \frac{\text{mass of 1 mol of molecule}}{\text{mass of 1 mol of } CO_2F_2} = \frac{82 \text{ g}}{82.0063 \text{ g}} = 1.00 \approx 1$$

The molecular formula is $(CO_2F_2)_1 = CO_2F_2$

✓ *Reasonable Answer Check:* The simplest ratios of C, O, and F in the sample end up very close to whole number values, so the empirical formula makes sense. The molar mass is the same as the mass of one mole of the empirical formula, so the molecular formula CO_2F_2 makes sense.

68. *Answer/Explanation:*

    (a)   A crystal of sodium chloride has alternating lattice of Na^+ and Cl^- ions.

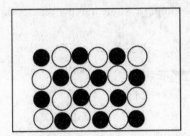

    (b)   The sodium chloride after it is melted has separated ions randomly distributed in the bottom of the box.

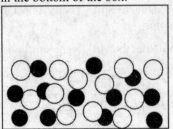

Applying Concepts

70. *Answer:* **(a) three (b) see identical pairs, below.**

    Strategy and Explanation: There are three isomers given for C_3H_8O. Each of these isomers is represented by two of the structures. The following pairs are identical:

    Isomer number one: Isomer number two: Isomer number three:

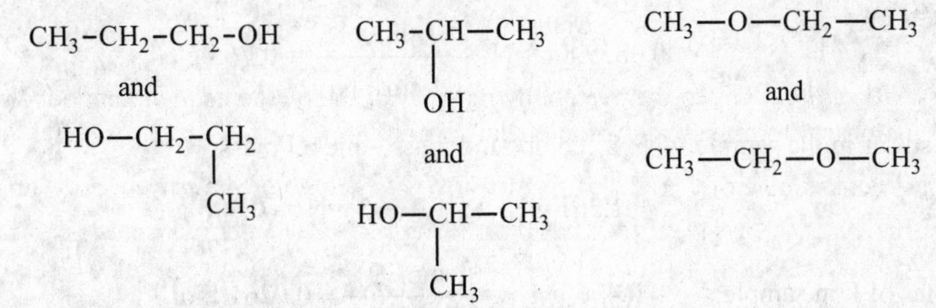

$$CH_3-CH_2-CH_2-OH$$

and

$$HO-CH_2-CH_2 \mid CH_3$$

$$CH_3-CH-CH_3 \mid OH$$

and

$$HO-CH-CH_3 \mid CH_3$$

$$CH_3-O-CH_2-CH_3$$

and

$$CH_3-CH_2-O-CH_3$$

    The first pair has an OH bonded to an end carbon. The second pair has an OH bonded to the middle carbon. The third pair has an O bonded between two carbon atoms.

71. *Answer:* **Tl_2CO_3, Tl_2SO_4**

    Strategy and Explanation: Thallium nitrate is $TlNO_3$. Since NO_3^- has a –1 charge, that means that thallium ion has a +1 charge, and is represented by Tl^+. The carbonate compound containing thallium will be a combination of Tl^+ and CO_3^{2-}, and the compound's formula will look like this: Tl_2CO_3. The sulfate compound containing thallium will be a combination of Tl^+ and SO_4^{2-}, and the compound's formula will look like this: Tl_2SO_4.

73. *Answer:* **(a) perbromate, bromate, bromite, hypobromite (b) selenate, selenite**
    Strategy and Explanation:

    (a)   Based on the guidelines for naming oxyanions in a series, relate the chloro-oxyanions to the bromo-oxyanions, since chlorine and bromine are in the same group (Group 7A).

        BrO_4^- is perbromate (like ClO_4^- is perchlorate) BrO_3^- is bromate (like ClO_3^- is chlorate)

        BrO_2^- is bromite (like ClO_2^- is chlorite) BrO^- is hypobromite (like ClO^- is hypochlorite)

    (b)   Based on the guidelines for naming oxyanions in a series, relate the sulfur-oxyanions to the selenium-oxyanions.

        SeO_4^{2-} is selenate (like SO_4^{2-} is sulfate) SeO_3^{2-} is selenite (like SO_3^{2-} is sulfite)

More Challenging Questions

76. *Answer:* **(a) 0.0130 mol Ni (b) NiF_2 (c) nickel(II) fluoride**

    Strategy and Explanation: Given the dimensions of a sample of nickel foil, the density of nickel metal, and the mass of a nickel fluoride compound (Ni_xF_y) produced using that foil as a source of nickel, determine the moles of nickel, and the formula of the nickel fluoride compound.

Always start with the sample – in this case, the dimensions of the Ni foil sample. Use metric conversion factors and the relationship between length, width, and height and volume to determine the sample's volume. Then use the density to determine the mass of Ni in the sample. Subtract the mass of Ni from the mass of the compound produced to find the mass of F in the sample. Then use the molar masses of Ni and F to find the moles of Ni and moles of F in the sample. Then set up a mole ratio and find the simplest whole number relationship between the atoms.

The foil is 0.550 mm thick, 1.25 cm long and 1.25 cm wide.

$$V = (\text{thickness}) \times (\text{length}) \times (\text{width})$$

$$V = \left(0.550 \text{ mm} \times \frac{1 \text{ m}}{1000 \text{ mm}} \times \frac{100 \text{ cm}}{1 \text{ m}}\right) \times (1.25 \text{ cm}) \times (1.25 \text{ cm}) = 0.0859 \text{ cm}^3$$

(a) $0.0859 \text{ cm}^3 \text{ Ni} \times \dfrac{8.908 \text{ g Ni}}{1 \text{ cm}^3 \text{ Ni}} \times \dfrac{1 \text{ mol Ni}}{58.69 \text{ g Ni}} = 0.0130 \text{ mol Ni}$

(b) $0.0859 \text{ cm}^3 \text{ Ni} \times \dfrac{8.908 \text{ g Ni}}{1 \text{ cm}^3 \text{ Ni}} = 0.765 \text{ g Ni}$

Mass of sulfur in the sample = Ni_xS_y compound mass – nickel mass

$$1.261 \text{ g } Ni_xS_y - 0.765 \text{ g Ni} = 0.496 \text{ g F}$$

Find moles of F in sample: $0.496 \text{ g F} \times \dfrac{1 \text{ mol F}}{18.9984 \text{ g F}} = 0.0261 \text{ mol F}$

Set up mole ratio: $0.0130 \text{ mol Ni} : 0.0261 \text{ mol F}$

Divide all the numbers in the ratio by the smallest number of moles, 0.0130 mol, and round to whole numbers: 1 Ni : 2 F. That gives an empirical formula of NiF_2

(c) NiF_2 is called nickel(II) fluoride. (It looks composed of nickel(II) ion, Ni^{2+}, and fluoride ion, F^-)

✓ *Reasonable Answer Check:* The mole ratio was very close to a whole number ratio, so the empirical formula of NiF_2 makes sense. One of the common ionic forms of nickel is nickel(II) ion, so this also confirms that the result is sensible.

77. *Answer:* CO_2

Strategy and Explanation: Given the mass of one molecule of a compound, the mass percentage of carbon, and the fact that the rest of the compound is oxygen, determine the identity of the compound.

Pick a convenient sample size, such as 100.0 g of compound, C_xO_y. Use the percent by mass of carbon in the compound to determine the mass of carbon in the sample. Subtract the mass of carbon from the mass of the compound to find the mass of oxygen in the sample. Then use the molar masses of C and O to find the moles of C and moles of O in the sample. Then set up a mole ratio and find the simplest whole number relationship between the atoms. Use the mass of one molecule to determine the molar mass of the compound. Use the molar mass and the empirical formula to determine the molecular formula.

The compound is 27.3% C and the rest is O. That means that a 100.0-gram sample of the compound will contain 27.3 grams of carbon. The rest is oxygen:

$$100.0 \text{ g } C_xO_y - 27.3 \text{ g C} = 72.7 \text{ g O}$$

Determine the moles of C and O in the sample.

$$27.3 \text{ g C} \times \frac{1 \text{ mol C}}{12.0107 \text{ g C}} = 2.27 \text{ mol C} \qquad 72.7 \text{ g O} \times \frac{1 \text{ mol O}}{15.9994 \text{ g O}} = 4.54 \text{ mol O}$$

Set up mole ratio: 2.27 mol C : 4.54 mol O

Divide all the numbers in the ratio by the smallest number, 2.27 mol, and round to whole numbers: 1 C : 2 O

Empirical formula: CO_2 Molecular formula: $(CO_2)_n$

Mass of 1 mol CO_2 = 12.0107 g + 2 × (15.9994 g) = 44.010 g/mol

Molar mass of compound: $\dfrac{7.308 \times 10^{-23} \text{ g}}{1 \text{ molecule}} \times \dfrac{6.022 \times 10^{23} \text{ molecule}}{1 \text{ mol}} = 44.0095 \dfrac{\text{g}}{\text{mol}}$

Set up a ratio to find n: $n = \dfrac{\text{mass of 1 mol compound}}{\text{mass of 1 mol emp. formula}} = \dfrac{44.01 \text{ g compound}}{44.0095 \text{ g } CO_2} = 1.000 \approx 1$

Molecular formula is $(CO_2)_1$ or CO_2

✓ *Reasonable Answer Check:* The mole ratio is very close to a whole number, and we recognize this compound as carbon dioxide, a common compound of carbon and oxygen.

79. *Answer:* **5.0 lb N, 4.4 lb P, and 4.2 lb K**

Strategy and Explanation: Given the percent by mass of some elements and compounds in a fertilizer, determine the mass (in pounds, lb) of three elements in a given mass of the fertilizer.

Use the given mass and the definition of percent (parts per hundred) to determine the mass of each compound in the sample. Then determine the mass of each element in one mole of each of the compounds (as is done when determining mass percent of the element), and use that ratio to determine the mass of the elements in the sample.

Assuming at least two significant figures precision, the fertilizer is 5.0% N, 10.% P_2O_5, and 5.0% K_2O. That means, a sample of 100. lb of the fertilizer mixture, has 5.0 lb N, 10. lb P_2O_5, and 5.0 lb K_2O.

Mass of P in one mole of P_2O_5 = 2 × (30.9738 g) = 61.9476 g P

Molar mass P_2O_5 = 61.9476 g + 5 × (15.9994 g) = 141.9446 g P_2O_5

$$10. \text{ lb } P_2O_5 \times \frac{453.59237 \text{ g}}{1 \text{ lb}} \times \frac{61.9476 \text{ g P}}{141.9446 \text{ g } P_2O_5} \times \frac{1 \text{ lb}}{453.59237 \text{ g}} = 4.4 \text{ lb P}$$

Mass of K in one mole of K_2O = 2 × (39.0983 g) = 78.1966 g K

Molar mass K_2O = 78.1966 g + 15.9994 g = 94.196 g K_2O

$$5.0 \text{ lb } K_2O \times \frac{453.59237 \text{ g}}{1 \text{ lb}} \times \frac{78.1966 \text{ g K}}{94.196 \text{ g } K_2O} \times \frac{1 \text{ lb}}{453.59237 \text{ g}} = 4.2 \text{ lb K}$$

So, 100. lb of the fertilizer contains 5.0 lb N, 4.4 lb P, and 4.2 lb K.

✓ *Reasonable Answer Check:* It makes sense that the percentage of P_2O_5 had to be higher to get approximately the same mass of P, since that compound has a significant mass of oxygen.

Chapter 4: Quantities of Reactants and Products

Chapter Contents:

Introduction

Balancing chemical equations is another one of those skills best learned as an algorithm. Random trial-and-error methods can cause more errors and much frustration. Mastering balancing chemical equations is critical in introductory chemistry. Teachers will expect you to be able to do it fast and flawlessly.

As in Chapter 3 with dissolving ionic compounds, focus on what you expect the usual products of a chemical reaction to be so you can avoid making up things. You should not have to guess what the form of the products are. With the identity of the reactants and some practiced anticipation of what a

One way to improve your comprehension of balanced chemical equations is to think about them with the corresponding word statements on both a particulate and molar level. For example, when looking at the equation: $O_2(g) + 2 H_2 (g) \longrightarrow 2 H_2O(\ell)$ you should see it as "one oxygen molecule reacts with two hydrogen molecules to produce two water molecules" or "one mole of oxygen gas reacts with two moles of hydrogen gas to produce two moles of liquid water." Remember, we *write* chemical equations and *observe* chemical reactions.

The word "Stoichiometry" may be intimidating, but it really just means the process of comparing quantities of counted things. You can get a sense for how stoichiometry works looking at Active Figure 4.5 showing the decomposition of water and Figure 4.6 showing a displacement reaction. Figure 4.8 shows a generic sequence of stoichiometry calculations.

Solutions to Blue-Numbered Questions
for Review and Thought in Chapter 4

Topical Questions

Stoichiometry

6. *Answer:*

	NH_3	O_2	NO	H_2O
No. molecules	4	5	4	6
No. atoms	16	10	8	18
No. moles of molecules	4	5	4	6
Mass (g)	68.1216	159.9940	120.0244	108.0912
Total mass of reactants (g)	228.1156		–	
Total mass of products (g)	–		228.1156	

Strategy and Explanation: Given the equation for the reaction, find the related molecules, atoms, moles, masses of each of the reactants and products, and total masses of the reactants and the products.

The stoichiometric coefficients (the numbers in front of the molecules' formulas) in the balanced equation can be interpreted as the related number of molecules or the related number of moles of molecules. The molar mass of each molecule is multiplied by the number of moles of that molecule represented in the equation to find its mass in grams. Add the mass of NH_3 and O_2 for the total reactant mass. Add the mass of NO and H_2O for the total product mass.

$$4 NH_3(g) + 5 O_2(g) \longrightarrow 4 NO(g) + 6 H_2O(g)$$

The stoichiometric coefficients are 4 for NH_3, 5 for O_2, 4 for NO and 6 for H_2O; these numbers represent relative numbers of molecules and relative numbers of moles of molecules.

	NH_3	O_2	NO	H_2O
No. molecules	4	5	4	6
No. atoms	$4 \times (1\ N + 3\ H) = 16$	$5 \times (2\ O) = 10$	$5 \times (1\ N + 1\ O) = 8$	$6 \times (2\ H + 1\ O) = 18$
No. moles of molecules	4	5	4	6
Mass	$4\ mol \times$ $(1 \times 14.0067g/mol +$ $3 \times 1.0079g/mol)$ $= 68.1216\ g$	$5\ mol \times$ $(2 \times 15.9994g/mol)$ $= 159.9940\ g$	$4\ mol \times$ $(1 \times 14.0067g/mol +$ $1 \times 15.9994g/mol) =$ $120.0244\ g$	$6\ mol \times$ $(2 \times 1.0079\ g/mol +$ $1 \times 15.9994g/mol) =$ $108.0912\ g$
Total mass of reactants	$68.1216\ g + 159.9940\ g = 228.1156\ g$		–	
Total mass of products		–	$120.0244\ g + 108.0912\ g = 228.1156\ g$	

✓ *Reasonable Answer Check:* A properly balanced equation has the same numbers of atoms of each type (4 N, 12 H, and 10 O) in the products and reactants, so the totals on each side should be equal, also (16+10=8+18). The law of conservation of mass says that mass of the reactants must equal the mass of the products, 228.1156g.

8. *Answer:* **Equation (b)**

Strategy and Explanation: In the first box, three A_2 molecules are combined with six B atoms. After the reaction occurs, the box contains only six AB molecules. That means symbolically: $3\ A_2 + 6\ B \longrightarrow 6\ AB$

We might be tempted to choose (a) after this observation. However, stoichiometric coefficients are supposed to be the smallest whole numbers relating the reactants and products and we find that all of these coefficients are divisible by 3. Therefore, the equation given in (b), $A_2 + 2\ B \longrightarrow 2\ AB$ is the best description of the proper stoichiometry of this reaction.

Classification of Chemical Reactions

11. *Answer:* **(a) combination   (b) decomposition (c) exchange (d) displacement**

Strategy and Explanation: Use the patterns of chemical reactions described in Section 4.2 to determine the type of reaction identified in each equation. The four major types of reactions (combination, decomposition, displacement, and exchange) can be identified by these four general equations:

Combination: $X + Y \longrightarrow XY$ Decomposition: $XY \longrightarrow X + Y$

Displacement: $A + BX \longrightarrow B + AX$ Exchange: $AX + BY \longrightarrow AY + BX$

(a) Two substances (Cu and O_2) combine to make a single product (CuO) in this equation, so it is a "Combination" reaction.

(b) Two substances (N_2O and H_2O) are produces when the reaction (NH_4NO_3) decomposes in this equation, so it is a "Decomposition" reaction.

(c) Two compounds ($BaCl_2$ and K_2SO_4) exchange ion partners to form two new compounds ($BaSO_4$ and KCl) in this equation, so it is an "Exchange" reaction.

(d) An element (Mg) replaces another element (H_2) in a compound in this equation, so it is a "Displacement" reaction.

✓ *Reasonable Answer Check:* All of these reactions follows one of the predicted patterns.

13. *Answer:* **(a) decomposition   (b) displacement   (c) combination   (d) exchange**

Strategy and Explanation: Use the patterns of chemical reactions described in Section 4.2 as summarized in the solution to Question 11.

(a) Two substances (PbO and CO_2) are produces when the reactant ($PbCO_3$) decomposes in this equation, so it is a "Decomposition" reaction.

(b) An element (Cu) replaces another element (H) in a compound in this equation, so it is a "Displacement"

(c) Two substances (Zn and O_2) combine to make a single product (ZnO) in this equation, so it is a "Combination" reaction.

(d) Two compounds ($BaCl_2$ and K_2SO_4) exchange ion partners to form new compounds ($BaSO_4$ and KCl) in this equation, so it is an "Exchange" reaction.

✓ *Reasonable Answer Check:* All of these reactions follows one of the predicted patterns.

Balancing Equations

15. *Answer:* **(a) $2\,Mg(s) + O_2(g) \longrightarrow 2\,MgO(s)$, magnesium oxide   (b) $2\,Ca(s) + O_2(g) \longrightarrow 2\,CaO(s)$, calcium oxide (c) $4\,In(s) + 3\,O_2(g) \longrightarrow 2\,In_2O_3(s)$, indium(III) oxide**

Strategy and Explanation: Balance the equation for the reaction between a metal and oxygen.

The products are metal oxides. Write the product in the form of a neutral metal oxide using the periodic table to predict the metal ion's charge. Balance the equations. Notice that the oxide anion has a –2 charge. Name the ionic compound product.

(a) Magnesium is in Group 2A. Its cation will have +2 charge.

$$2\,Mg(s) + O_2(g) \longrightarrow 2\,MgO(s) \qquad \text{The product is called magnesium oxide.}$$

(b) Calcium is in Group 2A. Its cation will have +2 charge.

$$2\,Ca(s) + O_2(g) \longrightarrow 2\,CaO(s) \qquad \text{The product is called calcium oxide.}$$

(c) Indium is in group 3A. Its cation will have +3 charge.

$$4\,In(s) + 3\,O_2(g) \longrightarrow 2\,In_2O_3(s) \qquad \text{The product is called indium(III) oxide.}$$

✓ *Reasonable Answer Check:* (a) 2 Mg & 2 O (b) 2 Ca & 2 O (c) 4 In & 6 O

17. *Answer:* **(a) $UO_2(s) + 4\,HF(\ell) \longrightarrow UF_4(s) + 2\,H_2O(\ell)$ (b) $B_2O_3(s) + 6\,HF(\ell) \longrightarrow 2\,BF_3(s) + 3\,H_2O(\ell)$ (c) $BF_3(g) + 3\,H_2O(\ell) \longrightarrow 3\,HF(\ell) + H_3BO_3(s)$**

When balancing equations, select a specific order in which the elements are balanced. Select first the atoms that are only in one product and one reactant, since they will be easier. Select next those elements that are in more than one reactant or product (such as H or O) and then those which are present in elemental form in either the reactants or products. Be systematic. If any of the coefficients end up fractional, multiply every coefficient by the same constant to eliminate the fraction.

(a) $?\,UO_2(s) + ?\,HF(\ell) \longrightarrow ?\,UF_4(s) + ?\,H_2O(\ell)$ Select order: U, F, H, O

     $UO_2(s) + ?\,HF(\ell) \longrightarrow UF_4(s) + ?\,H_2O(\ell)$ 1 U

     $UO_2(s) + 4\,HF(\ell) \longrightarrow UF_4(s) + ?\,H_2O(\ell)$ 4 F

     $UO_2(s) + 4\,HF(\ell) \longrightarrow UF_4(s) + 2\,H_2O(\ell)$ 4 H and 2 O

(b) $?\,B_2O_3(s) + ?\,HF(\ell) \longrightarrow ?\,BF_3(s) + ?\,H_2O(\ell)$ Select order: B, F, H, O

     $B_2O_3(s) + ?\,HF(\ell) \longrightarrow 2\,BF_3(s) + ?\,H_2O(\ell)$ 2 B

     $B_2O_3(s) + 6\,HF(\ell) \longrightarrow 2\,BF_3(s) + ?\,H_2O(\ell)$ 6 F

     $B_2O_3(s) + 6\,HF(\ell) \longrightarrow 2\,BF_3(s) + 3\,H_2O(\ell)$ 6 H and 3 O

(c) $?\,BF_3(g) + ?\,H_2O(\ell) \longrightarrow ?\,HF(g) + ?\,H_3BO_3(s)$ Select order: B, F, H, O

     $BF_3(g) + ?\,H_2O(\ell) \longrightarrow ?\,HF(\ell) + H_3BO_3(s)$ 1 B

     $BF_3(g) + ?\,H_2O(\ell) \longrightarrow 3\,HF(\ell) + H_3BO_3(s)$ 3 F

     $BF_3(g) + 3\,H_2O(\ell) \longrightarrow 3\,HF(\ell) + H_3BO_3(s)$ 6 H and 3 O

✓ *Reasonable Answer Check:* (a) 1 U, 2 O, 4 H & 4 F (b) 2 B, 3 O, 6 H & 6 F (c) 1 B, 3 F, 6 H & 3 O

19. *Answer:* **(a) H$_2$NCl(aq) + 2 NH$_3$(g) $\longrightarrow$ NH$_4$Cl(aq) + N$_2$H$_4$(aq)**

   (b) (CH$_3$)$_2$N$_2$H$_2$(ℓ) + 2 N$_2$O$_4$(g) $\longrightarrow$ 3 N$_2$(g) + 4 H$_2$O(g) + 2 CO$_2$(g)

   (c) CaC$_2$(s) + 2 H$_2$O(ℓ) $\longrightarrow$ Ca(OH)$_2$(s) + C$_2$H$_2$(g)

   Strategy and Explanation: Follow the method described in the solution for Question 17.

   (a) ? H$_2$NCl(aq) + ? NH$_3$(g) $\longrightarrow$? NH$_4$Cl(aq) + ? N$_2$H$_4$(aq)   Select order: Cl, N, H

   H$_2$NCl(aq) + ? NH$_3$(g) $\longrightarrow$ NH$_4$Cl(aq) + ? N$_2$H$_4$(aq)   1 Cl

   H$_2$NCl(aq) + 2 NH$_3$(g) $\longrightarrow$ NH$_4$Cl(aq) + N$_2$H$_4$(aq)   3 N and 8 H

   (b) ? (CH$_3$)$_2$N$_2$H$_2$(ℓ) + ? N$_2$O$_4$(g) $\longrightarrow$? N$_2$(g) + ? H$_2$O(ℓ) + ? CO$_2$(g)   Select order: H, C, O, N

   (CH$_3$)$_2$N$_2$H$_2$(ℓ) + ? N$_2$O$_4$(g) $\longrightarrow$? N$_2$(g) + 4 H$_2$O(g) + ? CO$_2$(g)   8 H

   (CH$_3$)$_2$N$_2$H$_2$(ℓ) + ? N$_2$O$_4$(g) $\longrightarrow$? N$_2$(g) + 4 H$_2$O(g) + 2 CO$_2$(g)   2 C

   (CH$_3$)$_2$N$_2$H$_2$(ℓ) + 2 N$_2$O$_4$(g) $\longrightarrow$? N$_2$(g) + 4 H$_2$O(g) + 2 CO$_2$(g)   8 O

   (CH$_3$)$_2$N$_2$H$_2$(ℓ) + 2 N$_2$O$_4$(g) $\longrightarrow$ 3 N$_2$(g) + 4 H$_2$O(g) + 2 CO$_2$(g)   6 N

   (c) ? CaC$_2$(s) + ? H$_2$O(ℓ) $\longrightarrow$? Ca(OH)$_2$(s) + ? C$_2$H$_2$(g)   Select order: Ca, C, O, H

   CaC$_2$(s) + ? H$_2$O(ℓ) $\longrightarrow$ Ca(OH)$_2$(s) + ? C$_2$H$_2$(g)   1 Ca

   CaC$_2$(s) + ? H$_2$O(ℓ) $\longrightarrow$ Ca(OH)$_2$(s) + C$_2$H$_2$(g)   2 C

   CaC$_2$(s) + 2 H$_2$O(ℓ) $\longrightarrow$ Ca(OH)$_2$(s) + C$_2$H$_2$(g)   2 O and 4 H

   ✓ *Reasonable Answer Check:* (a) 8 H, 3 N & 1 Cl (b) 2 C, 8 H, 6 N & 8 O (c) 1 Ca, 2 C, 4 H & 2 O

21. *Answer:* **(a) C$_6$H$_{12}$O$_6$ + 6 O$_2$ $\longrightarrow$ 6 CO$_2$ + 6 H$_2$O (b) C$_5$H$_{12}$ + 8 O$_2$ $\longrightarrow$ 5 CO$_2$ + 6 H$_2$O**

   (c) 2 C$_7$H$_{14}$O$_2$ + 19 O$_2$ $\longrightarrow$ 14 CO$_2$ + 14 H$_2$O (d) C$_2$H$_4$O$_2$ + 2 O$_2$ $\longrightarrow$ 2 CO$_2$ + 2 H$_2$O

   Strategy and Explanation: Balance the given combustion equations.

   When balancing combustion equations, select a C, H, O as the order in which the elements are balanced. Be systematic. If the coefficient of O$_2$ ends up a fraction, multiply every coefficient by 2 to make all of the coefficients whole numbers.

   (a) ? C$_6$H$_{12}$O$_6$ + ? O$_2$ $\longrightarrow$? CO$_2$ + ? H$_2$O

   C$_6$H$_{12}$O$_6$ + ? O$_2$ $\longrightarrow$ 6 CO$_2$ + ? H$_2$O   6 C

   C$_6$H$_{12}$O$_6$ + ? O$_2$ $\longrightarrow$ 6 CO$_2$ + 6 H$_2$O   12 H

   C$_6$H$_{12}$O$_6$ + 6 O$_2$ $\longrightarrow$ 6 CO$_2$ + 6 H$_2$O   18 O

   (b) ? C$_5$H$_{12}$ + ? O$_2$ $\longrightarrow$? CO$_2$ + ? H$_2$O

   C$_5$H$_{12}$ + ? O$_2$ $\longrightarrow$ 5 CO$_2$ + ? H$_2$O   5 C

   C$_5$H$_{12}$ + ? O$_2$ $\longrightarrow$ 5 CO$_2$ + 6 H$_2$O   12 H

   C$_5$H$_{12}$ + 8 O$_2$ $\longrightarrow$ 5 CO$_2$ + 6 H$_2$O   16 O

   (c) ? C$_7$H$_{14}$ + ? O$_2$ $\longrightarrow$? CO$_2$ + ? H$_2$O

   C$_7$H$_{14}$O$_2$ + ? O$_2$ $\longrightarrow$ 7 CO$_2$ + ? H$_2$O   7 C

   C$_7$H$_{14}$O$_2$ + ? O$_2$ $\longrightarrow$ 7 CO$_2$ + 7 H$_2$O   14 H

   C$_7$H$_{14}$O$_2$ + $\frac{19}{2}$ O$_2$ $\longrightarrow$ 7 CO$_2$ + 7 H$_2$O   21 O

   2 C$_7$H$_{14}$O$_2$ + 19 O$_2$ $\longrightarrow$ 14 CO$_2$ + 14 H$_2$O

   (d) ? C$_2$H$_4$O$_2$ + ? O$_2$ $\longrightarrow$? CO$_2$ + ? H$_2$O

   C$_2$H$_4$O$_2$ + ? O$_2$ $\longrightarrow$ 2 CO$_2$ + ? H$_2$O   2 C

   C$_2$H$_4$O$_2$ + ? O$_2$ $\longrightarrow$ 2 CO$_2$ + 2 H$_2$O   4 H

$$C_2H_4O_2 + 2\ O_2 \longrightarrow 2\ CO_2 + 2\ H_2O \qquad\qquad\qquad 6\ O$$

✓ *Reasonable Answer Check:* (a) 6 C, 12 H & 18 O, (b) 5 C, 12 H & 16 O, (c) 14 C, 28 H, 42 O (d) 2 C, 4 H, 6 O

The Mole and Chemical Reactions

23. *Answer:* **50.0 mol HCl**

Strategy and Explanation: Given a balanced chemical equation and the number of moles of a product formed, determine the moles of reactant that was needed.

Use the stoichiometry of the balanced equation as a conversion factor to convert the moles of product to moles of reactant.

The balanced equation says: 4 mol HCl are needed to make 1 mol Cl_2.

$$12.5\ \text{mol Cl}_2 \times \frac{4\ \text{mol HCl}}{1\ \text{mol Cl}_2} = 50.0\ \text{mol HCl}$$

✓ *Reasonable Answer Check:* More HCl molecules are needed than Cl_2 molecules are formed. It makes sense that moles of HCl are greater.

25. *Answer:* **1.1 mol O_2, 35 g O_2, 1.0 × 10^2 g NO_2**

Strategy and Explanation: Given a balanced chemical equation and the number of moles of a reactant, determine the moles and grams of another reactant needed and the grams of product produced.

Use the stoichiometry of the balanced equation as a conversion factor to convert the moles of the one reactant to moles of the other reactant. Use the molar mass to convert from moles to grams. Use the stoichiometry of the equation to find the moles of product formed, then use the molar mass of the product to find the grams.

The balanced equation says: 2 mol NO react with 1 mol O_2.

$$2.2\ \text{mol NO} \times \frac{1\ \text{mol O}_2}{2\ \text{mol NO}} = 1.1\ \text{mol O}_2$$

$$1.1\ \text{mol O}_2 \times \frac{31.9988\ \text{g O}_2}{1\ \text{mol O}_2} = 35\ \text{g O}_2$$

The balanced equation says: 2 mol NO produces 2 mol NO_2.

$$2.2\ \text{mol NO} \times \frac{2\ \text{mol NO}_2}{2\ \text{mol NO}} \times \frac{46.0055\ \text{g NO}_2}{1\ \text{mol NO}_2} = 1.0 \times 10^2\ \text{g NO}_2\ \text{produced}$$

✓ *Reasonable Answer Check:* Fewer moles of O_2 are needed than moles of NO. The mass of NO used is 2.2 mol × (30.0061 g/mol) = 66 g. The sum of the reactant masses (66 g + 35 g) is equal to the products mass (1.0 × 10^2 g), within known significant figures.

27. *Answer:* **(a) 12.7 g Cl_2 (b) 0.179 mol $FeCl_2$, 22.7 g $FeCl_2$ expected**

Strategy and Explanation: Given a balanced chemical equation and the number of grams of a reactant, determine the grams of another reactant needed and the moles and grams of product produced.

Use the molar mass of the first reactant to find the moles of that substance. Then use the stoichiometry of the balanced equation as a conversion factor to convert the moles of the one reactant to moles of the other reactant. Use the molar mass to convert from moles to grams. Use the stoichiometry of the equation to find moles of product formed, then use the molar mass of the product to find grams.

(a) The balanced equation says: 1 mol Fe reacts with 1 mol Cl_2.

$$10.0\ \text{g Fe} \times \frac{1\ \text{mol Fe}}{55.845\ \text{g Fe}} = 0.179\ \text{mol Fe}$$

$$0.179\ \text{mol Fe} \times \frac{1\ \text{mol Cl}_2}{1\ \text{mol Fe}} \times \frac{70.906\ \text{g Cl}_2}{1\ \text{mol Cl}_2} = 12.7\ \text{g Cl}_2$$

(b) The balanced equation says: 1 mol Fe produces 1 mol $FeCl_2$.

$$0.179 \text{ mol Fe} \times \frac{1 \text{ mol } FeCl_2}{1 \text{ mol Fe}} = 0.179 \text{ mol } FeCl_2 \text{ expected}$$

$$0.179 \text{ mol } FeCl_2 \times \frac{126.751 \text{ g } FeCl_2}{1 \text{ mol } FeCl_2} = 22.7 \text{ g } FeCl_2 \text{ expected}$$

✓ *Reasonable Answer Check:* The sum of the masses of the reactants must add up to the total mass of the product. 10.0 g + 12.7 g = 22.7 g. This is the same product mass as that calculated above.

29. *Answer:* The complete table looks like this:

$(NH_4)_2PtCl_6$	Pt	HCl
12.35 g	**5.428** g	**5.410** g
0.02782 mol	**0.02782** mol	**0.1484** mol

Strategy and Explanation: Given a balanced chemical equation and the number of grams of a reactant, determine the moles of the reactant and the mass and moles of two products.

Use the molar mass of the reactant to find the moles of that substance. Then use the stoichiometry of the balanced equation as a conversion factor to convert the moles of reactant to moles of each product formed, then use the molar mass of each product to find the grams.

Molar mass of $(NH_4)_2PtCl_6$ =
$$2 \times [1 \times (14.0067 \text{ g}) + 4 \times (1.0079 \text{ g})] + (195.078 \text{ g}) + 6 \times (35.453 \text{ g}) = 443.873 \text{ g/mol}$$

$$12.35 \text{ g } (NH_4)_2PtCl_6 \times \frac{1 \text{ mol } (NH_4)_2PtCl_6}{443.873 \text{ g } (NH_4)_2PtCl_6} = 2.782 \times 10^{-2} \text{ mol } (NH_4)_2PtCl_6$$

The balanced equation says: 3 mol $(NH_4)_2PtCl_6$ react to form 3 mol Pt.

$$2.782 \times 10^{-2} \text{ mol } (NH_4)_2PtCl_6 \times \frac{1 \text{ mol Pt}}{1 \text{ mol } (NH_4)_2PtCl_6} = 2.782 \times 10^{-2} \text{ mol Pt}$$

$$2.782 \times 10^{-2} \text{ mol Pt} \times \frac{195.078 \text{ g Pt}}{1 \text{ mol Pt}} = 5.428 \text{ g Pt}$$

The balanced equation says: 3 mol $(NH_4)_2PtCl_6$ react to form 16 mol HCl.

$$2.782 \times 10^{-2} \text{ mol } (NH_4)_2PtCl_6 \times \frac{16 \text{ mol HCl}}{3 \text{ mol } (NH_4)_2PtCl_6} = 0.1484 \text{ mol HCl}$$

$$0.1484 \text{ mol HCl} \times \frac{36.461 \text{ mol HCl}}{1 \text{ mol HCl}} = 5.410 \text{ g HCl}$$

✓ *Reasonable Answer Check:* The moles are smaller than the grams. This is appropriate. The moles of HCl are larger than the moles of $(NH_4)_2PtCl_6$ and Pt.

31. *Answer:* **2.0 mol, 36.0304 g**

Strategy and Explanation: Given the identity of a reactant undergoing a named reaction and the number of moles of a reactant, determine the number of moles and the mass of one product formed.

Balanced the equation for the reaction. Using the moles of reactant and the stoichiometry of the balanced equation, determine the moles of the product produced. Then use the molar mass of the product to find the grams.

Combustion of the hydrocarbon, C_3H_8, involves its reaction O_2 to produce CO_2 and H_2O.

$$C_3H_8 + 5 O_2 \longrightarrow 3 CO_2 + 4 H_2O$$

The balanced equation says: 4 mol H_2O is produced when 5 mol O_2 react.

$$2.5 \text{ mol } O_2 \times \frac{4 \text{ mol } H_2O}{5 \text{ mol } O_2} = 2.0 \text{ mol } H_2O$$

$$2.0 \text{ mol } H_2O \times \frac{18.0152 \text{ g } H_2O}{1 \text{ mol } H_2O} = 36.0304 \text{ g } H_2O$$

✓ *Reasonable Answer Check:* The moles of H_2O are is smaller than the moles of O_2. The number representing the mass of a molecule will always be larger than the number of moles.

32. *Answer:* **0.699 g Ga and 0.751 g As**

Strategy and Explanation: Given the formula of a compound and its mass, determine the mass of the elements required to make it.

Use the molar mass of the compound to find the moles of that substance. Then use the stoichiometry of the balanced equation to determine the moles of each element needed. Then use the molar masses of these elements to find the grams.

The formula gives: 1 mol GaAs is produced from 1 mol Ga and 1 mol As.

$$1.45 \text{ g GaAs} \times \frac{1 \text{ mol GaAs}}{144.6446 \text{ g GaAs}} \times \frac{1 \text{ mol Ga}}{1 \text{ mol GaAs}} \times \frac{69.723 \text{ g Ga}}{1 \text{ mol Ga}} = 0.699 \text{ g Ga}$$

$$1.45 \text{ g GaAs} \times \frac{1 \text{ mol GaAs}}{144.6446 \text{ g GaAs}} \times \frac{1 \text{ mol As}}{1 \text{ mol GaAs}} \times \frac{74.9215 \text{ g As}}{1 \text{ mol As}} = 0.751 \text{ g As}$$

✓ *Reasonable Answer Check:* The sum of the reactant masses (0.699 g + 0.751 g = 1.450 g) is the same as the total mass of the compound.

34. *Answer:* **(a) 699 g (b) 526 g**

Strategy and Explanation: Given a balanced chemical equation and the mass of a reactant in kilograms, determine the mass of one of the products produced and the mass of another reactant required.

Use metric conversion factors to convert kilograms to grams. Use the molar mass of the reactant to find the moles of that substance. Then use the stoichiometry of the equation to determine the moles of the product produced and the other reactant required. Then use molar masses to find the grams.

$$1.00 \text{ kg } Fe_2O_3 \times \frac{1000 \text{ g } Fe_2O_3}{1 \text{ kg } Fe_2O_3} \times \frac{1 \text{ mol } Fe_2O_3}{159.688 \text{ g } Fe_2O_3} = 6.26 \text{ mol } Fe_2O_3$$

(a) The balanced equation says: 1 mol Fe_2O_3 produces 1 mol Fe.

$$6.26 \text{ mol } Fe_2O_3 \times \frac{2 \text{ mol Fe}}{1 \text{ mol } Fe_2O_3} \times \frac{55.845 \text{ g Fe}}{1 \text{ mol Fe}} = 699 \text{ g Fe}$$

(b) The balanced equation says: 1 mol Fe_2O_3 requires 3 mol CO.

$$6.26 \text{ mol } Fe_2O_3 \times \frac{3 \text{ mol CO}}{1 \text{ mol } Fe_2O_3} \times \frac{28.0101 \text{ g CO}}{1 \text{ mol CO}} = 526 \text{ g CO}$$

✓ *Reasonable Answer Check:* The mass of iron produced is less than the mass of iron(III) oxide it was produced from. The mass of CO used is less than the mass of iron(III) oxide even with a larger stoichiometric coefficient because CO contains lighter-weight atoms.

Limiting Reactant

35. *Answer:* **BaCl$_2$, 1.12081 g BaSO$_4$**

Strategy and Explanation: Given a balanced chemical equation and the masses of both reactants, determine the limiting reactant and the mass of the product produced.

Here, we use a slight variation of what the text calls "the mole method." We will calculate a directly comparable quantity, the moles of the desired product. Use the molar mass of the reactants to find the moles of the reactant substances. Then use the stoichiometry of the equation to determine the moles of the product produced in each case. Identify the limiting reactant from the reactant that produces the least number of products. Then, use the moles of product produced from the limiting reactant and the molar mass of the product to find the grams.

Exactly 1 gram of each reactant is present initially.

The balanced equation says: 1 mol Na_2SO_4 produces 1 mol $BaSO_4$.

$$1 \text{ g } Na_2SO_4 \times \frac{1 \text{ mol } Na_2SO_4}{142.042 \text{ g } Na_2SO_4} \times \frac{1 \text{ mol } BaSO_4}{1 \text{ mol } Na_2SO_4} = 0.00704016 \text{ mol } BaSO_4$$

The balanced equation says: 1 mol $BaCl_2$ produces 1 mol $BaSO_4$.

$$1 \text{ g } BaCl_2 \times \frac{1 \text{ mol } BaCl_2}{208.233 \text{ g } BaCl_2} \times \frac{1 \text{ mol } BaSO_4}{1 \text{ mol } BaCl_2} = 0.00480231 \text{ mol } BaSO_4$$

The number of $BaSO_4$ moles produced from $BaCl_2$ is smaller (0.00480231 mol < 0.00704016 mol), so $BaCl_2$ is the limiting reactant.

Find the mass of 0.00480231 mol $BaSO_4$:

$$0.00480231 \text{ mol } BaSO_4 \times \frac{233.390 \text{ g } BaSO_4}{1 \text{ mol } BaSO_4} = 1.12081 \text{ g } BaSO_4$$

✓ *Reasonable Answer Check:* There are fewer moles of $BaSO_4$ present than $BaCl_2$, and the equation needs the same amount of $BaSO_4$ as $BaCl_2$, so it makes sense that $BaCl_2$ is the limiting reactant.

37. *Answer:* **(a) Cl_2 is limiting. (b) 5.08 g Al_2Cl_6 (c) 1.67 g Al unreacted**

Strategy and Explanation: Given a balanced chemical equation and the masses of both reactants, determine the limiting reactant, the mass of the product produced, and the mass remaining of the excess reactant when the reaction is complete.

Here, we use a slight variation of what the text calls "the mole method." We will calculate a directly comparable quantity, the moles of product. (a) Use the molar mass of the reactants to find the moles of the reactant substances. Then use the stoichiometry of the equation to determine the moles of the product produced in each case. Identify the limiting reactant from the reactant that produces the least number of products. (b) Use the moles of product produced from the limiting reactant and the molar mass of the product to find the grams. (c) From the limiting reactant quantity, determine the moles of the other reactant needed for complete reaction. Convert that number to grams using the molar mass. Then subtract the quantity used from the initial mass given to get the mass of excess reactant.

(a) The balanced equation says: 2 mol Al produces 1 mol Al_2Cl_6.

$$2.70 \text{ g } Al \times \frac{1 \text{ mol } Al}{26.9815 \text{ g } Al} \times \frac{1 \text{ mol } Al_2Cl_6}{2 \text{ mol } Al} = 0.0500 \text{ mol } Al_2Cl_6$$

The balanced equation says: 3 mol Cl_2 produces 1 mol Al_2Cl_6.

$$4.05 \text{ g } Cl_2 \times \frac{1 \text{ mol } Cl_2}{70.906 \text{ g } Cl_2} \times \frac{1 \text{ mol } Al_2Cl_6}{3 \text{ mol } Cl_2} = 0.0190 \text{ mol } Al_2Cl_6$$

The number of Al_2Cl_6 moles produced from Cl_2 is smaller (0.0190 mol < 0.0500 mol), so Cl_2 is the limiting reactant and Al is the excess reactant.

(b) Find the mass of 0.0190 mol Al_2Cl_6:

$$0.0190 \text{ mol } Al_2Cl_6 \times \frac{266.682 \text{ g } Al_2Cl_6}{1 \text{ mol } Al_2Cl_6} = 5.08 \text{ g } Al_2Cl_6$$

(c) The balanced equation says: 3 mol Cl_2 react with 2 mol Al.

$$4.05 \text{ g } Cl_2 \times \frac{1 \text{ mol } Cl_2}{70.906 \text{ g } Cl_2} \times \frac{2 \text{ mol Al}}{3 \text{ mol } Cl_2} \times \frac{26.9815 \text{ g Al}}{1 \text{ mol Al}} = 1.03 \text{ g Al}$$

2.70 g Al initial – 1.03 g Al used up = 1.67 g Al remains unreacted

✓ *Reasonable Answer Check:* There are fewer moles of Cl_2 present than Al, and the equation needs more Cl_2 than Al, so it makes sense that Cl_2 is the limiting reactant. In addition, the calculation in (c) proved that the initial mass of Al was larger than required to react with all of the Cl_2. The sum of the masses of the reactants that reacted (4.05 g + 1.03 g = 5.08 g) equals the mass of the product produced (5.08 g).

38. *Answer:* **(a) CO (b) 1.3 g H_2 (c) 85.2 g**

Strategy and Explanation: Given a balanced chemical equation and the masses of both reactants, determine the limiting reactant, the mass of the product produced, and the mass remaining of the excess reactant when the reaction is complete.

To solve (a): use the molar mass of the reactants to find the moles of the reactant substances. Then use the stoichiometry of the equation to determine the moles of the product produced in each case. Identify the limiting reactant from the reactant that produces the least number of products. To solve (b): from the limiting reactant quantity determine the moles of the other reactant needed for complete reaction. Convert that number to mass using the molar mass. Then subtract the quantity used from the initial mass given, to get the mass of excess reactant. (c) Use the moles of product produced from the limiting reactant and the molar mass of the product to find the grams.

(a) The balanced equation says: 1 mol CO produces 1 mol CH_3OH.

$$74.5 \text{ g CO} \times \frac{1 \text{ mol CO}}{28.0101 \text{ g CO}} \times \frac{1 \text{ mol } CH_3OH}{1 \text{ mol CO}} = 2.66 \text{ mol } CH_3OH$$

The balanced equation says: 2 mol H_2 produces 1 mol CH_3OH.

$$12.0 \text{ g } H_2 \times \frac{1 \text{ mol } H_2}{2.0158 \text{ g } H_2} \times \frac{1 \text{ mol } CH_3OH}{2 \text{ mol } H_2} = 2.98 \text{ mol } CH_3OH$$

The number of CH_3OH moles produced from CO is smaller (2.66 mol < 2.98 mol), so CO is the limiting reactant and H_2 is the excess reactant.

(b) The balanced equation says: 2 mol H_2 react with 1 mol CO.

$$74.5 \text{ g CO} \times \frac{1 \text{ mol CO}}{28.0101 \text{ g CO}} \times \frac{2 \text{ mol } H_2}{1 \text{ mol CO}} \times \frac{2.0158 \text{ g } H_2}{1 \text{ mol } H_2} = 10.7 \text{ g } H_2$$

12.0 g H_2 initial – 10.7 g H_2 used up = 1.3 g H_2 remains unreacted

(c) Find the mass of 2.66 mol CH_3OH:

$$2.66 \text{ mol } CH_3OH \times \frac{32.0417 \text{ g } CH_3OH}{1 \text{ mol } CH_3OH} = 85.2 \text{ g } CH_3OH$$

✓ *Reasonable Answer Check:* The calculation in (b) proved that the initial mass of H_2 was larger than required to react with all the CO. The sum of the masses of the reactants that reacted (74.5 g + 10.7 g = 85.2 g) equal the mass of the product produced (85.2 g).

39. *Answer:* **0 mol CaO, 0.19 mol NH_4Cl, 2.00 mol H_2O, 4.00 mol NH_3, 2.00 mol $CaCl_2$**

Strategy and Explanation: Given a balanced chemical equation and the masses of both reactants, determine the moles of reactants and products present when the reaction is finished.

Use the molar mass of the reactants to find the moles of the reactant substances present initially. Then use the stoichiometry of the equation to determine the moles of one of the products. Identify the limiting reactant from the reactant that produces the least number of products. From the limiting reactant quantity, and using the stoichiometry of the balanced equation, determine the moles of the other reactant and products.

The balanced equation says: 1 mol CaO produces 1 mol H_2O.

$$112 \text{ g CaO} \times \frac{1 \text{ mol CaO}}{56.077 \text{ g CaO}} \times \frac{1 \text{ mol H}_2\text{O}}{1 \text{ mol CaO}} = 2.00 \text{ mol H}_2\text{O}$$

The balanced equation says: 2 mol NH_4Cl produces 1 mol H_2O.

$$224 \text{ g NH}_4\text{Cl} \times \frac{1 \text{ mol NH}_4\text{Cl}}{53.4913 \text{ g NH}_4\text{Cl}} \times \frac{1 \text{ mol H}_2\text{O}}{2 \text{ mol NH}_4\text{Cl}} = 2.09 \text{ mol H}_2\text{O}$$

The number of H_2O moles produced from CaO is smaller (2.00 mol < 2.09 mol), so CaO is the limiting reactant and NH_4Cl is the excess reactant.

The balanced equation says: 1 mol CaO produces 2 mol NH_3.

$$112 \text{ g CaO} \times \frac{1 \text{ mol CaO}}{56.077 \text{ g CaO}} \times \frac{2 \text{ mol NH}_3}{1 \text{ mol CaO}} = 4.00 \text{ mol NH}_3$$

The balanced equation says: 1 mol CaO produces 1 mol $CaCl_2$.

$$112 \text{ g CaO} \times \frac{1 \text{ mol CaO}}{56.077 \text{ g CaO}} \times \frac{1 \text{ mol CaCl}_2}{1 \text{ mol CaO}} = 2.00 \text{ mol CaCl}_2$$

The balanced equation says: 1 mol CaO reacts with 2 mol NH_4Cl.

$$112 \text{ g CaO} \times \frac{1 \text{ mol CaO}}{56.077 \text{ g CaO}} \times \frac{2 \text{ mol NH}_4\text{Cl}}{1 \text{ mol CaO}} = 4.00 \text{ mol NH}_4\text{Cl reacted}$$

$$224 \text{ g NH}_4\text{Cl} \times \frac{1 \text{ mol NH}_4\text{Cl}}{53.4913 \text{ g NH}_4\text{Cl}} = 4.19 \text{ mol NH}_4\text{Cl present initially}$$

$$4.19 \text{ mol NH}_4\text{Cl initial} - 4.00 \text{ mol NH}_4\text{Cl reacted} = 0.19 \text{ mol NH}_4\text{Cl left}$$

✓ *Reasonable Answer Check:* The excess moles of NH_4Cl prove that the right limiting reactant was determined, since there are zero moles of CaO left. The moles of products are stoichiometrically appropriate multiples of 2.00 moles, as they should be.

Percent Yield

42. *Answer:* **56.0%**

Strategy and Explanation: Given the theoretical yield and the actual yield, determine the percent yield.

Divide the actual yield by the theoretical yield and multiply by 100% to get percent yield. *(Note that the balanced equation is given, too, but you don't need to use it to answer this Question.)*

$$\frac{36.7 \text{ g CaO actual}}{65.5 \text{ g CaO theoretical}} \times 100\% = 56.0\% \text{ yield}$$

✓ *Reasonable Answer Check:* A little more than half the maximum quantity of quicklime was produced, so it makes sense that the percent yield is a little more than 50%.

44. *Answer:* **8.8%**

Strategy and Explanation: Given the balanced chemical equation, the mass of the limiting reactant and the actual yield, determine the percent yield.

First, we need to calculate the theoretical yield. We get this by determining the maximum amount of product that could have been made from the given quantities of reactant: Take the mass of the limiting reactant and convert it to moles. Then use stoichiometry to find the moles of product. Then convert to grams using molar mass. Take the actual yield and divide by the calculated theoretical yield and multiply by 100% to get percent yield.

The limiting reactant is H_2. From its mass, find the maximum grams of CH_3OH that could be made. The mole ratio comes from the balanced equation.

$$5.0 \times 10^3 \text{ g } H_2 \times \frac{1 \text{ mol } H_2}{2.0158 \text{ g } H_2} \times \frac{1 \text{ mol } CH_3OH}{2 \text{ mol } H_2} \times \frac{32.0417 \text{ g } CH_3OH}{1 \text{ mol } CH_3OH} = 4.0 \times 10^4 \text{ g } CH_3OH$$

The given mass of CH_3OH is the actual yield. Use these two masses to calculate the percent yield.

$$\frac{3.5 \times 10^3 \text{ g } CH_3OH \text{ actual}}{4.0 \times 10^4 \text{ g } CH_3OH \text{ theoretical}} \times 100\% = 8.8\% \text{ yield}$$

✓ *Reasonable Answer Check:* The actual mass produced is roughly a factor of ten less than the maximum quantity of ethanol produced, so it makes sense that the percent yield is around 10%.

Empirical Formulas

48. *Answer:* **SO_3**

Strategy and Explanation: Given percent mass of elements in a compound, determine the empirical formula.

Choose a convenient sample mass of S_xO_y, such as 100.0 g. Find the mass of S and O in the sample, using the given mass percent. Use the molar mass of the elements to find their moles, then use a whole-number mole ratio to determine the empirical formula.

100.0 g of S_xO_y contains 40.0 g S and 60.0 g O.

$$40.0 \text{ g } S \times \frac{1 \text{ mol } S}{32.065 \text{ g } S} = 1.25 \text{ mol } S \qquad 60.0 \text{ g } O \times \frac{1 \text{ mol } O}{15.9994 \text{ g } O} = 3.75 \text{ mol } O$$

Set up mole ratio and simplify by dividing by the smallest number of moles:

1.25 mol S : 3.75 mol O

1 S : 3 O

Use the whole number ratio for the subscripts in the formula. The empirical formula is SO_3.

✓ *Reasonable Answer Check:* The percent by mass of oxygen in the compound is somewhat greater than the percent by mass of sulfur. Since sulfur atoms weigh more than oxygen atoms, it makes sense that there are several O atom per S atom present in this formula. SO_3 is a common oxide of sulfur.

50. *Answer:* **CH**

Strategy and Explanation: Given the mass of a compound, the identity of the elements in the compound, and the identity and masses of all the products produced, determine the empirical formula.

Use the molar mass of the products to find their moles and use the stoichiometry of their formulas to determine the moles of the elements in the compound. Then use a whole-number mole ratio to determine the empirical formula.

The compound is a hydrocarbon and contains only C and H: C_iH_j. Its combustion produced H_2O and CO_2. Use molar mass and formula stoichiometry to determine the moles of C and H.

$$1.481 \text{ g } CO_2 \times \frac{1 \text{ mol } CO_2}{44.0095 \text{ g } CO_2} \times \frac{1 \text{ mol } C}{1 \text{ mol } CO_2} = 0.03365 \text{ mol } C$$

$$0.303 \text{ g } H_2O \times \frac{1 \text{ mol } H_2O}{18.0152 \text{ g } H_2O} \times \frac{2 \text{ mol } H}{1 \text{ mol } H_2O} = 0.0336 \text{ mol } H$$

Set up mole ratio and simplify by dividing by the smallest number of moles:

0.03365 mol C : 0.0336 mol H

1 C : 1 H

Use the whole number ratio for the subscripts in the formula. The empirical formula is CH.

✓ *Reasonable Answer Check:* These are not necessary calculations, but we can calculate the masses of each element, C and H.

$$0.03365 \text{ mol C} \times \frac{12.0107 \text{ g C}}{1 \text{ mol C}} = 0.4041 \text{ g C} \qquad 0.0336 \text{ mol H} \times \frac{1.0079 \text{ g H}}{1 \text{ mol H}} = 0.0339 \text{ g H}$$

The sum of the masses (0.4041 g + 0.0339 g = 0.4380 g) adds up to the mass of the compound (0.438 g).

52. *Answer:* $C_3H_6O_2$

Strategy and Explanation: Given the mass of a compound, the identity of the elements in the compound, and the identity and masses of all the products produced, determine the empirical formula.

Combustion uses oxygen. When the compound also contains oxygen, determine the amount of oxygen after the other elements. Use the molar mass of the products to find their moles and use the stoichiometry of their formulas to determine the moles of the elements that are not oxygen in the compound. Find the masses of those elements, and subtract them from the total mass of the compound to get the mass of oxygen in the compound. Then use the molar mass to calculate the moles of oxygen in the compound. Then use a whole-number mole ratio to determine the empirical formula.

The compound contains C, H, and O: $C_iH_jO_k$. Its combustion produced H_2O and CO_2. Use molar mass and formula stoichiometry to determine the moles of C and H. Use the whole number ratio for the subscripts in the formula.

$$0.421 \text{ g CO}_2 \times \frac{1 \text{ mol CO}_2}{44.0095 \text{ g CO}_2} \times \frac{1 \text{ mol C}}{1 \text{ mol CO}_2} = 9.56 \times 10^{-3} \text{ mol C}$$

$$0.172 \text{ g H}_2\text{O} \times \frac{1 \text{ mol H}_2\text{O}}{18.0152 \text{ g H}_2\text{O}} \times \frac{2 \text{ mol H}}{1 \text{ mol H}_2\text{O}} = 1.91 \times 10^{-2} \text{ mol H}$$

Calculate the masses of C and H.

$$9.56 \times 10^{-3} \text{ mol C} \times \frac{12.0107 \text{ g C}}{1 \text{ mol C}} = 0.115 \text{ g C} \qquad 0.0191 \text{ mol H} \times \frac{1.0079 \text{ g H}}{1 \text{ mol H}} = 0.0192 \text{ g H}$$

Calculate the masses of O by subtracting the masses of C and H from the given total compound mass.

$$0.236 \text{ g C}_i\text{H}_j\text{O}_k - 0.115 \text{ g C} - 0.0192 \text{ g H} = 0.102 \text{ g O}$$

Calculate the moles of O: $\qquad 0.102 \text{ g O} \times \frac{1 \text{ mol O}}{15.9994 \text{ g O}} = 6.37 \times 10^{-3} \text{ mol O}$

Set up mole ratio and simplify by dividing by the smallest number of moles:

$$9.56 \times 10^{-3} \text{ mol C} : 1.91 \times 10^{-2} \text{ mol H} : 6.37 \times 10^{-3} \text{ mol O}$$

$$1.5 \text{ C} : 3 \text{ H} : 1 \text{ O}$$

Multiply by 2, to get a whole number ratio: 3 C : 6 H : 2 O

Use the whole number ratio for the subscripts in the formula.

The empirical formula is $C_3H_6O_2$.

✓ *Reasonable Answer Check:* The mole ratio came out very close to whole number values.

General Questions

54. *Answer:* **21.6 g N_2**

Strategy and Explanation: Given the unbalanced chemical equation and the mass of the limiting reactant, determine the mass of a product that can be isolated.

Balance the equation. Use the molar mass of the reactant to find moles of reactant, then use the stoichiometry of the equation to determine moles of product, then use the molar mass of the product to get the mass of the product.

Select a balance order: H, O, N, Cu

$$2 \text{ NH}_3 + 3 \text{ CuO} \longrightarrow \text{N}_2 + 3 \text{ Cu} + 3 \text{ H}_2\text{O}$$

Check the balance: 2 N, 6 H, 3 Cu, 3 O

$$26.3 \text{ g NH}_3 \times \frac{1 \text{ mol NH}_3}{17.0304 \text{ g NH}_3} \times \frac{1 \text{ mol N}_2}{2 \text{ mol NH}_3} \times \frac{28.0134 \text{ g N}_2}{1 \text{ mol N}_2} = 21.6 \text{ g N}_2$$

✓ *Reasonable Answer Check:* It makes sense that the initial mass of NH_3 is nearly the same as the mass of N_2 created, since the NH_3 molar mass is about half that of N_2 and the stoichiometry is 2:1.

56. *Answer:* **Element (b)**

Strategy and Explanation: Given the mass of the only reactant, the formula of the reactant with one element unknown, a balanced equation showing its decomposition, and the mass and identity of one of the products, determine the identity of the unknown element from a given list.

Use the molar mass of the product to find moles of product, then use the stoichiometry of the equation to determine moles of reactant. Then divide the grams of reactant by the calculated moles of reactant to determine the molar mass of the reactant. Subtract the molar masses of the known elements in the compound to determine the molar mass of the unknown element. Compare the molar masses of the elements on the list and find the one that most closely matches it.

Use mass of CO_2 to find moles of MCO_3

$$0.376 \text{ g CO}_2 \times \frac{1 \text{ mol CO}_2}{44.0095 \text{ g CO}_2} \times \frac{1 \text{ mol MCO}_3}{1 \text{ mol CO}_2} = 8.54 \times 10^{-3} \text{ mol MCO}_3$$

Divide given mass by moles:

$$\frac{1.056 \text{ g MCO}_3}{8.54 \times 10^{-3} \text{ mol MCO}_3} = 124 \frac{\text{g}}{\text{mol}} = \text{molar mass of MCO}_3$$

Subtract the molar mass of C and three times the molar mass of O from this molar mass to find the molar mass of M:

$$\frac{124 \text{ g MCO}_3}{\text{mol MCO}_3} - \frac{12.0107 \text{ g C}}{\text{mol MCO}_3} - \frac{3 \times 15.9994 \text{ g O}}{\text{mol MCO}_3} = 64 \text{ g M}$$

The element in the given list with the closest molar mass to 64 g/mol is (b) Cu.

✓ *Reasonable Answer Check:* The molar mass of Cu (63.5 g/mol) is 64. None of the others in the list (Ni at 58.7 g/mol, Zn at 65.4 g/mol, or Ba at 137.2 g/mol) are this close, though within ±1 g/mol Zinc almost qualifies. If we carry more decimal places than strictly allowed by the rules of significant figures, the molar mass of MCO_3 is 123.65 g/mol and the molar mass of M is 63.6 g/mol. It might make us feel better picking Cu, although the additional significant figures are unknown with the limited data provided.

57. *Answer:* **12.5 g Pt(NH$_3$)$_2$Cl$_2$**

Strategy and Explanation: Given the balanced chemical equation and the mass of one reactant and the moles of the other reactant, determine the maximum theoretical mass of a product.

Find the limiting reactant by determining moles of a product made from each reactant. Then using the molar mass of the product, determine the mass from the moles produced by the limiting reactant.

$$15.5 \text{ g (NH}_4)_2\text{PtCl}_4 \times \frac{1 \text{ mol (NH}_4)_2\text{PtCl}_4}{372.967 \text{ g (NH}_4)_2\text{PtCl}_4} \times \frac{1 \text{ mol Pt(NH}_3)_2\text{Cl}_2}{1 \text{ mol (NH}_4)_2\text{PtCl}_4} = 0.0416 \text{ mol Pt(NH}_3)_2\text{Cl}_2$$

$$0.15 \text{ mol NH}_3 \times \frac{1 \text{ mol Pt(NH}_4)_2\text{Cl}_2}{2 \text{ mol NH}_3} = 0.075 \text{ mol Pt(NH}_4)_2\text{Cl}_2$$

$(NH_4)_2PtCl_4$ is the limiting reactant, since fewer moles of $Pt(NH_3)_2Cl_2$ are produced from the $(NH_4)_2PtCl_4$ reactant (0.0416 mol) than produced from the NH_3 reactant (0.075 mol). Therefore, use 0.0416 mol $Pt(NH_3)_2Cl_2$ and the molar mass to determine the grams of $Pt(NH_3)_2Cl_2$ produced:

$$0.0416 \text{ mol Pt(NH}_3)_2\text{Cl}_2 \times \frac{300.0448 \text{ g Pt(NH}_3)_2\text{Cl}_2}{1 \text{ mol Pt(NH}_3)_2\text{Cl}_2} = 12.5 \text{ g Pt(NH}_3)_2\text{Cl}_2$$

✓ *Reasonable Answer Check:* It is satisfying that $(NH_4)_2PtCl_4$ is the limiting reactant, because that reactant contains the expensive platinum metal and the experimenter would want to make sure all of that got used up. Since the molar mass of the product is somewhat smaller than the molar mass of the limiting reactant and their stoichiometry is 1:1, it makes sense that a slightly smaller mass of product is formed in this reaction.

58. *Answer:* **SiH₄**

Strategy and Explanation: Given the mass of a compound, the identity of the elements in the compound, and the identity and masses of all the products produced, determine the empirical formula.

Use the molar mass of the products to find their moles and use the stoichiometry of their formulas to determine the moles of the elements in the compound. Then use a whole-number mole ratio to determine the empirical formula.

The compound is Si_xO_y. Its combustion produced SiO_2 and H_2O. Use molar mass and formula stoichiometry to determine the moles of Si and H.

$$11.64 \text{ g SiO}_2 \times \frac{1 \text{ mol SiO}_2}{60.0843 \text{ g SiO}_2} \times \frac{1 \text{ mol Si}}{1 \text{ mol SiO}_2} = 0.1937 \text{ mol Si}$$

$$6.980 \text{ g H}_2\text{O} \times \frac{1 \text{ mol H}_2\text{O}}{18.0152 \text{ g H}_2\text{O}} \times \frac{2 \text{ mol H}}{1 \text{ mol H}_2\text{O}} = 0.7749 \text{ mol H}$$

Set up mole ratio and simplify by dividing by the smallest number of moles: 0.1937 mol Si : 0.7749 mol H

1 Si : 4.000 H

Use the whole number ratio for the subscripts in the formula. The empirical formula is SiH_4.

✓ *Reasonable Answer Check:* These are not necessary calculations, but we can calculate the masses of each element, Si and H.

$$0.1937 \text{ mol Si} \times \frac{28.0855 \text{ g Si}}{1 \text{ mol Si}} = 5.440 \text{ g Si}$$

$$0.7749 \text{ mol H} \times \frac{1.0079 \text{ g H}}{1 \text{ mol H}} = 0.7810 \text{ g H}$$

The sum of these masses (5.440 g + 0.7810 g = 6.221 g) adds up to the mass of the original compound (6.22 g), to the given significant figures.

Applying Concepts

62. *Answer:* **Equation (b)**

Strategy and Explanation: Four XY_3 molecules are made from two diatomic X_2 molecules and six diatomic Y_2 molecules. So, symbolically, the reaction is $2 X_2 + 6 Y_2 \longrightarrow 4 XY_3$, and the stoichiometric equation representing that reaction is (b) $X_2 + 3 Y_2 \longrightarrow 2 XY_3$.

63. *Answer:* **When the metal mass is less than 1.2 g, the metal is the limiting reactant. When the metal mass is greater than 1.2 g, the bromine is the limiting reactant.**

Strategy and Explanation: When masses smaller than 1.2 gram of the metal are added, the metal is the limiting reactant, so the mass of the compound produced is directly proportional to the mass of metal present (shown by the straight line rising up to the right on the graph). More metal makes more products when the mass of metal is less than 1.2 gram.

When the mass larger than 1.2 g of metal are added, the bromine is the limiting reactant, so the particular mass of the metal is independent of how much compound is made. Since the amount of bromine is held constant, the mass of compound formed is also a constant (shown by a horizontal line on the graph).

More Challenging Questions

65. *Answer:* $H_2(g) + 3 Fe_2O_3(s) \longrightarrow H_2O(\ell) + 2 Fe_3O_4(s)$

Strategy and Explanation: Given the formulas of the reactants and one product and the percent mass of elements in the second product, determine the balanced chemical equation.

Choose a convenient sample mass of product, Fe_xO_y, such as 100.0 g. Find the mass of Fe and O in the sample, using the given mass percent. Use the molar mass of the elements to find their moles, then use a whole-number mole ratio to determine the empirical formula. Using the formulas of the reactants and products balance the equation.

100.0 g of Fe_xO_y contains 72.3 g Fe and 27.7 g O.

$$72.3 \text{ g Fe} \times \frac{1 \text{ mol Fe}}{55.845 \text{ g Fe}} = 1.29 \text{ mol Fe} \qquad 27.7 \text{ g O} \times \frac{1 \text{ mol O}}{15.9994 \text{ g O}} = 1.73 \text{ mol O}$$

Set up mole ratio and simplify by dividing by the smallest number of moles:

$$1.29 \text{ mol Fe} : 1.73 \text{ mol O}$$
$$1 \text{ Fe} : 1.34 \text{ O}$$

Multiply by 3 to get whole numbers: $3 \text{ Fe} : 4 \text{ O}$

Use the whole number ratio for the subscripts in the formula. The empirical formula is Fe_3O_4.

$? \ H_2(g) + ? \ Fe_2O_3(s) \longrightarrow ? \ H_2O(\ell) + ? \ Fe_3O_4(s)$	Select order: Fe, O, H
$? \ H_2(g) + 3 \ Fe_2O_3(s) \longrightarrow ? \ H_2O(\ell) + 2 \ Fe_3O_4(s)$	6 Fe
$? \ H_2(g) + 3 \ Fe_2O_3(s) \longrightarrow 1 \ H_2O(\ell) + 2 \ Fe_3O_4(s)$	9 O
$1 \ H_2(g) + 3 \ Fe_2O_3(s) \longrightarrow 1 \ H_2O(\ell) + 2 \ Fe_3O_4(s)$	2 H

✓ *Reasonable Answer Check:* Fe_3O_4 is a common oxide of iron. 6 Fe, 9 O and 2 H on each side.

67. *Answer:* **86.3 g**

Strategy and Explanation: Given a balanced chemical equation, the volume of a liquid limiting reactant, and the density of the liquid, determine the maximum theoretical yield of a product.

Use the density of the reactant to find the grams of the reactant. Then use the molar mass of the reactant to find the moles of reactant. Then use the stoichiometry of the equation to determine moles of product. Then use the molar mass of the product to get the mass of the product.

$$25.0 \text{ mL Br}_2 \times \frac{3.1023 \text{ g Br}_2}{1 \text{ mL Br}_2} \times \frac{1 \text{ mol Br}_2}{159.808 \text{ g Br}_2} \times \frac{1 \text{ mol Al}_2Br_6}{3 \text{ mol Br}_2} \times \frac{533.387 \text{ g Al}_2Br_6}{1 \text{ mol Al}_2Br_6} = 86.3 \text{ g Al}_2Br_6$$

✓ *Reasonable Answer Check:* The units all cancel, and the numerical multipliers are larger than the dividers.

70. *Answer:* **44.9 amu**

Strategy and Explanation: Given the mass of a sample of a compound X_2S_3 that is then roasted to form a given mass of X_2O_3, determine the atomic mass of X.

Because all the X atoms from X_2S_3 end up in X_2O_3, then it must be true that each sample must contains the same number of moles of X. Because each compound has the same number of X atoms, it must be true that each sample must contain the same number of moles of compound. Using a variable, x, for the molar mass of X, write an expression that shows the calculation of moles for each compound and equate these two expressions. Solve for x.

Let x = molar mass of X and y = molar mass of Y

$$\text{mol of } X_2Y_3 = \text{mass of } X_2Y_3 \times \frac{1 \text{ mol } X_2Y_3}{(2x + 3y)}$$

$$10.00 \text{ g } X_2S_3 \times \frac{1 \text{ mol } X_2S_3}{\left[2x + 3(32.065)\text{ g S}\right]} = 7.410 \text{ g } X_2O_3 \times \frac{1 \text{ mol } X_2O_3}{\left[2x + 3(15.9994)\text{ g O}\right]}$$

$$10.00 \times \frac{1}{\left(2x + 96.195\right)} = 7.410 \times \frac{1}{\left(2x + 47.9982\right)}$$

$$10.00 \,(2x + 47.9982) = 7.410 \,(2x + 96.195)$$

$$20.00\, x + 479.982 = 14.82\, x + 712.805$$

$$20.00\, x - 14.82\, x = 712.805 - 479.982$$

$$5.18\, x = 232.813$$

$$x = 44.9 \text{ g/mol}$$

$$\text{Atomic mass} = 44.9 \text{ amu}$$

✓ *Reasonable Answer Check:* The calculated atomic mass is very close to that of Sc, which does form 3+ ions and would combine with 2- ions like O^{2-} and S^{2-} to form Sc_2S_3 and Sc_2O_3.

72. *Answer:* **0 g $AgNO_3$, 9.82 g Na_2CO_3, 6.79 g Ag_2CO_3, 4.19 g $NaNO_3$**

Strategy and Explanation: Given the names of products and reactants of a chemical reaction, and the masses of the two reactants, determine the masses of all the products and reactants after the reaction is complete.

Determine the formulas of the reactants and products and set up and balance the chemical equation. Use the molar mass of the reactants to find the moles of the reactant substances. Then use the stoichiometry of the equation to determine the moles of the product produced in each case. Identify the limiting reactant from the reactant that produces the least number of products. Use the moles of product produced from the limiting reactant, the molar masses of the products to find the grams of products. Use the moles of product produced, determine the moles of the other reactant needed for complete reaction. Convert that number to grams using the molar mass. Then subtract the quantity used from the initial mass given to get the mass of excess reactant.

Silver nitrate is $AgNO_3$. Sodium carbonate is Na_2CO_3. Silver carbonate is Ag_2CO_3. Sodium nitrate is $NaNO_3$.

$? \text{ AgNO}_3 + ? \text{ Na}_2\text{CO}_3 \longrightarrow ? \text{ Ag}_2\text{CO}_3 + ? \text{ NaNO}_3$	Order: Ag, Na, N, C, O
$2 \text{ AgNO}_3 + ? \text{ Na}_2\text{CO}_3 \longrightarrow 1 \text{ Ag}_2\text{CO}_3 + ? \text{ NaNO}_3$	2 Ag
$2 \text{ AgNO}_3 + 1 \text{ Na}_2\text{CO}_3 \longrightarrow 1 \text{ Ag}_2\text{CO}_3 + 2 \text{ NaNO}_3$	2 Na, 2 N, 1 C, 9 O

The balanced equation says: 1 mol Na_2CO_3 produces 1 mol Ag_2CO_3.

$$12.43 \text{ g } Na_2CO_3 \times \frac{1 \text{ mol } Na_2CO_3}{105.9885 \text{ g } Na_2CO_3} \times \frac{1 \text{ mol } Ag_2CO_3}{1 \text{ mol } Na_2CO_3} = 0.1173 \text{ mol } Ag_2CO_3$$

The balanced equation says: 2 mol $AgNO_3$ produces 1 mol Ag_2CO_3.

$$8.37 \text{ g } AgNO_3 \times \frac{1 \text{ mol } AgNO_3}{169.8731 \text{ g } AgNO_3} \times \frac{1 \text{ mol } Ag_2CO_3}{2 \text{ mol } AgNO_3} = 0.0246 \text{ mol } Ag_2CO_3$$

The number of Ag_2CO_3 moles produced from $AgNO_3$ is smaller (0.0246 mol < 0.1173 mol), so $AgNO_3$ is the limiting reactant and Na_2CO_3 is the excess reactant. Therefore, at the end of the reaction, the mass of $AgNO_3$ present is zero grams.

Find the mass Ag_2CO_3: $0.0246 \text{ mol } Ag_2CO_3 \times \dfrac{275.7453 \text{ g } Ag_2CO_3}{1 \text{ mol } Ag_2CO_3} = 6.79 \text{ g } Ag_2CO_3$

Find the mass $NaNO_3$: $0.0246 \text{ mol } Ag_2CO_3 \times \dfrac{2 \text{ mol } NaNO_3}{1 \text{ mol } Ag_2CO_3} \times \dfrac{84.9947 \text{ g } NaNO_3}{1 \text{ mol } NaNO_3} = 4.19 \text{ g } NaNO_3$

Find the mass Na_2CO_3 used up, then subtract from initial for mass unreacted:

$$0.0246 \text{ mol Ag}_2\text{CO}_3 \times \frac{1 \text{ mol Na}_2\text{CO}_3}{1 \text{ mol Ag}_2\text{CO}_3} \times \frac{105.9885 \text{ g Na}_2\text{CO}_3}{1 \text{ mol Na}_2\text{CO}_3} = 2.61 \text{ g Na}_2\text{CO}_3 \text{ used up}$$

$$12.43 \text{ g Na}_2\text{CO}_3 \text{ initial} - 2.61 \text{ g Na}_2\text{CO}_3 \text{ used up} = 9.82 \text{ g Na}_2\text{CO}_3 \text{ remains unreacted}$$

✓ *Reasonable Answer Check:* The total mass before the reaction = 12.43 g + 8.37 g = 20.80 g is equal to the total mass after the reaction = 6.79 g + 4.19 g + 9.82 g = 20.80 g, in accordance with the conservation of mass.

74. *Answer:* **99.7% CH$_3$OH, 0.3% C$_2$H$_5$OH**

Strategy and Explanation: Given the mass of a sample of a liquid containing unknown amounts of two organic compounds and given the mass of one of the products of a chemical reaction, determine the composition of the liquid.

First, balance the combustion reactions for ethyl alcohol and methyl alcohol. Set two variables X = the mass in grams of methyl alcohol and Y = the mass in grams of ethyl alcohol. Then establish two equations, one describing the total mass of the sample related to the two components and one describing the moles of carbon dioxide produced when burning the sample. Algebraically solve for X and Y, then use those masses and the total sample mass to find the percentage by mass of the two compounds.

Unbalanced:	? C$_2$H$_5$OH(ℓ) + ? O$_2$(g) ⟶ ? CO$_2$(g) + ? H$_2$O(ℓ)	Order: C, H, then O
	1 C$_2$H$_5$OH(ℓ) + ? O$_2$(g) ⟶ 2 CO$_2$(g) + ? H$_2$O(ℓ)	2 C's
	1 C$_2$H$_5$OH(ℓ) + ? O$_2$(g) ⟶ 2 CO$_2$(g) + 3 H$_2$O(ℓ)	6 H's
	1 C$_2$H$_5$OH(ℓ) + 6 O$_2$(g) ⟶ 2 CO$_2$(g) + 3 H$_2$O(ℓ)	7 O's
Unbalanced:	? CH$_3$OH(ℓ) + ? O$_2$(g) ⟶ ? CO$_2$(g) + ? H$_2$O(ℓ)	Order: C, H, then O
	1 CH$_3$OH(ℓ) + ? O$_2$(g) ⟶ 1 CO$_2$(g) + ? H$_2$O(ℓ)	1 C's
	1 CH$_3$OH(ℓ) + ? O$_2$(g) ⟶ 1 CO$_2$(g) + 2 H$_2$O(ℓ)	4 H's
	1 CH$_3$OH(ℓ) + $\frac{3}{2}$ O$_2$(g) ⟶ 1 CO$_2$(g) + 2 H$_2$O(ℓ)	4 O's

Let X = grams of ethyl alcohol and Y = grams of methyl alcohol

Total mass of sample = 0.280 g = X + Y

Moles of carbon dioxide produced when burned = $0.385 \text{ g CO}_2 \times \dfrac{1 \text{ mol CO}_2}{44.0095 \text{ g CO}_2} = 0.00875 \text{ mol CO}_2$

$$0.00875 \text{ mol CO}_2 = X \text{ g C}_2\text{H}_5\text{OH} \times \frac{1 \text{ mol C}_2\text{H}_5\text{OH}}{46.0682 \text{ g C}_2\text{H}_5\text{OH}} \times \frac{2 \text{ mol CO}_2}{1 \text{ mol C}_2\text{H}_5\text{OH}}$$

$$+ \; Y \text{ g CH}_3\text{OH} \times \frac{1 \text{ mol CH}_3\text{OH}}{32.0417 \text{ g CH}_3\text{OH}} \times \frac{1 \text{ mol CO}_2}{1 \text{ mol CH}_3\text{OH}}$$

0.280 = X + Y   and   0.00875 = 0.0434139 X + 0.0312093 Y

Solve for X: 0.00875 = 0.0434139 X + 0.0312093(0.280 – X)

0.00875 = 0.0434139 X + 0.000874 – 0.0312093 X

0.00875 – 0.000874 = (0.0434139 – 0.0312093) X

0.00001 = 0.0122046 X

X = 0.0008 g ethyl alcohol

Y = 0.280 – X = 0.280 – 0.0008 = 0.279 g methyl alcohol

$$\frac{0.279 \text{ g CH}_3\text{OH}}{0.280 \text{ g liquid}} \times 100 \% = 99.7 \% \text{ CH}_3\text{OH} \qquad \frac{0.001 \text{ g C}_2\text{H}_5\text{OH}}{0.280 \text{ g liquid}} \times 100 \% = 0.3 \% \text{ C}_2\text{H}_5\text{OH}$$

✓ *Reasonable Answer Check:* Assuming the liquid is pure methyl alcohol gives the following mass of CO$_2$:

$$0.280 \text{ g CH}_3\text{OH} \times \frac{1 \text{ mol CH}_3\text{OH}}{32.0417 \text{ g CH}_3\text{OH}} \times \frac{1 \text{ mol CO}_2}{1 \text{ mol CH}_3\text{OH}} \times \frac{44.0095 \text{ g CO}_2}{1 \text{ mol CO}_2} = 0.385 \text{ g CO}_2$$

So it makes sense that the liquid is almost one hundred percent methanol.

75. *Answer:* **(a) $C_9H_{11}NO_4$ (b) $C_9H_{11}NO_4$**

Strategy and Explanation: Given the percent mass of elements in an organic compound and the compounds molar mass, determine the empirical formula and the molecular formula.

Choose a convenient sample mass of product, $C_xH_yN_zO_w$, such as 100.00 g. Find the mass of C and H in the sample, using the given mass percent, then subtract those masses from the total sample mass to get the mass of O. Use the molar mass of the elements to find their moles, then use a whole-number mole ratio to determine the empirical formula. For combustion, use the formula of the organic compound and O_2 as reactants and CO_2 and H_2O as products, then balance the equation.

(a) 100.00 g of $.C_xH_yN_zO_w$ contains 54.82 g C, 7.10 g N, and 32.46 g O.

$$100.00 \text{ g of } .C_xH_yN_zO_w - 54.82 \text{ g C} - 7.10 \text{ g N} - 32.46 \text{ g O} = 5.62 \text{ g H}$$

$$54.82 \text{ g C} \times \frac{1 \text{ mol C}}{12.0107 \text{ g C}} = 4.564 \text{ mol C} \qquad 7.10 \text{ g N} \times \frac{1 \text{ mol N}}{14.0067 \text{ g N}} = 0.507 \text{ mol N}$$

$$32.46 \text{ g O} \times \frac{1 \text{ mol O}}{15.9994 \text{ g O}} = 2.029 \text{ mol O} \qquad 5.62 \text{ g H} \times \frac{1 \text{ mol H}}{1.0079 \text{ g H}} = 5.58 \text{ mol H}$$

Set up mole ratio and simplify by dividing by the smallest number of moles:

$$4.564 \text{ mol C} : 5.58 \text{ mol H} : 0.507 \text{ mol N} : 2.029 \text{ mol O}$$

$$9 \text{ C} : 11 \text{ H} : 1 \text{ N} : 4 \text{ O}$$

Use the whole number ratio for the subscripts in the formula. The empirical formula is $C_9H_{11}NO_4$.

(b) The molar mass of the empirical formula = 197.1875 g/mol $C_9H_{11}NO_4$

$$\frac{197.19 \text{ g / mol compound}}{197.1875 \text{ g / mol emp. formula}} = 1 \text{ emp. formula / compound}$$

The molecular formula is $C_9H_{11}NO_4$.

✓ *Reasonable Answer Check:* The mole ratio is quite close to whole number values. The empirical formula is very close to the molecular formula.

Chapter 5: Chemical Reactions

Chapter Contents:

Introduction:

In Chapter 4, you learned how to balance equations. In Chapter 5, the concept of net ionic equations is introduced. Here you must balance the charges, as well as the atoms. Chunk information when possible, i.e., balance polyatomic ions as a whole unit instead of individual elements. When an ionic compound becomes aqueous, the common ions of which they are composed are recreated. Refer to Chapter 3 (especially Figure 3.2 and Table 3.7) if you need to refresh your memory.

Important new stoichiometric relationships are introduced in this chapter. Figure 5.15 on page 167 shows how the stoichiometric relationships interrelate.

Concentration, molarity, is very important. You need to be able to do problems that relate both to dilution and to volumetric reactions. The dilution equation described on page 163 should NOT be used for reactions.

Oxidation-reduction can be confusing because of all the opposite processes that occur, e.g., oxidation is loss of electrons whereas reduction is a gain, or the substance that is oxidized is the reducing agent. Keep in mind that only the reactants are oxidized or reduced, or can act as the oxidizing or reducing agent. There are two common mnemonics for remembering redox processes: "oil rig" which stands for **o**xidation **i**s **l**oss and **r**eduction **i**s **g**ain and "LEO the lion says GER", **l**osing **e**lectrons is **o**xidation and **g**aining **e**lectrons is **r**eduction.

Solutions to Blue-Numbered Questions
for Review and Thought for Chapter 5

Topical Questions

Solubility

7. *Answer:* **(a) soluble, Fe^{2+} and ClO_4^- (b) soluble, Na^+ and SO_4^{2-} (c) soluble, K^+ and Br^- (d) soluble, Na^+ and CO_3^{2-}**

   Strategy and Explanation: Given a compound's formula, determine whether it is water-soluble.

   Identify the cation and the anion in the salt, then use Table 5.1 on page 137.

   Notice: Any rule that applies is sufficient to determine the compound's solubility. For example, if Rule 1 applies, there is no need to look for other rules related to the anion.

   Notice: The Question asks which ions are present, not how many. When answering this Question, numbers that were part of the compound's formula stoichiometry are not included.

   (a) $Fe(ClO_4)_2$ contains iron(II) ion and perchlorate ion

       Rule 6: "All perchlorates are soluble." $Fe(ClO_4)_2$ is soluble.

       $Fe(ClO_4)_2$ ionizes to form Fe^{2+} and ClO_4^-.

   (b) Na_2SO_4 contains sodium ion (Group 1A) and sulfate ion

       Rule 1: "All ... sodium ... salts are soluble." Na_2SO_4 is soluble.

       Na_2SO_4 ionizes to form Na^+ and SO_4^{2-}.

   (c) KBr contains potassium ion (Group 1A) and bromide ion

       Rule 1: "All ... potassium ... salts are soluble." KBr is soluble.

       KBr ionizes to form K^+ and Br^-.

(d) Na_2CO_3 contains sodium ion (Group 1A) and carbonate ion

Rule 1: "All ... sodium ... salts are soluble." Na_2CO_3 is soluble.

Na_2CO_3 ionizes to form Na^+ and CO_3^{2-}.

9. *Answer:* **(a) soluble, K^+ and HPO_4^{2-} (b) soluble, Na^+ and ClO^- (c) soluble, Mg^{2+} and Cl^- (d) soluble, Ca^{2+} and OH^- (e) soluble, Al^{3+} and Br^-**

Strategy and Explanation: Given a compound's formula, determine whether it is water-soluble.

Identify the cation and the anion in the salt, then use Table 5.1 on page 137.

Notice: Any rule that applies is sufficient to determine the compound's solubility. For example, if Rule 1 applies, there is no need to look for other rules related to the anion.

Notice: The Question asks which ions are present, not how many. When answering this Question, numbers that were part of the compound's formula stoichiometry are not included.

(a) Potassium hydrogen phosphate is K_2HPO_4.

Rule 1: "All ... group 1A ... salts are soluble." Potassium is in group 1A, so potassium hydrogen phosphate is soluble. It ionizes to form potassium ion, K^+, and hydrogen phosphate ion, HPO_4^{2-}.

(b) Sodium hypochlorite is $NaOCl$.

Rule 1: "All ... group 1A ... salts are soluble." Sodium is in group 1A, so sodium hypochlorite is soluble.

It ionizes to form sodium ion, Na^+, and hypochlorite ion, ClO^-.

(c) Magnesium chloride is $MgCl_2$.

Rule 3: "All common... chlorides ... are soluble, except $AgCl$, Hg_2Cl_2, and $PbCl_2$..." Magnesium chloride is soluble.

It ionizes to form magnesium ion, Mg^{2+}, and chloride ion, Cl^-.

(d) Calcium hydroxide is $Ca(OH)_2$.

Rule 10: " ... $Ca(OH)_2$ is slightly soluble." That means it dissolves only slightly, but the ions that will be formed are calcium ion, Ca^{2+}, and hydroxide ion, OH^-.

(e) Aluminum bromide is $AlBr_3$.

Rule 3: "All common... bromides ... are soluble, except $AgBr$, Hg_2Br_2, and $PbBr_2$..." Aluminum bromide is soluble. It ionizes to form aluminum ion, Al^{3+}, and bromide ion, Br^-.

Exchange Reactions

11. *Answer:* **(a) $MnCl_2(aq) + Na_2S(aq) \longrightarrow MnS(s) + 2\,NaCl(aq)$ (b) no precipitate ("NP")**
(c) no precipiate ("NP") (d) $Hg(NO_3)_2(aq) + Na_2S(aq) \longrightarrow HgS(s) + 2\,NaNO_3(aq)$

(e) $Pb(NO_3)_2(aq) + 2\,HCl(aq) \longrightarrow PbCl_2(s) + 2\,HNO_3(aq)$

(f) $BaCl_2(aq) + H_2SO_4(aq) \longrightarrow BaSO_4(s) + 2\,HCl(aq)$

Strategy and Explanation: Given the reactants' formulas, determine if a precipitation reaction occurs. If so, write the balanced equation for the reaction. Create appropriate products of the possible exchange reaction, then check Table 5.1 on page 137 for insoluble products that would form a precipitate, then balance the equation.

Notice: Common acids (listed in Table 5.2 on page 144) are soluble.

(a) Possible exchange products are: MnS (Rule 13) Insoluble solid precipitate

NaCl (Rule 1) Soluble

$$MnCl_2(aq) + Na_2S(aq) \longrightarrow MnS(s) + 2\,NaCl(aq)$$

(b) Possible exchange products are: $Cu(NO_3)_2$ (Rule 2) Soluble

H_2SO_4 (Table 5.2) Soluble strong electrolyte

$$HNO_3(aq) + CuSO_4(aq) \longrightarrow \text{“NP.”}$$

(c) Possible exchange products are: H_2O Molecular liquid; no significant ionization

$NaClO_4$ (Rule 1) Soluble

$$NaOH(aq) + HClO_4(aq) \longrightarrow \text{“NP.”}$$

(d) Possible exchange products are: HgS (Rule 13) Insoluble solid precipitate

$NaNO_3$ (Rule 1) Soluble

$$Hg(NO_3)_2(aq) + Na_2S(aq) \longrightarrow HgS(s) + 2\,NaNO_3(aq)$$

(e) Possible exchange products are: $PbCl_2$ (Rule 3 exception) Insoluble solid precipitate

HNO_3 (Table 5.2) Soluble strong electrolyte

$$Pb(NO_3)_2(aq) + 2\,HCl(aq) \longrightarrow PbCl_2(s) + 2\,HNO_3(aq)$$

(f) Possible exchange products are: $BaSO_4$ (Rule 4 exception) Insoluble solid precipitate

HCl (Table 5.2) Soluble strong electrolyte

$$BaCl_2(aq) + H_2SO_4(aq) \longrightarrow BaSO_4(s) + 2\,HCl(aq)$$

✓ *Reasonable Answer Check:* All equations with insoluble products are identified as precipitation reactions. Those equations with all products soluble are identified with “NP.” Notice that, in both of the “NP” cases above, a reaction **does** occur; they are, however, not precipitation reactions.

13. *Answer:* **(a) CuS insoluble; $Cu^{2+} + H_2S(aq) \longrightarrow CuS(s) + 2\,H^+$; spectator ion is Cl^-. (b) $CaCO_3$ insoluble; $Ca^{2+} + CO_3^{2-} \longrightarrow CaCO_3(s)$; spectator ions are K^+ and Cl^-. (c) AgI insoluble;**

$Ag^+ + I^- \longrightarrow AgI(s)$; spectator ions are Na^+ and NO_3^-.

Strategy and Explanation: Given overall chemical equations, identify the water-insoluble products, write the net ionic equations, and determine the spectator ions.

Use the solubility rules to determine the actual physical state of the products. Identify the water-insoluble product. Use the solubility rules to determine the actual physical state of the reactants. Write the complete ionic equation by writing ionized forms for any soluble ionic compound and all strong acids and bases. Ions that remain completely unchanged in the complete ionic equation (same physical phase and same ionized form), i.e., any ions that are the same on the product side as on the reactant side, are eliminated to produce the net ionic equation. The ions eliminated are the spectator ions.

Notice: All soluble ionic compounds and strong acids and bases are ionized in the complete ionic equation. All solids, weak acids and bases, and molecular compounds remain un-ionized.

All ions have aqueous phase in the equations below.

(a) CuS is insoluble (Table 5.1, Rule 13), and HCl is a common strong acid (Table 5.2).

The water-insoluble product is CuS.

$CuCl_2$ is soluble (Table 5.1, Rule 3). H_2S is not a common strong acid (Table 5.2), so assume it is weak.

$$Cu^{2+} + 2\,Cl^- + H_2S(aq) \longrightarrow CuS(s) + 2\,H^+ + 2\,Cl^- \quad \text{complete ionic eq.}$$

Eliminate Cl^- ions on each side:

$$Cu^{2+} + H_2S(aq) \longrightarrow CuS(s) + 2\,H^+ \qquad \text{net ionic equation}$$

The spectator ion is Cl^-.

(b) KCl is soluble (Table 5.1, Rule 1). $CaCO_3$ is insoluble (Table 5.1, Rule 9).

The water-insoluble product is $CaCO_3$.

$CaCl_2$ is soluble (Table 5.1, Rule 3). K_2CO_3 is soluble (Table 5.1, Rule 1).

$$Ca^{2+} + 2\ Cl^- + 2\ K^+ + CO_3{}^{2-} \longrightarrow CaCO_3(s) + 2\ K^+ + 2\ Cl^- \qquad \text{complete ionic eq.}$$

Eliminate K^+ and Cl^-

$$Ca^{2+} + CO_3{}^{2-} \longrightarrow CaCO_3(s) \ \text{ net ionic equation}$$

The spectator ions are K^+ and Cl^-.

(c) AgI is insoluble (Table 5.1, Rule 3). $NaNO_3$ is soluble (Table 5.1, Rule 1).

The water-soluble product is AgI.

$AgNO_3$ is soluble (Table 5.1, Rule 2). NaI is soluble (Table 5.1, Rule 1).

$$Ag^+ + NO_3{}^- + Na^+ + I^- \longrightarrow AgI(s) + Na^+ + NO_3{}^- \ \text{ complete ionic equation}$$

Eliminate Na^+ and $NO_3{}^-$.

$$Ag^+ + I^- \longrightarrow AgI(s) \qquad \text{net ionic equation}$$

The spectator ions are Na^+ and $NO_3{}^-$.

✓ *Reasonable Answer Check:* All spectator ions have been identified and eliminated to make the net ionic equation. The atoms are all balanced and the charges are all balanced in each equation.

15. *Answer:* **(a) $Zn(s) + 2\ HCl(aq) \longrightarrow H_2(g) + ZnCl_2(aq)$; complete ionic: $Zn(s) + 2\ H^+ + 2\ Cl^-$ $\longrightarrow H_2(g) + Zn^{2+} + 2\ Cl^-$; net ionic: $Zn(s) + 2\ H^+ \longrightarrow H_2(g) + Zn^{2+}$ (b) $Mg(OH)_2(s) +$ $2\ HCl(aq) \longrightarrow MgCl_2(aq) + 2\ H_2O(\ell)$; complete ionic: $Mg(OH)_2(s) + 2\ H^+ + 2\ Cl^- \longrightarrow Mg^{2+} +$ $2\ Cl^- + 2\ H_2O(\ell)$; net ionic: $Mg(OH)_2(s) + 2\ H^+ \longrightarrow Mg^{2+} + 2\ H_2O(\ell)$ (c) $2\ HNO_3(aq) +$ $CaCO_3(s) \longrightarrow Ca(NO_3)_2(aq) + H_2O(\ell) + CO_2(g)$; complete ionic: $2\ H^+ + 2\ NO_3{}^- +$ $CaCO_3(s) \longrightarrow Ca^{2+} + 2\ NO_3{}^- + H_2O(\ell) + CO_2(g)$; net ionic: $2\ H^+ + CaCO_3(s) \longrightarrow Ca^{2+} + H_2O(\ell)$ $+ CO_2(g)$ (d) $4\ HCl(aq) + MnO_2(s) \longrightarrow MnCl_2(aq) + Cl_2(g) + 2\ H_2O(\ell)$; complete ionic: $4\ H^+ + 4\ Cl^-$ $+ MnO_2(s) \longrightarrow Mn^{2+} + 2\ Cl^- + Cl_2(g) + 2\ H_2O(\ell)$; net ionic: $4\ H^+ + 2\ Cl^- + MnO_2(s) \longrightarrow Mn^{2+} +$ $Cl_2(g) + 2\ H_2O(\ell)$**

Strategy and Explanation: Use the method described in the solution to Question 13. Given overall chemical equations, balance them, then write complete and net ionic equations.

Use standard balancing techniques to balance the overall equation. Use the solubility rules to determine the actual physical state of the reactants and products. Write complete ionic equations using that information. Ions that remain completely unchanged (same physical phase and same ionized form) are the spectator ions. Any ions that are the same on the product side as on the reactant side are eliminated to produce the net ionic equation.

Notice: Just like with balancing elements, we can balance ions as units: Select a balancing order and follow it.

Notice: All soluble ionic compounds and strong acids and bases are ionized in the complete ionic equation. All solids, weak acids and bases, and molecular compounds remain un-ionized.

All ions have aqueous phase in the equations below.

(a) $$Zn(s) + 2\ HCl(aq) \longrightarrow H_2(g) + ZnCl_2(aq)$$

$$Zn(s) + 2\ H^+ + 2\ Cl^- \longrightarrow H_2(g) + Zn^{2+} + 2\ Cl^- \ \text{ complete ionic equation}$$

Eliminate spectator ion, Cl^-. $Zn(s) + 2\ H^+ \longrightarrow H_2(g) + Zn^{2+}$ net ionic equation

(b)
$$Mg(OH)_2(s) + 2\,HCl(aq) \longrightarrow MgCl_2(aq) + 2\,H_2O(\ell)$$

$Mg(OH)_2(s) + 2\,H^+ + 2\,Cl^- \longrightarrow Mg^{2+} + 2\,Cl^- + 2\,H_2O(\ell)$ complete ionic equation

Eliminate spectator ion, Cl^-. $Mg(OH)_2(s) + 2\,H^+ \longrightarrow Mg^{2+} + 2\,H_2O(\ell)$ net ionic equation

(c)
$$2\,HNO_3(aq) + CaCO_3(s) \longrightarrow Ca(NO_3)_2(aq) + H_2O(\ell) + CO_2(g)$$

$2\,H^+ + 2\,NO_3^- + CaCO_3(s) \longrightarrow Ca^{2+} + 2\,NO_3^- + H_2O(\ell) + CO_2(g)$ complete ionic equation

Eliminate spectator ion, NO_3^-. $2\,H^+ + CaCO_3(s) \longrightarrow Ca^{2+} + H_2O(\ell) + CO_2(g)$ net ionic equation

(d)
$$4\,HCl(aq) + MnO_2(s) \longrightarrow MnCl_2(aq) + Cl_2(g) + 2\,H_2O(\ell)$$

$4\,H^+ + 4\,Cl^- + MnO_2(s) \longrightarrow Mn^{2+} + 2\,Cl^- + Cl_2(g) + 2\,H_2O(\ell)$ complete ionic equation

Eliminate spectator ions, two of the four Cl^-.

$4\,H^+ + 2\,Cl^- + MnO_2(s) \longrightarrow Mn^{2+} + Cl_2(g) + 2\,H_2O(\ell)$ net ionic equation

✓ *Reasonable Answer Check:* All of the equations are balanced. The net ionic equations have no spectator ions. The atoms are all balanced and the charges are all balanced in each equation.

17. *Answer:* (a) $Ca(OH)_2(s) + 2\,HNO_3(aq) \longrightarrow Ca(NO_3)_2(aq) + 2\,H_2O(\ell)$; complete ionic: $Ca(OH)_2(s) + 2\,H^+ + 2\,NO_3^- \longrightarrow Ca^{2+} + 2\,NO_3^- + 2\,H_2O(\ell)$; net ionic: $Ca(OH)_2(s) + 2\,H^+ \longrightarrow Ca^{2+} + 2\,H_2O(\ell)$
(b) $BaCl_2(aq) + Na_2CO_3(aq) \longrightarrow BaCO_3(s) + 2\,NaCl(aq)$; complete ionic: $Ba^{2+} + 2\,Cl^- + 2\,Na^+ + CO_3^{2-} \longrightarrow BaCO_3(s) + 2\,Na^+ + 2\,Cl^-$; net ionic: $Ba^{2+} + CO_3^{2-} \longrightarrow BaCO_3(s)$ (c) $2\,Na_3PO_4(aq) + 3\,Ni(NO_3)_2(aq) \longrightarrow Ni_3(PO_4)_2(s) + 6\,NaNO_3(aq)$; complete ionic: $6\,Na^+ + 2\,PO_4^{3-} + 3\,Ni^{2+} + 6\,NO_3^- \longrightarrow Ni_3(PO_4)_2(s) + 6\,Na^+ + 6\,NO_3^-$; net ionic: $2\,PO_4^{3-} + 3\,Ni^{2+} \longrightarrow Ni_3(PO_4)_2(s)$

Strategy and Explanation: Given overall chemical equations, balance them, then write complete and net ionic equations. Use the method described in the solution to Question 13. All ions have aqueous phase in the equations below.

(a) $Ca(OH)_2$ is slightly soluble (Table 5.1, Rule 10). HNO_3 is a strong acid (Table 5.2). $Ca(NO_3)_2$ is soluble (Table 5.1, Rule 2). H_2O is a liquid molecular compound.

$$Ca(OH)_2(s) + 2\,HNO_3(aq) \longrightarrow Ca(NO_3)_2(aq) + 2\,H_2O(\ell)$$ balanced

$Ca(OH)_2(s) + 2\,H^+ + 2\,NO_3^- \longrightarrow Ca^{2+} + 2\,NO_3^- + 2\,H_2O(\ell)$ complete ionic equation
Eliminate spectator ion, NO_3^-. $Ca(OH)_2(s) + 2\,H^+ \longrightarrow Ca^{2+} + 2\,H_2O(\ell)$ net ionic equation

(b) $BaCl_2$ is soluble (Table 5.1, Rule 3). Na_2CO_3 is soluble (Table 5.1, Rule 1).

$BaCO_3$ is insoluble (Table 5.1, Rule 9), $NaCl$ is soluble (Table 5.1, Rule 1).

$$BaCl_2(aq) + Na_2CO_3(aq) \longrightarrow BaCO_3(s) + 2\,NaCl(aq)$$ balanced

$Ba^{2+} + 2\,Cl^- + 2\,Na^+ + CO_3^{2-} \longrightarrow BaCO_3(s) + 2\,Na^+ + 2\,Cl^-$ complete ionic equation
Eliminate spectator ions, Cl^- and Na^+. $Ba^{2+} + CO_3^{2-} \longrightarrow BaCO_3(s)$ net ionic equation

(c) Na_3PO_4 is soluble (Table 5.1, Rule 1). $Ni(NO_3)_2$ is soluble (Table 5.1, Rule 2). $Ni_3(PO_4)_2$ is insoluble (Table 5.1, Rule 8). $NaNO_3$ is soluble (Table 5.1, Rule 1).

$$Na_3PO_4(aq) + Ni(NO_3)_2(aq) \longrightarrow Ni_3(PO_4)_2(s) + NaNO_3(aq)$$ unbalanced

Select balancing order: Ni^{2+}, then PO_4^{3-}, then NO_3^-, and Na^+

$$2\,Na_3PO_4(aq) + 3\,Ni(NO_3)_2(aq) \longrightarrow Ni_3(PO_4)_2(s) + 6\,NaNO_3(aq)$$ balanced

$6\,Na^+ + 2\,PO_4^{3-} + 3\,Ni^{2+} + 6\,NO_3^- \longrightarrow Ni_3(PO_4)_2(s) + 6\,Na^+ + 6\,NO_3^-$ complete ionic equation

Eliminate spectator ions, NO_3^- and Na^+.

$$2\ PO_4^{3-} + 3\ Ni^{2+} \longrightarrow Ni_3(PO_4)_2(s)\quad \text{net ionic equation}$$

✓ *Reasonable Answer Check:* All of the balanced equations are balanced. The net ionic equations have no spectator ions. The atoms are all balanced and the charges are all balanced in each equation.

19. *Answer:* **$CdCl_2$(aq) + 2 NaOH(aq) $\longrightarrow$ $Cd(OH)_2$(s) + 2 NaCl(aq); complete ionic: Cd^{2+} + 2 Cl^- +**

2 Na^+ + 2 OH^- $\longrightarrow$ $Cd(OH)_2$(s) + 2 Na^+ + 2 Cl^-; net ionic: Cd^{2+} + 2 OH^- $\longrightarrow$ $Cd(OH)_2$(s)

Strategy and Explanation: Given an overall chemical equation for a precipitation, balance it, then write complete and net ionic equations. Use the method described in the solution to Question 13. All ions have aqueous phase in the equations below.

$CdCl_2$ is soluble (Table 5.1, Rule 3). NaOH is soluble (Table 5.1, Rule 1). $Cd(OH)_2$ is insoluble (Table 5.1, Rule 10). NaCl is soluble (Table 5.1, Rule 1)

$$CdCl_2(aq) + 2\ NaOH(aq) \longrightarrow Cd(OH)_2(s) + 2\ NaCl(aq)\quad \text{balanced overall equation}$$

$$Cd^{2+} + 2\ Cl^- + 2\ Na^+ + 2\ OH^- \longrightarrow Cd(OH)_2(s) + 2\ Na^+ + 2\ Cl^- \qquad \text{complete ionic equation}$$

Eliminate spectator ions, Cl^- and Na^+.

$$Cd^{2+} + 2\ OH^- \longrightarrow Cd(OH)_2(s)\quad \text{net ionic equation}$$

✓ *Reasonable Answer Check:* The atoms and the charges in each equation are balanced. This precipitation reaction produces an insoluble solid.

21. *Answer:* **(a) base, strong, K^+ and OH^- (b) base, strong, Mg^{2+} and OH^- (c) acid, weak, H^+ and ClO^- (d) acid, strong, H^+ and Br^- (e) base, strong, Li^+ and OH^- (f) acid, weak; H^+, HSO_3^-, and SO_3^{2-}**

Strategy and Explanation: Given some chemical formulas, identify if they are acids or bases, identify whether they are weak or strong, and determine what ions produce with they dissolve in water.

Acids produce H^+ in aqueous solution. Bases produce OH^- in aqueous solution. In several of these cases, Table 5.1 on page 137 or Table 5.2 on page 144 can be used to determine whether a large or small amount of ions are produced, by determining if the compound is soluble and/or weak or strong. In some cases, however, it is not possible to look up that information in Chapter 5. However, it is almost always true that, if an acid is **not** one of the common strong acids listed in Table 5.2, it is a **weak** acid. That is the case with the weak acids in this Question. If the acid or base is strong, it will ionize. If the acid or base is weak it will not ionize to a great extent, remaining primarily in the molecular form.

(a) KOH is a strong base. (Given in Table 5.2), producing K^+ and OH^- ions when dissolved in water.

(b) $Mg(OH)_2$ is an insoluble ionic compound (Table 5.1, Rule 10) so few ions are produced in water, though the $Mg(OH)_2$ that dissolves does ionize completely. So, practically, it may be considered weak, because the OH^- ion concentration will never be very large. Technically, it may be considered to be strong, because all the dissolved $Mg(OH)_2$ is ionized, producing Mg^{2+} and OH^- ions when dissolved in water.

(c) HClO is a weak acid. (It's not listed as a common strong acid in Table 5.2.). It will not ionize very much, remaining mostly in the HClO(aq) form. The small amount that does ionize will form H^+ and ClO^- ions.

(d) HBr is a strong acid (Given in Table 5.2), producing H^+ and Br^- ions when dissolved in water.

(e) LiOH is a strong base (Given in Table 5.2), producing Li^+ and OH^- ions when dissolved in water.

(f) H_2SO_3 is a weak acid (It's not listed as a common strong acid in Table 5.2.). It will not ionize very much, remaining mostly in the H_2SO_3(aq) form. The small amount that does ionize will form H^+, HSO_3^-, and SO_3^{2-} ions.

✓ *Reasonable Answer Check:* All of the ions produced are common ions, most of them are found in Figure 3.2 and Table 3.7. Most of these are given specifically in Table 5.2. If you want to look far into your chemistry future, Table 15.2 on page 554 can also be used to confirm that the two weak acids given here in (c) and (f) are indeed weak.

23. *Answer:* **(a) HNO_2; NaOH; complete ionic: $HNO_2(aq) + Na^+ + OH^- \longrightarrow H_2O(\ell) + Na^+ + NO_2^-$; net ionic: $HNO_2(aq) + OH^- \longrightarrow H_2O(\ell) + NO_2^-$ (b) H_2SO_4; $Ca(OH)_2$; complete ionic & net ionic: $H^+ + HSO_4^- + Ca(OH)_2(s) \longrightarrow 2\,H_2O(\ell) + CaSO_4(s)$ (c) HI; NaOH; complete ionic: $H^+ + I^- + Na^+ + OH^- \longrightarrow H_2O(\ell) + Na^+ + I^-$; net ionic: $H^+ + OH^- \longrightarrow H_2O(\ell)$ (d) H_3PO_4; $Mg(OH)_2$; complete ionic & net ionic: $2\,H_3PO_4(aq) + 3\,Mg(OH)_2(s) \longrightarrow 6\,H_2O(\ell) + Mg_3(PO_4)_2(s)$**

Strategy and Explanation: Given some chemical formulas for salts, identify the acids and bases that would react to form them, then write the overall neutralization reaction both in complete and net ionic form.

The formation of a salt using neutralization comes from reacting an acid containing the salt's anion and a base containing the salt's cation. Set up the complete equation, then use the method described in previous Questions to find the complete and net ionic equations.

(a) $NaNO_2$ is formed from the neutralization of HNO_2 (to supply the nitrite anion) and NaOH (to supply the sodium cation). HNO_2 is a weak acid. NaOH is a strong base. $NaNO_2$ is a soluble compound.

$$HNO_2(aq) + NaOH(aq) \longrightarrow H_2O(\ell) + NaNO_2(aq) \quad \text{balanced overall equation}$$

$$HNO_2(aq) + Na^+ + OH^- \longrightarrow H_2O(\ell) + Na^+ + NO_2^- \quad \text{complete ionic equation}$$

Eliminate spectator ion, Na^+. $\quad HNO_2(aq) + OH^- \longrightarrow H_2O(\ell) + NO_2^- \quad$ net ionic equation

(b) $CaSO_4$ is formed from the neutralization of H_2SO_4 (to supply the sulfate anion) and $Ca(OH)_2$ (to supply the calcium cation). The first ionization of H_2SO_4 is strong, but HSO_4^- is a weak acid. The reactant $Ca(OH)_2$ is only slightly soluble. The product, $CaSO_4$, is insoluble.

$$H_2SO_4(aq) + Ca(OH)_2(s) \longrightarrow 2\,H_2O(\ell) + CaSO_4(s) \quad \text{balanced overall equation}$$

$$H^+ + HSO_4^- + Ca(OH)_2(s) \longrightarrow 2\,H_2O(\ell) + CaSO_4(s) \quad \text{complete ionic and net ionic equations}$$

(c) NaI is formed from the neutralization of HI (to supply the iodide anion) and NaOH (to supply the sodium cation). HI is a strong acid and NaOH is a strong base. The product, NaI, is soluble.

$$HI(aq) + NaOH(aq) \longrightarrow H_2O(\ell) + NaI(aq) \quad \text{balanced overall equation}$$

$$H^+ + I^- + Na^+ + OH^- \longrightarrow H_2O(\ell) + Na^+ + I^- \quad \text{complete ionic equation}$$

Eliminate spectator ions, Na^+ and I^-. $\quad H^+ + OH^- \longrightarrow H_2O(\ell) \quad$ net ionic equation

(d) $Mg_3(PO_4)_2$ is formed from the neutralization of H_3PO_4 (to supply the phosphate anion) and $Mg(OH)_2$ (to supply the magnesium cation). H_3PO_4 is a weak acid. The reactant $Mg(OH)_2$ is insoluble. The product, $Mg_3(PO_4)_2$, is insoluble.

$$2\,H_3PO_4(aq) + 3\,Mg(OH)_2(s) \longrightarrow 6\,H_2O(\ell) + Mg_3(PO_4)_2(s)$$

balanced overall, complete ionic, and net ionic equations

✓ *Reasonable Answer Check:* These acids and bases undergo neutralization to produce the appropriate salt and water. The net ionic equation does not always includes the whole salt, when one or both of the ions of the salt are found to be spectator ions.

25. *Answer:* **(a) precipitation reaction; products are NaCl and MnS; $MnCl_2(aq) + Na_2S(aq) \longrightarrow 2\,NaCl(aq) + MnS(s)$ (b) precipitation reaction; products are NaCl and $ZnCO_3$; $Na_2CO_3(aq) + ZnCl_2(aq) \longrightarrow 2\,NaCl(aq) + ZnCO_3(s)$ (c) gas-forming reaction; products are $KClO_4$, H_2O and CO_2; $K_2CO_3(aq) + 2\,HClO_4(aq) \longrightarrow 2\,KClO_4(aq) + H_2O(\ell) + CO_2(g)$**

Strategy and Explanation: Given the reactants of reactions, classify the reaction that occurs, identify the products, and balance the equations.

To classify these reactions, determine the exchange products and check their solubility and/or strength using Tables 5.1 on page 137 and 5.2 on page 144. Remember that carbonate compounds reacting with acids produce CO_2 gas.

(a) When $MnCl_2$ reacts with Na_2S, the exchange products are $NaCl$ and MnS. Checking their solubility, we find that $NaCl$ is soluble, but MnS is insoluble. That makes this a **precipitation reaction**.

$$MnCl_2(aq) + Na_2S(aq) \longrightarrow 2\ NaCl(aq) + MnS(s) \quad \text{balanced overall equation}$$

(b) When Na_2CO_3 reacts with $ZnCl_2$, the exchange products are $NaCl$ and $ZnCO_3$. Checking the solubility, we see that $NaCl$ is soluble, but $ZnCO_3$ is insoluble, so this is a **precipitation reaction**.

$$Na_2CO_3(aq) + ZnCl_2(aq) \longrightarrow 2\ NaCl(aq) + ZnCO_3(s) \quad \text{balanced overall equation}$$

(c) When K_2CO_3 reacts with $HClO_4$, the exchange products are $KClO_4$ and H_2CO_3, which decomposes into liquid H_2O and CO_2 gas. That makes this a **gas-forming reaction**. $KClO_4$ is soluble.

$$K_2CO_3(aq) + 2\ HClO_4(aq) \longrightarrow 2\ KClO_4(aq) + H_2O(\ell) + CO_2(g) \quad \text{balanced overall equation}$$

✓ *Reasonable Answer Check:* Precipitation reactions form insoluble compounds and gas-forming reactions produce gases. The atoms are all balanced and the charges are all balanced in each equation.

Oxidation-Reduction Reactions

27. *Answer:* **(a) Ox. # O = –2, Ox. # S = +6 (b) Ox. # O = –2, Ox. # H = +1, Ox. # N = +5 (c) Ox. # K = +1, Ox. # O = –2, Ox. # Mn = +7 (d) Ox. # O = –2, Ox. # H = +1 (e) Ox. # Li = +1, Ox. # O = –2, Ox. # H = +1 (f) Ox. # Cl is –1, Ox. # H = +1, Ox. # C = 0**

Strategy and Explanation: Given several formulas of compounds, determine the oxidation numbers of the atoms in each of them.

For this Question, the rules spelled out in Section 5.4 on page 156 are used extensively. Several elements in compounds have predictable oxidation numbers (Rules 1 to 3). The oxidation numbers of the remaining atom(s) can be determined using the sum rule (Rule 4). The term "oxidation number" is abbreviated below as "Ox. #".

(a) SO_3 contains oxygen. Rule 3 gives us Ox. # O = –2. This is a molecule, so the sum of the oxidation numbers is zero.

$$0 = 1 \times (\text{Ox. \# S}) + 3 \times (-2)$$

Therefore, Ox. # S = +6.

(b) HNO_3 contains oxygen and hydrogen. Rule 3 gives us Ox. # O = –2 and Ox. # H = +1. We use the sum rule to find the Ox. # N. This is a molecule, so the sum of the oxidation numbers is zero.

$$0 = 1 \times (+1) + 1 \times (\text{Ox. \# N}) + 3 \times (-2)$$

Therefore, Ox. # N = +5.

(c) $KMnO_4$ contains a monatomic cation, K^+, and a polyatomic anion, MnO_4^-. Rule 2 gives us Ox. # K = +1. Rule 3 gives us Ox. # O = –2. We use the sum rule to find the Ox. # Mn. For a polyatomic anion, the sum of the oxidation numbers is equal to the ion's charge of 1–.

$$-1 = 1 \times (\text{Ox. \# Mn}) + 4 \times (-2)$$

Therefore, Ox. # Mn = +7.

(d) H_2O contains oxygen and hydrogen. Rule 3 gives us Ox. # O = –2 and Ox. # H = +1.

(e) $LiOH$ contains a monatomic cation, Li^+, and a diatomic anion, OH^-. Rule 2 gives us Ox. # Li = +1. Rule 3 gives us Ox. # O = –2 and Ox. # H = +1.

(f) CH_2Cl_2 contains carbon, hydrogen, and chlorine. Since chlorine makes a –1 ion when part of ionic compounds, we will assume its oxidation number is –1 (as was described in Section 5.4 on page 156 for the compound PCl_3). Rule 3 gives us Ox. # H = +1. We use the sum rule to find the Ox. # C. This is a molecule, so the sum of the oxidation numbers is zero.

$$0 = (\text{Ox. \# C}) + 2 \times (+1) + 2 \times (-1)$$

Therefore, Ox. # C = 0.

✓ *Reasonable Answer Check:* The non-metal elements farther to the right on the periodic table have consistently more negative oxidation numbers. The metallic elements and nonmetals farther to the left on the periodic table have more positive oxidation numbers.

29. *Answer:* **(a) Ox. # O = –2, Ox. # S = +6 (b) Ox. # O = –2, Ox. # N = +5 (c) Ox. # O = –2, Ox. # Mn = +7 (d) Ox. # O = –2, Ox. # H = +1, Ox. # Cr = +3 (e) Ox. # O = –2, Ox. # H = +1, Ox. # P = +5 (f) Ox. # O = –2, Ox. # S = +2**

Strategy and Explanation: Given several formulas of ions, determine the oxidation numbers of the atoms in each one.

For this Question, the rules spelled out in Section 5.4 on page 156 are used extensively. Several elements in compounds have predictable oxidation numbers (Rules 1-3). The oxidation numbers of the remaining atom(s) can be determined using the sum rule (Rule 4). The term "oxidation number" is abbreviated below as "Ox. #".

(a) SO_4^{2-}. Rule 3 gives us Ox. # O = –2. We use the sum rule to find the Ox. # S. This is a polyatomic anion, so the sum of the oxidation numbers is equal to the ion's charge of 2–.

$$-2 = 1 \times (Ox. \# S) + 4 \times (-2)$$

Therefore, Ox. # S = +6.

(b) NO_3^- contains oxygen. Rule 3 gives us Ox. # O = –2. We use the sum rule to find the Ox. # N. This is a polyatomic anion, so the sum of the oxidation numbers is equal to the ion's charge of 1–:

$$-1 = 1 \times (Ox. \# N) + 3 \times (-2)$$

Therefore, Ox. # N = +5.

(c) MnO_4^- contains oxygen. Rule 3 gives us Ox. # O = –2. We use the sum rule to find the Ox. # Mn. For a polyatomic anion, the sum of the oxidation numbers is equal to the ion's charge of 1–.

$$-1 = 1 \times (Ox. \# Mn) + 4 \times (-2)$$

Therefore, Ox. # Mn = +7.

(d) $Cr(OH)_4^-$ contains oxygen and hydrogen. Rule 3 gives us Ox. # O = –2 and Ox. # H = +1. We use the sum rule to find the Ox. # Cr. For a polyatomic anion, the sum of the oxidation numbers is equal to the ion's charge of 1–.

$$-1 = 1 \times (Ox. \# Cr) + 4 \times [1 \times (-2) + 1 \times (+1)]$$

Therefore, Ox. # Cr = +3.

(e) $H_2PO_4^-$ contains oxygen and hydrogen. Rule 3 gives us Ox. # O = –2 and Ox. # H = +1. We use the sum rule to find the Ox. # P. For a polyatomic anion, the sum of the oxidation numbers is equal to the ion's charge of 1–.

$$-1 = 2 \times (+1) + 1 \times (Ox. \# P) + 4 \times (-2)$$

Therefore, Ox. # P = +5.

(f) $S_2O_3^{2-}$ contains oxygen. Rule 3 gives us Ox. # O = –2. We use the sum rule to find the Ox. # S. For a polyatomic anion, the sum of the oxidation numbers is equal to the ion's charge of 2–.

$$-2 = 2 \times (Ox. \# S) + 3 \times (-2)$$

Therefore, 2 × (Ox. # S) = +4 and Ox. # S = +2.

✓ *Reasonable Answer Check:* The non-metal elements farther to the right on the periodic table have consistently more negative oxidation numbers. The metallic elements and nonmetals farther to the left on the periodic table have more positive oxidation numbers.

32. *Answer:* **Only reaction (b) is an oxidation-reduction reaction; oxidation numbers change. Reaction (a) is a precipitation reaction; reaction (c) is an acid-base neutralization reaction.**

Strategy and Explanation: Given several reactions, determine if they are oxidation-reduction reactions and classify the remaining reactions.

To decide if a reaction is an oxidation-reduction reaction, we need to see if any of the elements change oxidation state. Oxidation-reduction reactions are ones in which the atoms have different oxidation states before and after the reaction. If no change in oxidation state is observed, then it is not an oxidation-reduction reaction.

(a) The ionic compounds representing reactants and products in this reaction all contain Cd^{2+}, Cl^-, Na^+, and S^{2-} ions. Therefore, this is **not an oxidation-reduction reaction**. The formation of insoluble CdS classifies this reaction as a **precipitation reaction.**

(b) The elements Ca and O_2, both with zero oxidation states, are combined into an ionic compound, CaO, with Ca^{2+} and O^{2-} ions. That means the oxidation numbers did change and this is an **oxidation-reduction reaction**.

(c) The ionic compounds representing reactants contain Ca^{2+}, OH^-, H^+, and Cl^- ions. The product ionic compound contains Ca^{2+} and Cl^-. The other product, water, has O in the –2 oxidation state and H in the +1 oxidation state. Therefore, this is **not an oxidation-reduction reaction**. The formation of water from the reaction of OH^- and H^+ classifies this reaction as an **acid-base neutralization** reaction.

✓ *Reasonable Answer Check:* The reactions that are not oxidation-reduction reactions are easily classified as one of the other reactions we have studied in this chapter.

34. *Answer:* **Substances (b), (c), and (d)**

Strategy and Explanation: (b) O_2, (c) HNO_3 and (d) MnO_4^- are common oxidizing agents. They are all good at oxidizing other chemicals because they are readily reduced.

36. *Answer:* **(a) CO_2 or CO (b) PCl_3 or PCl_5 (c) $TiCl_2$ or $TiCl_4$ (d) Mg_3N_2 (e) Fe_2O_3 (f) NO_2**

Strategy and Explanation: Given reactants of oxidation-reduction reactions, identify the combination reactions' products.

A combination reaction is the production of a product by combining two other chemicals. In some cases, more than one product might form. In those cases, all possibilities have been included.

(a) $$C(s) + O_2(g) \longrightarrow CO_2(g)$$
$$2\,C(s) + O_2(g) \longrightarrow 2\,CO(g)$$

(b) $$P_4(s) + 6\,Cl_2(g) \longrightarrow 4\,PCl_3(g)$$
$$P_4(s) + 10\,Cl_2(g) \longrightarrow 4\,PCl_5(g)$$

(c) $$Ti(s) + Cl_2(g) \longrightarrow TiCl_2(s)$$
$$Ti(s) + 2\,Cl_2(g) \longrightarrow TiCl_4(s)$$

(d) $$3\,Mg(s) + N_2(g) \longrightarrow Mg_3N_2(s)$$

(e) $$4\,FeO(s) + O_2(g) \longrightarrow 2\,Fe_2O_3(s)$$

(f) $$2\,NO + O_2(g) \longrightarrow 2\,NO_2(g)$$

✓ *Reasonable Answer Check:* The single product formed is a combination of the two substances given.

Activity Series

38. *Answer/Explanation:* An example of a displacement reaction that is also a redox reaction:

$$Fe(s) + 2\,HCl(aq) \longrightarrow FeCl_2(aq) + H_2(g)$$

(a) The species that is oxidized is Fe, since iron loses electrons in the half reaction:

$$Fe(s) \longrightarrow Fe^{2+}(aq) + 2\,e^-$$

(b) The species that is reduced is HCl, since hydrogen gains electrons in the half reaction:

$$2\,H^+(aq) + 2\,e^- \longrightarrow H_2(g)$$

(c) The species that is oxidized functions as the reducing agent, see (a): Fe(s)

(d) The species that is reduced functions as the oxidizing agent, see (b): HCl(aq)

Notice, many other answers are possible.

40. *Answer:* **(a) "NR" (b) "NR" (c) "NR" (d) 3 Ag$^+$(aq) + Au(s)**

Strategy and Explanation:

(a) Na is more active than Zn, so: $Na^+(aq) + Zn(s) \longrightarrow$ "NR."

(b) H_2 is more active than Pt, so: $HCl(aq) + Pt(s) \longrightarrow$ "NR."

(c) Ag is more active than Au, so: $Ag^+(aq) + Au(s) \longrightarrow$ "NR."

(d) Au is less active than Ag, so: $Au^{3+}(aq) + 3\ Ag(s) \longrightarrow 3\ Ag^+(aq) + Au(s)$

Solution Concentrations

42. *Answer:* **(a) 0.254 M Na$_2$CO$_3$ (b) 0.508 M Na$^+$, 0.254 M CO$_3^{2-}$**

Strategy and Explanation: Given the mass of the solute and the volume of the solution, find the molarity of the solute, and the concentrations of the ions.

Use the molar mass to find moles of solute, convert the volume to liters from milliliters, and divide the moles of solute by the volume in liters to get molarity. Use formula stoichiometry to find the concentrations of the ions.

(a) Find moles of solute: $6.73 \text{ g Na}_2\text{CO}_3 \times \dfrac{1 \text{ mol Na}_2\text{CO}_3}{105.9885 \text{ g Na}_2\text{CO}_3} = 6.35 \times 10^{-2} \text{ mol Na}_2\text{CO}_3$

$$250. \text{ mL solution} \times \frac{1 \text{ L}}{1000 \text{ mL}} = 0.250 \text{ L solution}$$

$$\text{Molarity} = \frac{6.35 \times 10^{-2} \text{ mol Na}_2\text{CO}_3}{0.250 \text{ L solution}} = 0.254 \text{ M Na}_2\text{CO}_3$$

Notice that the three sequential calculations shown above can be combined into one calculation:

$$\frac{6.73 \text{ g Na}_2\text{CO}_3}{250. \text{ mL solution}} \times \frac{1 \text{ mol Na}_2\text{CO}_3}{105.9885 \text{ g Na}_2\text{CO}_3} \times \frac{1000 \text{ mL}}{1 \text{ L}} = 0.254 \text{ M Na}_2\text{CO}_3$$

This combined version prevents having to write unnecessary intermediate answers and helps cut down on round-off errors in significant figures. Both ways give the right answer, but it is helpful to consolidate your work, when you can.

(b) Na_2CO_3 has a formula stoichiometry that looks like this:

$$1 \text{ mol Na}_2\text{CO}_3 : 2 \text{ mol Na}^+ \text{ ions} : 1 \text{ mol CO}_3^{2-} \text{ ions}.$$

So, the 0.254 M Na_2CO_3 contains $2 \times (0.254 \text{ M})$ Na$^+$ ion = 0.508 M Na$^+$ and 0.254 M CO$_3^{2-}$.

✓ *Reasonable Answer Check:* The number 0.0635 is about one quarter of .250; this value looks right. The concentration of Na$^+$ is twice than the concentration of CO$_3^{2-}$.

44. *Answer:* **0.494 g KMnO$_4$**

Strategy and Explanation: Given volume of a solution and its molarity, find the mass of solute in the solution.

Convert the volume of solution to liters from milliliters. Then use the molarity as a conversion factor to determine moles of solute. Then use the molar mass to find grams of solute.

$$250. \text{ mL solution} \times \frac{1 \text{ L}}{1000 \text{ mL}} \times \frac{0.0125 \text{ mol KMnO}_4}{1 \text{ L solution}} \times \frac{158.0339 \text{ g KMnO}_4}{1 \text{ mol KMnO}_4} = 0.494 \text{ g KMnO}_4$$

✓ *Reasonable Answer Check:* The units cancel appropriately. The relative size of the mass seems appropriate.

46. *Answer:* **5.08 × 10^3 mL**

Strategy and Explanation: Given the mass of the solute and the solution's molarity, find the volume of the solution.

Use the molar mass to find determine moles of solute. Then use the molarity as a conversion factor to determine volume of the solute in liters. Then convert the volume of solution to milliliters from liters.

$$25.0 \text{ g NaOH} \times \frac{1 \text{ mol NaOH}}{39.9971 \text{ g NaOH}} \times \frac{1 \text{ L solution}}{0.123 \text{ mol NaOH}} \times \frac{1000 \text{ mL}}{1 \text{ L}} = 5.08 \times 10^3 \text{ mL solution}$$

✓ *Reasonable Answer Check:* The units cancel appropriately. The relatively large number of milliliters seems appropriate, since this relatively dilute solution contains a relatively large mass of solute.

48. *Answer:* **0.0150 M CuSO$_4$**

Strategy and Explanation: Given the volume of a concentrated solution, the concentrated solution's molarity, and the final volume of the dilute solution, find the concentration of the dilute solution.

Moles of solute do not change when water is added. So we can equate the moles of solute in the concentrated solution with the moles of solute in the dilute solution. Get the moles of solute in the concentrated solution by using the molarity of the concentrated solution as a conversion factor to convert the volume into moles. Divide the moles of solute by the new dilute solution's volume to get the dilute solution's concentration. The logical plan outlined here is built into the equation given in Section 5.6 on page 163 for doing dilution calculations:

$$Molarity(\text{conc}) \times V(\text{conc}) = Molarity(\text{dil}) \times V(\text{dil})$$

$$\frac{Molarity(\text{conc}) \times V(\text{conc})}{V(\text{dil})} = \frac{\left(0.0250 \text{ M CuSO}_4 \text{ conc}\right) \times (6.00 \text{ mL conc})}{(10.0 \text{ mL dil})} = 0.0150 \text{ M CuSO}_4$$

✓ *Reasonable Answer Check:* The dilute solution has a smaller concentration than the concentrated solution. That is a sensible result, since water was added to make the new solution.

That is a sensible result, since water was added to make the new solution.

Calculations for Reactions in Solution

50. *Answer:* **0.205 g Na$_2$CO$_3$**

Strategy and Explanation: Given the volume and molarity of a solution containing one reactant and the balanced chemical equation for a reaction, determine the mass of another reactant required for complete reaction.

The stoichiometry of a balanced chemical equation dictates how the moles of reactants combine, so we will commonly look for ways to calculate moles. Here, the volume and molarity can be used to find the moles of one reactant. Then we will use the equation stoichiometry to find out moles of the other reactant needed. Finally, we will use the molar mass to find the grams.

Notice: It is NOT appropriate to use the dilution equation when working with reactions!

We learn from the balanced equation that 2 mol of HNO$_3$ reacts with 1 mol Na$_2$CO$_3$.

$$25.0 \text{ mL HNO}_3 \text{ solution} \times \frac{1 \text{ L}}{1000 \text{ mL}} \times \frac{0.155 \text{ mol HNO}_3}{1 \text{ L HNO}_3 \text{ solution}}$$

$$\times \frac{1 \text{ mol Na}_2\text{CO}_3}{2 \text{ mol HNO}_3} \times \frac{105.9885 \text{ g Na}_2\text{CO}_3}{1 \text{ mol Na}_2\text{CO}_3} = 0.205 \text{ g Na}_2\text{CO}_3$$

52. *Answer:* **121 mL HNO$_3$**

Strategy and Explanation: Given the mass of one reactant, the balanced chemical equation for a reaction, and the molarity of a solution containing the other reactant, determine the volume of the second solution for a complete reaction.

The stoichiometry of a balanced chemical equation dictates how the moles of reactants combine, so we will commonly look for ways to calculate moles. Here, the mass and molar mass can be used to find the moles of one reactant. Then we will use the equation stoichiometry to find out moles of the other reactant needed. Then we will use the moles and molarity to find volume in liters and convert liters into milliliters.

Notice: It is NOT appropriate to use the dilution equation when working with reactions!

We learn from the balanced equation that 1 mol of Ba(OH)$_2$ reacts with 2 mol HNO$_3$.

$$1.30 \text{ g Ba(OH)}_2 \times \frac{1 \text{ mol Ba(OH)}_2}{171.3416 \text{ g Ba(OH)}_2} \times \frac{2 \text{ mol HNO}_3}{1 \text{ mol Ba(OH)}_2}$$

$$\times \frac{1 \text{ L HNO}_3 \text{ solution}}{0.125 \text{ mol HNO}_3} \times \frac{1000 \text{ mL}}{1 \text{ L}} = 121 \text{ mL HNO}_3 \text{ solution}$$

54. *Answer:* **0.18 g AgCl; NaCl is the excess reactant; 0.0080 M NaCl**

Strategy and Explanation: Given the volumes and molarities of separate solutions containing each reactant of a precipitation reaction, determine the maximum mass of product produced and the identity and concentration of the excess reactant.

First, complete and balance the precipitation equation. Then use the volumes, molarities, and the equation stoichiometry to find the moles of product produced for each reactant. The reactant that produces the smallest number of moles is the limiting reactant. Use the moles of product produced by the limiting reactant to determine the maximum mass of product formed. Determine the number of moles of excess reactant in the solution, using stoichiometry. Determine the excess reactant's final concentration by dividing by the new solution's volume in liters.

Notice: It is NOT appropriate to use the dilution equation when working with reactions!

The precipitation equation looks like this:

$$\text{AgNO}_3\text{(aq)} + \text{NaCl(aq)} \longrightarrow \text{AgCl(s)} + \text{NaNO}_3\text{(aq)}$$

We learn from this balanced equation that 1 mol of NaCl produces 1 mol AgCl.

$$50.0 \text{ mL AgNO}_3 \text{ solution} \times \frac{1 \text{ L}}{1000 \text{ mL}} \times \frac{0.025 \text{ mol AgNO}_3}{1 \text{ L AgNO}_3 \text{ solution}} \times \frac{1 \text{ mol AgCl}}{1 \text{ mol AgNO}_3} = 0.0013 \text{ mol AgCl}$$

$$100.0 \text{ mL NaCl solution} \times \frac{1 \text{ L}}{1000 \text{ mL}} \times \frac{0.025 \text{ mol NaCl}}{1 \text{ L NaCl solution}} \times \frac{1 \text{ mol AgCl}}{1 \text{ mol NaCl}} = 0.0025 \text{ mol AgCl}$$

AgCl is the limiting reactant, and NaCl is the excess reactant. Use 0.0013 mol AgCl to determine the maximum grams of AgCl produced:

$$0.0013 \text{ mol AgCl} \times \frac{143.321 \text{ g AgCl}}{1 \text{ mol AgCl}} = 0.18 \text{ g AgCl}$$

If 0.0013 mol AgCl was formed and 0.0025 mol AgCl was expected from NaCl, the difference is how much AgCl was not formed:

$$0.0025 \text{ mol AgCl expected} - 0.0013 \text{ mol AgCl formed} = 0.0012 \text{ mol AgCl not formed}$$

$$0.0012 \text{ mol AgCl not formed} \times \frac{1 \text{ mol NaCl}}{1 \text{ mol AgCl}} = 0.0012 \text{ mol NaCl not reacted}$$

$$\text{Total Volume} = (50.0 \text{ mL} + 100.0 \text{ mL}) \times \frac{1 \text{ L}}{1000 \text{ mL}} = 0.1500 \text{ L}$$

Molarity of NaCl after the AgCl has completely precipitated:

$$\text{Molarity} = \frac{0.013 \text{ mol NaCl}}{0.1500 \text{ L solution}} = 0.0080 \text{ M NaCl}$$

✓ *Reasonable Answer Check:* The concentration of the excess reactant after the reaction is over is smaller than it was at the beginning. This is an expected result, because some of it reacted, and what was left over got diluted. The maximum mass of AgCl produced is a reasonable number.

56. *Answer:* **1.192 M HCl**

Strategy and Explanation: Given the mass of one reactant, the balanced chemical equation for a reaction, and the volume of a solution containing the other reactant needed for a complete reaction, determine the molarity of the second solution.

The mass and molar mass can be used to find the moles of one reactant. Then we will use the equation stoichiometry to find out moles of the other reactant needed. Then we will use the moles and volume in liters to determine the molarity. *Note: It is NOT appropriate to use the dilution equation when working with reactions!*

We learn from the balanced equation that 1 mol of Na_2CO_3 reacts with 2 mol HCl.

$$2.050 \text{ g } Na_2CO_3 \times \frac{1 \text{ mol } Na_2CO_3}{105.9885 \text{ g } Na_2CO_3} \times \frac{2 \text{ mol HCl}}{1 \text{ mol } Na_2CO_3} = 3.868 \times 10^{-2} \text{ mol HCl}$$

$$32.45 \text{ mL HCl solution} \times \frac{1 \text{ L}}{1000 \text{ mL}} = 0.03245 \text{ L HCl solution}$$

$$\frac{3.868 \times 10^{-2} \text{ mol HCl}}{0.03245 \text{ L HCl solution}} = 1.192 \text{ M HCl solution}$$

The three separate calculations above can be consolidated into one calculation as follows:

$$\frac{2.050 \text{ g } Na_2CO_3}{32.45 \text{ mL HCl solution}} \times \frac{1 \text{ mol } Na_2CO_3}{105.9885 \text{ g } Na_2CO_3} \times \frac{2 \text{ mol HCl}}{1 \text{ mol } Na_2CO_3} \times \frac{1000 \text{ mL}}{1 \text{ L}} = 1.192 \text{ M HCl solution}$$

Both methods will give the right answer, however the consolidated calculation eliminates the need to write down unnecessary intermediate answers and helps eliminate round-off errors.

✓ *Reasonable Answer Check:* The units cancel appropriately, and the moles and liters are nearly the same value so it makes sense that the molarity is near 1.

General Questions

59. *Answer/Explanation:*

(a) Using Table 5.1 page 137, we find that ammonium compounds and nitrate compounds are soluble. The HgS compound is insoluble.

$$\textbf{(NH}_4\textbf{)}_2\textbf{S(aq) + Hg(NO}_3\textbf{)}_2\textbf{(aq)} \longrightarrow \textbf{HgS(s) + 2 NH}_4\textbf{NO}_3\textbf{(aq)} \qquad \text{balanced equation}$$

(b) The reactants are **ammonium sulfide** and **mercury(II) nitrate**. The products are **mercury(II) sulfide** and **ammonium nitrate**.

(c) The aqueous compounds are ionized. The solid remains unionized. Spectator ions are NH_4^+ and NO_3^-.

$$\textbf{S}^{2-}\textbf{(aq) + Hg}^{2+}\textbf{(aq)} \longrightarrow \textbf{HgS(s)} \qquad \text{balanced net ionic equation}$$

(d) This is a **precipitation** reaction.

60. *Answer/Explanation:*

(a) Combination reaction: $SO_3(g) + H_2O(\ell) \longrightarrow H_2SO_4(aq)$

(b) Combination reaction: $Sr(s) + H_2(g) \longrightarrow SrH_2(s)$

(c) Displacement reaction: $Mg(s) + H_2SO_4(aq) \longrightarrow MgSO_4(aq) + H_2(g)$

(d) Exchange (precipitation) reaction: $Na_3PO_4(aq) + 3 \, AgNO_3(aq) \longrightarrow Ag_3PO_4(s) + 3 \, NaNO_3(aq)$

(e) Decomposition and gas-forming reaction: $Ca(HCO_3)_2(s) \longrightarrow CaO(s) + H_2O(\ell) + 2 \, CO_2(g)$

(f) Oxidation-reduction reaction: $2 \, Fe^{3+}(aq) + Sn^{2+}(aq) \xrightarrow{\text{heat}} 2 \, Fe^{2+}(aq) + Sn^{4+}(aq)$

61. *Answer:* **(a) H_2O and NH_3 molecules, plus a small amount of NH_4^+ and OH^- (b) H_2O and CH_3CO_2H molecules, plus a small amount of $CH_3CO_2^-$ and H^+ (c) H_2O, Na^+, and OH^- (d) H_2O, H^+, and Br^-**

Strategy and Explanation:

(a) NH_3(aq) is a weak base (according to Table 5.2). That means the solution's primary components are H_2O and NH_3 molecules. A small amount of NH_4^+ and OH^- also exists in the solution.

(b) $CH_3CO_2H(aq)$ is a weak acid (according to Table 5.2). That means the solution's primary components are H_2O and CH_3CO_2H molecules. A small amount of $CH_3CO_2^-$ and H^+ also exists in the solution.

(c) NaOH(aq) is a strong base (according to Table 5.2). That means the solution's primary components are H_2O, Na^+, and OH^-.

(d) HBr(aq) is a strong acid (according to Table 5.2). That means the solution's primary components are H_2O, H^+, and Br^-.

63. *Answer:* **The only redox reaction is reaction (c); oxidizing agent is Ti; reducing agent is Mg.**

Strategy and Explanation: Given several reactions, determine if they are redox (oxidation-reduction) reactions and identify the oxidizing and reducing agents in those that are redox reactions.

To decide if a reaction is a redox reaction, we need to see if any of the elements change oxidation state. In redox reactions, the atoms change oxidation states during the reaction. If no change in oxidation state is observed, then the reaction is not a redox reaction. The oxidizing agent is the reactant that assists an oxidation by being reduced, so it will be the reactant whose atoms gain electrons, and end up with a lower (more negative or less positive) oxidation number. The reducing agent is the reactant that assists a reduction by being oxidized, so it will be the reactant whose atoms lose electrons, and end up with a higher (more positive or less negative) oxidation number.

(a) Look at the oxidation numbers for the reactants:

NaOH contains a monatomic cation, Na^+, and a diatomic anion, OH^-. Rule 2 gives us Ox. # Na = +1. Rule 3 gives us Ox. # O = –2 and Ox. # H = +1.

H_3PO_4 contains oxygen and hydrogen. Rule 3 gives us Ox. # O = –2 and Ox. # H = +1. We use the sum rule to find the Ox. # P. For a molecule, the sum of the oxidation numbers is equal to zero.

$$0 = 3 \times (+1) + 1 \times (Ox. \# P) + 4 \times (-2) \qquad \text{Therefore, Ox. \# P = +5.}$$

Look at the oxidation numbers for the products:

NaH_2PO_4 contains a monatomic cation, Na^+, and a polyatomic anion, $H_2PO_4^-$. Rule 2 gives us Ox. # Na = +1. $H_2PO_4^-$ contains oxygen and hydrogen. Rule 3 gives us Ox. # O = –2 and Ox. # H = +1. We use the sum rule to find the Ox. # P. For a polyatomic anion, the sum of the oxidation numbers is equal to the ion's charge of 1–.

$$-1 = 2 \times (+1) + 1 \times (Ox. \# P) + 4 \times (-2) \qquad \text{Therefore, Ox. \# P = +5.}$$

H_2O contains oxygen and hydrogen. Rule 3 gives us Ox. # O = –2 and Ox. # H = +1.

The oxidation numbers don't change, so this is NOT a redox reaction.

(b) Look at the oxidation numbers for the reactants:

NH_3 contains hydrogen. Rule 3 gives us Ox. # H = +1. We use the sum rule to find the Ox. # N. For a molecule, the sum of the oxidation numbers is equal to zero.

$$0 = 1 \times (Ox. \# N) + 3 \times (+1) \qquad \text{Therefore, Ox. \# N = -3}$$

CO_2 contains oxygen. Rule 3 gives us Ox. # O = –2. We use the sum rule to find the Ox. # C. For a molecule, the sum of the oxidation numbers is equal to zero.

$$0 = 1 \times (Ox. \# C) + 2 \times (-2) \qquad \text{Therefore, Ox. \# C = +4.}$$

H_2O contains oxygen and hydrogen. Rule 3 gives us Ox. # O = –2 and Ox. # H = +1.

Look at the oxidation numbers for the products:

NH_4HCO_3 contains a polyatomic cation, NH_4^+, and a polyatomic anion, HCO_3^-. NH_4^+ contains hydrogen. Rule 3 gives us Ox. # H = +1. We use the sum rule to find the Ox. # N. For a polyatomic cation, the sum of the oxidation numbers is equal to the ion's charge of 1+.

$$+1 = 1 \times (Ox. \# N) + 4 \times (+1) \qquad \text{Therefore, Ox. \# N = -3.}$$

HCO_3^- contains oxygen and hydrogen. Rule 3 gives us Ox. # O = –2 and Ox. # H = +1. We use the sum rule to find the Ox. # C. For a polyatomic anion, the sum of the oxidation numbers is equal to the ion's charge of 1–.

$$-1 = 1 \times (+1) + 1 \times (\text{Ox. \# C}) + 3 \times (-2)$$

Therefore, Ox. # C = +4.

The oxidation numbers don't change, so this is NOT a redox reaction.

(c) Look at the oxidation numbers for the reactants:

$TiCl_4$ contains a monatomic cation, Ti^{4+}, and a monatomic anion, Cl^-. Rule 2 gives us Ox. # Ti = +4 and Ox. # Cl = –1.

Rule 1 indicates the oxidation number of elements is zero, so Ox. # Mg = 0.

Look at the oxidation numbers for the products:

Rule 1 indicates the oxidation number of elements is zero, so Ox. # Ti = 0.

$MgCl_2$ contains a monatomic cation, Mg^{2+}, and a monatomic anion, Cl^-. Rule 2 gives us Ox. # Mg = +2 and Ox. # Cl = –1.

The oxidation numbers of Mg and Ti do change, so this IS a redox reaction.

The oxidizing agent is Ti since its Ox. # goes from +4 to zero.

The reducing agent is Mg since its Ox. # goes from zero to +2.

(d) Look at the oxidation numbers for the reactants:

NaCl contains a monatomic cation, Na^+, and a monatomic anion, Cl^-. Rule 2 gives us Ox. # Na = +1 and Ox. # Cl = –1.

$NaHSO_4$ contains a monatomic cation, Na^+, and a polyatomic anion, HSO_4^-. Rule 2 gives us Ox. # Na = +1. HSO_4^- contains oxygen and hydrogen. Rule 3 gives us Ox. # O = –2 and Ox. # H = +1. We use the sum rule to find the Ox. # S. For a polyatomic anion, the sum of the oxidation numbers is equal to the ion's charge of 1–.

$$-1 = 1 \times (+1) + 1 \times (\text{Ox. \# S}) + 4 \times (-2)$$

Therefore, Ox. # S = +6

Look at the oxidation numbers for the products:

HCl contains hydrogen. Rule 3 gives us Ox. # H = +1. We use the sum rule to find the Ox. # Cl. For a molecule, the sum of the oxidation numbers is equal to zero.

$$0 = 1 \times (+1) + 1 \times (\text{Ox. \# Cl})$$

Therefore, Ox. # Cl = –1

Na_2SO_4 contains a monatomic cation, Na^+, and a polyatomic anion, SO_4^{2-}. Rule 2 gives us Ox. # Na = +1. SO_4^{2-} contains oxygen. Rule 3 gives us Ox. # O = –2. We use the sum rule to find the Ox. # S. For a polyatomic anion, the sum of the oxidation numbers is equal to the ion's charge of 2–.

$$-2 = 1 \times (\text{Ox. \# S}) + 4 \times (-2)$$

Therefore, Ox. # S = +6

The oxidation numbers don't change, so this is NOT a redox reaction.

✓ *Reasonable Answer Check:* The oxidation numbers are consistent with typical oxidation states of these elements. The reactants and products are common acids, ionic compounds and ions.

Applying Concepts

64. *Answer/Explanation:* For the first case: LiCl(aq) and $AgNO_3$(aq)

(a) The two separate solutions of soluble salts would be clear and colorless like water. Once combined, insoluble white AgCl (Table 5.1, Rule 3) would precipitate. We would probably see it eventually sink to the bottom of the beaker, leaving a clear, probably colorless liquid containing aqueous $LiNO_3$ above it.

(b) Notice, for proper proportions, these diagrams really need many more water molecules.

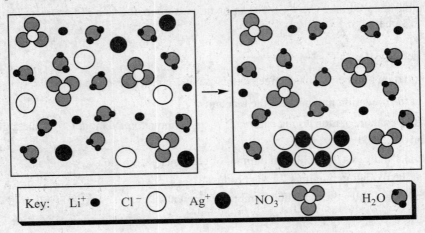

Key: Li$^+$ ●   Cl$^-$ ◯   Ag$^+$ ●   NO$_3^-$ ⬤   H$_2$O ⬤

(c) Li$^+$(aq) + Cl$^-$(aq) + Ag$^+$(aq) + NO$_3^-$(aq) ⟶ AgCl(s) + Li$^+$(aq) + NO$_3^-$(aq)

For the second case: NaOH(aq) and HCl(aq)

(a) The separate acid and base solutions would be clear and colorless like water. Once combined, the solution would still be clear and colorless.

(b) Notice, for proper proportions, these diagrams really need many more water molecules.

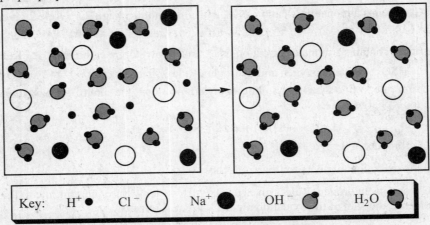

Key: H$^+$ ●   Cl$^-$ ◯   Na$^+$ ●   OH$^-$ ⬤   H$_2$O ⬤

(c) Na$^+$(aq) + OH$^-$(aq) + H$^+$(aq) + Cl$^-$(aq) ⟶ H$_2$O(ℓ) + Na$^+$(aq) + Cl$^-$(aq)

66. *Answer:* **(a) combine H$_2$SO$_4$ and Ba(OH)$_2$ (b) combine Na$_2$SO$_4$ and Ba(NO$_3$)$_2$ (c) combine H$_2$SO$_4$(aq) and BaCO$_3$(s)**

Strategy and Explanation: Prepare barium sulfate from a given list of chemicals by various means.

(a) To make BaSO$_4$ from an acid-base reaction, use a base with the cation and an acid with the anion:

$$H_2SO_4(aq) + Ba(OH)_2(aq) \longrightarrow BaSO_4(s) + 2\ H_2O(\ell)$$

(b) To make BaSO$_4$ from a precipitation reaction, use a soluble salt containing the anion and a soluble salt containing the cation:

$$Na_2SO_4(aq) + Ba(NO_3)_2(aq) \longrightarrow BaSO_4(s) + 2\ NaNO_3(aq)$$

(c) To make BaSO$_4$ from a gas-forming reaction, use an acid with the anion and the carbonate salt of the cation:

$$H_2SO_4(aq) + BaCO_3(s) \longrightarrow BaSO_4(s) + H_2O(\ell) + CO_2(g)$$

✓ *Reasonable Answer Check:* All of these reactants were provided in the list of chemicals. The neutralization reaction produces the solid and water. The precipitation reaction produces the solid and a soluble salt. The gas-forming reaction produces the solid and carbon dioxide gas.

68. *Answer:* **(d)**

Strategy and Explanation: Too much water was added, making the solution too dilute. So, (d) the concentration of the solution is less than 1 M because you added more solvent than necessary.

70. *Answer:* **(a) and (d) are correct.**

Strategy and Explanation:

(a) Since the solution is acidic, there are more H^+ ions than OH^- ions in the mixture. This statement is TRUE.

(b) This statement is FALSE, for the same reason (a) was true.

(c) Only equal quantities of strong acid and strong base make a neutral solution. If either the acid or the base are weak or if the molarities are unequal, then the resulting solution will be basic or acidic, so this statement is FALSE.

(d) Since the resulting solution was acidic, and equal volumes were added, that means the acid's concentration must have been greater than the base's concentration. This statement is TRUE.

(e) While the concentration of H_2SO_4 might have been greater, it is not necessarily true that it MUST have been greater, since only half as many moles of the diprotic acid is needed to neutralize the NaOH base; therefore, it's concentration need only have been anything more than half as large. This is TRUE.

72. *Answer:* **(a) Groups C&D: $Ag^+ + Cl^- \longrightarrow AgCl$ (s); Groups A&B: $Ag^+ + Br^- \longrightarrow AgBr$ (s) (b) silver halide product is the same for A&B and different from C or D (c) Curve has upward slope while the Ag^+ is the limiting reactant. Bromide is heavier than chloride, so the curve levels out at different masses of product.**

Strategy and Explanation:

(a) Na^+ and NO_3^- are spectator ions.

Net Ionic for groups C and D

$$Ag^+ + Cl^- \longrightarrow AgCl \ (s)$$

Net Ionic for groups A and B

$$Ag^+ + Br^- \longrightarrow AgBr \ (s)$$

(b) The silver halide solid (AgCl) produced by groups C and D is the same. The silver halide solid (AgBr) produced by groups A and B is the same.

(c) The "crossover" point, where the line changes from linearly rising to horizontal, describes the stoichiometric equivalence of the reactant and the product. To find the crossover point, (1) extrapolate the linearly-rising part of the line upward and to the right, (2) extrapolate the horizontal part of the line to the "Mass of product" axis, (3) determine where those two straight lines cross each other, and (4) draw a vertical line down from that crossing point to the "Mass of $AgNO_3$" axis:

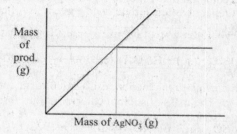

The cross-over point here occurs at 0.75 g $AgNO_3$. Below the mass of 0.75 g, the graph line increases with increasing masses of $AgNO_3$. In this region, Ag^+ is the limiting reactant. These reactions both require the same mass of $AgNO_3$ to make their respective silver halide. However, since bromide ion is heavier than chloride ion, the mass of the product will be different and the mass of product where the graph levels out will be different because AgBr is heavier than AgCl. That means the products in groups A and B (AgCl) weigh less than the products in groups C and D (AgBr).

More Challenging Questions

73. *Answer:* **104 g/mol**

Strategy and Explanation: Given an unknown diprotic acid, the volume and concentration of a solution containing the base that is used to neutralize it, the balanced chemical equation for the neutralization reaction, determine the molar mass of the acid.

The volume and molarity of the first reactant are used to calculate the moles. Equation stoichiometry can be used to find the moles of the acid. Then we will divide the mass by the moles to get molar mass. *Note: It is NOT appropriate to use the dilution equation when working with reactions!*

We learn from the balanced equation that 1 mol of H_2A reacts with 2 mol NaOH.

$$36.04 \text{ mL NaOH solution} \times \frac{1 \text{ L}}{1000 \text{ mL}} \times \frac{0.509 \text{ mol NaOH}}{1 \text{ L NaOH solution}} \times \frac{1 \text{ mol } H_2A}{2 \text{ mol NaOH}} = 0.00917 \text{ mol } H_2A$$

$$\frac{0.954 \text{ g } H_2A}{0.00917 \text{ mol } H_2A} = 104 \frac{\text{g } H_2A}{\text{mol } H_2A}$$

The two separate calculations above can be consolidated into one calculation as follows:

$$\frac{0.954 \text{ g } H_2A}{36.04 \text{ mL NaOH solution}} \times \frac{1000 \text{ mL}}{1 \text{ L}} \times \frac{1 \text{ L NaOH solution}}{0.509 \text{ mol NaOH}} \times \frac{2 \text{ mol NaOH}}{1 \text{ mol } H_2A} = 104 \frac{\text{g } H_2A}{\text{mol } H_2A}$$

Both methods will give the right answer, however the consolidated calculation eliminates the need to write down unnecessary intermediate answers and helps eliminate round-off errors.

✓ *Reasonable Answer Check:* The moles are nearly 0.01 times the mass value so it makes sense that the molar mass is near 100.

76. *Answer:* **2.26 g NaCl, 1.45 g NaCl**

Strategy and Explanation: Given the volumes of two different chicken soup solutions containing a reactant of a precipitation reaction and given the mass of the precipitate formed when excess amount of the second reactant is added, determine the mass of the first reactant present in the solution.

Using molar mass, determine the number of moles of product, AgCl, formed, then using stoichiometry determine the number of reactant, NaCl, in the solution. Determine the mass of the reactant, NaCl, using the molar mass.

The precipitation reaction looks like this:

$$AgNO_3(aq) + NaCl(aq) \longrightarrow AgCl(s) + NaNO_3(aq)$$

We learn from this balanced equation that 1 mol of AgCl is produced from 1 mol NaCl.

$$5.55 \text{ g AgCl} \times \frac{1 \text{ mol AgCl}}{143.321 \text{ g AgCl}} \times \frac{1 \text{ mol NaCl}}{1 \text{ mol AgCl}} \times \frac{58.443 \text{ g NaCl}}{1 \text{ mol NaCl}} = 2.26 \text{ g NaCl}$$

$$3.55 \text{ g AgCl} \times \frac{1 \text{ mol AgCl}}{143.321 \text{ g AgCl}} \times \frac{1 \text{ mol NaCl}}{1 \text{ mol AgCl}} \times \frac{58.443 \text{ g NaCl}}{1 \text{ mol NaCl}} = 1.45 \text{ g NaCl}$$

✓ *Reasonable Answer Check:* The mass of sodium chloride in the "less salt" solution is 64% of the mass of the sodium chloride in the regular soup.

77. *Answer:* **0.0154 M CaSO₄; 0.341 g CaSO₄ undissolved**

Strategy and Explanation: Given the maximum amount of an ionic solid that will dissolve in a fixed volume of a solution, the mass of the ionic solid added to a specific volume of a solution, determine the molarity of the solute in the solution and the mass of the solid that does not dissolve.

Use the molar mass to determine moles of solute dissolved in the solution, and determine the number of liters of solution. Divide these two numbers to get the molarity. Subtract the mass of solid that will dissolve from the mass of the solid added to the solution to determine how much does not dissolve.

The most $CaSO_4$ that can dissolve in 100.0 mL of water is 0.209 grams. Since more than that amount of solid was added, only 0.209 grams of it will dissolve.

$$\frac{0.209 \text{ g } CaSO_4}{100.0 \text{ mL}} \times \frac{1 \text{ mol } CaSO_4}{136.141 \text{ g } CaSO_4} \times \frac{1000 \text{ mL}}{1 \text{ L}} = 0.0154 \text{ M } CaSO_4$$

0.550 g $CaSO_4$ added – 0.209 g $CaSO_4$ dissolved = 0.341 g $CaSO_4$ remain undissolved

✓ *Reasonable Answer Check:* There is more than the maximum amount of solid added, so it makes sense that there would be some left over.

78. *Answer:* **184 mL $K_2Cr_2O_7$**

Strategy and Explanation: Given the molarity of a reactant in a solution and the moles of product generated, determine the volume of the solution for a complete reaction with excess amounts of the other reactants available.

First, use the moles of product and the equation stoichiometry to find out moles of the reactant needed. Then use the moles and molarity to find volume in liters, and convert liters into milliliters. *Note: It is NOT appropriate to use the dilution equation when working with reactions!*

The balanced equation says 3 mol of CH_3COOH are produced from 2 mol $K_2Cr_2O_7$.

$$0.166 \text{ mol } CH_3COOH \times \frac{2 \text{ mol } K_2Cr_2O_7}{3 \text{ mol } CH_3COOH} \frac{1 \text{ L } K_2Cr_2O_7 \text{ solution}}{0.600 \text{ mol } K_2Cr_2O_7} \times \frac{1000 \text{ mL}}{1 \text{ L}} = 184 \text{ mL } K_2Cr_2O_7$$

80. *Answer:* **6.28% impurity**

Strategy and Explanation: Given the mass of a tablet containing vitamin C, the balanced neutralization equation, and the volume and molarity of a base solution used for neutralization, determine what percentage of the tablet is impurity.

Use the volume, molarity, stoichiometry of the chemical equation and molar mass of the compound to calculate the mass of the vitamin C in the sample. Subtract the mass of vitamin C from the mass of the tablet, to determine that mass of the impurity. Divide the mass of the impurity into the mass of the soil sample and multiply by 100% to get percentage mass of the impurity.

$$21.30 \text{ mL } NaOH \times \frac{1 \text{ L } NaOH}{1000 \text{ mL } NaOH} \times \frac{0.1250 \text{ mol } NaOH}{1 \text{ L } NaOH} \times \frac{1 \text{ mol } HC_6H_7H_6}{1 \text{ mol } HCl}$$

$$\times \frac{176.1238 \text{ g } HC_6H_7O_6}{1 \text{ mol } HC_6H_7O_6} \times \frac{1000 \text{ mg } HC_6H_7O_6}{1 \text{ g } HC_6H_7O_6} = 468.6 \text{ mg } HC_6H_7O_6$$

$$500.0 \text{ mg tablet} - 468.6 \text{ mg } HC_6H_7O_6 = 468.6 \text{ mg impurity}$$

$$\frac{31.4 \text{ g impurity}}{500.0 \text{ g tablet}} \times 100\% = 6.28\% \text{ impurity}$$

✓ *Reasonable Answer Check:* A vitamin C tablet should be mostly vitamin C, so it makes sense that the mass percent if the impurity is small.

Chapter 6: Energy and Chemical Reactions

Introduction

Heat and temperature are obviously related, but they are not interchangeable terms. To distinguish between heat and temperature, imagine a bowl, a bucket, and a barrel of boiling water. If a thermometer were placed in each container, which one would cause the mercury to raise the highest? Which container would be the better heat source on a cold winter night?

Enthalpy change is a chemist's way of quantifying of the general concept of heat energy being released or absorbed in a chemical reaction. The heat energy change is expressed by a ΔH value. Whether the enthalpy change is called enthalpy of formation, enthalpy of combustion, enthalpy of sublimation, enthalpy of reaction, standard molar enthalpy of formation, enthalpy of decomposition, enthalpy of fusion, etc., don't lose sight of the concept: it is still just referring to the heat energy released or absorbed for the specific named process.

Solutions to Blue-Numbered Questions
for Review and Thought for Chapter 6

Topical Questions

The Nature of Energy

7. *Answer:* **(a) 399 Cal (b) 5.0×10^6 J/day**

    Strategy and Explanation: Convert a quantity of kilojoules (provided by a piece of cake) into food Calories, and convert a quantity of food Calories into joules.

    Use metric and energy conversion factors to achieve the conversions.

    (a)
    $$1670 \ \text{kJ} \times \frac{1000 \ \text{J}}{1 \ \text{kJ}} \times \frac{1 \ \text{cal}}{4.184 \ \text{J}} \times \frac{1 \ \text{kcal}}{1000 \ \text{cal}} \times \frac{1 \ \text{Cal}}{1 \ \text{kcal}} = 399 \ \text{Cal}$$

    (b)
    $$\frac{1200 \ \text{Cal}}{1 \ \text{day}} \times \frac{1 \ \text{kcal}}{1 \ \text{Cal}} \times \frac{1000 \ \text{cal}}{1 \ \text{kcal}} \times \frac{4.184 \ \text{J}}{1 \ \text{cal}} = \frac{5.0 \times 10^6 \ \text{J}}{1 \ \text{day}}$$

    ✓ *Reasonable Answer Check:* The food Calorie is about four times bigger than a kilojoule. So, the energy quantity in Calories should be about four times smaller than in kilojoules.

9. *Answer:* **3×10^8 J**

    Strategy and Explanation: Given the rate of solar energy reaching the earth per square centimeter of area and the dimensions of a house's roof, determine how much energy reaches the roof in one hour.

    Determine the roof's area in units of square centimeters and the time in hours, then multiply them by the given rate.

    $$A = (12 \ \text{m}) \times (12 \ \text{m}) \times \left(\frac{100 \ \text{cm}}{1 \ \text{m}}\right)^2 = 1.4 \times 10^6 \ \text{cm}^2$$

    $$1.4 \times 10^6 \ \text{cm}^2 \times 1 \ \text{h} \times \frac{60 \ \text{min}}{1 \ \text{hr}} \times \frac{4.0 \ \text{J}}{\text{min} \times \text{cm}^2} = 3 \times 10^8 \ \text{J}$$

✓ *Reasonable Answer Check:* This is a large amount of energy, as implied in the Question's parenthetical remark. This is about 100 kW-hr of energy and at the current rate of energy costs, represents energy that would cost about $80. That explains why many homeowners in sunny places have elected to put solar panels on their roofs.

Conservation of Energy

11. *Answer/Explanation:*

(a) In the process of lighting the match, the kinetic energy of moving the match across the striking surface is converted into thermal energy due to friction. This thermal energy causes the match to "light." This "lighting" is the result of a combustion process, where the chemical energy stored in the reactants is converted into heat energy and light energy. The heat energy of the match is used to light the fuse, and the chemical energy of the fuse is converted into heat energy and light energy. The heat from the fuse ignites the chemical propellants in the rocket. The chemical energy here is converted into heat, light, and the kinetic energy and potential energy of the rocket as the rocket's speed and altitude increase. When the rocket explodes, more chemical energy is converted into light, heat, and kinetic energy.

(b) As the fuel is pumped from the underground storage tank, its potential energy is increased by the mechanical energy of the pump. By using the fuel to drive 25 miles, some of the chemical potential energy stored in the fuel is converted into kinetic energy that moves the car and into heat energy energy that warms the engine and passenger compartment, when necessary.

13. *Answer/Explanation:* Describe and explain your choice of the system and the surroundings, describe transfer of energy and materials into and out of the system, and determine if the process is exothermic or endothermic.

The system is identified as precisely what we are studying. The surroundings are everything else. The important aspects of the surroundings are usually those things in contact with the system or in close proximity. The process is exothermic if the system loses energy. The process is endothermic if the system gains energy.

(a) The System: NH_4Cl

This choice was made following the discussion in Section 6.2 (page 184) when describing a reaction, "the system is usually defined as all the atoms that make up the reactants." Here the equation for the reaction being studied is: $NH_4Cl(s) \longrightarrow NH_4^+(aq) + Cl^-(aq)$

The Surroundings: Anything not NH_4Cl, including the water.

The choice to consider water as part of the surroundings instead of part of the system is based upon the fact that it is neither a reactant nor a product in given the dissolving equation. Although water undeniably has a strong interaction with the products of the reaction, there are still no H_2O molecules in this equation.

Defining the system to include the water can also be done. Data collected from dissolving experiments will not actually be able to functionally separate the water from the material dissolving.

(b) To study the release of energy during the phase change of this ionic compound, we must isolate it and see how it interacts with the surroundings.

(c) The system's interaction with the surroundings causes heat energy to be transferred into the surroundings and out of the system. There is no material transfer in this process, but there is a change in the specific interaction between the water and system.

(d) Since changes in the system cause energy to be gained by the system, the process is endothermic.

✓ *Reasonable Answer Check:* This definition of the system is restrictive enough to allow us to learn more about the relationship between the energy required to break the ionic bonds in the solid and the energy involved with the products' increased interaction with the surrounding solvent molecules.

15. *Answer:* **$\Delta E = +32$ J**

Strategy and Explanation: Given descriptions and numerical values for work and heat energy changes, determine the ΔE for the system.

The change in energy of the system, ΔE, is calculated using Equation 6.1 on page 185: $\Delta E = q + w$, where q and w (work energy and heat energy) cause energy to flow across the boundary between the system and the surroundings. In chemisty, both q and w are chosen to be positive when the direction of energy flow is from the surroundings into the system. Alternatively, w and q will carry a negative sign, if energy flows out of the system into the surroundings.

The system does work, so energy flows out of the system and w is negative: $w = -75.4$ J.

Heat energy is transferred into the system, so energy flows into the system and q is positive: $q = +25.7$ cal

Notice: Make sure that the units of energy are the same before adding the energy values.

$$\Delta E = (+25.7 \text{ cal}) \times \frac{4.184 \text{ J}}{1 \text{ cal}} + (-75.4 \text{ J}) = 32 \text{ J}$$

✓ *Reasonable Answer Check:* The heat energy input is more than the work energy output, so it is reasonable that ΔE is positive.

Heat Capacity

17. *Answer:* **Process (a) requires more energy than (b)**

Strategy and Explanation: Given the mass and temperature change for two samples, determine which process requires a greater transfer of energy.

Get the specific heat capacity of water from Table 6.1 on page 188, then use Equation 6.2′ on page 187 to calculate the heat energy required in each scenario.

Table 6.1 tells us that the specific heat capacity of liquid water is 4.184 J g^{-1}°C^{-1}.

$$q = c \times m \times \Delta T.$$

(a) $\Delta T = T_f - T_i = 50 \text{ °C} - 20\text{°C} = 30 \text{ °C}$ *(must be rounded to tens place)*

$$q_{H_2O} = (4.184 \text{ J g}^{-1}\text{°C}^{-1}) \times (10.0 \text{ g}) \times (30 \text{ °C}) = 1 \times 10^3 \text{ J}$$

(b) $\Delta T = T_f - T_i = 37 \text{ °C} - 25\text{°C} = 12 \text{ °C}$

$$q_{Cu} = (0.385 \text{ J g}^{-1}\text{°C}^{-1}) \times (20.0 \text{ g}) \times (12 \text{ °C}) = 92 \text{ J}$$

The cooling described in (a) requires a greater transfer of energy than that described in (b).

✓ *Reasonable Answer Check:* The specific heat capacity of water is more that ten times the specific heat capacity of copper and the mass and the temperature change for the copper sample was also smaller. The water sample should thus require much greater transfer of energy.

19. *Answer:* **It takes less time to raise the Cu sample to body temperature.**

Strategy and Explanation: Given the mass and initial temperature of two samples made out of two different metals, also given the assumption that they take up heat energy at the same rate when held in your hand, determine which metal will warm up to body temperature faster.

First, look up the specific heat capacities of the substances being heated in Table 6.1. The rate of absorption of heat energy can be related to heat capacity, mass, temperature increase, and elapsed time using Equation 6.2. The elapsed time can then be related to the specific heat capacity, the mass, the change in temperature, and the rate of absorption of heat energy. The fastest elapsed time will reach body temperature first.

The specific heat capacity (c) of Al is 0.902 J g^{-1}°C^{-1} and of Cu is 0.385 J g^{-1}°C^{-1} from Table 6.1. Use Equation 6.2′, then divide by Δt to write an equation relating the rate of heat energy absorption to the specific heat capacity, mass, temperature increase, and elapsed time:

$$\text{rate of heat energy absorption} = \frac{q}{\Delta t} = c \times m \times \frac{\Delta T}{\Delta t}$$

Solve this equation for the elapsed time, Δt:

$$\Delta t = c \times \frac{m \times \Delta T}{\text{rate}}$$

Comparing the two samples, their mass, change in temperature, and the rate of heat energy absorption are all the same, simplifying the equation above to the following: $\Delta t = c \times$ constant. This equation demonstrates that the elapsed heating time is proportional to the specific heat capacity. That means the copper sample will warm to body temperature before the aluminum sample at a constant rate over a common temperature range.

✓ *Reasonable Answer Check:* Specific heat capacity describes the amount of energy needed to increase the temperature of a 1-gram sample by 1 °C. Since these samples have the same mass and temperature change, the object with the smaller specific heat capacity will be faster to heat.

21. *Answer:* **More energy (1.48 × 10⁶ J) is absorbed by the water sample than by the ethylene glycol sample (9.56 × 10⁵ J).**

Strategy and Explanation: Given the volume of a cooling system, the densities of water and ethylene glycol, and the temperature change for two different samples, compare the thermal energy increase in the two samples.

First convert the volumes into cubic centimeters, then use the densities to determine the masses of each sample. Then use Equation 6.2′ to calculate the heat energy required in each process and compare them.

Volume of the cooling system in cubic centimeters:

$$5.00 \text{ quarts} \times \frac{0.946 \text{ L}}{1 \text{ quart}} \times \frac{1000 \text{ mL}}{1 \text{ L}} \times \frac{1 \text{ cm}^3}{1 \text{ mL}} = 4730 \text{ cm}^3$$

Mass of the two samples: $4730 \text{ cm}^3 \times \dfrac{1.113 \text{ g}}{1 \text{ cm}^3} = 5260 \text{ g ethylene glycol}$

$$4730 \text{ cm}^3 \times \frac{1.00 \text{ g}}{1 \text{ cm}^3} = 4730 \text{ g water}$$

In both samples, the change in temperature is the same:

$$\Delta T = T_f - T_i = 100.0 \text{ °C} - 25.0 \text{ °C} = 75.0 \text{ °C}$$

$$q_{\text{ethylene glycol}} = (2.42 \text{ J g}^{-1}\text{°C}^{-1}) \times (5260 \text{ g}) \times (75.0 \text{ °C}) = 9.56 \times 10^5 \text{ J}$$

$$q_{\text{water}} = (4.184 \text{ J g}^{-1}\text{°C}^{-1}) \times (4730 \text{ g}) \times (75.0 \text{ °C}) = 1.48 \times 10^6 \text{ J}$$

More energy is absorbed (1.48 × 10⁶ J) by the water sample than is absorbed (9.56 × 10⁵ J) by the ethylene glycol sample.

✓ *Reasonable Answer Check:* Water has a much larger specific heat capacity than ethylene glycol, so a somewhat smaller mass of water will still absorb more thermal energy than a larger mass of ethylene glycol.

23. *Answer:* **Gold**

Strategy and Explanation: Given the mass, initial and final temperatures, and energy needed to heat an unknown element, determine its most probable identity using Table 6.1.

Adapt the method described in the solution to Questions 17 and 21.

$$c = \frac{q}{m \times \Delta T} = \frac{34.7 \text{ J}}{23.4 \text{ g} \times \left(28.9 \text{ °C} - 17.3 \text{ °C}\right)} = 0.128 \ \frac{\text{J}}{\text{g °C}}$$

Looking at Table 6.1, the element whose specific heat capacity is closest to this is Au.

✓ *Reasonable Answer Check:* The specific heat capacity of Au (0.128 J g⁻¹°C⁻¹) matches the calculated value to three significant figures (0.128 J g⁻¹°C⁻¹). The similarity in the two values gives us confidence in the answer.

25. *Answer:* **$\Delta T_{\text{surroundings}}$ = positive, ΔE_{system} = negative**

Strategy and Explanation: Given the direction of transfer of thermal energy between a system and the surroundings with no work done, determine the algebraic sign of $\Delta T_{\text{surroundings}}$ and ΔE_{system}.

If thermal energy enters the surroundings as heat energy, then the temperature in the surroundings, $T_{\text{surroundings}}$, rises. If thermal energy enters the system as heat energy, then the temperature in the surroundings, $T_{\text{surroundings}}$, drops. If the system gets energy from the surroundings, then the internal energy of the system, E_{system}, rises. If the surroundings gets energy from the system, then the internal energy of the system, E_{system}, lowers.

The thermal energy enters the surroundings as heat energy, so the temperature in the surroundings, $T_{\text{surroundings}}$, rises:

$$T_{f,\text{surroundings}} > T_{i,\text{surroundings}}$$

$$\Delta T_{\text{surroundings}} = T_{f,\text{surroundings}} - T_{i,\text{surroundings}} = \text{positive}$$

Here, the surroundings gets energy from the system, so the internal energy of the system, E_{system}, lowers:

$$E_{f,system} < E_{i,system}$$

$$\Delta E_{system} = E_{f,system} - E_{i,system} = \text{negative}$$

✓ *Reasonable Answer Check:* Energy leaves the system, so a negative ΔE_{system} makes sense. The energy must show up in the surroundings, so the fact that these two algebraic signs are opposite signs also makes sense.

Energy and Enthalpy

27. *Answer:* **4.13×10^5 J**

Strategy and Explanation: Given the number of cubes in a tray of ice, the mass of each cube, and the thermal energy required to melt a given mass of ice, determine the energy required to melt the whole tray of ice cubes. The tray of ice cubes is the sample. Use the rest of the information as unit factors to determine the energy.

$$1 \text{ tray} \times \frac{20 \text{ cubes}}{1 \text{ tray}} \times \frac{62.0 \text{ g ice}}{1 \text{ cube}} \times \frac{333 \text{ J}}{1.00 \text{ g ice}} = 4.13 \times 10^5 \text{ J}$$

✓ *Reasonable Answer Check:* The tray has over 1000 grams of ice, so this large number of joules makes sense. Notice, the tray and the number of cubes are countable objects; hence, those numbers are exact with infinite significant figures, so their values do not limit the significant figures of the answer.

29. *Answer:* **5.00×10^5 J**

Strategy and Explanation: Given the mass and initial temperature of a liquid substance, the boiling point of the liquid, the final temperature of the gaseous substance, the specific heat capacity and the enthalpy of vaporization, determine the energy (in joules) required to complete the transition.

Rearranging Equation 6.2, determine the heat energy required to raise the temperature to the boiling point. Then use the enthalpy of vaporization (ΔH_{vap}) to determine the energy required to boil the liquid at the boiling point.

Define the system as the liquid benzene. To change the temperature of the system, use $q = c \times m \times \Delta T$

$$q_{T\text{-rise}} = c_{benzene} \times m_{benzene} \times \Delta T_{benzene}$$

$$= (1.74 \text{ J g}^{-1}°\text{C}^{-1}) \times (1.00 \text{ kg}) \times \frac{1000 \text{ g}}{1 \text{ kg}} \times (80.1 \text{ °C} - 20.0 \text{ °C}) = 1.05 \times 10^5 \text{ J}$$

To change a phase in the system, use $q = m \times \Delta H_{vap}$.

$$q_{boil} = m_{benzene} \times \Delta H_{vap,benzene} = (1.00 \text{ kg}) \times \frac{1000 \text{ g}}{1 \text{ kg}} \times (395 \text{ J/g}) = 3.95 \times 10^5 \text{ J}$$

The sum gives the total heat energy required.

$$q_{benzene} = q_{T\text{-rise}} + q_{boil} = (1.05 \times 10^5 \text{ J}) + (3.95 \times 10^5 \text{ J}) = 5.00 \times 10^5 \text{ J}$$

✓ *Reasonable Answer Check:* To make the final product, the liquid's temperature needed to be raised, and then the liquid needed to be vaporized. Both of these changes require energy to be added to the system.

31. *Answer/Explanation:* A cooling curve shows how the temperature drops as the heat energy is removed from the system. In that respect, the lower part of this graph will look like the reverse of Figure 6.11.

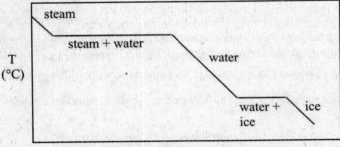

Quantity of heat transferred out of the system

Thermochemical Expressions

33. *Answer:* **endothermic**

Strategy and Explanation: Given a thermochemical expression, determine if the reaction is exothermic or endothermic.

Find the sign of ΔH, and use that to determine if the reaction is exothermic or endothermic. When ΔH is negative, the reaction is exothermic. When ΔH is positive, the reaction is endothermic. Notice: If students can't always remember the Greek origins of these words and they find themselves mixing which is positive and which is negative, ask them to think of the words "exhale" and "inhale." When you **ex**hale, air goes **out** of your lungs (exo = negative); when you **in**hale, air goes **into** your lungs (endo = positive).

$\Delta H = +38$ kJ, so this reaction is **endothermic**.

✓ *Reasonable Answer Check:* Endothermic reactions have positive ΔH.

35. *Answer:* **(a) – 2.1 × 10² kJ (b) –33 kJ**

Strategy and Explanation: Given a thermochemical expression for a phase change, determine the quantity of energy transferred to the surroundings when two different samples undergo that phase change.

When necessary, convert the sample quantity into moles. Then use the thermochemical expression to create a unit conversion factor relating energy to moles of reactant.

Reversing the given balanced thermochemical expression to show the freezing reaction, causes the $\Delta H°$ to change sign:

$$H_2O(\ell) \longrightarrow H_2O\ (s) \qquad\qquad \Delta H° = -6.0 \text{ kJ}$$

The thermochemical expression tells us when 1 mol $H_2O(\ell)$ is frozen, that 6.0 kJ are transferred to the surroundings.

(a)
$$34.2 \text{ mol } H_2O(s) \times \frac{-6.0 \text{ kJ}}{1 \text{ mol } H_2O(s)} = -2.1 \times 10^2 \text{ kJ}$$

(b)
$$100.0 \text{ g } H_2O(s) \times \frac{1 \text{ mol } H_2O(s)}{18.0152 \text{ g } H_2O(s)} \times \frac{-6.0 \text{ kJ}}{1 \text{ mol } H_2O(s)} = -33 \text{ kJ}$$

✓ *Reasonable Answer Check:* The freezing reaction is exothermic, since heat energy must be removed from the reactants to make the products. Also, 34 mol of water weigh more than 100 grams, so it makes sense that the answer in (a) is larger than the answer in (b).

Enthalpy Changes for Chemical Reactions

37. *Answer:* **(a) 0.500 $C_8H_{18}(\ell)$ + 6.25 O_2(g) $\longrightarrow$ 4.00 CO_2(g) + 4.50 $H_2O(\ell)$; $\Delta H° = -2.75 \times 10^3$ kJ**
    (b) 100. $C_8H_{18}(\ell)$ + 1250 O_2(g) $\longrightarrow$ 800. CO_2(g) + 900. $H_2O(\ell)$; $\Delta H° = -5.50 \times 10^5$ kJ
    (c) 1.00 $C_8H_{18}(\ell)$ + 12.5 O_2(g) $\longrightarrow$ 8.00 CO_2(g) + 9.00 $H_2O(\ell)$; $\Delta H° = -5.50 \times 10^3$ kJ

Strategy and Explanation: Given a thermochemical expression for a reaction, write a thermochemical expression for multiples of the reaction.

Determine what number the expression must be multiplied by to get the desired number of reactants or products given. To determine the multiplier, divide the stoichiometric coefficient we want to have in the new expression by the stoichiometric coefficient given in the original expression. Multiply every coefficient in the expression and the $\Delta H°$ by that multiplier to set up the new thermochemical expression.

In each of the samples, a number of moles is given. Assume these are exact numbers.

(a) The original expression makes 16 mol CO_2. We want to make 4.00 mol CO_2. So,

$$\text{Multiplier} = \frac{4.00 \text{ mol}}{16 \text{ mol}} = 0.250$$

$$0.250 \times [2 \text{ } C_8H_{18}(\ell) + 25 \text{ } O_2(g) \longrightarrow 16 \text{ } CO_2(g) + 18 \text{ } H_2O(\ell)] \qquad \Delta H° = 0.250 \times (-10{,}992 \text{ kJ})$$

$$0.500 \ C_8H_{18}(\ell) + 6.25 \ O_2(g) \longrightarrow 4.00 \ CO_2(g) + 4.50 \ H_2O(\ell) \qquad \Delta H° = -2.75 \times 10^3 \ kJ$$

(b) The original expression burns 2 mol isooctane. We want to burn 100. mol isooctane. So,

$$\text{Multiplier} = \frac{100. \ mol}{2 \ mol} = 50.0$$

$$50.0 \times [2 \ C_8H_{18}(\ell) + 25 \ O_2(g) \longrightarrow 16 \ CO_2(g) + 18 \ H_2O(\ell)] \qquad \Delta H° = 50.0 \times (-10,992 \ kJ)$$

$$100. \ C_8H_{18}(\ell) + 1250 \ O_2(g) \longrightarrow 800. \ CO_2(g) + 900. \ H_2O(\ell) \qquad \Delta H° = -5.50 \times 10^5 \ kJ$$

(c) The original expression burns 2 mol isooctane. We want to burn 1.00 mol isooctane. So,

$$\text{Multiplier} = \frac{1.00 \ mol}{2 \ mol} = 0.500$$

$$0.500 \times [2 \ C_8H_{18}(\ell) + 25 \ O_2(g) \longrightarrow 16 \ CO_2(g) + 18 \ H_2O(\ell)] \qquad \Delta H° = 0.500 \times (-10,992 \ kJ)$$

$$1.00 \ C_8H_{18}(\ell) + 12.5 \ O_2(g) \longrightarrow 8.00 \ CO_2(g) + 9.00 \ H_2O(\ell) \qquad \Delta H° = -5.50 \times 10^3 \ kJ$$

✓ *Reasonable Answer Check:* It makes sense that when more moles are involved, the $\Delta H°$ is larger, and when fewer moles are involved, the $\Delta H°$ is smaller.

39. *Answer:* **see thermostoichiometric factors below**

Strategy and Explanation: Given a thermochemical expression for a reaction, write all the thermostoichiometric factors that can be derived.

The expression's stoichiometric coefficients are interpreted in units of moles. The thermostoichiometric factors are ratios between the enthalpy of reaction and the moles of all the different reactants and products

The balanced expression says that 1 mol CaO reacts with 3 mol C to form 1 mol CaC_2 and 1 mol CO with the consumption of 464.8 kJ of thermal energy.

$$\frac{464.8 \ kJ}{1 \ mol \ CaO} \qquad \frac{464.8 \ kJ}{3 \ mol \ C} \qquad \frac{464.8 \ kJ}{1 \ mol \ CaC_2} \qquad \frac{464.8 \ kJ}{1 \ mol \ CO}$$

The reciprocal of the four factors above are also appropriate factors.

$$\frac{1 \ mol \ CaO}{464.8 \ kJ} \qquad \frac{3 \ mol \ C}{464.8 \ kJ} \qquad \frac{1 \ mol \ CaC_2}{464.8 \ kJ} \qquad \frac{1 \ mol \ CO}{464.8 \ kJ}$$

✓ *Reasonable Answer Check:* In each factor, the enthalpy change is related to the moles of one of the reactants or products. There are four different reactants and products and two ways to set up the ratio, so it makes sense that there are eight factors total.

41. *Answer:* **-1.45×10^3 kJ/mol**

Strategy and Explanation: Given a chemical equation for the combustion of a fuel, the mass of fuel burned, and the thermal energy evolved at constant pressure for the reaction, determine the molar enthalpy of combustion of the fuel.

Molar enthalpy change ($\Delta H°$) is identical to the thermal energy released at constant pressure per mol of substance. So, convert the mass into moles, then divide the thermal energy by the moles to get the molar enthalpy of combustion.

The balanced thermochemical expression tells us that exactly 1 mol C_2H_5OH burns to produce heat energy.

$$q = -3.62 \ kJ \ \text{(It is negative since heat energy is evolved, rather than absorbed.)}$$

$$n = 0.115 \ g \ C_2H_5OH \times \frac{1 \ mol \ C_2H_5OH}{46.0682 \ g \ C_2H_5OH} = 0.00250 \ mol \ C_2H_5OH$$

$$\Delta H° = \frac{q}{n} = \frac{-3.62 \ kJ}{0.00250 \ mol} = -1.45 \times 10^3 \ kJ/mol$$

✓ *Reasonable Answer Check:* Ethanol is used as a fuel, so it makes sense that a large amount of heat energy is evolved.

43. *Answer:* **6×10^4 kJ released**

Strategy and Explanation: Given a thermochemical expression for a reaction and a specific quantity of a reactant, determine the quantity of energy released.

Convert the mass into moles. Then use the thermochemical expression to create a unit conversion factor relating energy to moles of reactant.

The balanced chemical equation tells us that when exactly 1 mol CH_2O react, 425 kJ of thermal energy are evolved by the reaction.

$$10 \text{ lb } CH_2O \times \frac{454 \text{ g}}{1 \text{ lb}} \times \frac{1 \text{ mol } CH_2O}{30.0259 \text{ g } CH_2O} \times \frac{425 \text{ kJ}}{1 \text{ mol } CH_2O} = 6 \times 10^4 \text{ kJ evolved}$$

✓ *Reasonable Answer Check:* About 150 moles should produce about 150 times the molar enthalpy change.

Where Does the Energy Come From?

45. *Answer:* **HF**

Strategy and Explanation: Refer to the chart given above Question 45 on page 216. The bond with the largest bond enthalpy is the strongest, so H–F (566 kJ/mol) is the strongest of the four hydrogen halide bonds. The others are weaker with smaller bond enthalpies: H–Cl (431 kJ/mol), H–Br (366 kJ/mol), and H–I (299 kJ/mol).

47. *Answer:* **For reaction with fluorine: (a) 594 kJ (b) –1132 kJ (c) –538 kJ; For reaction with chlorine: (a) 678 kJ (b) –862 kJ (c) –184 kJ (d) Reaction of fluorine with hydrogen is more exothermic.**

Strategy and Explanation: Given a table of bond enthalpies (above Question 45) and the description of two chemical reactions, determine for each reaction (a) the enthalpy change for breaking all the bonds in the reactants, (b) the enthalpy change for forming all the bonds in the products, (c) the enthalpy change for the reaction, and (d) determine which reaction is most exothermic.

Balance the equations and determine how many moles of bonds are broken and formed. Use each bond's bond enthalpy (ΔH_{bond}) as a conversion factor to determine energy per bond type, then add up all the energies.

The balanced chemical equations look like this:

$$H_2 + F_2 \longrightarrow 2 \text{ HF} \qquad\qquad H_2 + Cl_2 \longrightarrow 2 \text{ HCl}$$

(a) In both reactions, one H–H bond and one halogen–halogen bond is broken.

$\Delta H_{reactants} = 1 \text{ mol} \times (\Delta H_{H-H \text{ bond}}) + 1 \text{ mol} \times (\Delta H_{halogen-halogen \text{ bond}})$

For fluorine, $\Delta H_{reactants} = 1 \text{ mol} \times (436 \text{ kJ/mol}) + 1 \text{ mol} \times (158 \text{ kJ/mol}) = 594 \text{ kJ}$

For chlorine, $\Delta H_{reactants} = 1 \text{ mol} \times (436 \text{ kJ/mol}) + 1 \text{ mol} \times (242 \text{ kJ/mol}) = 678 \text{ kJ}$

(b) In both reactions, two H–halogen bonds form. The enthalpy of forming a bond is the opposite sign of the enthalpy for breaking a bond, so we add a minus sign in the equation.

$$\Delta H_{products} = -2 \text{ mol} \times (\Delta H_{H-halogen \text{ bond}})$$

For fluorine, $\Delta H_{products} = -2 \text{ mol} \times (566 \text{ kJ/mol}) = -1132 \text{ kJ}$

For chlorine, $\Delta H_{products} = -2 \text{ mol} \times (431 \text{ kJ/mol}) = -862 \text{ kJ}$

(c) To get the enthalpy change for the reaction, add the enthalpy change of the reactants to the enthalpy change of the products

$$\Delta H_{total} = \Delta H_{reactants} + \Delta H_{products}$$

For fluorine, $\Delta H_{total} = (594 \text{ kJ}) + (-1132 \text{ kJ}) = -538 \text{ kJ}$

For chlorine, $\Delta H_{total} = (678 \text{ kJ}) + (-862 \text{ kJ}) = -184 \text{ kJ}$

(d) The reaction of fluorine with hydrogen is more exothermic (–538 kJ is more negative) than the reaction of chlorine with hydrogen (–184 kJ is less negative).

✓ *Reasonable Answer Check:* We expect the fluorine to be more reactive than chlorine, so more energy would be released.

Measuring Enthalpy Changes: Calorimetry

49. *Answer:* **18 °C**

Strategy and Explanation: Given the mass of a piece of ice, and the volume and temperature of a sample of water, determine the final temperature after the two are combined, the ice is melted and thermal equilibrium is established.

First, look up the specific heat capacity of water and the enthalpy of fusion of ice. Heat energy from the drop in temperature of the water is used for two things: to melt the ice and to raise the temperature of the resulting ice water to the final temperature. At the end, all the water reaches the same temperature. Using Equation 6.2, set up one equation showing the heat energy gained by the water equal to the heat energy used to melt the ice and to raise its temperature.

The specific heat capacity (c) water is $4.184 \text{ J g}^{-1}°\text{C}^{-1}$ according to Table 6.1. The enthalpy of fusion of ice (ΔH_{fus}) is 333 J/g, according to Section 6.4 and Question 28.

For a temperature change, rearrange Equation 6.2: $q = c \times m \times \Delta T$ and with a phase change use $q = m \times \Delta H_{fus}$. Heat energy is gained in the conversion of ice to ice water and when the ice water increases in temperature, so q_{ice} is positive. Heat energy is lost by the original water, so q_{water} is negative. The quantity of heat energy lost by the water is absorbed by the ice and ice water. So,

$$q_{ice} + q_{ice\,water} = -q_{water}$$

$$m_{ice} \times \Delta H_{fus,ice} + c_{ice\,water} \times m_{ice\,water} \times \Delta T_{ice\,water} = -(c_{water} \times m_{water} \times \Delta T_{water})$$

$$\Delta T_{ice\,water} = T_f - T_{i,ice\,water} = T_f - 0\ °\text{C} \qquad \Delta T_{water} = T_f - T_{i,water} = T_f - 25\ °\text{C}$$

$$(15.0\text{ g}) \times (333\text{ J/g}) + (4.184\text{ J g}^{-1}°\text{C}^{-1}) \times (15.0\text{ g}) \times (T_f - 0\ °\text{C})$$

$$= -[(4.184\text{ J g}^{-1}°\text{C}^{-1}) \times (200.\text{ mL}) \times \frac{1.00\text{ g H}_2\text{O}}{1\text{ mL H}_2\text{O}} \times (T_f - 25\ °\text{C})]$$

$$5.00 \times 10^3 + 62.8T_f = -837.T_f + 2.1 \times 10^4$$

$$9.00 \times 10^2 \times T_f = 1.6 \times 10^4$$

$$T_f = 18\ °\text{C}$$

✓ *Reasonable Answer Check:* The final temperature is less than the initial temperature of the water and more than the presumed initial temperature of the ice. Quantitatively, we can use $q = c \times m \times \Delta T$ to confirm that the water is losing 6×10^3 J at the same time as the water is gaining 6.1×10^3 J. They are the same within the uncertainty limit of $\pm 1 \times 10^3$ J, so this final temperature makes sense.

51. *Answer:* **6.6 kJ**

Strategy and Explanation: Given the mass of water in a bomb calorimeter, the initial and final temperatures, and the heat capacity of the bomb, determine the energy evolved by a reaction in the bomb.

First, look up the specific heat capacity of water and rearrange Equation 6.2, to find heat energy gained by the water. Calculate the heat energy gained by the bomb using the heat capacity (C_{bomb}). Relate the heat energy gained by the water and the bomb to that evolved by the reaction.

The specific heat capacity (c) water is $4.184 \text{ J g}^{-1}°\text{C}^{-1}$ according to Table 6.1.

For a temperature change, rearrange Equation 6.2: $q = c \times m \times \Delta T$. For a change in temperature in the bomb with the heat capacity: $q = C_{bomb} \times \Delta T$.

$$q_{total} = c_{water} \times m_{water} \times \Delta T + C_{bomb} \times \Delta T$$

$$\Delta T = T_f - T_i = 22.83\ °\text{C} - 19.50\ °\text{C} = 3.33\ °\text{C}$$

$$q_{total} = (4.184\text{ J g}^{-1}°\text{C}^{-1}) \times (320.\text{ g}) \times (3.33\ °\text{C}) + (650\text{ J/}°\text{C}) \times (3.33\ °\text{C})$$

$$q_{total} = 4.46 \times 10^3\text{ J} + 2.26 \times 10^3\text{ J}$$

$$q_{total} = 6.6 \times 10^3\text{ J}$$

$$\text{Heat energy evolved by the reaction} = 6.6 \times 10^3 \text{ J} \times \frac{1 \text{ kJ}}{1000 \text{ J}} = 6.6 \text{ kJ}$$

✓ *Reasonable Answer Check:* Heat energy is positive, since heat energy is gained by the water and the bomb.

Hess's Law

53. *Answer:* $\Delta H° = -1220.$ kJ/mol

Strategy and Explanation: Given three thermochemical expressions, determine the enthalpy change of a reaction.

Identify unique occurrences of the reactants and/or products in the given equations to help you determine which and how many of a given equation to use. If an equation's reactant shows in the products of the desired equation (or vice versa) then reverse the equation and change the sign of $\Delta H°$. If the substance has a different stoichiometric coefficient than the desired coefficient, determine an appropriate multiplier for the equation and multiply the $\Delta H°$ by the same multiplier (as was done in Question 37).

Look at the first reactant: Sr(s). The only given equation that has Sr(s) is the first one, where Sr(s) is also a reactant, so we must include that equation, as written, with its given $\Delta H°$. Look at the second reactant: C(graphite). The only given equation that has C(graphite) is the third one, where C(graphite) is also a reactant, so we must include that equation, as written, with its given $\Delta H°$. The third reactant is O_2. This chemical is found in several of the given reactions, so let's skip it. Look at the product: $SrCO_3(s)$. The only given equation that has $SrCO_3(s)$ is the second one, so we must include that equation, as written, with its given $\Delta H°$. We have now planned to use all three equations, so let's do it:

Add all three equations, and eliminate reactants that also show up as products: SrO(s) and $CO_2(g)$. The remaining reactants and products should form the net equation for the reaction, and the sum of all the individual $\Delta H°$ values will give us the $\Delta H°$ for that reaction:

$$Sr(s) + \frac{1}{2} O_2(g) \longrightarrow SrO(s) \qquad \Delta H° = -592 \text{ kJ}$$

$$SrO(s) + CO_2(g) \longrightarrow SrCO_3(s) \qquad \Delta H° = -234 \text{ kJ}$$

$$+ \quad C(graphite) + O_2(g) \longrightarrow CO_2(g) \qquad + \quad \Delta H° = -394 \text{ kJ}$$

$$Sr(s) + C(graphite) + \frac{3}{2} O_2(g) \longrightarrow SrCO_3(s) \qquad \Delta H° = -1220. \text{ kJ}$$

✓ *Reasonable Answer Check:* The net equation adds up the equation we are looking for. The enthalpy of formation of strontium carbonate is not in the table of formation enthalpies; however, the enthalpy of formation of group-related calcium carbonate is –1206.92 kJ/mol (from Table 6.2). The values are similar, so this seems reasonable.

Standard Molar Enthalpies of Formation

55. *Answer:* (a) $2 \text{ Al (s)} + \frac{3}{2} O_2 \text{ (g)} \longrightarrow Al_2O_3 \text{ (s)}$ $\Delta H° = -1675.7$ kJ (b) $Ti(s) + 2 Cl_2 \text{ (g)} \longrightarrow TiCl_4(\ell)$

$\Delta H° = -804.2$ kJ (c) $N_2(g) + 2 H_2 \text{ (g)} + \frac{3}{2} O_2 \text{ (g)} \longrightarrow NH_4NO_3(s)$ $\Delta H° = -365.56$ kJ

Strategy and Explanation:

(a) The formation of $Al_2O_3(s)$ is written with the reactants as standard state elements, Al(s) and $O_2(g)$, and the product as one mole of standard state compound:

$$2 \text{ Al (s)} + \frac{3}{2} O_2 \text{ (g)} \longrightarrow Al_2O_3 \text{ (s)} \qquad \Delta H° = -1675.7 \text{ kJ}$$

(b) The formation of $TiCl_4(\ell)$ is written with the reactants as standard state elements, Ti(s) and $Cl_2(g)$, and the product as one mole of standard state compound:

$$Ti(s) + 2 Cl_2 \text{ (g)} \longrightarrow TiCl_4(\ell) \qquad \Delta H° = -804.2 \text{ kJ}$$

(c) The formation of $NH_4NO_3(s)$ is written with the reactants as standard state elements, $N_2(g)$, $H_2(g)$, and $O_2(g)$, and the product as one mole of standard state compound:

$$N_2(g) + 2 H_2(g) + \frac{3}{2} O_2(g) \longrightarrow NH_4NO_3(s) \qquad \Delta H° = -365.56 \text{ kJ}$$

57. *Answer:* **(a) $\Delta H° = 1372.5$ kJ (b) endothermic**

Strategy and Explanation: Given a balanced chemical equation for a reaction and a table of molar enthalpies of formation, determine the enthalpy change of the reaction and whether the reaction is endothermic or exothermic.

Using molar enthalpies of formation is a common variation of Hess's Law. We will form all the products (using their molar enthalpies, as written) and we will destroy – the opposite of formation – the reactants (using their molar enthalpies, with opposite sign). That logic is the basis of Equation 6.11:

$$\Delta H° = \sum \left[(\text{moles of product}) \times \Delta H_f°(\text{product}) \right] - \sum \left[(\text{moles of reactant}) \times \Delta H_f°(\text{reactant}) \right]$$

Use the stoichiometric coefficient of the balanced equation to describe the moles of each of the reactants and products. Set up a specific version of Equation 6.11. Look up the $\Delta H_f°$ for each; remember to check the physical phase and remember that $\Delta H_f°$ for standard state elements is exactly zero. Plug them into the equation and solve for $\Delta H°$. Check the sign of $\Delta H°$ to assess endothermic or exothermic.

(a) The products are O_2 and $C_6H_{12}O_6$. The reactants are CO_2 and $H_2O(\ell)$. O_2 is the elemental form for oxygen.

$$\Delta H° = [(1 \text{ mol}) \times \Delta H_f°(C_6H_{12}O_6) + (6 \text{ mol}) \times \Delta H_f°(O_2)]$$
$$- [(6 \text{ mol}) \times \Delta H_f°(CO_2) + (1 \text{ mol}) \times \Delta H_f°(H_2O(\ell))]$$

Look up the $\Delta H_f°$ values in Table 6.2.

$$\Delta H° = [(1 \text{ mol}) \times (-1274.4 \text{ kJ/mol}) + (6 \text{ mol}) \times (0 \text{ kJ/mol})]$$
$$- [(6 \text{ mol}) \times (-393.509 \text{ kJ/mol}) + (1 \text{ mol}) \times (-285.830 \text{ kJ/mol})] = 1372.5 \text{ kJ}$$

(b) $\Delta H°$ is positive, so the reaction is endothermic.

✓ *Reasonable Answer Check:* It makes sense that the formation of an organic molecule is endothermic.

59. *Answer:* **41.2 kJ evolved**

Strategy and Explanation: Given the mass of a reactant, the description of a reaction, and molar enthalpies of formation, determine the thermal energy transferred out of the system at constant pressure.

Balance the equation and use it to describe the moles of each of the reactants and products. Set up a specific version of Equation 6.11. Look up the $\Delta H_f°$ for each. Plug them into the equation and solve for the molar enthalpy ($\Delta H°$) – this represents the molar heat energy evolved at constant pressure. Convert mass to moles, and use the molar enthalpy as a conversion factor to get heat energy evolved at constant pressure for the reactant sample.

The product is Fe_2O_3. The reactants are both elements: Fe and O_2.

$$2 Fe(s) + \frac{3}{2} O_2(g) \longrightarrow Fe_2O_3(s)$$

$$\Delta H° = (1 \text{ mol}) \times \Delta H_f°(Fe_2O_3) - [(2 \text{ mol}) \times \Delta H_f°(Fe) + (\frac{3}{2} \text{ mol}) \times \Delta H_f°(O_2)]$$

Look up the $\Delta H_f°$ value in Table 6.2.

$$\Delta H° = (1 \text{ mol}) \times (-824.2 \text{ kJ/mol}) - (2 \text{ mol}) \times (0 \text{ kJ/mol}) - (\frac{3}{2} \text{ mol}) \times (0 \text{ kJ/mol}) = -824.2 \text{ kJ}$$

That means the thermochemical expression looks like this:

$$2 Fe(s) + \frac{3}{2} O_2(g) \longrightarrow Fe_2O_3(s) \qquad \Delta H° = -824.2 \text{ kJ}$$

For 2 mol of Fe, 824.2 kJ needs to be transferred out. Now, we'll start the calculation for how much thermal energy is transferred out using the sample given:

$$5.58 \text{ g Fe} \times \frac{1 \text{ mol Fe}}{55.845 \text{ g Fe}} \times \frac{824.2 \text{ kJ}}{2 \text{ mol Fe}} = 41.2 \text{ kJ is transferred out}$$

✓ *Reasonable Answer Check:* The sample is a tenth of a mole. The equation shows two mol of Fe are needed. So the resulting heat energy is half of a tenth of the original molar enthalpy.

General Questions

61. *Answer:* **Gold reaches 100 °C first.**

Strategy and Explanation: The metal with the smaller specific heat capacity will increase in temperature faster than a metal with a larger specific heat capacity. From Table 6.1, $c_{Cu} = 0.385$ J g^{-1}°C^{-1} and $c_{Au} = 0.128$ J g^{-1}°C^{-1}, so the Au will reach 100 °C first.

63. *Answer:* $\Delta H^\circ_f(B_2H_6) = 36$ **kJ/mol**

Strategy and Explanation: Given a balanced thermochemical expression for a reaction and some molar enthalpies of formation, determine an unknown molar enthalpy of formation.

Use the stoichiometric coefficient of the balanced equation to describe the moles of each of the reactants and products. Set up a specific version of Equation 6.11, and solve it for the unknown ΔH°_f. Look up the ΔH°_f for the rest of the reactants and products. Plug them into the equation and solve for ΔH°_f.

The products are B_2O_3 and $H_2O(\ell)$. The reactants are B_2H_6 and O_2. O_2 is the elemental form for oxygen.

$$\Delta H^\circ = [(1 \text{ mol}) \times \Delta H^\circ_f(B_2O_3) + (3 \text{ mol}) \times \Delta H^\circ_f(H_2O(\ell))] - (1 \text{ mol}) \times \Delta H^\circ_f(B_2H_6)$$

$$\Delta H^\circ_f(B_2H_6) = -\frac{\Delta H^\circ}{(1 \text{ mol})} + \Delta H^\circ_f(B_2O_3) + 3 \times \Delta H^\circ_f(H_2O(\ell))$$

Look up the ΔH°_f value for $H_2O(\ell)$ in Table 6.2.

$$\Delta H^\circ_f(B_2H_6) = -\frac{(-2166 \text{ kJ})}{(1 \text{ mol})} + (-1273 \text{ kJ/mol}) + 3 \times (-285.830) = 36 \text{ kJ/mol}$$

✓ *Reasonable Answer Check:* The high exothermicity of the reaction can be almost completely accounted for by the formation of the products. So, it makes sense that the molar enthalpy of formation of the reactant B_2H_6 is small.

65. *Answer:* $\Delta H^\circ_f(C_2H_4Cl_2(\ell)) = -165.2$ **kJ/mol**

Strategy and Explanation: Given two thermochemical expressions, determine the molar enthalpy of another formation reaction. Write the equation for the desired net reaction. Follow the procedure described in Question 53.

The formation of $C_2H_4Cl_2$ is written with the standard state elements as reactants and one mole of compound as the product:

$$2 \text{ C(graphite)} + 2 \text{ H}_2(g) + Cl_2(g) \longrightarrow C_2H_4Cl_2(\ell)$$

Look at the first reactant: C(graphite). The only given equation that has C(graphite) is the first one, so we must include that equation, as written, with its ΔH°. Look at the second reactant: $H_2(g)$. This reactant is taken care of by using the first equation, also. Look at the third reactant: $Cl_2(g)$. The only given equation that has $Cl_2(g)$ is the second one, but it is a product, so the equation must be reversed. Therefore, we must change the sign of its ΔH°. We have now planned how we will use all the given equations, so let's do it:

Add all the equations as planned, eliminate reactants that also show up as products: the $C_2H_4(g)$ molecules.

Add all the remaining reactants and products to form the net equation for the reaction, and add all the individual ΔH° values to get the ΔH° for the reaction:

$$2\text{ C(graphite)} + 2\text{ H}_2(g) \longrightarrow C_2H_4(g) \qquad\qquad \Delta H° = 52.3 \text{ kJ}$$

$$+\ C_2H_4(g) + Cl_2(g) \longrightarrow C_2H_4Cl_2(g) \qquad + \ \Delta H° = -217.5 \text{ kJ}$$

$$\overline{2\text{ C(graphite)} + 2\text{ H}_2(g) + Cl_2(g) \longrightarrow C_2H_4Cl_2(\ell) \qquad \Delta H° = -165.2 \text{ kJ}}$$

So, $\Delta H_f°(C_2H_4Cl_2(\ell)) = -165.2$ kJ/mol

An alternative approach to this Question is to notice that the first equation is also a formation reaction. Hence the second reaction's $\Delta H°$ can be related to the two formation reactions using Equation 6.11.

$$\Delta H°_{\text{second reaction}} = (1 \text{ mol}) \times \Delta H_f°(C_2H_4) - (1 \text{ mol}) \times \Delta H_f°(C_2H_4Cl_2)$$

$$\Delta H_f°(C_2H_4Cl_2) = (1 \text{ mol}) \times \Delta H_f°(C_2H_4) - \frac{\Delta H°_{\text{second reaction}}}{(1 \text{ mole})}$$

$$\Delta H_f°(C_2H_4Cl_2) = 52.3 \text{ kJ/mol} - 217.5 \text{ kJ/mol} = -165.2 \text{ kJ}$$

✓ *Reasonable Answer Check:* Both methods result in the same answers.

67. *Answer:* **(a) 36.03 kJ evolved (b) 1.18 × 10⁴ kJ evolved**

Strategy and Explanation: Given a balanced chemical equation for two reactions and a table of molar enthalpies of formation, determine the enthalpy change of the first reaction and the thermal energy evolved by the second.

Use the stoichiometric coefficient of the balanced equation to describe the moles of each of the reactants and products. Set up a specific version of Equation 6.11. Look up the $\Delta H_f°$ for each. Plug them into the equation and solve for $\Delta H°$.

(a) The products of the first reaction are N_2O and $H_2O(g)$. The reactant is $NH_4NO_3(s)$.

$$\Delta H° = (1 \text{ mol}) \times \Delta H_f°(N_2O) + (2 \text{ mol}) \times \Delta H_f°(H_2O(g)) - (1 \text{ mol}) \times \Delta H_f°(NH_4NO_3(s))$$

Look up the $\Delta H_f°$ values in Table 6.2 and Appendix J.

$\Delta H° = (1 \text{ mol}) \times (82.05 \text{ kJ/mol}) + (2 \text{ mol}) \times (-241.818 \text{ kJ/mol}) - (1 \text{ mol}) \times (-365.56 \text{ kJ/mol}) = -36.03$ kJ

(b) The products of the second reaction are N_2, $H_2O(g)$, and O_2. The reactant is $NH_4NO_3(s)$. N_2 and O_2 are elemental forms, with $\Delta H_f°$ exactly zero.

$$\Delta H° = (4 \text{ mol}) \times \Delta H_f°(H_2O(g)) - (2 \text{ mol}) \times \Delta H_f°(NH_4NO_3(s))$$

Look up the $\Delta H_f°$ values in Table 6.2 and Appendix J.

$$\Delta H° = (4 \text{ mol}) \times (-241.818 \text{ kJ/mol}) - (2 \text{ mol}) \times (-365.56 \text{ kJ/mol}) = -236.15 \text{ kJ}$$

The balanced chemical equation says that 2 mol NH_4NO_3 react with the evolution of 236.15 kJ of thermal energy.

$$8.00 \text{ kg } NH_4NO_3 \times \frac{1000 \text{ g}}{1 \text{ kg}} \times \frac{1 \text{ mol } NH_4NO_3}{80.0432 \text{ g } NH_4NO_3} \times \frac{236.15 \text{ kJ}}{2 \text{ mol } NH_4NO_3} = 1.18 \times 10^4 \text{ kJ}$$

✓ *Reasonable Answer Check:* These explosive reactions produce large amounts of heat energy. The sign and size of these answers make sense.

69. *Answer:* **Step 1: −137.23 kJ, Step 2: 275.341 kJ, Step 3: 103.71 kJ, $H_2O(g) \longrightarrow H_2(g) + \frac{1}{2}O_2(g)$;**
$\Delta H° = 241.82$ kJ; endothermic

Strategy and Explanation: Given a balanced chemical equation for two reactions and a table of molar enthalpies of formation, determine the enthalpy change of the first reaction and the thermal energy evolved by the second.

Use the stoichiometric coefficient of the balanced equations to describe the moles of each of the reactants and products. Set up specific versions of Equation 6.11. Look up the ΔH_f° for each species. Plug them into the equations and solve for the three ΔH°. Then add (as described in Question 53) the three chemical equations together to make the overall equation for the reaction and determine the overall ΔH°.

Step 1: The products of the reaction are $H_2SO_4(\ell)$ and $HBr(g)$. The reactants are SO_2, $H_2O(g)$, and $Br_2(g)$. $Br_2(g)$ is NOT elemental bromine.

$$\Delta H^\circ = (1 \text{ mol}) \times \Delta H_f^\circ(H_2SO_4(\ell)) + (2 \text{ mol}) \times \Delta H_f^\circ(HBr(g))$$
$$- [(1 \text{ mol}) \times \Delta H_f^\circ(SO_2) + (2 \text{ mol}) \times \Delta H_f^\circ(H_2O(g)) + (1 \text{ mol}) \times \Delta H_f^\circ(Br_2(g))]$$

Look up the ΔH_f° values in Table 6.2 and Appendix J.

$$\Delta H^\circ = (1 \text{ mol}) \times (-813.989 \text{ kJ/mol}) + (2 \text{ mol}) \times (-36.40 \text{ kJ/mol}) - (1 \text{ mol}) \times (-296.830 \text{kJ/mol})$$
$$- (2 \text{ mol}) \times (-241.818 \text{kJ/mol}) - (1 \text{ mol}) \times (30.907 \text{kJ/mol}] = -137.23 \text{ kJ}$$

Step 2: The products of the reaction are $H_2O(g)$, SO_2, and O_2. The reactant is $H_2SO_4(\ell)$. $O_2(g)$ is elemental.

$$\Delta H^\circ = (1 \text{ mol}) \times \Delta H_f^\circ(H_2O(g)) + (1 \text{ mol}) \times \Delta H_f^\circ(SO_2) - (1 \text{ mol}) \times \Delta H_f^\circ(H_2SO_4(\ell))$$

Look up the ΔH_f° values in Table 6.2 and Appendix J.

$$\Delta H^\circ = (1 \text{ mol}) \times (-241.818 \text{ kJ/mol}) + (1 \text{ mol}) \times (-296.830 \text{ kJ/mol}) -$$
$$(1 \text{ mol}) \times (-813.989 \text{ kJ/mol}) = 275.341 \text{ kJ}$$

Step 3: The products of the reaction are $H_2(g)$ and $Br_2(g)$. The reactant is $HBr(g)$. $H_2(g)$ is in elemental form, but $Br_2(g)$ is not.

$$\Delta H^\circ = (1 \text{ mol}) \times \Delta H_f^\circ(Br_2(g)) - (2 \text{ mol}) \times \Delta H_f^\circ(HBr(g))$$

Look up the ΔH_f° values in Table 6.2 and Appendix J.

$$\Delta H^\circ = (1 \text{ mol}) \times (30.907 \text{ kJ/mol}) - (2 \text{ mol}) \times (-36.40 \text{ kJ/mol}) = 103.71 \text{ kJ}$$

Add the three thermochemical expressions to get the net equation:

$$\cancel{SO_2(g)} + \cancel{2\,H_2O(g)} + \cancel{Br_2(g)} \longrightarrow \cancel{H_2SO_4(\ell)} + 2\,\cancel{HBr(g)} \qquad \Delta H^\circ = -137.23 \text{ kJ}$$

$$\cancel{H_2SO_4(\ell)} \longrightarrow \cancel{H_2O(g)} + \cancel{SO_2(g)} + \tfrac{1}{2}O_2(g) \qquad \Delta H^\circ = 275.341 \text{ kJ}$$

$$+ \quad 2\,\cancel{HBr(g)} \longrightarrow H_2(g) + \cancel{Br_2(g)} \qquad\qquad + \quad \Delta H^\circ = 103.71 \text{ kJ}$$

$$H_2O(g) \longrightarrow H_2(g) + \tfrac{1}{2}O_2(g) \qquad\qquad\qquad \Delta H^\circ = 241.82 \text{ kJ}$$

The net equation is endothermic.

✓ *Reasonable Answer Check:* The net reaction is the reverse of the reaction for $\Delta H_f^\circ(H_2O(g))$, which Appendix J shows with a value of -241.818 kJ/mol. The positive value makes sense because the reaction is reversed, and the size of these two numbers is the same within given significant figures.

Applying Concepts

71. *Answer:* **Substance A**

Strategy and Explanation: Rearrange Equation 6.2: $q = m \times c \times \Delta T$. The equation says that energy transferred is proportional to ΔT for constant mass samples, using the specific heat capacity, c. So, the slower the temperature rises with the transfer of energy, the larger the specific heat capacity. On the graph, the shallowest line (the line with the least steep slope) has the highest specific heat capacity; here, that is Substance A.

73. *Answer/Explanation:* Thermal energy content is greater in Beaker 1 than in Beaker 2. A larger mass of water will contain larger quantity of thermal energy at a given temperature.

75. *Answer/Explanation:* Enthalpy change is an extensive property. The given equation produces 2 mol SO_3. Formation enthalpy from Table 6.2 is for the production of 1 mol SO_3, so the enthalpy values must be different by a factor of two.

More Challenging Questions

77. *Answer:* $\Delta H_f^\circ(OF_2) = 18$ kJ/mol

Strategy and Explanation: Given a balanced thermochemical expression for a reaction and some molar enthalpies of formation, determine an unknown molar enthalpy of formation.

Use the stoichiometric coefficient of the balanced equation to describe the moles of each of the reactants and products. Set up a specific version of Equation 6.11, and solve it for the unknown ΔH_f°. Look up the ΔH_f° for the rest of the reactants and products. Plug them into the equation and solve for ΔH_f°.

The products are HF and O_2. The reactants are OF_2 and $H_2O(g)$. O_2 is the elemental form for oxygen.

$$\Delta H^\circ = [(2 \text{ mol}) \times \Delta H_f^\circ(HF) + (1 \text{ mol}) \times \Delta H_f^\circ(O_2)] - [(1 \text{ mol}) \times \Delta H_f^\circ(OF_2) + (1 \text{ mol}) \times \Delta H_f^\circ(H_2O(g))]$$

$$\Delta H^\circ = (2 \text{ mol}) \times \Delta H_f^\circ(HF) + (1 \text{ mol}) \times \Delta H_f^\circ(O_2) - (1 \text{ mol}) \times \Delta H_f^\circ(OF_2) - (1 \text{ mol}) \times \Delta H_f^\circ(H_2O(g))$$

$$\Delta H_f^\circ(OF_2) = - \frac{\Delta H^\circ}{(1 \text{ mol})} + 2 \times \Delta H_f^\circ(HF) + \Delta H_f^\circ(O_2) - \Delta H_f^\circ(H_2O(g))$$

Look up the ΔH_f° values for $H_2O(g)$ and HF in Table 6.2.

$$\Delta H_f^\circ(OF_2) = -(-318 \text{ kJ/mol}) + 2 \times (-271.1 \text{ kJ/mol}) + (0 \text{ kJ/mol}) - (-241.818 \text{ kJ/mol}) = 18 \text{ kJ/mol}$$

✓ *Reasonable Answer Check:* The high exothermicity of the reaction can be almost completely accounted for with the destruction of the other reactant and formation of the products. So, it makes sense that the molar enthalpy of formation of OF_2, the reactant, is small.

79. *Answer:* **50.014 kJ/g methane, 47.484 kJ/g ethane, 46.354 kJ/g propane, 45.7140 kJ/g butane; methane > ethane > propane > butane**

Strategy and Explanation: Compare the quantity of thermal energy evolved per gram (also called the fuel value) when burning four different organic fuels.

Write the formulas for the four smallest hydrocarbons. Balance their combustion equations. Then use Equation 6.11, a table of molar enthalpies of formation, and the stoichiometry of the balanced equations to determine the molar enthalpy change for each combustion reaction. Then use the molar mass to determine the enthalpy per gram of each fuel and rank them.

The reaction for the combustion of propane is described in Problem-Solving Example 4.4 on page 109 and those describing the combustion of methane, ethane, and butane are similar to that one. We'll assume that all fuels are initially in the gas state. Look up the ΔH_f° value in Appendix J.

Methane: $$CH_4 + 2\,O_2 \longrightarrow CO_2 + 2\,H_2O$$

$$\Delta H^\circ_{methane} = (1 \text{ mol}) \times \Delta H_f^\circ(CO_2) + (2 \text{ mol}) \times \Delta H_f^\circ(H_2O(g)) - (1 \text{ mol}) \times \Delta H_f^\circ(CH_4) - (2 \text{ mol}) \times \Delta H_f^\circ(O_2)$$

$$= (1 \text{ mol}) \times (-393.509 \text{ kJ/mol}) + (2 \text{ mol}) \times (-241.818 \text{ kJ/mol})$$
$$- (1 \text{ mol}) \times (-74.81 \text{ kJ/mol}) - (2 \text{ mol}) \times (0 \text{ kJ/mol}) = -802.34 \text{ kJ}$$

$$\frac{802.34 \text{ kJ produced}}{1 \text{ mol } CH_4} \times \frac{1 \text{ mol } CH_4}{16.0423 \text{ g } CH_4} = 50.014 \frac{\text{kJ produced}}{\text{g } CH_4}$$

Ethane: $$C_2H_6 + \frac{7}{2}\,O_2 \longrightarrow 2\,CO_2 + 3\,H_2O$$

$$\Delta H^\circ_{ethane} = (2 \text{ mol}) \times \Delta H_f^\circ(CO_2) + (3 \text{ mol}) \times \Delta H_f^\circ(H_2O(g)) - (1 \text{ mol}) \times \Delta H_f^\circ(C_2H_6) - \left(\frac{7}{2} \text{ mol}\right) \times \Delta H_f^\circ(O_2)$$

$$= (2 \text{ mol}) \times (-393.509 \text{ kJ/mol}) + (3 \text{ mol}) \times (-241.818 \text{ kJ/mol})$$

$$- (1 \text{ mol}) \times (-84.68 \text{ kJ/mol}) - (\tfrac{7}{2} \text{ mol}) \times (0 \text{ kJ/mol}) = -1427.79 \text{ kJ}$$

$$\frac{1427.79 \text{ kJ produced}}{1 \text{ mol } C_2H_6} \times \frac{1 \text{ mol } C_2H_6}{30.0688 \text{ g } C_2H_6} = 47.484 \frac{\text{kJ produced}}{\text{g } C_2H_6}$$

Propane: $C_3H_8 + 5 O_2 \longrightarrow 3 CO_2 + 4 H_2O$

$$\Delta H^\circ = (3 \text{ mol}) \times \Delta H^\circ_f(CO_2) + (4 \text{ mol}) \times \Delta H^\circ_f(H_2O(g)) - (1 \text{ mol}) \times \Delta H^\circ_f(C_3H_8) - (5 \text{ mol}) \times \Delta H^\circ_f(O_2)$$

$$\Delta H^\circ = (3 \text{ mol}) \times (-393.509 \text{ kJ/mol}) + (4 \text{ mol}) \times (-241.818 \text{ kJ/mol})$$

$$- (1 \text{ mol}) \times (-103.8 \text{ kJ/mol}) - (5 \text{ mol}) \times (0 \text{ kJ/mol}) = -2044.0 \text{ kJ}$$

$$\frac{2044.0 \text{ kJ produced}}{1 \text{ mol } C_3H_8} \times \frac{1 \text{ mol } C_3H_8}{44.0953 \text{ g } C_3H_8} = 46.354 \frac{\text{kJ produced}}{\text{g } C_3H_8}$$

Butane: $C_4H_{10} + \tfrac{13}{2} O_2 \longrightarrow 4 CO_2 + 5 H_2O$

$$\Delta H^\circ_{butane} = (4 \text{ mol}) \times \Delta H^\circ_f(CO_2) + (5 \text{ mol}) \times \Delta H^\circ_f(H_2O(g)) - (1 \text{ mol}) \times \Delta H^\circ_f(C_4H_{10}) - (\tfrac{13}{2} \text{ mol}) \times \Delta H^\circ_f(O_2)$$

$$= (4 \text{ mol}) \times (-393.509 \text{ kJ/mol}) + (5 \text{ mol}) \times (-241.818 \text{ kJ/mol})$$

$$- (1 \text{ mol}) \times (-126.148 \text{ kJ/mol}) - (\tfrac{13}{2} \text{ mol}) \times (0 \text{ kJ/mol}) = -2656.978 \text{ kJ}$$

$$\frac{2656.978 \text{ kJ produced}}{1 \text{ mol } C_4H_{10}} \times \frac{1 \text{ mol } C_4H_{10}}{58.1218 \text{ g } C_4H_{10}} = 45.7140 \frac{\text{kJ produced}}{\text{g } C_4H_{10}}$$

Methane has the highest fuel value, followed by ethane, then propane, and then butane.

✓ *Reasonable Answer Check:* While we might not have predicted which of these would be the highest, seeing a consistent trend between these numbers is satisfying.

81. *Answer:* **(a) 26.6 °C, temperature leveled off from Exp. 3 onward (b) ascorbic acid is limiting in Exp. 1, 2, and 3; sodium hydroxide is limiting in Exp. 3, 4, and 5 (c) One; equal quantities of each reactant are present in Exp. 3 at the stoichiometric equivalence point**

Strategy and Explanation:

(a) We predict the temperature of experiment to be 26.6 °C, because, above $C_6H_8O_6$ masses of 8.81g, the other reactant NaOH appears to be the limiting reactant.

(b) $C_6H_8O_6$ limits in Experiments 1, 2, and 3, which is why the final temperature increases proportionally to the mass of $C_6H_8O_6$ from experiment to experiment in those three. NaOH limits in Experiments 3,4, and 5, as seen by the constant temperature increase in the available data for those three experiments, even though larger masses of $C_6H_8O_6$ are used.

(c) Experiment 3 seems to be the stoichiometric equivalence point, where both reactants run out at the same time. Before that experiment $C_6H_8O_6$ limits, and after that experiment, NaOH limits. Use the mass of $C_6H_8O_6$ and the volume and molarity of the NaOH to determine the moles present in that experiment.

$$100. \text{ mL NaOH} \times \frac{1 \text{ L}}{1000 \text{ mL}} \times \frac{0.500 \text{ mol NaOH}}{1 \text{ L NaOH}} = 0.0500 \text{ mol NaOH}$$

$$8.81 \text{ g } C_6H_8O_6 \times \frac{1 \text{ mol } C_6H_8O_6}{176.1238 \text{ g } C_6H_8O_6} = 0.0500 \text{ mol } C_6H_8O_6$$

Because equal quantities of each reactant are present at the stoichiometric equivalence point, ascorbic acid, $C_6H_8O_6$, must have just one hydrogen ion.

Chapter 7: Electron Configurations and the Periodic Table

Introduction

This chapter explains the basis for the charges on monatomic ions introduced in Chapter 3. Instead of memorizing the charges, you will be able to determine the charges from the electron configurations.

Table 7.3 is a good way to pull together all the ideas of shells, subshells, orbitals, and electron arrangement. For each shell, determine (1) what subshells are available, (2) how many orbitals each subshell has, (3) how many electrons can each subshell hold, and (4) how many electrons can the shell hold. Note that the sum of electrons in the subshells equals the number electrons in a given shell.

When drawing Lewis structures, the position and order of the dots is not particularly important, though typically, one does not put more than two on a side.

There are several devices for remembering the details of writing electron configurations. The maximum number of electrons in a subshell increases by four (s has 2, p has $2 + 4 = 6$, etc.). In addition to using the periodic table to determine the order, the electron-filling "diagonal rule" (shown in Figure 7.18) is a good mnemonic.

Solutions to Blue-Numbered Questions
for Review and Thought for Chapter 7

Topical Questions

Electromagnetic Radiation

5.   *Answer:* **(a) Radio waves (b) Microwaves**

    Strategy and Explanation: Use Figure 7.1

    (a) **Radio waves** are lower frequency, thus have less energy, than infrared light.

    (b) **Microwaves** are higher frequency than radio waves.

7.   *Answer:* **3.00×10^{-3} m, 6.63×10^{-23} J/photon, 39.9 J/mol**

    Strategy and Explanation: Given the frequency of electromagnetic radiation, determine the wavelength in meters, the energy of one photon, and the energy of one mole of photons. Use equations described in Section 7.1 and 7.2. $\nu = 1.00 \times 10^{11}$ s^{-1}

    (a) $$\lambda = \frac{c}{\nu} = \frac{2.998 \times 10^8 \text{ m/s}}{1.00 \times 10^{11} \text{ s}^{-1}} = 3.00 \times 10^{-3} \text{ m}$$

    (b) $E = h\nu = (6.626 \times 10^{-34} \text{ J·s}) \times (1.00 \times 10^{11} \text{ s}^{-1}) = 6.63 \times 10^{-23}$ J for one photon

    (c) $$\frac{6.63 \times 10^{-23} \text{ J}}{1 \text{ photon}} \times \frac{6.022 \times 10^{23} \text{ photons}}{1 \text{ mol photons}} = 39.9 \text{ J/mol}$$

    ✓ *Reasonable Answer Check:* High frequency has short wavelength. The energy for one photon is a tiny number, but a mole of photons has a sizable energy.

9.   *Answer:* **6.06×10^{14} Hz**

    Strategy and Explanation: Given the wavelength of light, determine the frequency. Use appropriate length conversions and the equation in Section 7.1. 1 Hz is defined as 1 s^{-1}. The wavelength is $\lambda = 495$ nm.

$$\nu = \frac{c}{\lambda} = \frac{2.998 \times 10^8 \, \text{m/s}}{495 \, \text{nm} \times \frac{1 \times 10^{-9} \, \text{m}}{1 \, \text{nm}}} \times \frac{1 \, \text{Hz}}{1 \, \text{s}^{-1}} = 6.06 \times 10^{14} \, \text{Hz}$$

✓ *Reasonable Answer Check*: Nanometer wavelengths are fairly small, so it makes sense that the frequency is high. Figure 7.1 also shows visible light in the 10^{14} - 10^{16} Hz frequency range.

11. *Answer:* **1.1×10^{15} Hz, 7.4×10^{-19} J/photon**

Strategy and Explanation: Given the wavelength of electromagnetic radiation, determine the frequency and energy of this radiation. Use equations in Sections 7.1 and 7.2, and appropriate metric conversion factors. The energy calculated is the energy of one photon of this light.

$$\lambda = 270 \, \text{nm} \qquad \nu = \frac{c}{\lambda} = \frac{2.998 \times 10^8 \, \text{m/s}}{270 \, \text{nm} \times \frac{1 \times 10^{-9} \, \text{m}}{1 \, \text{nm}}} = 1.1 \times 10^{15} \, \text{s}^{-1} = 1.1 \times 10^{15} \, \text{Hz}$$

$$E_{\text{photon}} = h\nu = \left(6.626 \times 10^{-34} \, \text{J} \cdot \text{s}\right) \times \left(1.1 \times 10^{15} \, \text{s}^{-1}\right) = 7.4 \times 10^{-19} \, \text{J}$$

✓ *Reasonable Answer Check*: The wavelength calculated coincides with that of ultraviolet light, according to Figure 7.1. This ultraviolet photon's energy is also more than those of visible light calculated in Section 7.2.

12. *Answer:* **8.42×10^{-19} J, it is larger**

Strategy and Explanation: Given the wavelength of electromagnetic radiation, determine the energy of one photon of this radiation and compare it to the given energy of another kind of photon. We use appropriate length conversions and the equation in Section 7.2. The wavelength is $\lambda = 2.36$ nm.

$$E_{\text{photon}} = \frac{hc}{\lambda} = \frac{6.626 \times 10^{-34} \, \text{J} \cdot \text{s} \times 2.998 \times 10^8 \, \text{m/s}}{2.36 \, \text{nm} \times \frac{1 \times 10^{-9} \, \text{m}}{1 \, \text{nm}}} = 8.42 \times 10^{-17} \, \text{J}$$

$$8.42 \times 10^{-17} \, \text{J} > 3.18 \times 10^{-19} \, \text{J}$$

✓ *Reasonable Answer Check*: The energy of an x-ray photon is more than 250 times more than that of a visible photon. That make sense, since the wavelength of the x-ray is about 250 times shorter than the wavelengths of visible light (400 nm - 700 nm).

Photoelectric Effect

15. *Answer/Explanation:* Photons of light with long wavelength are low in energy. It is clear from the description that the energy of these photons is insufficient to cause electrons to be ejected. Increasing the intensity only increases the number of photons, not their individual energy.

16. *Answer:* **no**

Strategy and Explanation: Given the wavelength of light and the minimum energy for ejecting electrons from a metal surface, determine if the light has sufficient energy to cause the photoelectric effect (i.e., eject electrons from that metal's surface). We use appropriate length conversions and the equation in Section 7.2 to find the energy of the photon. Compare the photon energy to the minimum energy. If the photon energy is larger, then the photon can eject electrons; if the energy is smaller, it cannot. The wavelength is $\lambda = 600.$ nm.

$$E_{\text{photon}} = \frac{hc}{\lambda} = \frac{6.626 \times 10^{-34} \, \text{J} \cdot \text{s} \times 2.998 \times 10^8 \, \text{m/s}}{600. \, \text{nm} \times \frac{1 \times 10^{-9} \, \text{m}}{1 \, \text{nm}}} = 3.31 \times 10^{-19} \, \text{J}$$

$$E_{\text{minimum}} = 3.69 \times 10^{-19} \, \text{J} > E_{\text{photon}}$$

Therefore it has insufficient energy to eject electrons.

✓ *Reasonable Answer Check*: The minimum energy is similar to ones calculated for visible light photons in Section 7.2. These two energies are close, so we probably would not have been able to predict this result before doing the calculation.

Atomic Spectra and the Bohr Atom

19. *Answer:* **(a) absorbed (b) emitted (c) absorbed (d) emitted**

Strategy and Explanation: Given the values of n for the initial and final states of an electron in hydrogen, determine whether energy is absorbed or emitted. The size of n indicates the relative energy of the state. If the final state has a larger n value than the initial state, then energy must be absorbed. If the final state has a smaller n value than the initial state, then energy must be emitted.

(a) n = 1 to n = 3. Energy absorbed.   (b) n = 5 to n = 1. Energy emitted.

(c) n = 2 to n = 4. Energy absorbed.   (d) n = 5 to n = 4. Energy emitted.

✓ *Reasonable Answer Check:* Energy needs to be used up to get the electron to a higher n value state. Energy is lost when moving the electron to a lower n value state.

21. *Answer:* **(a), (b), and (d)**

Strategy and Explanation: The difference between the energies of the two levels is the energy emitted. The smaller the energy, the longer the wavelength. When we look at Figure 7.8, we need to find which of the given levels is closer together than the energy levels represented by n = 1 and n = 4. The energies levels represented by (a) n = 2 and n = 4, (b) n = 1 and n = 3, and (d) n = 3 and n = 5, are closer together than n = 1 and n = 4. So, **(a), (b), and (d)** will require radiation with longer wavelength.

25. *Answer:* **4.576×10^{-19} J absorbed and 434.0 nm**

Strategy and Explanation: Given the values of n for the initial and final states of an electron in hydrogen, determine the energy and wavelength of the emission.

Using equations given in Section 7.3, calculate the energy of the photon emitted during the transition. Use the equation given in Section 7.1 and appropriate metric conversions to calculate the wavelength in nanometers.

Here, $n_i = 2$ and $n_f = 5$:

$$\Delta E_{e^-} = \left(2.179 \times 10^{-18}\ \text{J}\right)\left(\frac{1}{n_i^2} - \frac{1}{n_f^2}\right) = \left(2.179 \times 10^{-18}\ \text{J}\right)\left(\frac{1}{2^2} - \frac{1}{5^2}\right) = 4.576 \times 10^{-19}\ \text{J}$$

The energy of the photon absorbed is equal to the energy gained by the electron as it moves to a higher-energy state. $E_{photon} = \Delta E_{e^-}$. The wavelength of the photon absorbed relates to the energy:

$$\lambda = \frac{hc}{E_{photon}} = \frac{\left(6.626 \times 10^{-34}\ \text{J} \cdot \text{s}\right)\left(2.998 \times 10^8\ \text{m/s}\right)}{4.576 \times 10^{-19}\ \text{J}} \times \frac{1\ \text{nm}}{10^{-9}\ \text{m}} = 434.0\ \text{nm}$$

✓ *Reasonable Answer Check:* The wavelength of radiation produced from the reverse of this transition is reported in Figure 7.8.

de Broglie Wavelength

27. *Answer:* **0.05 nm**

Strategy and Explanation: Given the mass and speed of a particle, determine the de Broglie wavelength. Use the equation in Section 7.4, the method in Problem-Solving Example 7.5, and appropriate unit conversions.

Problem-Solving Example 7.5 shows how Planck's constant can be represented using a mass of kilograms: $h = 6.626 \times 10^{-34}\ \text{kg m}^2\text{s}^{-1}$

Find speed in meters per second: $v = (0.05) \times (2.998 \times 10^8\ \text{m/s}) = 1 \times 10^7\ \text{m/s}$

Now, find the de Broglie wavelength: $\lambda_{DeBroglie} = \dfrac{h}{mv} = \dfrac{6.626 \times 10^{-34}\ \text{kg} \cdot \text{m}^2\text{s}^{-1}}{\left(9.11 \times 10^{-31}\ \text{kg}\right)\left(1 \times 10^7\ \dfrac{\text{m}}{\text{s}}\right)} = 5 \times 10^{-11}\ \text{m}$

This wavelength is 0.05 nm.

✓ *Reasonable Answer Check:* The electron is a sub-nanoscale particle, so it makes sense that its de Broglie wavelength is similar (50 pm) to its size dimensions.

Quantum Numbers

30. *Answer:* **(a) cannot occur, m_ℓ must be between $-\ell$ and $+\ell$ (b) can occur (c) cannot occur, m_s cannot be 1, here (d) cannot occur, ℓ must be less than n (e) can occur**

Strategy and Explanation:

(a) $n = 2$, $\ell = 1$, $m_\ell = 2$, $m_s = +\frac{1}{2}$ could not occur because m_ℓ must be between $-\ell$ and $+\ell$.

(b) $n = 3$, $\ell = 2$, $m_\ell = 0$, $m_s = -\frac{1}{2}$ can occur.

(c) $n = 1$, $\ell = 0$, $m_\ell = 0$, $m_s = 1$ could not occur because m_s must be $+\frac{1}{2}$ or $-\frac{1}{2}$.

(d) $n = 3$, $\ell = 3$, $m_\ell = 2$, $m_s = -\frac{1}{2}$ could not occur because ℓ must be less than n.

(e) $n = 2$, $\ell = 0$, $m_\ell = 0$, $m_s = +\frac{1}{2}$ can occur.

Quantum Mechanics

32. *Answer/Explanation:* There are n subshells in the nth energy level; they have $\ell = n - 1, \ldots, 0$. That means **four** different subshells (designated $\ell = 3$, 2, 1, and 0) are found in the 4th energy level.

34. *Answer/Explanation:* The wave mechanical model of the atom tells us that electrons do not follow simple paths as do planets. Rather, they occupy regions of space having certain shapes and varying distances around the nucleus. Hence, subshells and orbitals were not explained by the Bohr model.

35. *Answer/Explanation:* Orbits have predetermined paths – position and momentum are both exactly known at all times. Heisenberg's uncertainty principle says that we cannot know both position and momentum simultaneously.

36. *Answer/Explanation:* The $n = 3$ shell has orbitals with $\ell = 2$, 1, and 0. That means it has **d, p, and s orbitals**. There are n^2 orbitals in the nth shell. That means there are ($3^2 =$) **nine** orbitals in the $n = 3$ shell.

Electron Configurations

39. *Answer/Explanation:* $_{32}$Ge electron configuration: $\mathbf{1s^2 2s^2 2p^6 3s^2 3p^6 3d^{10} 4s^2 4p^2}$

41. *Answer/Explanation:*

(a) The diagram in the book shows total of 28 electrons and six 3d sublevels, but Fe atom should have only 26 electrons and the 3d sublevel should have only five orbitals. Also, in the ground state of Fe atom, the lower-energy **4s orbital must be full**.

(b) The **orbital labels must be 3, not 2**. The electrons in 3p subshell of P should be in separate orbitals with parallel spin (Hund's Rule).

(c) The diagram in the book shows total of 50 electrons and six 3d sublevels, but the cation, Sn^{2+} should have only 48 electrons and the 3d sublevel should have only five orbitals. Also, electrons should be removed from the higher-energy valence 5p orbitals of Sn to make the cation, Sn^{2+}, not from the lower-energy 4d orbitals. The **4d orbital must be completely filled before the 5p orbitals start filling**.

43. *Answer/Explanation:* **18** elements are in the fourth period of the periodic table. It is not possible for there to be another element in this period because **all possible orbital electron combinations are already used**.

45. *Answer/Explanation:* When transition metals form cations, the electrons lost first are those from their valence (highest n) s orbitals.

$_{25}$Mn electron configuration: $1s^22s^22p^63s^23p^63d^54s^2$. Mn has **5 unpaired electrons**:

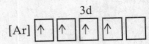

$_{25}$Mn^{2+} electron configuration: $1s^22s^22p^63s^23p^63d^5$. Mn^{2+} has **5 unpaired electrons**:

$$\begin{array}{c} 3d \\ [\text{Ar}]\ \boxed{\uparrow}\ \boxed{\uparrow}\ \boxed{\uparrow}\ \boxed{\uparrow}\ \boxed{\uparrow} \end{array}$$

$_{25}$Mn^{3+} electron configuration: $1s^22s^22p^63s^23p^63d^4$. Mn^{3+} has **4 unpaired electrons**:

$$\begin{array}{c} 3d \\ [\text{Ar}]\ \boxed{\uparrow}\ \boxed{\uparrow}\ \boxed{\uparrow}\ \boxed{\uparrow}\ \boxed{} \end{array}$$

48. *Answer/Explanation:*

(a) $_{63}$Eu electron configuration: **[Xe]4f^{7}6s^2**   or   **[Xe]4f^{6}5d^{1}6s^2**

(b) $_{70}$Yb electron configuration: **[Xe]4f^{14}6s^2**   or   **[Xe]4f^{13}5d^{1}6s^2**

Valence Electrons

50. *Answer/Explanation:*   (a) $\cdot$ Sr $\cdot$ (b) $:\overset{\displaystyle ..}{\underset{\displaystyle ..}{Br}}\cdot$ (c) $\cdot\overset{\displaystyle .}{Ga}\cdot$ (d) $\cdot\overset{\displaystyle ..}{Sb}\cdot$

52. *Answer/Explanation:*

(a) $_{20}$Ca^{2+} electron configuration: $1s^22s^22p^63s^23p^6$ or **[Ar]**

(b) $_{19}$K$^+$ electron configuration: $1s^22s^22p^63s^23p^6$ or **[Ar]**

(c) $_8$O^{2-} electron configuration: $1s^22s^22p^6$ or **[Ne]**

$_{20}$Ca^{2+} and $_{19}$K$^+$ are isoelectronic.   They have the same number of electrons.

54. *Answer/Explanation:*

$_{50}$Sn electron configuration: **[Kr]4d^{10}5s^{2}5p^2**

$_{50}$Sn^{2+} electron configuration: **[Kr]4d^{10}5s^2**

$_{50}$Sn^{4+} electron configuration: **[Kr]4d^{10}**

Paramagnetism and Unpaired Electrons

56. *Answer/Explanation:* In both paramagnetic and ferromagnetic substances, atoms have unpaired spins and so are attracted to magnets. Ferromagnetic substances retain their aligned spins after an external magnetic field has been removed, so they can function as magnets. Paramagnetic substances lose their aligned spins after a time and therefore cannot be used as permanent magnets.

Periodic Trends

58. *Answer:* **P < Ge < Ca < Sr < Rb**

Strategy and Explanation: Without using the table of numerical sizes, list elements in order of increasing size. Without using the table that gives numerical sizes, we are asked to list elements in order of increasing size. Periodic Trends in atom size are related to the radius of the atoms:

- Comparing atoms across the period (row) of the periodic table, atom sizes increase from right to left.

- Comparing atoms down a group (column) of the periodic table, atom sizes increase from top to bottom.

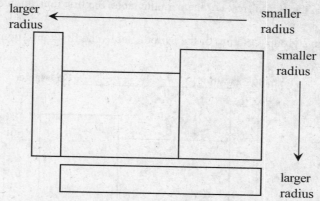

Looking at a periodic table, the smallest of the ions given is P, since its the closest to the top right corner. Next is Ge, diagonally down and to the left of P. Then comes Ca, in the same period as Ge, but further to the left. Then comes Sr, in Group 2A, below Ca. Then comes Rb, in the same period as Sr, but further to the left. So, the order is: P < Ge < Ca < Sr < Rb.

✓ *Reasonable Answer Check:* Using Figures 7.20 and 7.21, we can confirm these predictions.

60. *Answer:* **(a) Rb (b) O (c) Br (d) Ba^{2+} (e) Ca^{2+}**

Strategy and Explanation: Without using the table that gives numerical sizes, determine which element of a pair of atoms and ions has larger radius. Use the periodic table for atom size trends (as described in the solution to Question 58) and the following periodic trends.

Periodic trends comparing atoms to ions:

•   The cation of an element will always be smaller than the atom of the same element (since fewer electrons are attracted more closely to the nucleus).

•   The anion of an element will always be larger than the atom of the same element (since more electrons will be less attracted to the nucleus).

Periodic trend comparing ions to ions:

•   When comparing isoelectronic ions, the element with the larger atomic number is smaller (since it has more protons to attract the electrons).

•   When comparing ions of the same element, the ion with the larger positive charge is smaller (since fewer electrons are attracted more closely to the nucleus).

(a)   Using the periodic table, Rb has a smaller radius than Cs atom, because they are both in the same group and Cs is below Rb.

(b)   The O atom has a smaller radius than O^{2-} ion, since anions are larger than the neutral atom. (8 protons attract 8 electrons better than they can attract 10 electrons.)

(c)   Using the periodic table, Br has a smaller radius than As atom, because they are both in the same period and Br is further to the right.

(d)   The Ba^{2+} ion has a smaller radius than Ba atom, since cations are smaller than the neutral atom. (56 protons attract 54 electrons better than they can attract 56 electrons.)

(e)   The Ca^{2+} ion has a smaller radius than Cl^- ion, since they are isoelectronic and the atomic number of Ca (20) is larger than the atomic number of Cl (17). (20 protons can attract 18 electrons better than 17 protons can.)

✓ *Reasonable Answer Check:* Using Figures 7.20 - 7.22, we can confirm these predictions.

62. *Answer:* **Choice (c)**

Strategy and Explanation: List elements in order of increasing first ionization energy. Sometimes we will not have a chart with numbers like Figure 7.22, so let's get an idea of what periodic trends are exhibited in first ionization energies:

- Comparing atoms in the period (row) of the periodic table, the first ionization energy increases from left to right.

- Comparing atoms in a group (column) of the periodic table, the first ionization energy increases from bottom to top.

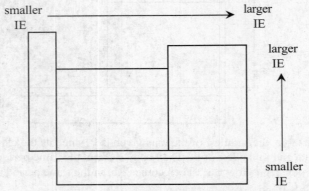

- Elements in group 2A, 2B, and 5A are exceptions to this general trend, having higher IE than elements to their immediate right.

Looking at a periodic table, the element the furthest to the left is Li, so, we'll predict that it has the lowest first ionization energy. Then comes Si, to the right of the metal. Then comes C, above Si in the same group. Then comes Ne, to the right of C in the same period. So the correct order is: Li < Si < C < Ne. That is choice (c).

✓ *Reasonable Answer Check*: Using Figure 7.22, we can confirm these predictions.

64. *Answer/Explanation:* Second ionization energies are all higher than the first ionization energies and once core electrons are being removed, the ionization energy rises dramatically. To have a large gap between the first and second ionization energies, the element **must have one valence electron**. The only element given that has one valence electron is **Na**, from Group 1A.

66. *Answer/Explanation:* Adding a free electron to the oxygen atom is exothermic and produces an anion. When the second electron is added, its negative charge is repelled by the anion's negative charge, according to Coulomb's Law. Additional energy is required to overcome the coulombic charge repulsion, making the second electron affinity endothermic.

Born-Haber Cycle

67. *Answer:* **–862 kJ**

Strategy and Explanation: Given sublimation, ionization energy, bond energy, electron affinity, and formation enthalpies, determine the lattice energy. Adapt the method in Section 7.13, using Hess's Law.

Construct a diagram similar to Figure 7.25, using the enthalpy values given.

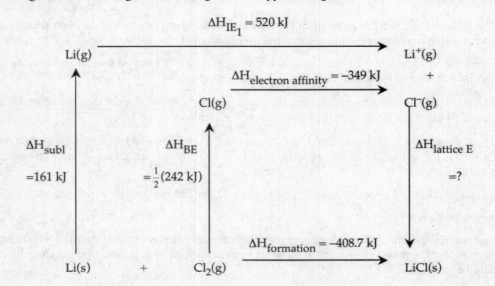

Use Hess's Law to combine the given chemical equations given to make the desired chemical equation:

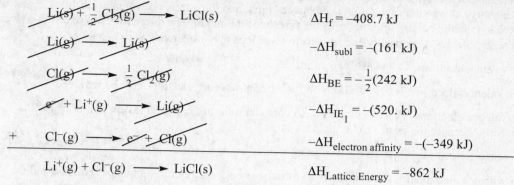

$$Li(s) + \tfrac{1}{2} Cl_2(g) \longrightarrow LiCl(s) \qquad \Delta H_f = -408.7 \text{ kJ}$$

$$Li(g) \longrightarrow Li(s) \qquad -\Delta H_{subl} = -(161 \text{ kJ})$$

$$Cl(g) \longrightarrow \tfrac{1}{2} Cl_2(g) \qquad \Delta H_{BE} = -\tfrac{1}{2}(242 \text{ kJ})$$

$$e^- + Li^+(g) \longrightarrow Li(g) \qquad -\Delta H_{IE_1} = -(520. \text{ kJ})$$

$$+ \quad Cl^-(g) \longrightarrow e^- + Cl(g) \qquad -\Delta H_{electron\ affinity} = -(-349 \text{ kJ})$$

$$Li^+(g) + Cl^-(g) \longrightarrow LiCl(s) \qquad \Delta H_{Lattice\ Energy} = -862 \text{ kJ}$$

✓ *Reasonable Answer Check*: Lithium ion is smaller than sodium ion and both chloride compounds are composed by combining a +1-cation and a –1-ion , so, as expected, this lattice energy is similar in size but slightly more negative than that seen for sodium chloride (which is given in the text (–786 kJ).

General Questions

69. *Answer/Explanation:* Selenium has the following electron configuration: $1s^2 2s^2 2p^6 3s^2 3p^6 3d^{10} 4s^2 4p^4$. That means the first two p sublevels are full, and the third one has 4 electrons. Look at the orbital diagrams of the p sublevels to find out the number of pairs:

From this, it is very easy to tell that there are **seven pairs** of electrons in p orbitals.

71. *Answer:* **(a) He (b) Sc (c) Na**

Strategy and Explanation:

(a) For a ground state element to be in Group 8A but have no p electrons, it must have fewer than five electrons, because the fifth electron goes into a 2p orbital (i.e., $1s^2 2s^2 2p^1$). This Group 8 element is **He** ($1s^2$).

(b) For a ground state element to have a single electron in the 3d subshell, it must have this electron configuration: $1s^2 2s^2 2p^6 3s^2 3p^6 3d^1$. This element is **Sc**.

(c) For a ground state element's cation to have a +1 charge and an electron configuration of $1s^2 2s^2 2p^6$, the element's electron configuration would have one more electron: $1s^2 2s^2 2p^6 3s^1$. This element is **Na**.

74. *Answer:* **(a) S (b) Ra (c) N (d) Ru (e) Cu**

Strategy and Explanation:

(a) The ground state element whose atoms have an electron configuration of $1s^2 2s^2 2p^6 3s^2 3p^4$ has 16 electrons and 16 protons. This is sulfur, **S**, with the atomic number of 16.

(b) The largest *known* element in the alkaline earth group (Group 2A) is the one at the bottom of the group on the periodic table. This is radium, **Ra**.

(c) The largest first ionization energy of elements in the group is the group member at the top of the group on the periodic table. In Group 5A, this is nitrogen, **N**.

(d) For a ground state element's cation to have a +2 charge and an electron configuration of $[Kr]4d^6$, the element's electron configuration would have two more electrons: $[Kr]4d^6 5s^2$. This element is ruthenium, **Ru**.

(e) The ground state element whose neutral atoms have an electron configuration of $[Ar]3d^{10}4s^1$ has 29 electrons and 29 protons. This is copper, **Cu**, with the atomic number of 29.

76. *Answer:* **In^{4+}, Fe^{6+}, Sn^{5+}, very high successive ionization energies**

Strategy and Explanation: Cations likely to form are ones that have lost only valence electrons. Any ion that has a more positive charge than its number of valence electrons is not likely to form. Anions likely to form are ones that add enough electrons to the atom to increase the number of valence electrons to eight. Any ion that has a more negative charge than the difference between eight and the number of valence electrons is not likely to form. Free ions with charges larger than ±3 are rare, because making the ion requires us to remove electrons from ions that are already very positive or add electrons to ions that are already very negative.

Cs has one valence electrons; Cs^+ **is** likely to form.

In has three valence electrons; In^{4+} is **unlikely** to form because a core electron must be removed.

Fe has two valence electrons and six d electrons, so core electrons are being removed; however, Fe^{6+} is **unlikely** to form, because the positive charge is too large.

Te has six valence electrons. It needs $8 - 6 = 2$ electrons. Te^{2-} **is** likely to form.

Sn has four valence electrons; Sn^{5+} is **unlikely** to form because a core electron must be removed and the positive charge is too large.

I has seven valence electrons. It needs $8 - 7 = 1$ electron. I^- **is** likely to form.

78. *Answer:* **(a) replace "inversely" with "directly" (b) it is inversely proportional to *the square of* n (c) replace "as soon as" with "before" (d) replace "frequency" with "wavelength"**

Strategy and Explanation:
(a) This statement: "The energy of a photon is inversely related to its frequency." is **false**. Actually, the energy of a photon is **directly** related to its frequency according to the equation: $E = h\nu$.

(b) This statement: "The energy of the hydrogen electron is inversely proportional to its principle quantum number n." is **false**. Actually, the energy of the hydrogen electron is **inversely proportional to the square of** its principle quantum number, n.

$$E = - \frac{2.179 \times 10^{-18} \text{ J}}{n^2}$$

(c) This statement: "Electrons start to fill the fourth energy level as soon as the third level is full." is **false**. Several different slight changes can make it into a true statement, for example: "Electrons start to fill the fourth energy level **before** the third level is full" This can be seen in the electron configuration for potassium and calcium: $1s^2 2s^2 2p^6 3s^2 3p^6 3d^4 4s^1$ and $1s^2 2s^2 2p^6 3s^2 3p^6 3d^4 4s^2$ where the 4s sublevel fills before the 3d sublevel starts to fill.

(d) This statement: "Light emitted by an n = 4 to n = 2 transition will have a longer frequency than that from an n = 5 to n = 2 transition." is **false**. Several different slight changes can make it into a true statement, for example: "Light emitted by an n = 4 to n = 2 transition will have a longer **wavelength** than that from an n = 5 to n = 2 transition." The energy difference is smaller in the first transition than the second, so the frequency of the emitted photon is smaller, and its wavelength is longer.

80. *Answer:* **ultraviolet; 91.18 nm or shorter is needed to ionize hydrogen.**

Strategy and Explanation: Given a list of types of radiation, determine which is needed to ionize hydrogen. Estimate the ionization energy of hydrogen, from Figure 7.22 in kJ/mol, and use Avogadro's number to determine the energy required to ionize one hydrogen atom, then determine the wavelength of light with that energy using the equation from Section 7.2.

The graph in Figure 7.22 suggests that the energy is between 1200-1300

$$\frac{1300 \text{ kJ}}{1 \text{ mol}} \times \frac{1000 \text{ J}}{1 \text{ kJ}} \times \frac{1 \text{ mol}}{6.022 \times 10^{23} \text{ atoms}} \times \frac{1 \text{ atom}}{1 \text{ photon}} = 2.2 \times 10^{-18} \text{ J / photon}$$

$$\lambda = \frac{hc}{E_{photon}} = \frac{6.626 \times 10^{-34} \text{ J} \cdot \text{s} \times 2.998 \times 10^8 \text{ m/s}}{2.2 \times 10^{-18} \text{ J}} = 9.2 \times 10^{-8} \text{ m}$$

To ionize hydrogen, we need light with a wavelength of 9.2×10^{-8} m or shorter. This is the ultraviolet region of the electromagnetic spectrum. (See Figure 7.1)

✓ *Reasonable Answer Check*: It is satisfying to get 2.179×10^{-18} J/photon, since that is the number used in the Bohr model calculations for electrons changing quantum states in hydrogen:

$$\Delta E_{e^-} = \left(2.179 \times 10^{-18} \ J\right)\left(\frac{1}{n_i^2} - \frac{1}{n_f^2}\right)$$

To ionize the electron, we must move it from n = 1 (the ground state) to above n = ∞ (the last state in the atom). This should be the minimum photon energy needed for ionization:

$$\Delta E_{e^-} = \left(2.179 \times 10^{-18} \ J\right)\left(\frac{1}{1^2} - \frac{1}{\infty^2}\right) = 2.179 \times 10^{-18} \ J = E_{\text{ionization photon}}$$

Applying Concepts

81. *Answer/Explanation:* The element, X, with electron configuration: $1s^2 2s^2 2p^6 3s^1$ has one valence electron ($3s^1$), placing it in Group 1A. Group 1A metals combine with nonmetals forming ionic compounds. The Group 1A elements form cations with a 1+ charge. They combine with the anion of chlorine (chloride, Cl^-) to form **XCl**.

83. *Answer:* **(a) ground state (b) either ground state or excited state (c) excited state (d) impossible (e) excited state (f) excited state**

Strategy and Explanation:

(a) $1s^2 2s^1$ is a ground state electron configuration. The 1s sublevel has the lowest energy it is full. The 2s sublevel has the second lowest energy and the remaining electron is there. This is the lowest energy electron configuration, so it would be called the **ground state**.

(b) $1s^2 2s^2 2p^3$ could be **ground state or excited state**. If each of the p electrons is in a separate orbital with the same spin, that is the ground state. However, if any of the spins are reversed or if any of the electrons are paired, this is a different and higher energy state, hence an excited state. We would need to see an orbital energy diagram of this state to determine an unambiguous answer:

Ground state 2p orbital energy diagram:

Some excited state 2p orbital energy diagrams:

(c) $[Ne]3s^2 3p^3 4s^2$ is an **excited state**. The lower-energy 3p sublevel needs to be full ($3p^6$) before the 4s sublevel gets any electrons; therefore, this possible electron configuration is not the lowest energy state, and would be called an excited state. The ground state electron configuration would be $[Ne]3s^2 3p^5$.

(d) $[Ne]3s^2 3p^6 4s^3 3d^2$ is **impossible**. The 4s sublevel has only one orbital, so the maximum number of electrons is 2. This electron configuration has three electrons in that orbital.

(e) $[Ne]3s^2 3p^6 4f^4$ is an **excited state**. Several lower-energy sublevels need to be full before the 4f sublevel gets any electrons; therefore, this possible electron configuration is not the lowest energy state, and would be called an excited state. The ground state electron configuration would be $[Ne]3s^2 3p^6 3d^2 4s^2$.

(f) $1s^2 2s^2 2p^4 3s^2$ is an **excited state**. The lower-energy 2p sublevel needs to be full ($2p^6$) before the 3s sublevel gets any electrons; therefore, this possible electron configuration is not the lowest energy state, and would be called an excited state. The ground state electron configuration would be $1s^2 2s^2 2p^6$.

85. *Answer:* **(a) increase, decrease (b) helium (c) 5 and 13 (d) no electrons left (e) It is a core electron. (f) $Mg^{2+}(g) \longrightarrow Mg^{3+}(g) + e^-$**

Strategy and Explanation:

(a) Based in the graphical data, ionization energies **increase** left to right and **decrease** top to bottom on the periodic table.

(b) The element with the largest first ionization energy is **helium** (atomic number = 2).

(c) The peaks of the fourth ionization energies will occur at the atomic number of boron (**5**) and at the atomic number of aluminum (**13**).

(d) He has only two electrons, so after two ionizations, there are **no electrons left**.

(e) First electron in lithium is a valence electron, but the **second electron is a core electron**.

(f) The atomic number of 12 belongs to magnesium, $_{12}Mg$.

$$Mg^{2+}(g) \longrightarrow Mg^{3+}(g) + e^-$$

More Challenging Questions

88. *Answer:* **(a) 4.34×10^{-19} J   (b) 6.54×10^{14} Hz**

Strategy and Explanation: Given wavelengths of light absorbed by an electron ejected from a metal and the kinetic energy of the ejected electron, determine the binding energy and the minimum frequency required for the photoelectric effect.

Use the equation in Section 7.2 to find the energy of each photon. Subtract the kinetic energy of the electron from the photon energy to find the binding energy. Use $E = h\nu$ to find the lowest frequency photon.

(a) $\lambda = 4.00 \times 10^{-7}$ m $E_{photon} = \dfrac{hc}{\lambda} = \dfrac{6.626 \times 10^{-34} \text{ J} \cdot \text{s} \times 2.998 \times 10^{8} \text{ m/s}}{4.00 \times 10^{-7} \text{ m}} = 4.97 \times 10^{-19}$ J

Binding Energy = Photon Energy – Kinetic Energy $= 4.97 \times 10^{-19}$ J $- 0.63 \times 10^{-19}$ J $= 4.34 \times 10^{-19}$ J

(b) $\nu_{photon} = \dfrac{E}{h} = \dfrac{4.34 \times 10^{-19} \text{ J}}{6.626 \times 10^{-34} \text{ J} \cdot \text{s}} = 6.55 \times 10^{14} \text{ s}^{-1} = 6.55 \times 10^{14}$ Hz

✓*Reasonable Answer Check*: The binding energy is lower than the photon energy, since the photon required to eject a zero kinetic energy electron needs to have less energy than one with 6.3×10^{-20} J.

90. *Answer:* **2.18×10^{-18} J**

Strategy and Explanation: Given an equation describing the relationship between the electron energy level, the atomic number, Z, and the quantum number, n, determine the ionization energy of the electron. Subtract the ionization energy state ($n = \infty$) of the electron from ground state ($n = 1$) of the electron in hydrogen.

Z = 1 for hydrogen

$$E_{ionization} = E_\infty - E_1 = -\frac{1^2}{(\infty)^2}\left(2.18 \times 10^{-18} \text{ J}\right) - \left[-\frac{1^2}{(1)^2}\left(2.18 \times 10^{-18} \text{ J}\right)\right] = 2.18 \times 10^{-18} \text{ J}$$

✓*Reasonable Answer Check*: The ionization of hydrogen is given in Figure 7.8 as the energy required to go from n = 1 to n= = ∞.

91. *Answer/Explanation:* To remove an electron from an N atom requires disruption of a half filled sub-shell (p^3), which is relatively stable, so the ionization of oxygen will be less energy than the ionization of nitrogen.

94. *Answer/Explanation:*

(a) Element 112 will reside in Group 2B, as one of the **transition metals**, directly below mercury.

(b) The ground state electron configuration of element 112 using the radon core is $[_{86}Rn]7s^2 5f^{14} 6d^{10}$.

Chapter 8: Covalent Bonding

Introduction

Teaching for Conceptual Understanding

Lewis structures are pictorial representations of the atoms and electrons involved in bonding. (They do not indicate the geometry of the molecule.) The process of drawing Lewis structures is rule-based and best taught (and learned) as an algorithm. It is important to follow the Guidelines for Writing Lewis structures found in Section 8.2. Although there are several approaches to drawing correct structures, you will succeed best when a systematic set of steps is done in order.

Hydrogen and the halogens generally have one bond, oxygen has two bonds, nitrogen has three bonds, and carbon has four. This information will provide you with a quick check for the Lewis structures you draw.

Periodic trends can help you see why the periodic table is valuable and how it was created. The periodicity of the properties of elements made it necessary to arrange elements this way.

Solutions to Blue-Numbered Questions
for Review and Thought for Chapter 8

Topical Questions

Lewis Structures

13. *Answer:* **(a)** **(b)** **(c)** **(d)**

Strategy and Explanation: Write Lewis structures for a list of ions and molecules.

Following a systematic plan will give you reliable results every time. Trial and error often works for small molecules, but as molecules get more complex, it's better to follow a procedure than to try to guess where electrons will end up. There are a number of methods. The one described here is the same as that described in the text and it works all the time.

[A]　Count the total number of valence electrons. If there is a nonzero charge, adjust the electron count appropriately. Add electrons for negative charges and subtract electrons for positive charges.

[B]　Determine the skeleton structure, which often means figure out which atom is the central atom. As a general rule, the central atom is usually the first element in the formula, and it will often have a smaller electronegativity (see Section 8.6) than the rest of the atoms in the formula. There will be exceptions, for example: H will **never** be the central atom. The tendency of certain atoms to make certain numbers of bonds will also help inform you of how to arrange the skeleton structure (e.g., we often see elements in Group 4A make four bonds, Group 5A make 3 bonds, Group 6A make two bonds, and Group 7A and H make one bond.) Connect each atom in the skeleton structure with a single bond.

[C]　Each single bond (–) is formed using two electrons. Subtract the electrons used for single bonds in the skeleton structure from the total number of electrons.

[D]　Complete the octets of all of the terminal "outer" atoms by adding pairs of dots (called lone pairs) so that each of them ends up with a total of eight electrons, including the two shared with the central atom. There is an exception: H only need two electrons, so **never** put dots on H.

[E]   Subtract the electrons used for lone pairs from the total.

[F]   If you still have unused electrons, distribute all the remaining electrons on the central atom (or the inner atoms), preferentially with a goal to complete all their octets. If you have extra electrons after completing all the octets, put the extra electrons as dots on the central atom so that only one atom gets more than eight electrons.

[G]   Check the octet of the inner, central atom(s).

   (i)   If each has eight or more electrons, then the structure is complete. The only reason an atom would have more than eight electrons is if there were too many electrons for single bonds and the octet rule to use up all the electrons.

   (ii)   If any atom has less than eight electrons, move a lone pair of electrons from one of the outer atoms to make a new shared pair, forming a multiple bond to the central/inner atom that needs electrons. Repeat this procedure until all atoms have an octet. Never make more than the minimum number of multiple bonds unless you have a really good reason you can explain.

[H]   If the structure is an ion, put it in brackets and designate the net charge outside the bracket at the upper right corner.

✓ *Reasonable Answer Check*: Check the octets of every atom in the structure (and make sure H atoms have only two electrons) then count the electrons and make sure the total is right.

(a) [A]   The atoms Cl and F are both in Group 7A: $7 + 7 = 14$ e$^-$

   [B]   Choose the Cl atom is the central atom with the F atom bonded to it, since it has the lower electronegativity.

Cl——F

   [C]   Connecting the two atoms uses two electrons, so subtract 2 electrons from the current electron count: 14 e$^-$ $- 2$ e$^-$ $= 12$ e$^-$

   [D]   F has two electrons already (in the single bond, so add six dots to the F atom: Cl——F̈:

Notice that F now has a complete octet:

8 e$^-$ inside the circle

   [E]   Completing the octet of the F atom uses 6 electrons, so subtract 6 electrons from the current electron count: : 12 e$^-$ $- 6$ e$^-$ $= 6$ e$^-$

   [F]   Put the last six electrons on the Cl: :C̈l——F̈:

   [G]   Cl now has a complete octet:

8 e$^-$ inside the circle

:C̈l——F̈:

So, the structure is complete.

✓ *Reasonable Answer Check*: The Cl has eight electrons (one pair in a single bond and six dots). The F has eight electrons (one pair in a single bond and six dots). Looking at the structure from left to right, the total number of electrons can be counted: 6 dots + 2 in a bond + 6 dots = 14 total.

(b) [A]   The H atoms are in Group 1A and the Se atom is in Group 6A: $2 \times (1) + 6 = 8$ e$^-$

   [B]   The Se atom is the central atom with the H atoms around it.

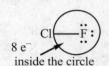

H——Se——H

[C]   Connecting the H and Se atoms uses four electrons: 8 e⁻ – 4 e⁻ = 4 e⁻

[D]   H atoms need no more electrons.

$$\text{H}\text{---Se---}\text{H}$$

[E]   No electrons used in this step. So we still have 4 e⁻.

[F]   Put the last four electrons on the Se.

[G]   Se has eight electrons, so the structure is complete.

(c) [A]   The B atom is in Group 3A and the F atoms are in Group 7A. The – charge **adds** one extra electron:
3 + 4 × (7) + 1 = 32 e⁻

[B]   The B atom is the central atom with the F atoms around it:

[C]   Connecting four F atoms uses eight electrons: 32 e⁻ – 8 e⁻ = 24 e⁻

[D]   Complete the octets of all the F atoms:

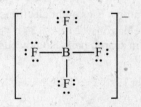

[E]   Completing the octets of the four F atoms uses 24 electrons: 24 e⁻ – 24 e⁻ = 0 e⁻

[F]   No more electrons are available.

[G]   B has eight electrons, so the structure is complete.

[H]   The structure is an ion, so put it in brackets and add the – charge.

(d) [A]   The P atom is in Group 5A and the O atoms are in Group 6A. The 3– charge **adds** three extra electrons: 5 + 4 × (6) + 3 = 32 e⁻

[B]   The P atom is the central atom with the O atoms around it.

[C]   Connecting four O atoms uses eight electrons: 32 e⁻ – 8 e⁻ = 24 e⁻

[D]   Complete the octets of all the O atoms:

[E]   Completing the octets of the four O atoms uses 24 electrons: 24 e⁻ – 24 e⁻ = 0 e⁻

[F]   No more electrons are available.

[G]   P has eight electrons, so the structure is complete.

[H]   The structure is an ion, so put it in brackets and add the 3– charge.

$$\begin{bmatrix} :\overset{..}{\underset{..}{O}}: \\ :\overset{..}{\underset{..}{O}}-P-\overset{..}{\underset{..}{O}}: \\ :\overset{..}{\underset{..}{O}}: \end{bmatrix}^{3-}$$

✓ *Reasonable Answer Check:* All the Period 2 and 3 elements have an octet of electrons. The H atoms have two electrons. All valance electrons are accounted for in the structures, either as members of shared pairs or of lone pairs.

15. *Answer:* **(a)**   H—C—Cl: (with H above and below C)   **(b)** $\begin{bmatrix} :\overset{..}{O}: \\ :\overset{..}{O}-Si-\overset{..}{O}: \\ :\overset{..}{O}: \end{bmatrix}^{4-}$ **(c)** $\begin{bmatrix} :\overset{..}{F}: \\ :\overset{..}{F}-Cl-\overset{..}{F}: \\ :\overset{..}{F}: \end{bmatrix}^{+}$ **(d)** H—C—C—H (with H above and below each C)

Strategy and Explanation: Write Lewis structures for a list of ions and molecules.

Follow the systematic plan given in the solution to Question 13. Use what you learned about organic molecules in Chapter 3 to determine how atoms are bonded in the organic structures.

(a) The C atom is in Group 4A, the H atom is in Group 1A, and the Cl atom is in Group 7A: $4 + 3 \times (1) + 7 = 14$ e$^-$. The C atom is the central atom with the other atoms around it. Connecting four atoms uses eight electrons: 14 e$^-$ $- 8$ e$^-$ $= 6$ e$^-$. H needs no more electrons. Complete the octet of the Cl atom using six electrons: 6 e$^-$ $- 6$ e$^-$ $= 0$ e$^-$. No more electrons are available. The central atom, C, has eight electrons in shared pairs, so it needs no more. We get the following structure:

$$H-\overset{\displaystyle H}{\underset{\displaystyle H}{C}}-\overset{..}{\underset{..}{Cl}}:$$

(b) The Si atom is in Group 4A and the O atoms are in Group 6A. The 4– charge **adds** four extra electrons: $4 + 4 \times (6) + 4 = 32$ e$^-$. The Si atom is the central atom with the O atoms around it. Connecting four O atoms uses eight electrons: 32 e$^-$ $- 8$ e$^-$ $= 24$ e$^-$. Complete the octets of all the O atoms using 24 electrons. 24 e$^-$ $- 24$ e$^-$ $= 0$ e$^-$. No more electrons are available. The central atom, Si, has eight electrons in shared pairs, so it needs no more. The structure is an ion, so put it in brackets and add the 4– charge:

$$\begin{bmatrix} :\overset{..}{O}: \\ :\overset{..}{O}-Si-\overset{..}{O}: \\ :\overset{..}{O}: \end{bmatrix}^{4-}$$

(c) The atoms Cl and F are both in Group 7A. The + charge **removes** one electron: $7 + 4 \times (7) - 1 = 34$ e$^-$. The Cl atom is the central atom with the F atoms around it. Connecting four F atoms uses eight electrons: 34 e$^-$ $- 8$ e$^-$ $= 26$ e$^-$. Complete the octets of all the F atoms, using 24 electrons: 26 e$^-$ $- 24$ e$^-$ $= 2$ e$^-$. Put the last two electrons on the Cl. The central atom, Cl, has ten electrons already, so it needs no more. The structure is an ion, so put it in brackets and add the + charge.

$$\begin{bmatrix} :\overset{..}{F}: \\ :\overset{..}{F}-Cl-\overset{..}{F}: \\ :\overset{..}{F}: \end{bmatrix}^{+}$$

(d) The C atom is in Group 4A and the H atoms are in Group 1A: $2 \times (4) + 6 \times (1) = 14$ e$^-$. As described in Chapter 3 Section 3.3, the ethane molecule has three H atoms bonded to each C, and the two carbons bonded to each other. Connecting eight atoms uses 14 electrons: 14 e$^-$ $- 14$ e$^-$ $= 0$ e$^-$ No more electrons are available. Each C has eight electrons in shared pairs, so they need no more.

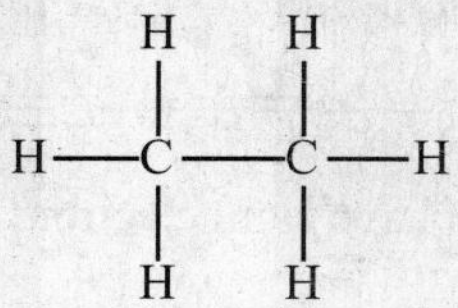

$$
\begin{array}{ccc}
& H & H \\
& | & | \\
H- & C- & C -H \\
& | & | \\
& H & H
\end{array}
$$

✓ *Reasonable Answer Check:* All the Period 2 elements have an octet of electrons. The H atoms have two electrons. All the Period 3 elements have eight or more electrons. All valence electrons are accounted for in the structures, either as members of shared pairs or of lone pairs.

17. *Answer:* **(a)**

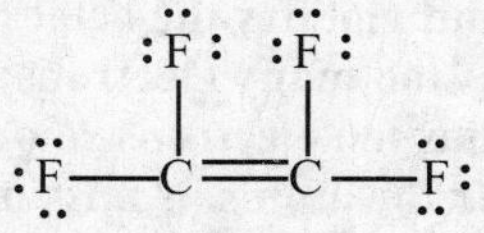

:F: :F:
 | |
:F—C=C—F:

(b)

H   H
| |
H—C=C—C≡N:

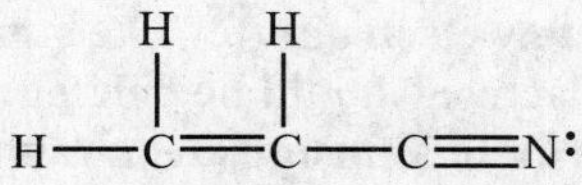

Strategy and Explanation: Write Lewis structures for a list of ions and molecules.

Follow the systematic plan given in the solution to Question 13. Use what you learned about organic molecules in Chapter 3 to determine how atoms are bonded in the organic structures.

(a) $2 \times (4) + 4 \times (7) = 36 \ e^-$. This organic molecule looks like ethene with all the H atoms changed to F atoms. The two carbon atoms are bonded to each other, and each has two F atoms bonded to it. Any other arrangement would make F–F bonds. F has the highest electronegativity (see Section 8.6) and won't bond with another F atom if any other element is present. Connecting six atoms with single bonds uses 10 electrons: $36 \ e^- - 10 \ e^- = 26 \ e^-$. For the purposes of this method, let's call the first C atom the central atom. Complete the octets of the F atoms and the second C atom, using 26 electrons: $26 \ e^- - 26 \ e^- = 0 \ e^-$. At this point, the structure looks like this:

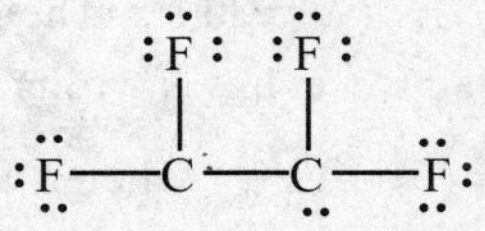

:F: :F:
 | |
:F—C—C—F:

The first C atom has only six electrons in shared pairs, so it needs two more, but there are no more electrons available for lone pairs. Therefore, we must move one lone pair of electrons from the C atom to make a new shared pair, forming a double bond to the C atom:

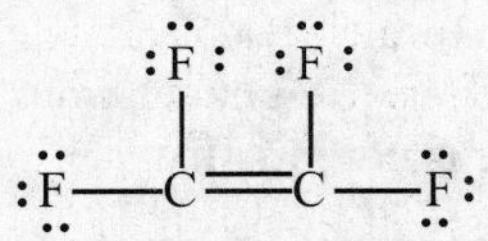

:F: :F:
 | |
:F—C=C—F:

The first C atom now has eight electrons in shared pairs and this is the proper Lewis structure.

(b) $3 \times (4) + 3 \times (1) + 5 = 20 \ e^-$. As described in Chapter 3 Section 3.1, the structural formula given here tells us that this organic compound has two H atoms and a C atom bonded to the first C atom, an H atom and a C atom bonded to the second C atom, and an N atom bonded to the third C atom. Connecting seven atoms uses 12 electrons: $20 \ e^- - 12 \ e^- = 8 \ e^-$. For the purposes of this method, let's call the second and third C atoms "central atoms" so we can start from the ends and work toward the middle. H needs no more electrons. Complete the octet of the N atom and the octet of the first carbon using eight electrons: $8 \ e^- - 8 \ e^- = 0 \ e^-$. At this point, we have the following structure:

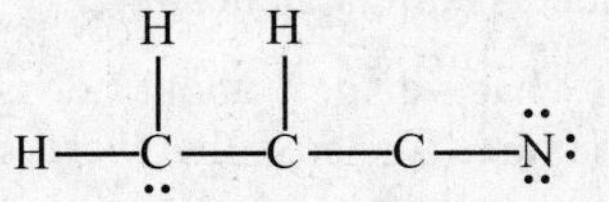

H   H
| |
H—C—C—C—N:

The second C atom has only six electrons in shared pairs and the third C atom has only four electrons in shared pairs, but there are no more electrons available. Therefore, we must move one lone pair of electrons from the first C atom to make a new shared pair between the first and second C atoms, making a double bond. We must also move two lone pairs of electrons from the N atom to the third C atom to make two new shared pairs, forming a triple bond between the N atom and C atom:

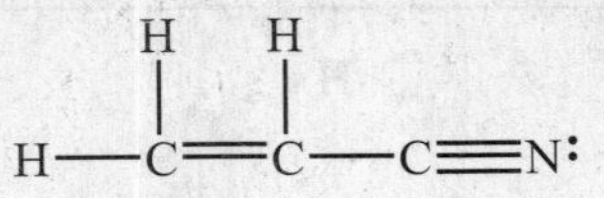

$$H \quad H$$
$$| \quad |$$
$$H - C = C - C \equiv N:$$

The second and third C atoms now both have eight electrons in shared pairs and this is the proper Lewis structure.

✓ *Reasonable Answer Check:* All the Period 2 elements have an octet of electrons. The H atoms have two electrons. All valance electrons are accounted for in the structures, either as members of shared pairs or of lone pairs.

21. *Answer:* **All are incorrect. (a) The structure has two few electrons and violates the octet rule on both F atoms. (b) The structure has too few electrons. (c) The structure has too many electrons; carbon atom has nine electrons, so the single electron should be deleted; oxygen has ten electrons so one of the lone pairs should be removed. (d) has an in appropriate arrangement for the atoms, is missing one H atom, and violates the rule of two for hydrogen. (e) NO_2^- has too few electrons and violates the octet rule.**

Strategy and Explanation: Determine if given Lewis structures are correct and explain what is wrong with the incorrect ones.

A Lewis structure is correct if it has the correct type and number of atoms, if all valance electrons are accounted for in the structures, either as members of shared pairs or of lone pairs; if all the Period 2 elements have an octet of electrons; if the H atoms have two electrons; and if all the Period 3 and higher elements have eight or more electrons. So, first count the atoms and compare with the formula. Then count the electrons in the structure and compare that with the total number of valence electrons. If the count is correct, then check the octets of all the elements and check H atoms for two electrons. Only elements in period 3 and higher are allowed to exceed eight electrons.

(a) OF_2 **Check electron count:** $6 + 2 \times (7) = 20$ e⁻. The given structure has 8 electrons (two lone pairs and two single bonds), so the total electron count is wrong. This structure is incorrect.

(b) O_2 **Check electron count:** $2 \times (6) = 12$ e⁻. The given structure has 10 electrons (two lone pairs and a triple bond), so the total electron count is wrong. This structure is incorrect.

(c) CCl_2O **Check electron count:** $4 + 2 \times (7) + 6 = 24$ e⁻. The given structure also has 27 electrons (nine lone pairs, one unpaired electron, two single bonds, and one double bond), so the total electron count is wrong.

Check octets, etc.: Both Cl atoms each have eight electrons, but the C atom has 9 electrons and the O atom has 10 electrons, so the octet rule is not satisfied.

(d) CH_3Cl **Check type and number of atoms:** The structure has only two H atoms, not three.

Check electron count: $4 + 3 \times (1) + 7 = 14$ e⁻. The given structure has 16 electrons (four lone pairs, and four single bonds), so the total electron count is wrong.

Check octets, etc.: The structure also has one hydrogen atom with too many electrons (4, not 2).

This structure is incorrect.

(e) NO_2^- **Check electron count:** $5 + 2 \times (6) + 1 = 18$ e⁻. The given structure has 16 electrons (five lone pairs, one single bond, and one double bond), so the total electron count is wrong. This structure is incorrect. It is also missing the brackets and the net charge.

✓ *Reasonable Answer Check:* We used what we know about Lewis structures to determine the incorrect structures. Let's draw correct Lewis structures for those that were incorrect, (a), (b), (d) and (e).

The correct structure for (a) OF_2 needs 12 more electrons; each F atom needs three more lone pairs:

$$:\ddot{F}:\ddot{O}:\ddot{F}:$$

The correct structure for (b) O_2 needs 2 more electrons, two more lone pairs, and one less shared pair:

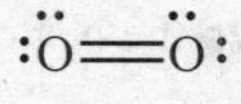

$$:\ddot{O} = \ddot{O}:$$

The correct structure for (d) has two fewer electrons and the carbon bonded to the chlorine without an H atom between them. Remember, H atom only shares two electrons:

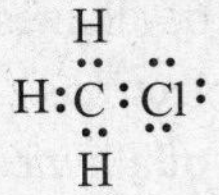

$$\begin{array}{c} H \\ | \\ H\!:\!\overset{\cdot\cdot}{C}\!:\!\overset{\cdot\cdot}{\underset{\cdot\cdot}{Cl}}\!: \\ | \\ H \end{array}$$

The correct structure for (e) has two more electrons on N (and brackets with a charge):

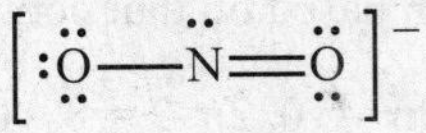

$$\left[\; :\!\overset{\cdot\cdot}{\underset{\cdot\cdot}{O}}\!\!-\!\!\overset{\cdot\cdot}{N}\!\!=\!\!\overset{\cdot\cdot}{\underset{\cdot\cdot}{O}} \;\right]^{-}$$

The corrected structures are different from the incorrect ones, so these answers look right.

Bonding in Hydrocarbons

22. *Answer:* **(see structures below)**

Strategy and Explanation: Write structural formulas for all the branched-chain compounds with a given formula.

Be systematic. Start with a long chain that has only one methyl branch. Move the methyl around (but don't put it on the end carbon and don't put it on a carbon past the first half of the chain). Then make two methyl branches, and move them around similarly. Continue this process until the main chain is too short for methyl branches. Then make an ethyl branch and move it around (but don't put it on the end carbon or the carbon next to the end carbon and don't put it on a carbon past the first half of the chain). Follow a similar pattern with ethyls as with methyls.

The straight chain isomer of C_6H_{14} has six carbons, so start with a five-carbon chain.

One methyl branch can go on the five-carbon chain in two different ways. The methyl branch can go on the second carbon or on the third carbon.

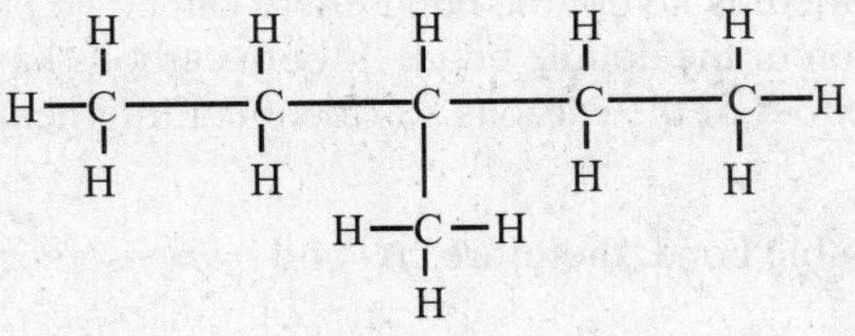

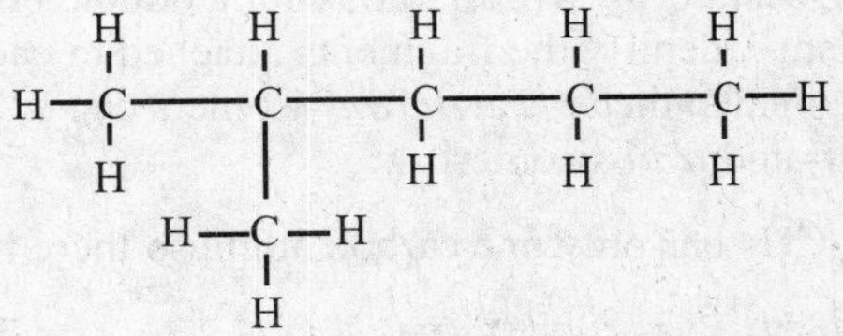

Two methyl branches can go on the four-carbon chain in two ways. They can both go on the second carbon or one can go on the second carbon and one can go on the third carbon.

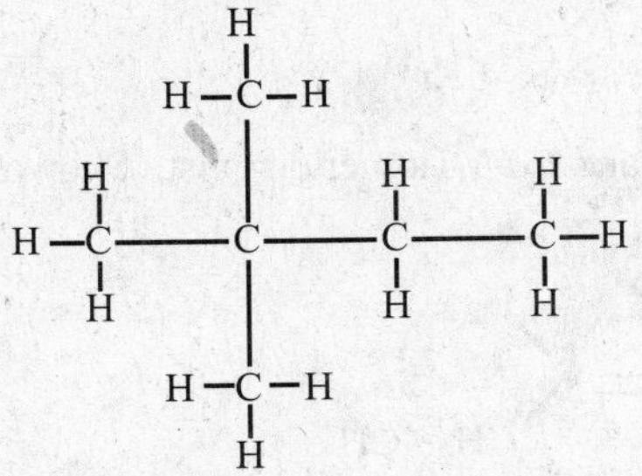

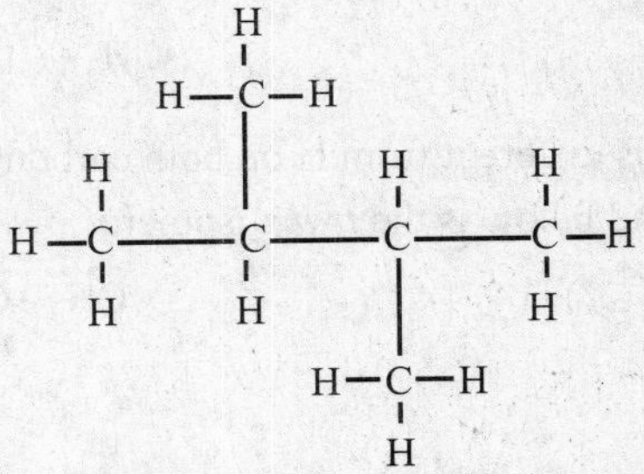

Three methyl branches cannot all three be attached to the one middle carbon in a three-carbon chain, so we're done with methyl branches.

We could try to make a four-carbon chain with an ethyl branch. However, ethyl branches can't be attached to chain-carbons within two carbons of the end of the chain. (If you do that, the molecules "longest" chain will actually include your branch!) A four-carbon chain isn't long enough and therefore we can't use ethyl branches. So we have found all the branched isomers of the hydrocarbon with the formula C_6H_{14}.

✓ *Reasonable Answer Check:* The molecules have one or more branches off the three-carbon chain. They also all have six C atoms and 14 H atoms.

24. *Answer:* **(a) alkyne (b) alkane (c) alkene**

Strategy and Explanation: Given formulas of hydrocarbons determine if they are alkane, alkene, or alkyne.

If we assume that each of these straight chain molecules has at most one multiple bond, we can use the formula pattern to determine the class. The formulas of alkanes are C_nH_{2n+2}, for n = 1 and higher. (See Section 3.3 on page 68.) Each multiple bond removes two electrons. The formulas of alkenes are C_nH_{2n}, for n = 2 and higher. (See Section 8.4 on page 280.) The formulas of alkynes are C_nH_{2n-2}, for n = 2 and higher.(See Section 8.4 on page 281.) Using the number of C atoms, set n. Then determine 2n+2, 2n, or 2n–2. Compare these numbers to the number of H atoms. Form a conclusion based on that comparison.

(a) C_5H_8 n = 5, with this n, 2n+2 = 12, 2n = 10, 2n–2 = 8. This hydrocarbon is an alkyne. Notice, that it could also be an alkene, if there are two double bonds in the molecule.

(b) $C_{24}H_{50}$ n = 24, with this n, 2n+2 = 50, 2n = 48, 2n–2 = 46. This hydrocarbon is an alkane.

(c) C_7H_{14} n = 7, with this n, 2n+2 = 16, 2n = 14, 2n–2 = 12. This hydrocarbon is an alkene.

✓ *Reasonable Answer Check:* One and only one of the three calculations using n gave a matching number.

26. *Answer:* **(a) no (b) yes; cis** 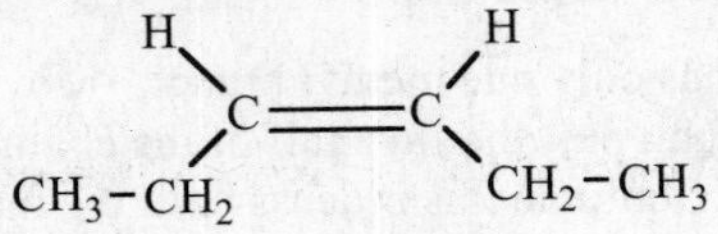 **and trans**

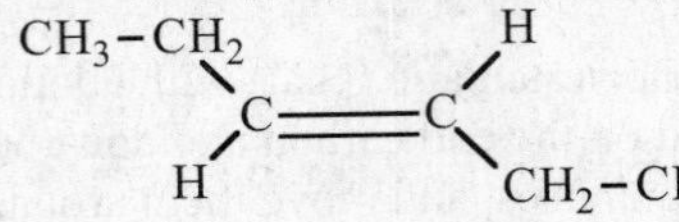

(c) yes; cis 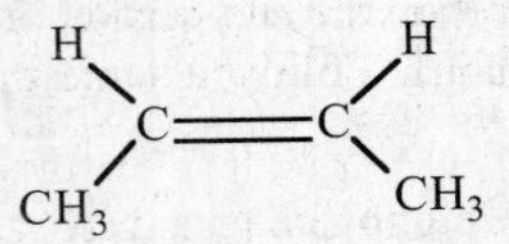 **and trans** 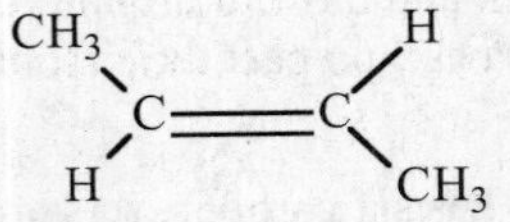   **(d) no**

Strategy and Explanation: Given the condensed structural formulas for some organic compounds, determine if *cis*- and *trans*-isomers exist.

In this Question we will first look for a double bond. If there is no double bond, there can be no *cis*- and *trans*-isomerism. Identify the fragments attached to each carbon of the double bond. If both carbons have different two fragments, then *cis* and *trans*-isomers can exist. If either of the carbons has identical fragments, then there is no *cis*- and *trans*-isomerism.

(a) Br_2CH_2 has only one carbon atom, so there is no double bond; therefore *cis*- and *trans*-isomer do not exist.

(b) $CH_3CH_2CH=CHCH_2CH_3$

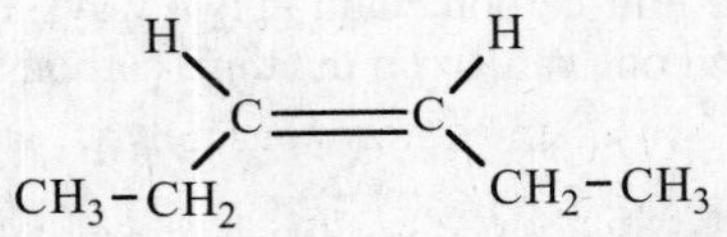

CH_3CH_2 is different from H on both carbons, so *cis*- and *trans*-isomer do exist. Shown above is the *cis*-isomer, and below is the *trans*-isomer:

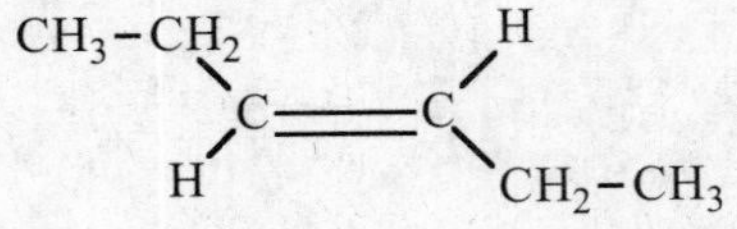

(c) $CH_3CH=CHCH_3$

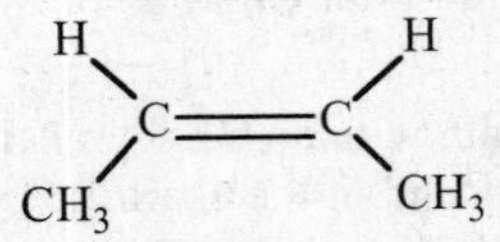

CH_3 is different from H on both carbons, so *cis*- and *trans*-isomer do exist. Shown above is the *cis*-isomer, and below is the *trans*-isomer:

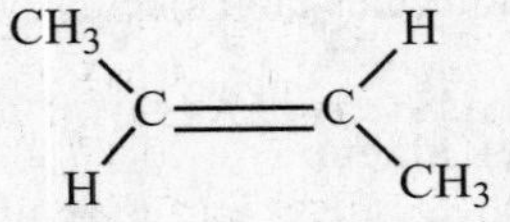

(d) CH$_2$=CHCH$_2$CH$_3$

It is a terminal alkene where that the terminal carbon has two of the same atoms (2 hydrogens), so *cis*- and *trans*-isomer do not exist.

✓ *Reasonable Answer Check:* The isomers shown in (b) and (c) are the *cis*-isomers. Switching the fragments on the left carbon, gives the *trans*-isomer. If that is done in (d) nothing changes.

28. *Answer/Explanation:* It is not possible for oxalic acid to have *cis-trans* isomerization since the double bonds are only to the oxygen atoms and the C–C bond is a single bond. That means free rotation about the C–C bond can occur.

Bond Properties

30. *Answer:* **(a) B–Cl (b) C–O (c) P–O (d) C=O**

Strategy and Explanation: Given a series of pairs of bonds, predict which of the bonds will be shorter.

Use periodic trends in atomic radii to identify the smaller atoms. The bond with the smaller atoms, will have a shorter bond. In cases where bonds between the same atoms are compared, triple bonds are shorter than double bonds, which are shorter than single bonds.

(a) Both bonds have Cl, so compare the sizes of B and Ga. B is smaller than Ga. (It is higher in the same group of the periodic table.) So, B–Cl is shorter than Ga–Cl.

(b) Both bonds have O, so compare the sizes of C and Sn. C is smaller than Sn. (It is higher in the same group of the periodic table.) So, C–O is shorter than Sn–O.

(c) Both bonds have P, so compare the sizes of O and S. O is smaller than S. (It is higher in the same group of the periodic table.) So, P–O is shorter than P–S.

(d) Both bonds have C, so compare the sizes of C and O. O is smaller than C. (It is further to the right in the same period of the periodic table.) So, C=O is shorter than C=C.

✓ *Reasonable Answer Check:* Several of these predictions are confirmed in Table 8.1.

32. *Answer:* **CO**

Strategy and Explanation: Given two chemical formulas, predict which has the shorter carbon-oxygen bond.

First, write Lewis structures for the two molecules. In cases where bonds between the same two atoms are compared, triple bonds are shorter and stronger than double bonds, which are shorter and stronger than single bonds. There is only one plausible Lewis structure for formaldehyde, H$_2$CO, with 12 valence electrons.

In formaldehyde, the carbon-oxygen bond is a double bond.

There is only one plausible Lewis structure for carbon monoxide, CO, with 10 valence electrons.

:C≡O:

In carbon monoxide, the carbon-oxygen bond is a triple bond.

That means the shorter bond is the bond in CO.

✓ *Reasonable Answer Check:* The Lewis structures obey the octet rule and have the right number of valance electrons. These predictions are upheld in general with values given in Table 8.1.

34. *Answer:* **CO$_3$$^{2-}$; it has more resonance structures with single C–O bonds, so the bond is longer.**

Strategy and Explanation: Given two chemical formulas, predict which has the shorter carbon-oxygen bond.

First, write resonance structures for the two molecules. In cases where bonds between the same two atoms are compared, triple bonds are shorter and stronger than double bonds, which are shorter and stronger than single bonds. If more than one resonance structure is possible, average their contribution.

Consider formate ion first. Two equivalent plausible Lewis structures exist for HCO_2^- with 18 valence electrons:

In the formate ion, one resonance structure has a single carbon-oxygen bond and one has a double carbon-oxygen bond. We predict that the carbon-oxygen bond will be halfway between a single and double bond.

There are three plausible equivalent Lewis structures for carbonate ion, CO_3^{2-}, with 24 valence electrons:

In the carbonate, one resonance structure has a double carbon-oxygen bond and two have single carbon-oxygen bond. We predict that the carbon-oxygen bond will be closer to a single bond than a double bond.

That means the longer bond is the bond in CO_3^{2-}.

✓ *Reasonable Answer Check:* The Lewis structures obey the octet rule and have the right number of valance electrons. These predictions are upheld in general with values given in Table 8.1.

Bond Energies and Enthalpy Changes

36. *Answer:* **−92 kJ; the reaction is exothermic.**

Strategy and Explanation: Given a description of a chemical equation for a reaction and a table of bond enthalpies, estimate the standard enthalpy change of the reaction and determine whether the reaction is exothermic or endothermic.

First, balance the chemical equation. Then use another variation of Hess's Law to estimate $\Delta H°$. We will break all the bonds in the reactants (by putting their bond enthalpies into the system) and then we will form all the bonds – the opposite of breaking – in the products (by removing their bond enthalpy from the system). That is the logic behind Equation 8.1.

$$\Delta H° = \sum\left[(\text{moles of bond}) \times D(\text{bond broken})\right] - \sum\left[(\text{moles of bond}) \times D(\text{bond formed})\right]$$

Get the balanced equation and use it to describe the moles of each of the reactants and products. Count the moles of bonds of each type that break and form. Set up a specific version of the above equation. Look up the D for each bond in Table 8.2, plug them into the equation and solve for $\Delta H°$.

The reactants are nitrogen, N_2, and hydrogen, H_2. The product is ammonia, NH_3. The balanced equation is

$$N_2 \quad + \quad 3\,H_2 \quad \longrightarrow \quad 2\,NH_3$$

Hence, we break one mol of N≡N bond and three mol of H–H bonds then form six mol of N–H bonds:

$$\Delta H° = (1 \text{ mol of } N≡N) \times D_{N≡N} + (3 \text{ mol } H–H) \times D_{H–H} - (6 \text{ mol } N–H) \times D_{N–H}$$

Look up the D values in Table 8.2.

$$\Delta H° = (1 \text{ mol}) \times (946 \text{ kJ/mol}) + (3 \text{ mol}) \times (436 \text{ kJ/mol}) - (6 \text{ mol}) \times (391 \text{ kJ/mol}) = -92 \text{ kJ}$$

This reaction is exothermic.

✓ *Reasonable Answer Check:* The reaction is twice the formation reaction for NH_3. Appendix J tells us that NH_3 has $\Delta H_f^\circ = -46.11$ kJ/mol. Twice this value produces a $\Delta H^\circ = -92.22$ kJ/mol. This is very close to the estimate calculated here.

Electronegativity and Bond Polarity

39. *Answer:* **(a) N, C, Br, and O (b) S–O is most polar.**

Strategy and Explanation:

(a) Look up electronegativity values (in Figure 8.6) for the atoms in these bonds.
$$EN_C = 2.5, \ EN_N = 3.0, \ EN_H = 2.1, \ EN_{Br} = 2.8, \ EN_S = 2.5, \ N_O = 3.5$$

N is more electronegative in C–N, C is more electronegative in C–H, Br is more electronegative in C–Br, and O is more electronegative in S–O.

(b) Bonds are more polar when the electronegativity difference is larger. Get ΔEN to find most polar.

$$\Delta EN_{C-N} = EN_N - EN_C = 3.0 - 2.5 = 0.5$$

$$\Delta EN_{C-H} = EN_C - EN_H = 2.4 - 2.1 = 0.4$$

$$\Delta EN_{C-Br} = EN_{Br} - EN_C = 2.8 - 2.5 = 0.3$$

$$\Delta EN_{S-O} = EN_O - EN_S = 3.5 - 2.5 = 1.0 \quad \text{most polar}$$

Formal Charge

41. *Answer:* **(a)** 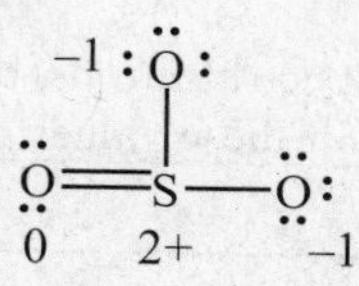   **(b)** 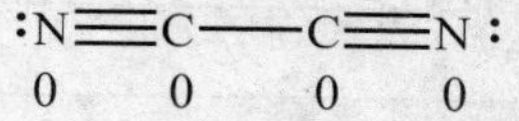   **(c)**

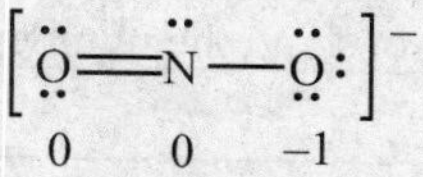

Strategy and Explanation: Given the formulas of molecules or ions, write the correct Lewis structure and assign formal charges to each atom.

Write the Lewis structures. Then determine the number of lone pair electrons and bonding electrons around each atom. Use the method described in Section 8.7 on page 291 to determine the formal charges on each atom.

Formal charge = (number of valence electrons in an atom)

$$- \ [(number \ of \ lone \ pair \ electrons) + (\tfrac{1}{2} \ number \ of \ bonding \ electrons)]$$

To get the *(number of lone pair electrons)* just count all the dots. To get *(number of bonding electrons)* just count two times the number lines representing covalent bonds.

(a) Lewis structure for SO_3 molecule: 24 electrons total

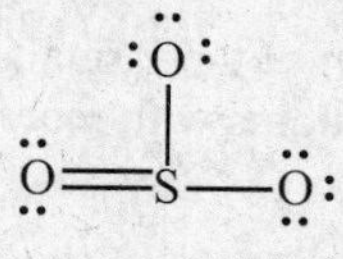

The formal charges are calculated for each atom using the number of valence electrons, the number of lone pair electrons and the number of bonding electrons. Let's set up a chart for these values:

	S	=O	–O
Valence electrons	6	6	6
Lone pair electrons	0	4	6
Bonding electrons	8	4	2
Formal charge	6 – (0 + 4) = +2	6 – (4 + 2) = 0	6 – (6 + 1) = –1

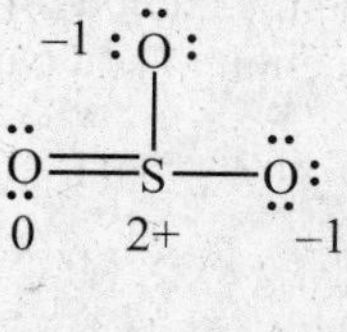

There are other resonance structures that could be written for SO_3 with the double bond moved to each of the other two O atoms and with more than one double bond to the sulfur atom. (A more involved description of writing and judging the feasibility of resonance structures is found in the solutions to Questions 45-51.)

(b) Lewis structure for NCCN molecule: 18 electrons total.

$$:N\equiv C—C\equiv N:$$

The formal charges are calculated for each atom using the number of valence electrons, the number of lone pair electrons and the number of bonding electrons. Let's set up a chart for these values:

	C	N
Valence electrons	4	5
Lone pair electrons	0	2
Bonding electrons	8	6
Formal charge	$4 - (0 + 4) = 0$	$5 - (2 + 3) = 0$

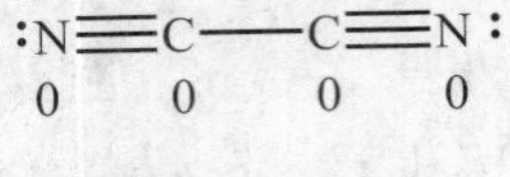

$$:N\equiv C—C\equiv N:$$
$$0000$$

(c) Lewis structure for NO_2^- ion: 18 electrons total

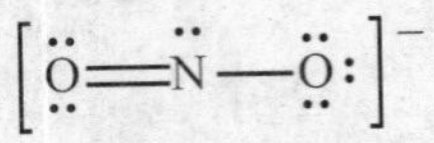

$$\left[\ddot{O}=\ddot{N}—\ddot{O}:\right]^-$$

The formal charges are calculated for each atom using the number of valence electrons, the number of lone pair electrons and the number of bonding electrons. Let's set up a chart for these values:

	N	=O	–O
Valence electrons	5	6	6
Lone pair electrons	2	4	6
Bonding electrons	6	4	2
Formal charge	$5 - (2 + 3) = 0$	$6 - (4 + 2) = 0$	$6 - (6 + 1) = -1$

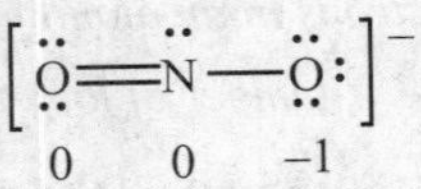

$$\left[\ddot{O}=\ddot{N}—\ddot{O}:\right]^-$$
$$0\phantom{\ddot{O}=}0\phantom{\ddot{N}—}-1\phantom{\ddot{O}:]}$$

One other resonance structure could be written for NO_2^- with the double bond moved to the other O atom. (A more involved description of writing resonance structures is found in the solutions to Questions 45-51.)

✓ *Reasonable Answer Check:* The sum of the formal charges is zero for the neutral molecules and the ionic charge for the charged ion. The atoms with more bonds have more positive formal charges than those with fewer bonds and more lone pairs.

43. *Answer:* **(a)**

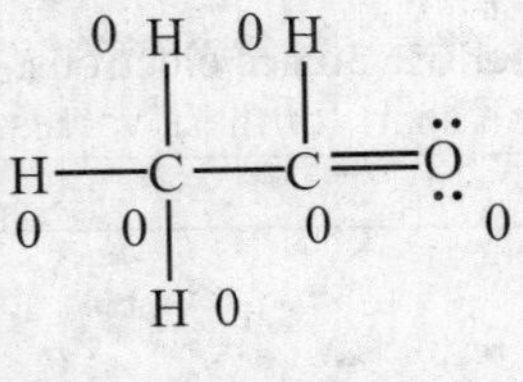

(b)

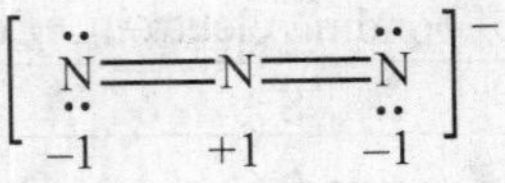

(c)

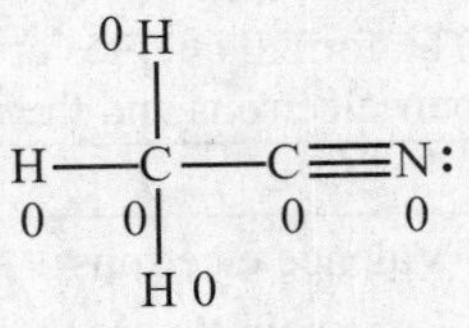

Strategy and Explanation: Given the formulas of molecules or ions, write the correct Lewis structure and assign formal charges to each atom.

Write the Lewis structures. Then determine the number of lone pair electrons and bonding electrons around each atom. Use the method described in Section 8.7 on page 291 and the solution to Question 41 to determine the formal charges on each atom.

(a) The Lewis structure for CH_3CHO molecule: 18 electrons total.

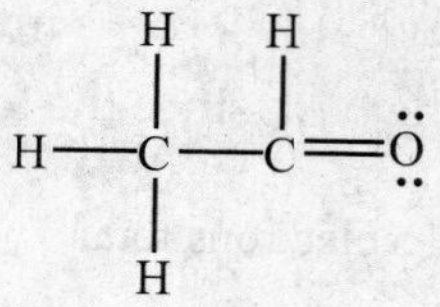

$$H-\underset{\underset{\displaystyle H}{|}}{\overset{\overset{\displaystyle H}{|}}{C}}-\underset{\overset{\displaystyle H}{|}}{C}=\ddot{\underset{\cdot\cdot}{O}}$$

Set up the chart:

	C–	C=	H	O
Valence electrons	4	4	1	6
Lone pair electrons	0	0	0	4
Bonding electrons	8	8	2	4
Formal charge	$4-(0+4)=0$	$4-(0+4)=0$	$1-(0+1)=0$	$6-(4+2)=0$

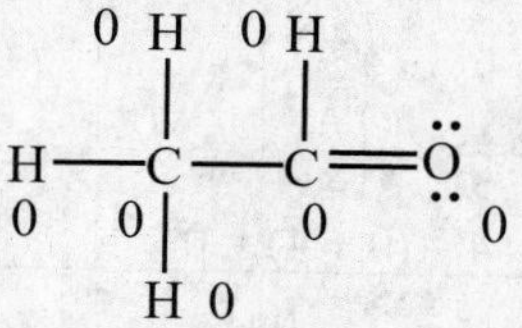

$$\overset{0}{H}-\underset{\underset{0}{\overset{|}{\underset{H}{0}}}}{\overset{\overset{0}{\overset{|}{H}}}{\underset{0}{C}}}-\underset{0}{\overset{\overset{0}{\overset{|}{H}}}{C}}=\ddot{\underset{\cdot\cdot}{O}}\ \ 0$$

(b) There are three possible Lewis structures for N_3^- ion: 16 electrons total.

First structure:

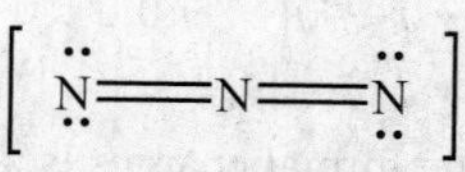

$$\left[\ddot{\underset{\cdot\cdot}{N}}=N=\ddot{\underset{\cdot\cdot}{N}}\right]^-$$

Set up the chart:

	N=	=N=
Valence electrons	5	5
Lone pair electrons	4	0
Bonding electrons	4	8
Formal charge	$5-(4+2)=-1$	$5-(0+4)=+1$

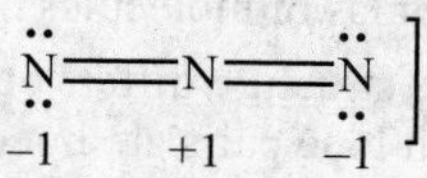

$$\left[\underset{-1}{\ddot{\underset{\cdot\cdot}{N}}}=\underset{+1}{N}=\underset{-1}{\ddot{\underset{\cdot\cdot}{N}}}\right]^-$$

Second structure:

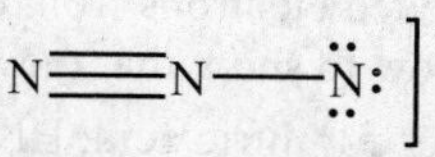

$$\left[:N\equiv N-\ddot{\underset{\cdot\cdot}{N}}:\right]^-$$

Set up the chart:

	Left-most N	Middle N	Right-most N
Valence electrons	5	5	5
Lone pair electrons	2	0	6
Bonding electrons	6	8	2
Formal charge	$5-(2+3)=0$	$5-(0+4)=+1$	$5-(6+1)=-2$

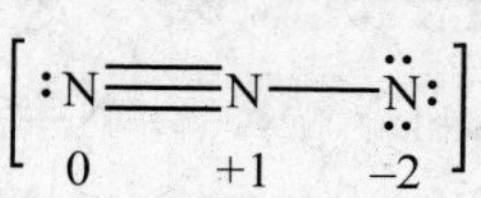

$$\left[:N\equiv\underset{+1}{N}-\underset{-2}{\ddot{\underset{\cdot\cdot}{N}}}:\right]^-$$
$$\ \ \ 0$$

Third structure:

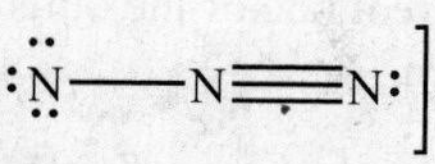

$$\left[:\ddot{\underset{\cdot\cdot}{N}}-N\equiv N:\right]^-$$

Set up the chart:

	Left-most N	Middle N	Right-most N
Valence electrons	5	5	5
Lone pair electrons	6	0	2
Bonding electrons	2	8	6
Formal charge	$5-(6+1)=-2$	$5-(0+4)=+1$	$5-(2+3)=0$

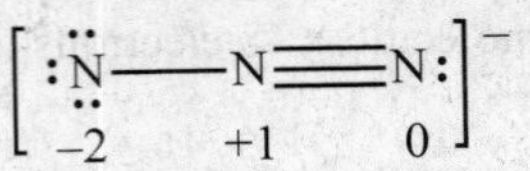

$$\left[:\overset{..}{\underset{..}{N}}\text{——}N\text{====}N: \right]^{-}$$
$$\quad -2 \qquad +1 \qquad 0$$

The first of these three structures is the best, since the formal charges on each atom are closer to zero.

(c) Lewis structure for CH_3CN molecule: 16 electrons total

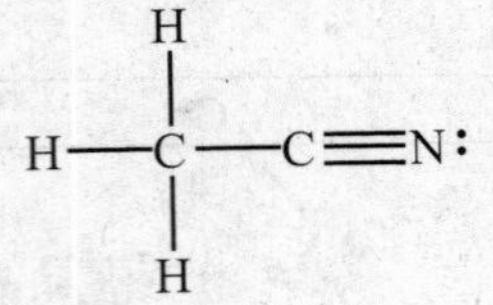

$$H\text{——}\underset{\underset{H}{|}}{\overset{\overset{H}{|}}{C}}\text{——}C\text{≡≡}N:$$

Set up the chart:

	C	H	N
Valence electrons	4	1	5
Lone pair electrons	0	0	2
Bonding electrons	8	2	6
Formal charge	$4 - (0 + 4) = 0$	$1 - (0 + 1) = 0$	$5 - (2 + 3) = 0$

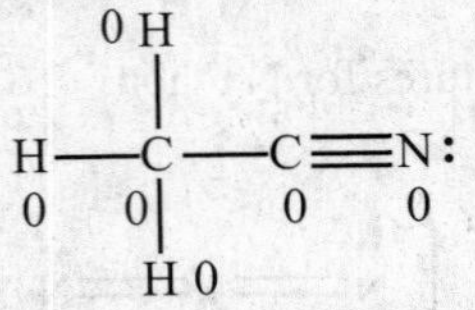

$$\underset{\underset{H}{|} \; 0}{0 \;\; 0}{\overset{0 \; H}{H\text{——}C\text{——}C\text{≡≡}N:}}$$
$$0 \quad 0 \qquad 0 \quad 0$$

✓ *Reasonable Answer Check:* The sum of the formal charges is zero for the neutral molecules and the ionic charge for the charged ion. The atoms with more bonds have more positive formal charges than those with fewer bonds and more lone pairs.

Resonance

45. *Answer:* **See structures below**

Strategy and Explanation: Given the formulas of molecules or ions, write all the resonance structures.

Write the Lewis structure. Each resonance structure differs only by where the electrons for a multiple bond come from. When two or more atoms with lone pairs are bonded to an atom that needs more electrons, any one of them can supply the electron pair for a multiple bond. To write all resonance structures, systematically and sequentially supply the central atom with needed electrons from each of the possible sources. Separate these different structures with a double-headed arrow to show that they are resonance structures.

(a) There are three plausible Lewis structures for nitric acid, HNO_3, with 24 valence electrons. They are formed using one lone pair from a different one of the outer O atoms to make the second bond in the double bond to complete the octet of the N atom.

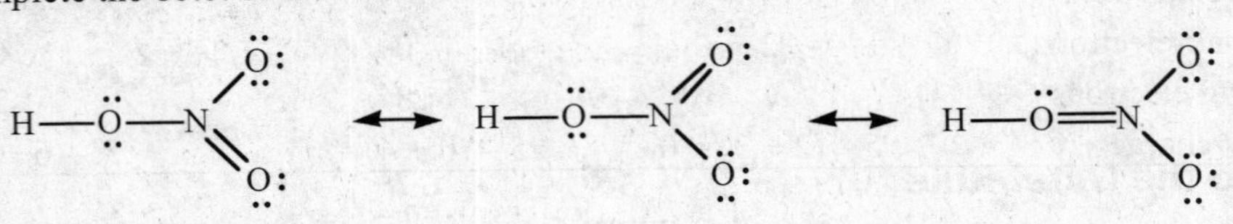

These structures are not equally plausible.

(b) There are three plausible Lewis structures for nitrate ion, NO_3^-, with 24 valence electrons. They are formed using one lone pair from a different one of the outer O atoms to make the second bond in the double bond to complete the octet of the N atom.

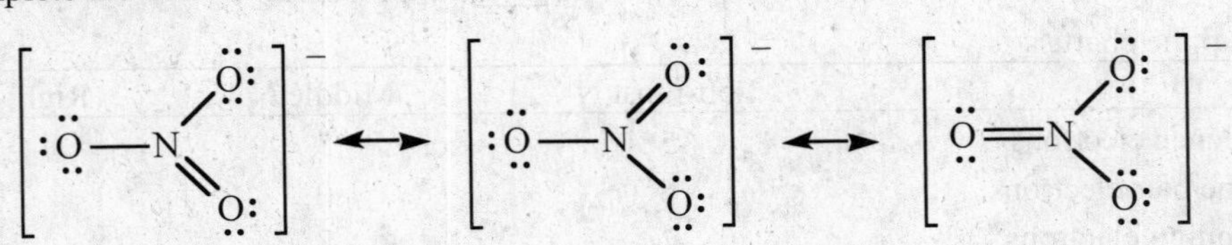

✓ *Reasonable Answer Check:* The structures drawn all follow the octet rule and have the right number of valence electrons. They differ by which outer atom is double bonded to the N atom.

47. *Answer:* **see structures below; the fourth one is the most plausible.**

Strategy and Explanation: Write resonance structures using all single bonds, one, two and three double bonds, and use formal charges to predict the most plausible one.

Write the Lewis structure. To write all resonance structures, systematically and sequentially supply the central atom with needed electrons from each of the possible sources. Keep in mind that Period 2 elements must have an octet, but Period 3 elements can have 8 or more electrons. Separate these different structures with a double-headed arrow to show that they are resonance structures. To determine the relative plausibility of the structures, determine the formal charges (as described in the answers to Question 41) then use the rules described in Section 8.7:

• Smaller formal charges are more favorable than larger ones.

• Negative formal charges should reside on the more electronegative atoms. Conversely, positive formal charges should reside on the least electronegative atoms.

• Like charges should not be on adjacent atoms.

BrO_4^- has 32 electrons. Formal charges for single bonded O atoms are always –1. Formal charges for double bonded O are always 0. Each time electrons are moved from lone pairs into bonding pairs the positive formal charge on Br goes down: The Lewis structure that follows the octet rule for all the atoms is the first one:

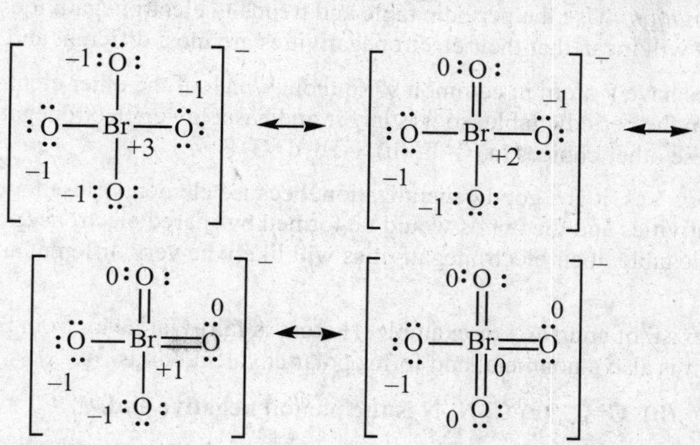

Since smaller formal charges are more favorable than larger ones, we predict that this **fourth resonance structure**, with three double bonds and the most zero formal charges, is the most plausible.

✓ *Reasonable Answer Check:* The Period 2 O atoms in all these structures follow the octet rule. The Period 3 Br atom has eight or more electrons. All structures have the right number of valence electrons. They differ by which outer atom forms the multiple bonds to the central atom. While it might feel strange to select a resonance structure that does not follow the octet rule as being the most plausible, it is clear from the formal charges that this structure is preferred over the one that does follow the octet rule.

Exceptions to the Octet Rule

50. *Answer:* **See structures below**

Strategy and Explanation: Follow the systematic plan given in the answers to Question 13.

(a) BrF_5 has 42 valance electrons. Use ten of the electrons to connect the six F atoms to the central Br atom. Use 30 more of them to fill the octets of the F atoms. Put the last two on Br.

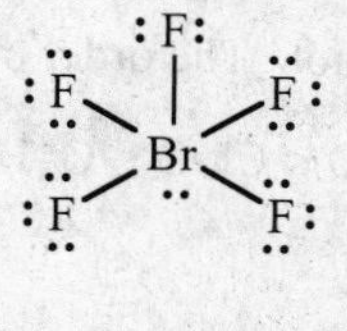

(b) IF_5 has 42 valance electrons. Use ten of the electrons to connect the five F atoms to the central I atom. Use 30 more of them to fill the octets of the F atoms. Put the last two on I.

$$\begin{array}{c}
:\overset{\cdot\cdot}{\underset{\cdot\cdot}{F}}: \\[4pt]
:\overset{\cdot\cdot}{\underset{\cdot\cdot}{F}} \!\!-\!\! \overset{|}{\underset{\cdot\cdot}{I}} \!\!-\!\! \overset{\cdot\cdot}{\underset{\cdot\cdot}{F}}: \\[4pt]
:\overset{\cdot\cdot}{\underset{\cdot\cdot}{F}} \qquad \overset{\cdot\cdot}{\underset{\cdot\cdot}{F}}:
\end{array}$$

(c) IBr_2^- has 22 valance electrons. Use four of the electrons to connect the Br atoms to the central I atom. Use 12 more of them to fill the octets of the Br atoms. Put the last six on I.

$$\left[:\overset{\cdot\cdot}{\underset{\cdot\cdot}{Br}} \!\!-\!\! \overset{\cdot\cdot}{\underset{\cdot\cdot}{I}} \!\!-\!\! \overset{\cdot\cdot}{\underset{\cdot\cdot}{Br}}: \right]^-$$

✓ *Reasonable Answer Check:* All the Period 2 elements have an octet of electrons. All the Period 3 elements have eight or more electrons. All valance electrons are accounted for in the structures, either as members of shared pairs or of lone pairs.

General Questions

54. *Answer:* **The bond in (c) Si–F, Si is farthest from F on the periodic table, so it is larger and has a lower electronegativity.**

Strategy and Explanation: Use the periodic table and trends in electronegativities to determine which pair is farthest apart. That will mean that their electronegativities are most different and the bond will be most polar.

Here, all the choices have F atom in common so find out which of the other elements is the farthest from F. Si is farthest from F on the periodic table, so it is larger and has a lower electronegativity. Therefore, (c) Si–F is more polar than these other choices (a) C–F, (b) S–F, (d) O–F

55. *Answer/Explanation:* **Yes**, it is a good generalization, because elements close together in the periodic table have similar electronegativities and the bonds would be formed by shared electrons (polar covalent). If they are far apart on the periodic table, their electronegativities will likely be very different and the bond more likely to be ionic.

A few exceptions exist, of course. For example, H atom is fairly far away from most of the nonmetals on the periodic table, yet it is also a nonmetal and forms polar covalent bonds.

56. *Answer:* **(a) C=C   (b) C=C   (c) C≡N; N is the partial negative end.**

Strategy and Explanation: The resonance structure given in the Question is the only resonance structure with zero formal changes, so it is the dominant form for the molecule.

(a) There are two carbon-carbon bonds, a C=C and a C–C. Double bonds are shorter than single bonds, so C=C is the shortest.

(b) Double bonds are stronger than single bonds, so C=C is the strongest.

(c) The most polar bond is the bond with atoms that have the largest electronegativity difference. The difference between the electronegativities of C (EN = 2.5) and H (EN = 2.1) is 0.4. The difference between the electronegativities of C and N (EN = 3.0) is 0.5. So, the carbon-nitrogen bond is slightly more polar. Nitrogen has the higher electronegativity, so it is the partial negative end of the bond.

$$\underset{\delta^+ \qquad \delta^-}{C\equiv N}$$

59. *Answer:* **O–O < Cl–O < O–H < O=O < O=C**

Strategy and Explanation: The strongest chemical bond has the largest bond enthalpy. Look up the bond enthalpies for the bonds in Table 8.2: $D_{O-H} = 467$ kJ/mol, $D_{O=O} = 498$ kJ/mol, $D_{O=C} = 695$ to 803 kJ/mol, $D_{O-O} = 146$ kJ/mol, and $D_{O-Cl} = 205$ kJ/mol. The order of increasing strength is:

$$O-O < Cl-O < O-H < O=O < O=C$$

60. *Answer:* **(d) and (e) are alkane. (a), (b), (c), and (f) do not fit into this category**

Strategy and Explanation: Given formulas of hydrocarbons, determine if they are alkanes or not. The formulas of alkanes are C_nH_{2n+2}, for n = 1 and higher.

	n	2n+2 (alkane)
$C_6H_?$	6	? = 14
$C_8H_?$	8	? = 18
$C_{10}H_?$	10	? = 22

Check the given formulas to the formulas for alkanes derived in this chart:

(a) A hydrocarbon with eight C atoms is an alkane, if it has the formula C_8H_{18}. Therefore, C_8H_{10} does not fit into this category.

(b) A hydrocarbon with ten C atoms is an alkane, if it has the formula $C_{10}H_{22}$. Therefore, $C_{10}H_8$ does not fit into this category.

(c) A hydrocarbon with six C atoms is an alkane, if it has the formula C_6H_{14}. Therefore, C_6H_{12} does not fit into this category (though it could be a cyclic alkane).

(d) C_6H_{14} is an alkane.

(e) C_8H_{18} is an alkane.

(f) A hydrocarbon with six C atoms is an alkane, if it has the formula C_6H_{14}. Therefore, $C_{10}H_8$ does not fit into this category.

Applying Concepts

62. *Answer:* **student may have forgotten to subtract one electron for the positive charge**

Strategy and Explanation: The number of valence electrons in SF_5^+ is $6 + 5 \times (7) - 1 = 40$ e$^-$. It looks like the student forgot to subtract one electron for the positive charge, since the given structure has $6 + 5 \times (7) = 41$ e$^-$.

63. *Answer:* **their bonding arrangements differ**

Strategy and Explanation: Resonance structures must only be different by where the electrons are. The bonding arrangement of the atoms (what atom is bonded to what atom) must be the same from structure to structure. In the first structure, the C atom has only one S atom bonded to it. In the second structure the C atom has an N atom and an S atom bonded to it. These bonding arrangements differ; therefore these structures represent different molecules, not resonance structures of one molecule.

66. *Answer:* **Cl: 3.0;   S: 2.5;   Br: 2.5;   Se: 2.4;   As: 2.1**

Strategy and Explanation: Using the periodic trend for electronegativity (EN)

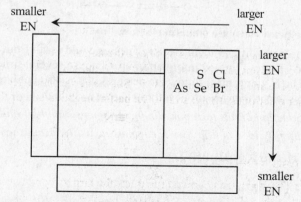

We certainly know that Cl's EN is the largest, so we'll assign it the value of 3.0. We certainly know that As's EN is the lowest, so we'll assign it value of 2.1 The trend shows that EN's of S and Br are both larger than that of Se, so we'll assign them both the value of 2.5. That leaves the EN value of 2.4 for Se. With the values of 2.5 and 2.4 so close together, the assignment of those values are uncertain.

69. *Answer:* **(a) C–O (b) C≡N (c) C–O**

Strategy and Explanation:

(a) Bond strength increases from single to double to triple, so the weakest bond will be a single bond. Table 8.2 gives the various bond enthalpies for single bonds. The C–C bond enthalpy is 356 kJ/mol. The C–O bond enthalpy is 336 kJ/mol. Therefore, the C–O bond is the weakest carbon-containing bond.

(b) The triple C≡N bond is the strongest carbon-containing bond.

(c) The electronegativity of O atom is greater than that of any other element in this structure, so the C–O bond is the most polar.

More Challenging Questions

70. *Answer:* **(a) see structure below (b)**   H—Ö̤—N̈=N̈—Ö—H

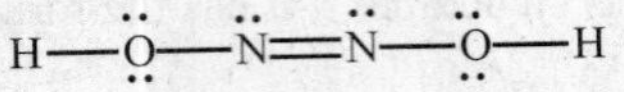

Strategy and Explanation:

(a) B_4Cl_4 cannot follow the octet rule; there are too few electrons. The structure has 40 valence electrons. We must use 8 electrons to attach the Cl atoms to the B atoms and we must use 24 electrons to fill the octets of the Cl atoms. If the B atoms cluster together and use a total of four electrons to bond with each other, then all the electrons could be used. Dashed lines (- - - - -) are used between B and B in the structure below because they do not represent typical two electron bonds.

(Notice: At the time of production of this textbook, the three-dimensional ball-and-stick model of the B_4H_4 molecule is provided at http://www.chem.ox.ac.uk/inorganicchemistry3/index.html, as are several of the others in this section of Questions.)

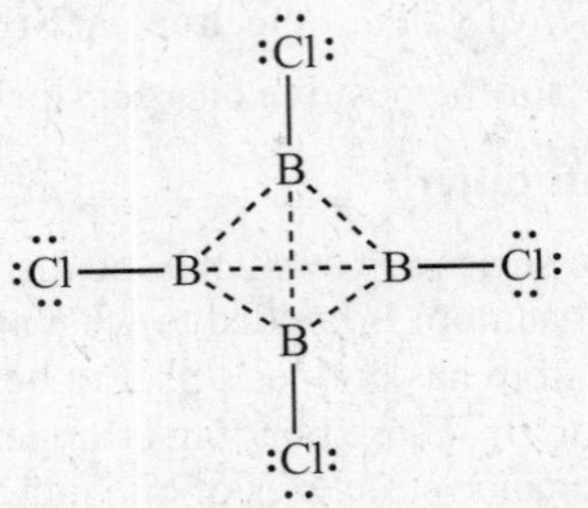

(b) This molecule is an acid so we will attach the H atoms to the O atom as we have seen with other acids. The molecule has 24 electrons. The best Lewis structure would look like this:

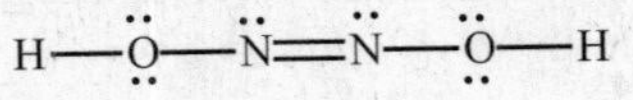

H—Ö—N̈=N̈—Ö—H

72. *Answer:* **(a) see structure below (b) see structure below**

Strategy and Explanation: Many different and creative guesses could result from attempting to write Lewis structures of this molecule. There is limited information provided, so several structures will be equally good at answering the Question. The N atoms must follow the octet rule. It is also not likely that the N atoms will bond to each other or to Cl atoms, since their electronegativity is relatively high.

(Notice: At the time of production of this textbook, the three-dimensional ball-and-stick model of the $(Cl_2PN)_3$ and $(Cl_2PN)_4$ molecules are provided at http://www.chem.ox.ac.uk/inorganicchemistry3/index.html.)

(a) There are 72 valence electrons in $(Cl_2PN)_3$.

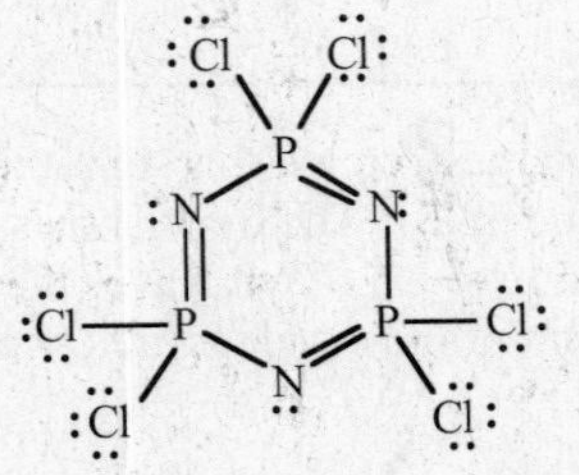

(b)   There are 96 valence electrons in $(Cl_2PN)_4$.

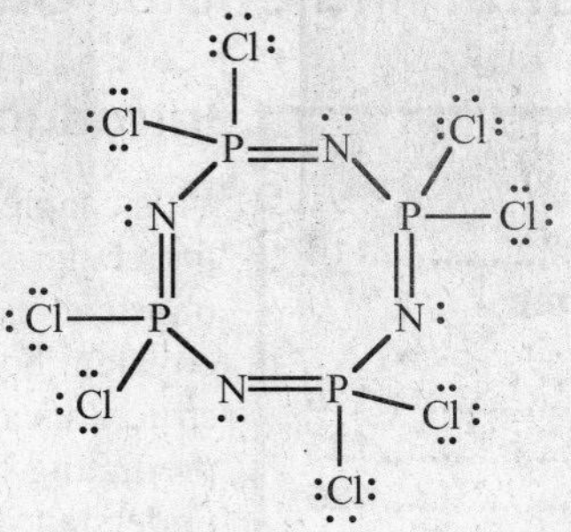

74. *Answer:*

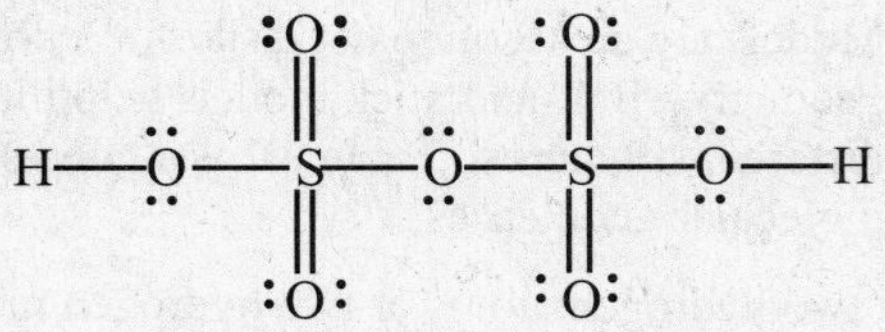

Strategy and Explanation: The molecule has 56 electrons. If we imagine an SO_3 molecule bonding with a sulfuric acid molecule, by having one of the O atoms on H_2SO_4 attached to the S atom of SO_3, then the molecule would look like this.

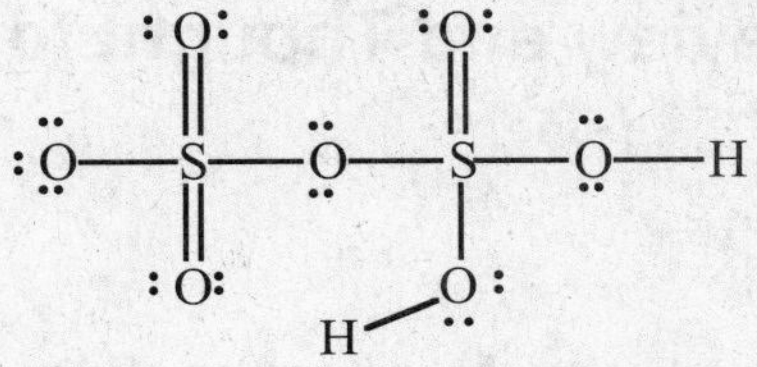

This structure has two atoms with non-zero formal charge (the left-most O and the right S), so it is likely that the H atom will shift to the other side of the molecule during the reaction to form the more stable (all zero-formal-charge) molecule.

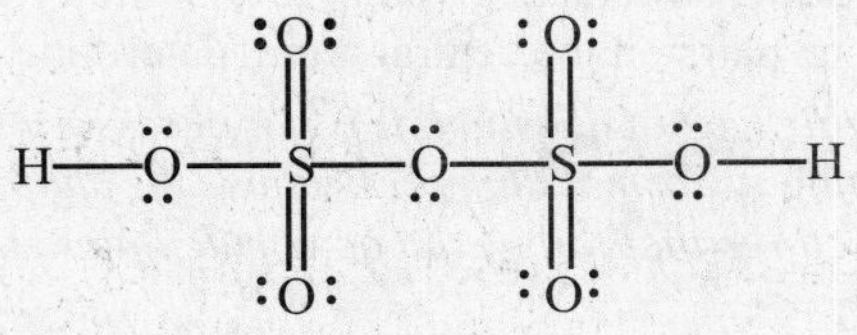

76. *Answer:* **see structure below**

Strategy and Explanation: The C and N atoms in this molecule must all follow the octet rule. Make a straight chain, and add as many multiple bonds to fulfill the octet rule for both period two atoms.

78. *Answer:* **see structure below**

Strategy and Explanation: Tetraphosphrous trisulfide is P_4S_3, with 38 valance electrons.

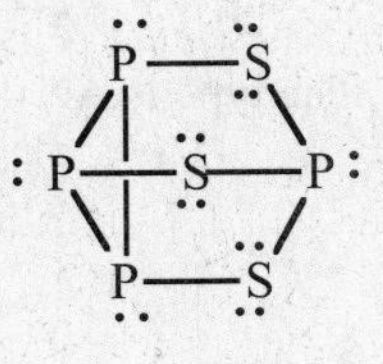

Chapter 9: Molecular Structures

Chapter Contents:

Introduction

This chapter explains the importance of understanding molecular shapes. From DNA, to designer drugs, to physical properties of molecules, the understanding covalent bonding and an ability to draw correct Lewis structures are prerequisites for understanding and predicting correct molecular shapes, which helps with understanding reactivity.

Models are an absolute necessity for learning molecular geometry. Ball-and-stick models help illustrate the different structures. Table 9.1 is a good summary of molecular geometries.

Hydrogen bonding is misnamed. It is not a bonding between two hydrogen atoms or two hydrogen molecules. Questions for Review and Thought 96-99 can help you improve your understanding of hydrogen bonding. If your major involves the study of living systems, it is most important that you have a solid understanding of the concept.

Solutions to Blue-Numbered Questions
for Review and Thought for Chapter 9

Topical Questions

Molecular Shape

8. *Answer:* **See structures below; (a) linear (b) triangular planar (c) octahedral (d) triangular bipyramidal**

Strategy and Explanation: Write Lewis structures for a list of formulas and identify their shape.

Follow the systematic plan for Lewis structures given in the solution to Question 8.12, then determine the number of bonded atoms and lone pairs on the central atom, determine the designated type (AX_nE_m) and use Table 9.1. *Notice: To answer this type of question, it is not necessary to expand the octet of a central atom solely for the purposes of lowering its formal charge, because the shape of the molecule will not change. Hence, we will write Lewis structures that follow the octet rule unless the atom needs more than eight electrons.*

(a) BeH_2 (4 e^-)   H—Be—H   The type is AX_2E_0, so it is linear.

(b) CH_2Cl_2 (20 e^-)   The type is AX_4E_0, so it is tetrahedral.

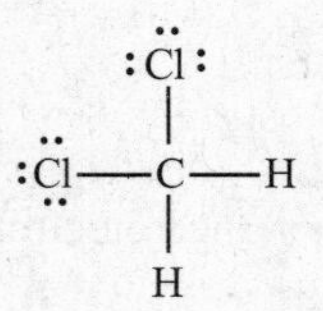

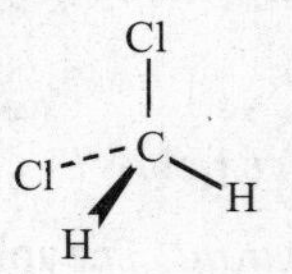

(c) BH_3 (6 e^-)   The type is AX_3E_0, so it is triangular planar.

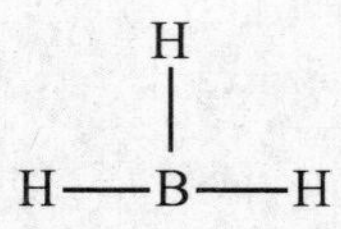

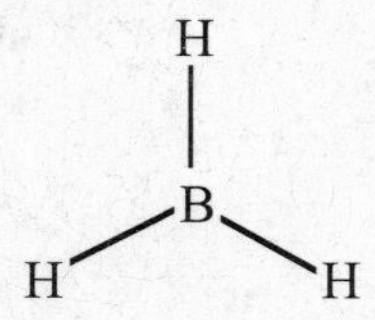

(d) SCl_6 (48 e$^-$)   The type is AX_6E_0, so it is octahedral.

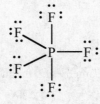

(e) PF_5 (40 e$^-$)   The type is AX_5E_0, so it is triangular bipyramidal.

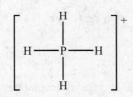

10. *Answer:* **(a) electron-pair geometry and molecular geometry are tetrahedral (b) electron-pair geometry is tetrahedral; molecular geometry is angular (109.5°) (c) electron-pair geometry and molecular geometry are both triangular planar (d) electron-pair geometry and molecular geometry are both triangular planar**

Strategy and Explanation: Adapt the method given in the solution to Question 9.

(a) PH_4^+ (8 e$^-$)   The type is AX_4E_0, so electron-pair geometry and the molecular geometry are both tetrahedral.

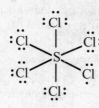

(b) OCl_2 (20 e$^-$)   The type is AX_2E_2, so the electron-pair geometry is tetrahedral and the molecular geometry is angular (109.5°).

(c) SO_3 (24 e$^-$)   The type is AX_3E_0, so electron-pair geometry and the molecular geometry are both triangular planar.

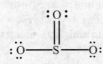

(d) H_2CO (12 e$^-$)   The type is AX_3E_0, so electron-pair geometry and the molecular geometry are both triangular planar.

12. *Answer:* **See structures below; (a) electron-pair geometry and molecular geometry are both triangular planar (b) electron-pair geometry and molecular geometry are both triangular planar (c) electron-pair geometry is tetrahedral and the molecular geometry is triangular pyramidal (d) electron-pair geometry is tetrahedral; molecular geometry is triangular pyramidal**

Strategy and Explanation: Adapt the method given in the solution to Question 9.

(a) BO_3^{3-} (24 e⁻)   The type is AX_3E_0, so both the electron-pair geometry and the molecular geometry are triangular planar.

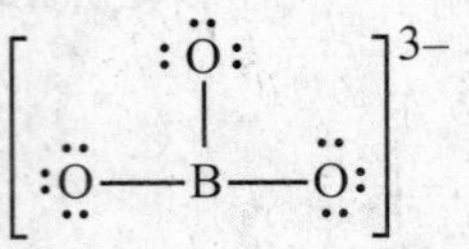

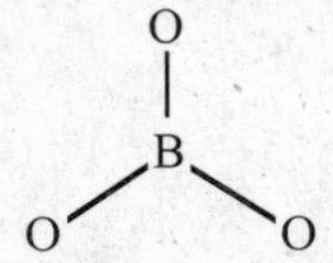

(b) CO_3^{2-} (24 e⁻) AX_3E_0, so both the electron-pair geometry and the molecular geometry are triangular planar.

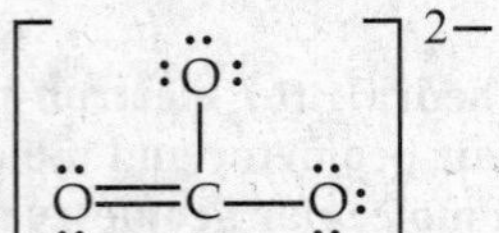

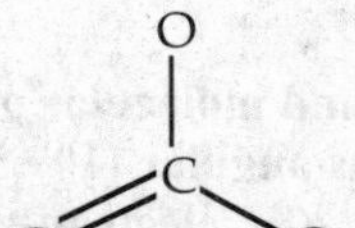

(c) SO_3^{2-} (26 e⁻) AX_3E_1, so the electron-pair geometry is tetrahedral and the molecular geometry is triangular pyramidal.

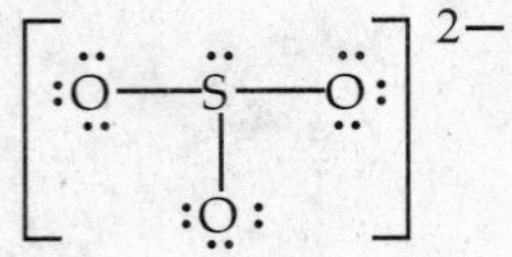

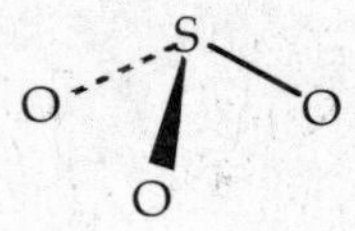

(d) ClO_3^{-} (26 e⁻) AX_3E_1, so the electron-pair geometry is tetrahedral and the molecular geometry is triangular pyramidal.

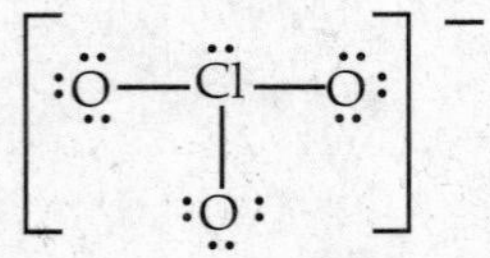

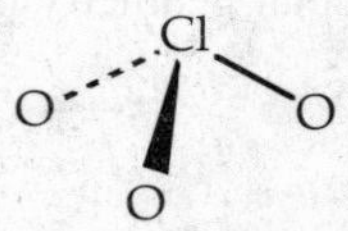

All of these ions and molecules have one central atom with three O atoms bonded to it. The number and type of bonded atoms is constant. The geometries vary depending on how many lone pairs are on the central atom. The structures with the same number of valance electrons all have the same geometry.

14. *Answer:* **See structures below; (a) electron-pair geometry and the molecular geometry are both octahedral (b) electron-pair geometry is triangular bipyramidal; molecular geometry is seesaw (c) electron-pair geometry and molecular geometry are both triangular bipyramidal (d) electron-pair geometry is octahedral; molecular geometry is square planar**

Strategy and Explanation: Adapt the method given in the solution to Question 9.

(a) SiF_6^{2-} (48 e⁻)   The type is AX_6E_0, so both the electron-pair geometry and the molecular geometry are octahedral.

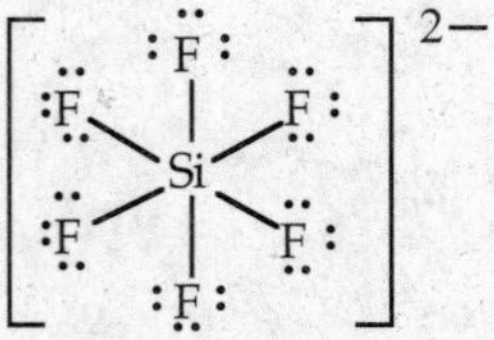

(b) SF₄ (34 e⁻)   The type is AX₄E₁, so the electron-pair geometry is triangular bipyramidal and the molecular geometry is seesaw.

(c) PF₅ (40 e⁻)   The type is AX₅E₀, so the electron-pair geometry is triangular bipyramidal and the molecular geometry is triangular bipyramidal.

(d) XeF₄ (36 e⁻)   The type is AX₄E₂, so the electron-pair geometry is octahedral and the molecular geometry is square planar.

16. *Answer:* **(a) O–S–O angle is 120° (b) F–B–F angle is 120° (c) H–C–H angle is 120°; C–C–N angle is 180° (d) N–O–H angle is 109.5°; O–N–O is 120°**

Strategy and Explanation: Adapt the method given in the solution to Question 9 to get the electron-pair geometry of the second atom in the bond. Use this to predict the approximate bond angle.

(a) SO₂ (18 e⁻)   The type is AX₂E₁, so the electron-pair geometry is triangular planar and the approximate O–S–O angle is 120°.

(b) BF₃ (24 e⁻)   The type is AX₃E₀, so the electron-pair geometry is triangular planar and the approximate F–B–F angle is 120°.

(c) CH₂CHCN (20 e⁻)   Look at the first C atom: The type is AX₃E₀, so the electron-pair geometry is triangular planar and the approximate H–C–H angle is 120°. Look at the third C atom: The type is AX₂E₀, so the electron-pair geometry is linear and the approximate C–C–N angle is 180°.

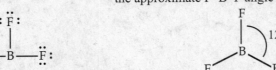

(d) HNO_3 (24 e⁻)

Look at the first O atom: The type is AX_2E_2, so the electron-pair geometry is tetrahedral and the approximate N–O–H angle is 109.5°.
Look at the N atom: The type is AX_3E_0, so the electron-pair geometry is triangular planar and the approximate O–N–O angle is 120°.

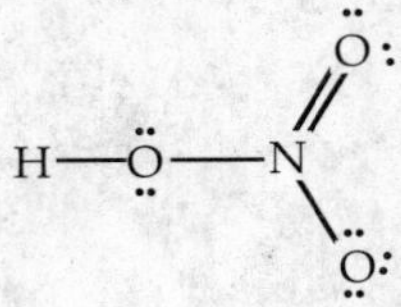

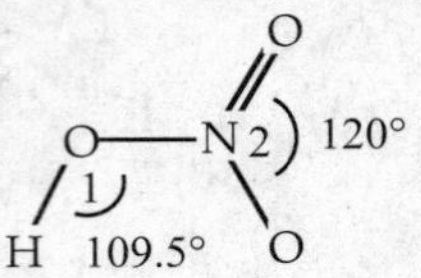

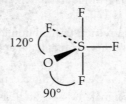

18. *Answer:* **(a) four at 90°, one at 120°, and one at 180° (b) 90° and 120° (c) eight at 90° and two at 180°**

Strategy and Explanation: Adapt the method given in the solution to Question 16.

(a) SeF_4 (34 e⁻)

The type is AX_4E_1, so the electron-pair geometry is triangular bipyramidal. The $F_{equitorial}$–Se–$F_{equitorial}$ angle is 120°, the $F_{equitorial}$–Se–F_{axial} angles are 90° and F_{axial}–Se–F_{axial} angle is 180°.

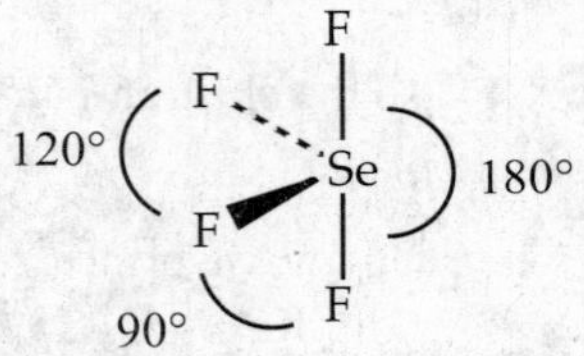

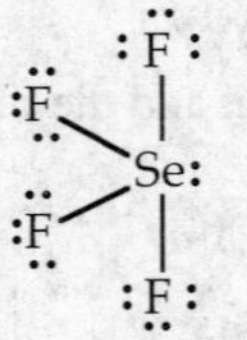

(b) SOF_4 (40 e⁻)

The type is AX_5E_0, so the electron-pair geometry is triangular bipyramidal, equatorial-F–S–O angles are 120° and the axial-F–S–O angles are 90°.

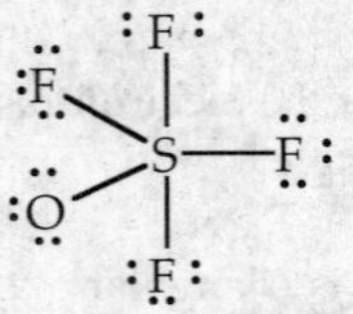

(c) BrF_5 (42 e⁻)

The type is AX_5E_1, so the electron-pair geometry is octahedral and all the F–Br–F angles are 90°. The angles across the bottom of the structure will be 180°.

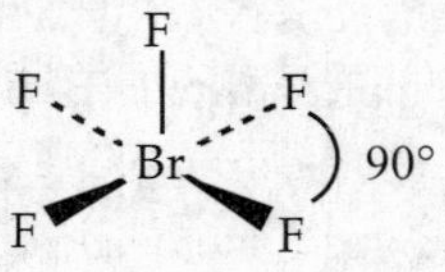

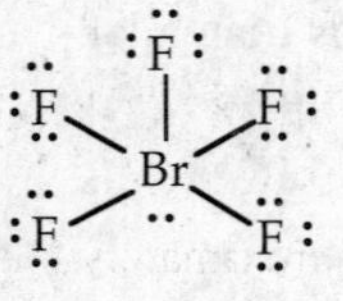

20. *Answer:* **NO_2^+**

Strategy and Explanation: NO_2 molecule and NO_2^+ ion differ by only one electron. NO_2 has 17 electrons and NO_2^+ has 16 electrons. Their Lewis structures are quite similar:

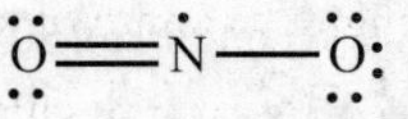

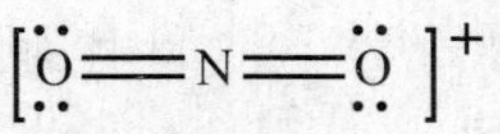

According to VESPR, NO_2 is triangular planar (AX_2E_1) and NO_2^+ is linear (AX_2). The predict O–N–O angle in NO_2 is approximately 120°. The predict O–N–O angle in NO_2^+ ion is 180°. So, **NO_2^+** has a greater O–N–O angle.

Hybridization

22. *Answer:* **(a)** sp^3 **(b)** sp^2

    Strategy and Explanation:

    (a) One s and three p orbitals combine to make sp^3-hybrid orbitals.

    (b) One s and two p orbitals combine to make sp^2-hybrid orbitals.

24. *Answer:* **tetrahedral, sp^3-hybridization**

    Strategy and Explanation: Write a Lewis structure for $HCCl_3$ and use VSEPR to determine the molecular geometry as described in the solution to Question 8. Use Table 9.2 to determine the hybridization of the central atom using the electron-pair geometry. The molecule is AX_4E_0 type, so its electron-pair geometry and its molecular geometry are tetrahedral.

    To make four equal bonds with an electron-pair geometry of tetrahedral, the C atom must be sp^3-hybridized, according to Table 9.2. The H atom and Cl atoms are not hybridized.

28. *Answer:* **N: sp^3, 109.5°;   first two C's (from the left): sp^3, 109.5°;   third C: sp^2, 120°;   top O: sp^2, 120°; right O: sp^3, 109.5°**

    Strategy and Explanation: Use the Lewis structure of alanine and the VSEPR model to determine the electron-pair geometry of each of the atoms. Use Table 9.2 to determine the hybridization of the central atom using the electron-pair geometry. The bond angles are associated with the electron-pair geometry also, so use Figure 9.4. For reference, the atoms present in multiple quantity are numbered here left to right: C_1, C_2, C_3, O_1, and O_2.

    - Look at C_1: The type is AX_4E_0, so the electron-pair geometry is tetrahedral and this C atom's hybridization is sp^3. The sp^3-hybridized C atom has tetrahedral bond angles of approximately 109.5°.

    - Look at C_2: The type is AX_4E_0, so the electron-pair geometry is tetrahedral and this C atom's hybridization is sp^3. The sp^3-hybridized C atom has tetrahedral bond angles of approximately 109.5°.

    - Look at the N atom: The type is AX_3E_1, so the electron-pair geometry is tetrahedral and its hybridization is

    - Look at C_3: The type is AX_3E_0, so the electron-pair geometry is triangular planar and this C atom's hybridization is sp^2. The sp^2-hybridized C atom has triangular planar bond angles of approximately 120°.

    - Look at the N atom: The type is AX_3E_1, so the electron-pair geometry is tetrahedral and its hybridization is sp^3. The sp^3-hybridized N atom has tetrahedral bond angles of approximately 109.5°.

    - Look at O_1: The type is AX_1E_2, so the electron-pair geometry is triangular planar and this C atom's hybridization is sp^2. The sp^2-hybridized C atom has triangular planar bond angles of approximately 120°.

    - Look at O_2: The type is AX_2E_2, so the electron-pair geometry is tetrahedral and this O atom's hybridization is sp^3. The sp^3-hybridized O atom has tetrahedral bond angles of approximately 109.5°.

30. *Answer:* **(a) first two C's (from the left):** sp^3, **109.5°;   third and fourth C's (with triple bond):** sp, **180°**
    (b) shortest: C≡C   (c) strongest: C≡C

Strategy and Explanation:

(a) Write the Lewis structure. Use the VSEPR model to determine the electron-pair geometry of each of the atoms. Use Table 9.2 to determine the hybridization of the central atom using the electron-pair geometry. The bond angles are associated with the electron-pair geometry also, so use Figure 9.4.

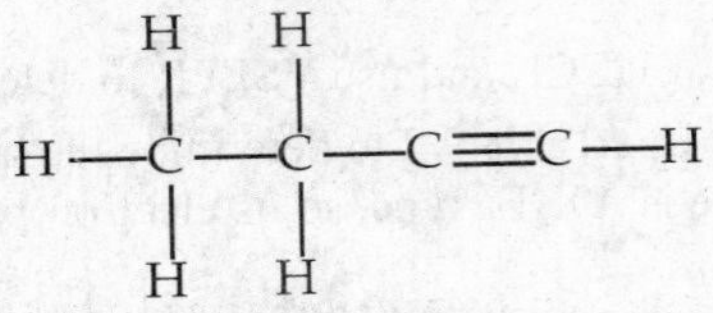

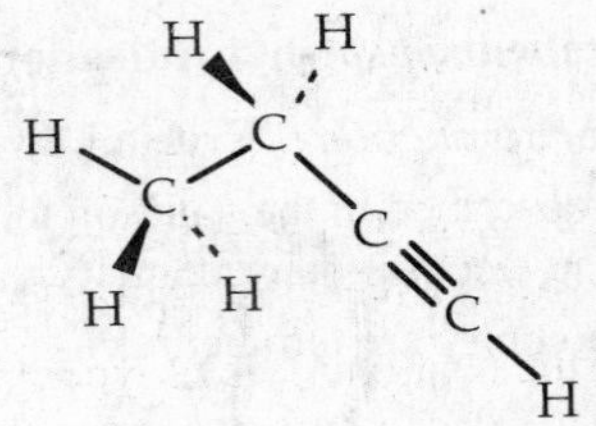

Look at the first two C atoms (the two farthest to the left): They are both of type AX_4E_0, so the electron-pair geometry is tetrahedral and these C atoms have sp^3-hybridization. The sp^3-hybridized C atoms have tetrahedral bond angles of approximately 109.5°.

Look at the second two C atoms (the two farthest to the right): They are both of type AX_2E_0, so the electron-pair geometry is linear and these C atoms have sp-hybridization. The sp-hybridized C atoms have linear bond angles of approximately 180°.

(b) The shortest carbon-carbon bond is the triple bond. That can be confirmed by looking up the bond lengths in Table 8.1.

(c) The strongest carbon-carbon bond is the triple bond. That can be confirmed by looking up the average bond enthalpies in Table 8.2.

32. *Answer:* **see structures with σ and π bonds designated below**

Strategy and Explanation: The first bond between two atoms must always be a σ bond. When more than one pair of electrons are shared between atoms, they are always part of π bonds. So the second bond in a double bond and the second and third bonds in a triple bond are always π bonds.

(a)

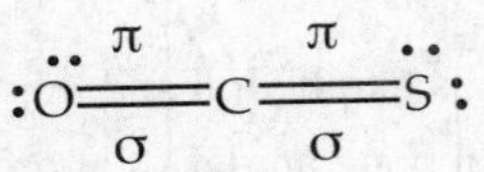

(b)

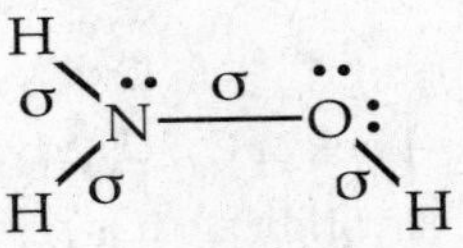

(c)

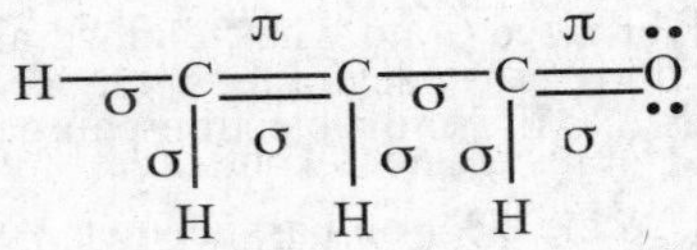

(d)

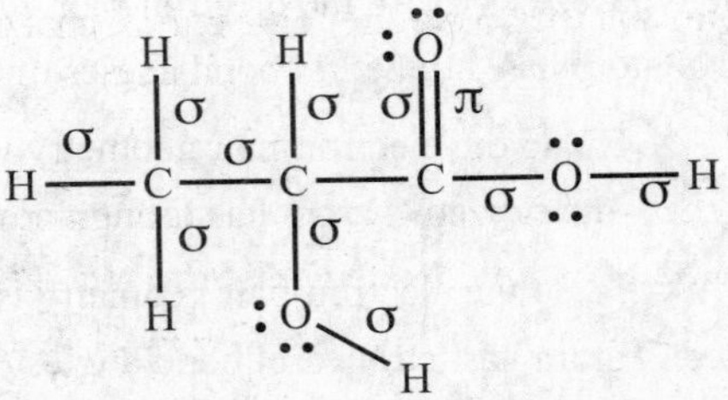

34. *Answer:* **see structure below for location of sigma and pi bonds; (a) 6 sigma bonds   (b) 3 pi bonds**
(c) *sp*-**hybridization   (d)** *sp*-**hybridization   (e) both have** *sp²*-**hybridization**

Strategy and Explanation: Use the method described in the solution to Question 32.

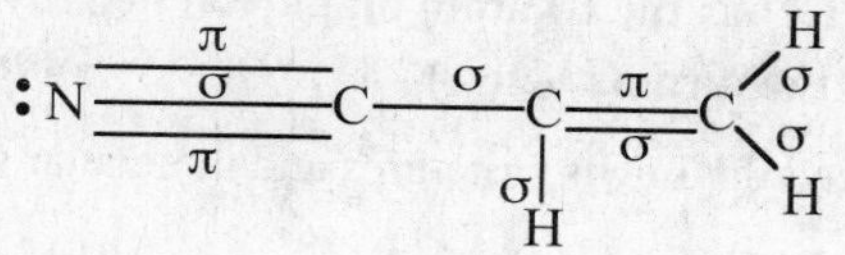

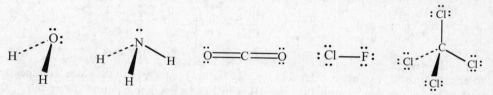

(a) As seen above, the structure has six sigma bonds.

(b) As seen above, the structure has three pi bonds.

(c) Look at the C atom bonded to the N atom: It is of type AX_2E_0, so the electron-pair geometry is linear and these C atoms have *sp*-hybridization.

(d) Look at the N atom: It is of type AX_1E_1, so the electron-pair geometry is linear and it has *sp*-hybridization.

(e) Both H-bearing C atoms are type AX_3E_0, so the electron-pair geometry is triangular planar and both of them have a hybridization of *sp²*.

Molecular Polarity

36. *Answer:* **(a) H_2O (b) CO_2 and CCl_4 are not polar   (c) F**

Strategy and Explanation: Find the Lewis structures and molecular shapes of H_2O, NH_3, CS_2, ClF, and CCl_4.

(a) The bond polarity is related to the difference in electronegativity. Use Figure 8.6 to get electronegativity (EN) values:

$$EN_O = 3.5, EN_H = 2.1, EN_N = 3.0, EN_C = 2.5, EN_{Cl} = 3.0$$

$$\Delta EN_{O-H} = EN_O - EN_H = 3.5 - 2.1 = 1.4 \qquad \leftarrow \text{O–H has largest } \Delta EN.$$

$$\Delta EN_{N-H} = EN_N - EN_H = 3.0 - 2.1 = 0.9$$

$$\Delta EN_{C-O} = EN_O - EN_C = 3.5 - 2.5 = 1.0$$

$$\Delta EN_{Cl-F} = EN_F - EN_{Cl} = 3.5 - 3.0 = 0.5$$

$$\Delta EN_{C-Cl} = EN_{Cl} - EN_C = 3.0 - 2.5 = 0.5$$

The most polar bonds are in H_2O.

(b) Use the description given in Section 9.5 to determine if the molecule is polar:

H_2O is polar, since the terminal atoms are not symmetrically arranged around the O atom.

NH_3 is polar, since the terminal atoms are not symmetrically arranged around the N atom.

CO_2 is not polar, since the two terminal atoms are all the same, they are symmetrically arranged (in a l shape) around the C atom, and they have the same partial charge.

$$C \!-\! O$$
$$\delta^+ \quad \delta^-$$

CCl_4 is not polar, since the four terminal atoms are all the same, they are symmetrically arranged (in a tetrahedral shape) around the C atom, and they all have the same partial charge.

$$C \!-\! Cl$$
$$\delta^+ \quad \delta^-$$

So, CO_2 and CCl_4 are the molecules in the list that are not polar.

(c) The more negatively charged atom of a pair of bonded atoms is the atom with the largest electronegativity. In ClF, the F atom is more negatively charged.

38. *Answer:* **Molecules (b) and (c) are polar; HBF$_2$ has the F atoms on partial negative end and the H atom on the partial positive end; CH$_3$Cl has the Cl atom on partial negative end and the H atoms on the partial positive end. (see dipole moment diagrams below)**

Strategy and Explanation: We need the Lewis structures and molecular shapes of the molecules:

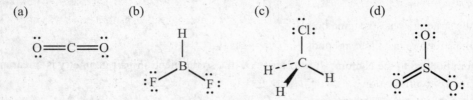

The molecules that are polar have asymmetrical atom arrangements; here, they are (b) HBF$_2$ and (c) CH$_3$Cl. The others, (a) CO$_2$ and (d) SO$_3$, have symmetrical arrangements of identical atoms with the same partial charge. (Notice: Three equivalent resonance structures can be written for SO$_3$, making each S–O bond the same length and strength.)

The bond polarities related to the difference in electronegativity. Use Figure 8.6 to get electronegativity (EN) values: $EN_B = 2.0$, $EN_H = 2.1$, $EN_F = 4.0$, $EN_C = 2.5$, $EN_{Cl} = 3.0$

$$\Delta EN_{B-H} = EN_H - EN_B = 2.1 - 2.0 = 0.1$$

$$\Delta EN_{B-F} = EN_F - EN_B = 4.0 - 2.0 = 2.0$$

$$\Delta EN_{C-H} = EN_C - EN_H = 2.5 - 2.1 = 0.4$$

$$\Delta EN_{C-Cl} = EN_{Cl} - EN_C = 3.0 - 2.5 = 0.5$$

The bond pole arrows' lengths are related to their ΔEN, and points toward the atom with the more negative EN. So, the B–F arrows are much longer than the HB arrow, and a net dipole points toward the F atoms' side of the molecule, making the F atoms' side of the molecule the partial negative end and the H atom's side of the molecule the partial positive end.

The C–Cl arrow points toward Cl, and the C–H points toward the C, so all the arrows point toward the C. (Left right, and back forward cancel due to the symmetry of the triangular orientation of the H atoms). That means a net dipole points toward the Cl atom's side of the molecule, making the Cl atom's side of the molecule the partial negative end and the H atoms' side of the molecule the partial positive end.

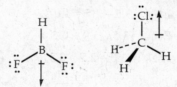

40. *Answer:* **(a) The Br–F bond has a larger electronegativity difference. (b) The H–O bond has a larger electronegativity difference.**

Strategy and Explanation: The dipole moment relates both to the strength of the bond poles and their directionality.

The bond polarities are related to the difference in electronegativity. Use Figure 8.6 to get electronegativity (EN) values: $EN_{Br} = 2.8$, $EN_F = 4.0$, $EN_{Cl} = 3.0$

$$\Delta EN_{Br-F} = EN_F - EN_{Br} = 4.0 - 2.8 = 1.2$$

$$\Delta EN_{Cl-F} = EN_F - EN_{Cl} = 4.0 - 3.5 = 0.5$$

The bond polarity of the bond in BrF is much greater than that in ClF, which explains the significant dipole moment difference. $EN_H = 2.1$, $EN_O = 3.5$, $EN_S = 2.5$

$$\Delta EN_{O-H} = EN_O - EN_H = 3.5 - 2.1 = 1.4$$

$$\Delta EN_{S-H} = EN_S - EN_H = 2.5 - 2.1 = 0.4$$

The bond polarity of the O–H bonds is greater than that of the S–H bonds, which explains the dipole moment difference.

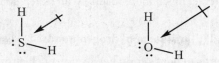

42. *Answer:* **Both molecules have dipole moments. (a) NH$_2$OH has the O atom on partial negative end and the H atoms on the partial positive end. (b) S$_2$F$_2$ has the F atoms on partial negative end and the S atoms on the partial positive end.**

Strategy and Explanation: Follow the method described in the solution to Question 38 - 40.

(a) 14 e$^-$. The Lewis structure looks like this:

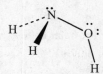

$$\Delta EN_{O-H} = EN_O - EN_H = 3.5 - 2.1 = 1.4$$

$$\Delta EN_{N-H} = EN_N - EN_H = 3.0 - 2.1 = 0.9$$

$$\Delta EN_{N-O} = EN_O - EN_N = 3.5 - 3.0 = 0.5$$

(b) 20 e$^-$. The Lewis structure looks like this:

$$\Delta EN_{Cl-S} = EN_{Cl} - EN_S = 3.5 - 2.5 = 1.0$$

Noncovalent Interactions

44. *Answer/Explanation:*

Interaction	Distance	Example
ion-ion	longest range	Na$^+$ interaction with Cl$^-$
ion-dipole	long range	Na$^+$ ions in H$_2$O
dipole-dipole	medium range	H$_2$O interaction with H$_2$O
dipole-induced dipole	short range	H$_2$O interaction with Br$_2$
induced dipole-induced dipole	shortest range	Br$_2$ interaction with Br$_2$

46. *Answer:* **Wax is nonpolar. Water droplets will form so that the polar water molecules can avoid interaction with the wax. Dirty, unwaxed cars have surface soils and salts that interact well with water.**

Strategy and Explanation: Wax is made up of nonpolar molecules that interact almost exclusively by London forces. The water molecules are highly polar and would have to give up hydrogen bonds between other water molecules to interact with wax using only much weaker London forces. These two substances would not interact very well. The water "beads up" in an attempt to have the smallest possible necessary surface interaction with the wax. On a dirty, unwaxed car, the soils and salts that are often found on cars contain ions and polar compounds. They interact much more readily with water, so the water would not "bead up" on the dirty surface.

48. *Answer:* **Molecules (c), (d) and (e) will form hydrogen bonds.**

Strategy and Explanation: Hydrogen bonds form between very electronegative atoms in one molecule to H atoms bonded to a very electronegative atom (EN ≥ 3.0) in another molecule. If H atoms are present in a molecule but they are bonded to lower electronegativity atoms, such as C atoms, the molecule cannot use those H atoms for hydrogen bonding.

(a) The H atoms are bonded to C atoms and the highest electronegativity atom in the molecule is Br (EN = 2.8), so this molecule cannot form hydrogen bonds.

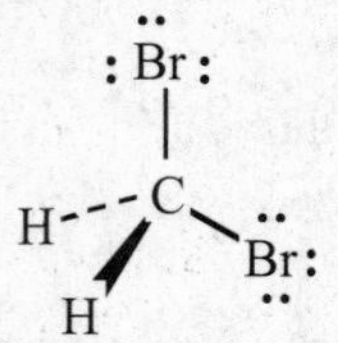

(b) The H atoms are bonded to C atoms, so this molecule cannot form hydrogen bonds with other molecules of the same compound. It does have a high electronegativity O atom (EN = 3.5), so this molecule could interact with other molecules, such as H_2O, which can provide the H atoms for hydrogen bonding to the O atom on this molecule.

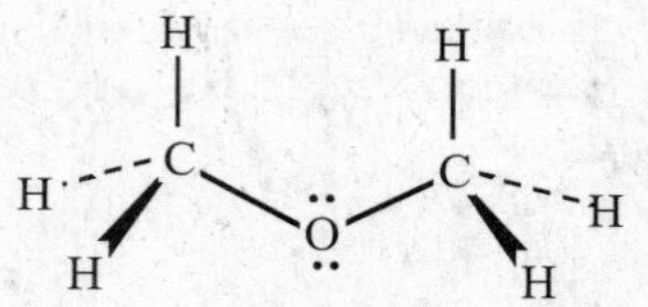

(c) Three H atoms are bonded to high electronegativity atoms (EN_O = 3.5 and EN_N = 3.0). Those H atoms (circled in the structure below) can form hydrogen bonds with the N and O atoms in neighboring molecules.

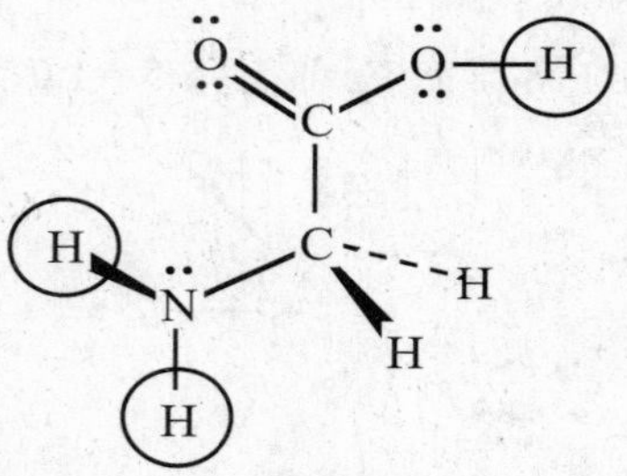

(d) Two H atoms are bonded to high electronegativity O atoms (EN = 3.5). Those H atoms (circled in the structure below) can form hydrogen bonds with the O atoms in neighboring molecules.

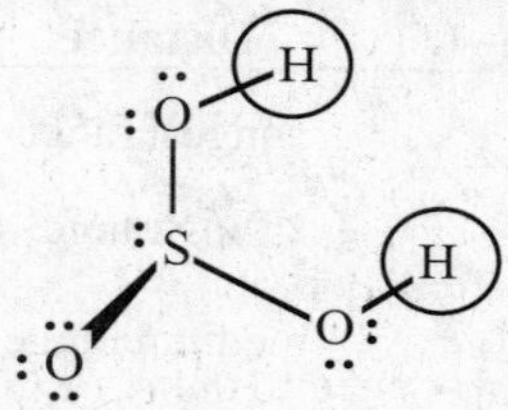

(e) One H atom is bonded to a high electronegativity O atom (EN = 3.5). That H atom (circled in the structure below) can form hydrogen bonds with the O atoms in neighboring molecules.

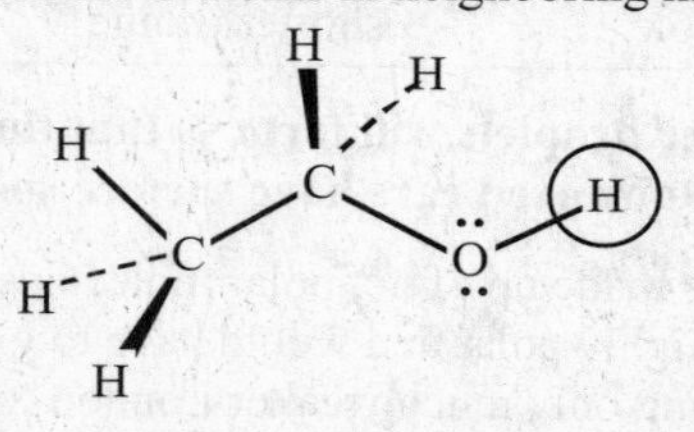

49. *Answer:* **Vitamin C is capable of forming hydrogen bonds with water, thus vitamin is miscible in H_2O.**

Strategy and Explanation: Four H atoms are bonded to high electronegativity O atoms (EN = 3.5). Those H atoms (circled in the structure below) can form hydrogen bonds with H_2O molecules.

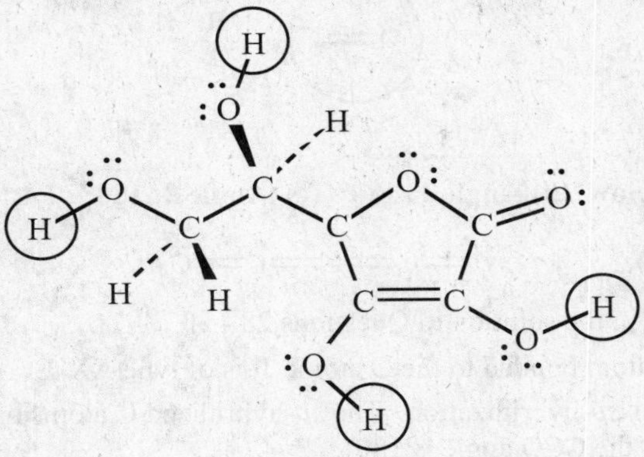

This molecule is quite polar and would not interact well with fats, because fats interact primarily using weaker London forces.

51. *Answer:* **(a) London forces (b) London forces (c) Covalent bonds (d) Dipole-dipole forces**

Strategy and Explanation:

(a) Hydrocarbons have no capability of forming hydrogen bonds and the bonds are close to non-polar. The molecules interact primarily with London forces. The London forces between the molecules must be overcome to sublime $C_{10}H_8$.

(b) As described in (a), propane molecules, a hydrocarbon, would need to overcome London forces to melt.

(c) Decomposing molecules of nitrogen and oxygen require breaking covalent bonds, so the forces that must be overcome are the intramolecular (covalent) forces.

(d) To evaporate polar PCl_3 requires that the molecules overcome dipole-dipole forces.

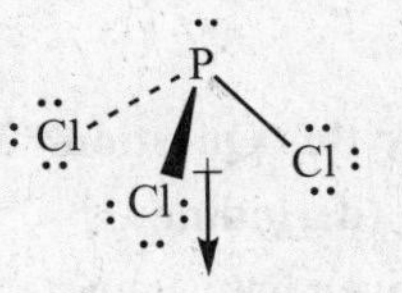

General Questions

54. *Answer:* **(a) Angle 1: 180° Angle 2: 109.5° Angle 3: 109.5° (b) C=O (c) C=O**

Strategy and Explanation:

(a) Follow the method described in the solution to Questions 28 - 30.

Angle 1: Look at the C atom bonded to the N atom: It is of type AX_2E_0, so the electron-pair geometry is linear and this C atom has *sp*-hybridization. The *sp*-hybridized C atoms have linear bond angles of approximately 180°. So, the CCN angle is 180°.

Angles 2 and 3: Look at the C atom bonded to only one O atom: The type is AX_4E_0, so the electron-pair geometry is tetrahedral this C atom has *sp³*-hybridization. The *sp³*-hybridized C atom has tetrahedral bond angles of approximately 109.5°. So, the HCH angle (Angle 2) and the OCH angle (Angle 3) are both 109.5°.

(b) Follow the method described in the solution to Question 36. The most polar bond is the C=O bond.

(c) Multiple bonds are shorter than single bonds, so the C=O bond is the shortest CO bond:

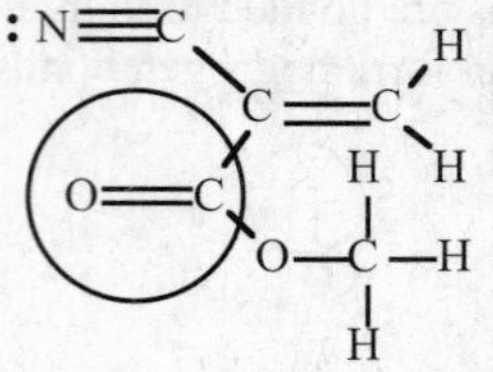

56. *Answer:* **(a) see structure below (b) Angle 1: 180° (c) Angle 2: 180°**

Strategy and Explanation: (a)   :Ö══C══C══C══Ö:

Follow the method described in the solution to Questions 28 - 30.

(b) Angle 1: Look at the C atom bonded to the O atom: It is of type AX_2E_0, so the electron-pair geometry is linear and this C atom has *sp*-hybridization. The *sp*-hybridized C atoms have linear bond angles of approximately 180°. So, the CCO angle is 180°.

(c) Angle 2: Look at the C atom bonded only to the C atoms: It is of type AX_2E_0, so the electron-pair geometry is linear and this C atom has *sp*-hybridization. The *sp*-hybridized C atoms have linear bond angles of approximately 180°. So, the CCO angle is 180°.

61. *Answer:* **If the polarity of the bonds exactly cancel each other, a molecule will be nonpolar.**

Strategy and Explanation:

Bonds are polar, if the atom's EN's are different. If equal bond poles point in opposite or symmetrically balanced directions such that they cancel each other to give a zero dipole moment for the molecule, the molecule will be nonpolar.

A molecule with polar bonds can have a dipole moment of zero if the bond poles are equal and point in opposite directions. CO_2 is a good example of a nonpolar molecule with polar bonds.

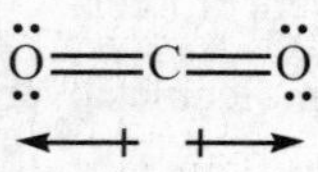

Applying Concepts

65. *Answer:* **Many correct answers exist for this Question. These are examples:**

(a) nitrogen (b) boron (c) phosphorus (d) iodine

Strategy and Explanation: The answers given here are not the only correct answers. They are only examples.

(a) XH_3 would have a central atom with one lone pair of electrons, if X = N, nitrogen.

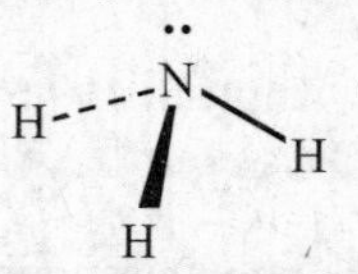

(b) XCl_3 would have no lone pairs on the central atom, if X = B, boron.

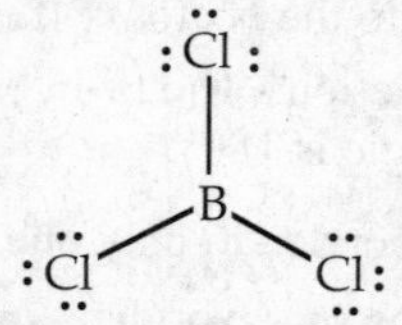

(c) XF_5 would have no lone pairs on the central atom, if X = P, phosphorus.

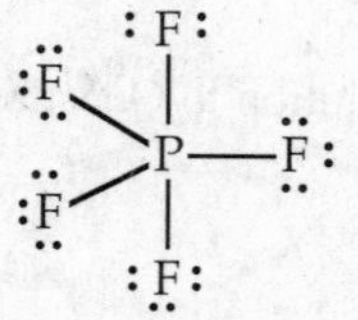

(d) XCl_3 would have two lone pairs on the central atom, if X = I, iodine.

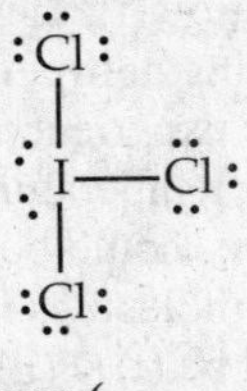

67. *Answer:* **Five; see diagram below with dashed lines (_ _ _ _ _) showing intermolecular interactions.**

Strategy and Explanation: Five water molecules could hydrogen bond to an acetic acid molecule: one to each of the four lone pairs on O atoms in acetic acid and one to the H atom bonded to an O atom in acetic acid.

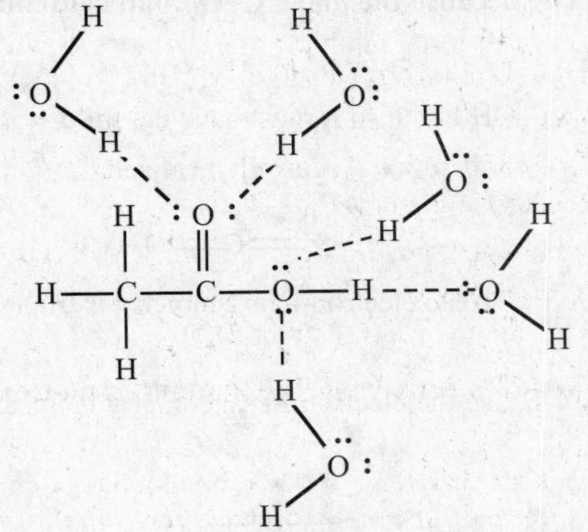

69. *Answer:* **Diagram (d)**

Strategy and Explanation:

(a) This is incorrect, because the H atoms are not bonded to highly electronegative atoms.

(b) This is incorrect. The H atoms are both bonded to O atom (EN = 3.5), but two partial-positive H atoms will not form a hydrogen bond to each other!

(c) This is incorrect. The H atom is bonded F. This is a covalent bond, not an example of the noncovalent interactive force called hydrogen bonding.

(d) This is correct. The H atom is bonded to an O atom (EN = 3.5) and is hydrogen bonding to another O atom.

The only answer that is correct is (d).

More Challenging Questions

71. *Answer:* **(a) See diagrams below   (b) (i) < (ii) < (iii)**

Strategy and Explanation:

(a) The Lewis structures for the three isomers of $C_6H_6O_2$ (each with one benzene ring and two –OH groups attached) look like this:

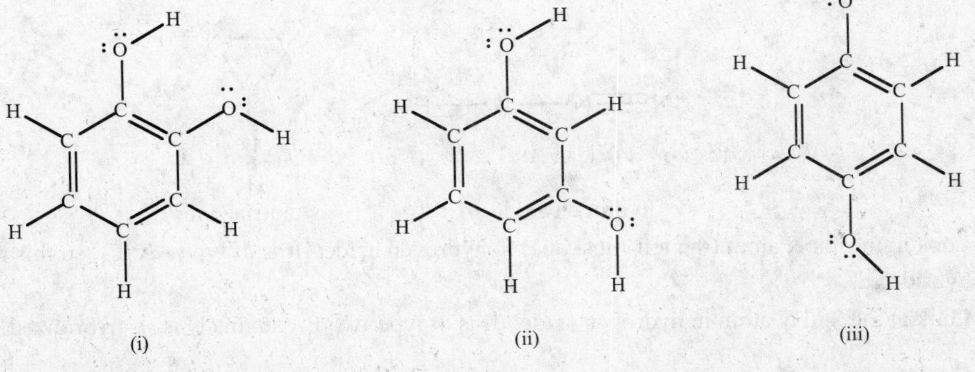

(b) All of these molecules have the same kind and strength of hydrogen bonding; however, because of the proximity of the two –OH groups in structure (i), these molecules are able to experience intramolecular hydrogen bonding (that is hydrogen bonding between H atoms and O atoms within the same molecule). Less hydrogen bonding disruption will be experienced when the molecules undergo a transition to the liquid state; hence, (i) will melt at a lower temperature than the other two. The asymmetry of structure (ii) might suggest that the intermolecular forces could be a little bit weaker in the solid, compared to those experienced among the higher-symmetry molecules in structure (iii). Therefore, the predicted order of for melting points of these three solids would be (i) < (ii) < (iii).

73. *Answer:* **(a) see structure below (b) First C: triangular planar, H–C–C and H–C–H bond angles are 120°; Second C: linear, C–C–O bond angle is 180° (c) First C: *sp²*-hybridized; Second C: *sp*-hybridized, O atom: *sp²*-hybridized (d) polar, because the polar C=O bond contributes to a nonzero dipole moment.**

Strategy and Explanation:

(a) Ketene, C_2H_2O (16 e⁻) has no –OH bond, so it must have the following Lewis structure.

$$H-\overset{\overset{\textstyle H}{|}}{C}=C=\ddot{O}:$$

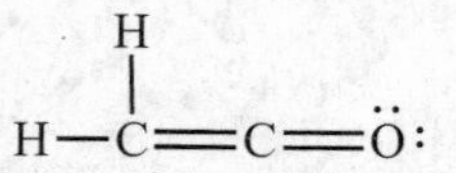

(b) 1st C atom: It is of type AX_3E_0, so the electron-pair geometry is triangular planar with H–C–C and H–C–H bond angles of 120°.

2nd C atom: It is of type AX_2E_0, so the electron-pair geometry is linear with a C–C–O bond angle of 180°.

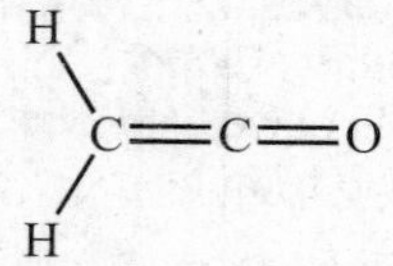

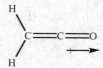

(c) Look at the 1st C atom: It is of type AX_3E_0, so the C atom must be *sp²*-hybridized.

Look at the 2nd C atom: It is of type AX_2E_0, so the C atom must be *sp*-hybridized.

Look at the O atom: It is of type AX_1E_2, so the O atom must be *sp²*-hybridized.

(d) The C=O bond in this asymmetric molecule is a significantly polar bond which makes the molecule polar.

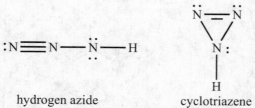

75. *Answer:* **(a) see diagrams below (b) *sp*; *sp*; *sp²* (c) *sp²*, *sp²*, *sp³* (d) hydrogen azide: 3 σ bonds, cyclotriazene: 4 σ bonds (e) hydrogen azide: 2 π bonds; cyclotriazene: 1 π bond (f) hydrogen azide: 180°; cyclotriazene: 60°.**

Strategy and Explanation:

(a) The two isomers of N_3H (16 e⁻) have the following Lewis structures.

$$:N\equiv N-\ddot{N}-H$$

hydrogen azide   cyclotriazene

(b) Look at the 1st N atom (the left-most one) in hydrogen azide: It is of type AX_1E_1, so this N must be *sp*-hybridized.

Look at the 2nd N atom in hydrogen azide: It is of type AX_2E_0, so this N is *sp*-hybridized.

Look at the 3rd N atom in hydrogen azide: It is of type AX_2E_2, so we'd think that the N is sp^3-hybridized. However, to accommodate the other resonance structure (shown in (c) with two double bonds between the N atoms), one p-orbital must remain unhybridized, so the N is sp^2-hybridized

(c) Look at the first two N atom (the two on top of the triangle) in cyclotriazene: Each N is of type AX_2E_1, so each N atom must be sp^2-hybridized.

Look at the 3rd N atom in cyclotriazene: It is of type AX_3E_1, so this N is sp^3-hybridized.

For answering Questions (d) and (e), recall that every single bond is a sigma-bond; every double bond is one sigma bond and one pi bond. So the sigma and pi bonds in this molecule are shown here:

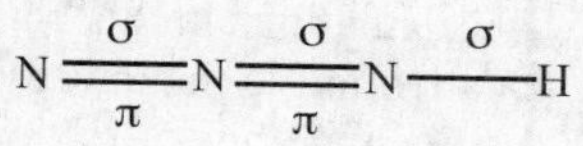

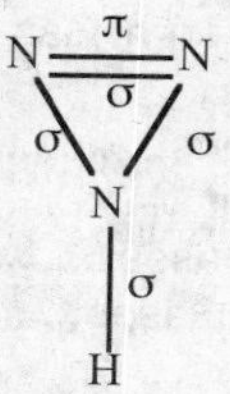

(d) In hydrogen azide, there are three sigma bonds. In cyclotriazene, there are four sigma bonds.

(e) In hydrogen azide, there are two pi bonds. In cyclotriazene, there is one pi bond.

(f) In hydrogen azide, the second N dictates the N–N–N bond angle. In (b), we ascertained that this N atom is sp-hybridized, so the bond angle must be 180°.

In cyclotriazene, the methods of this chapter indicate that the bond angles should be 120° (for the two angles represented by [left N-right N-bottom N], or [right N-left N- bottom N]) and 109.5° (for the angle represented by [left N-bottom N-right N]); however, since the three N atoms form a triangle, geometric constraints dictate that the bond angles will be approximately 60°.

77. *Answer:* **(a) Angle 1: 120°, Angle 2: 120°, Angle 3: 109.5° (b) sp^3 (c) single-bonded O: sp^3, double-bonded O: sp^2**

Strategy and Explanation:

(a) Angle 1: Look at the C atom: It is of type AX_3E_0, so it must have triangular planar electron geometry, and the bond angle is approximately 120°.

Angle 2: Look at the second of the three carbon atoms: It is of type AX_3E_0, so it must have triangular planar electron geometry, and the bond angle is approximately 120°.

Angle 3: Look at the C atom: It is of type AX_4E_0, so it must have tetrahedral electron geometry, and the bond angle is approximately 109.5°.

(b) The N atom is of type AX_3E_1, so it must be sp^3-hybridized.

(c) Look at the O atoms with two single bonds: Each of them are of type AX_2E_2, so they must each be sp^3-hybridized.

Look at the double bonded O atom: It is of type AX_1E_2, so it must be sp^2-hybridized.

Chapter 10: Gases and the Atmosphere

Introduction

Several named laws (Avogadro's, Boyles, Charles's, and Dalton's laws) describe the basis properties of gases. The *relationships* among pressure, volume, temperature, and amount of gas are more important than the name of the person who discovered them.

It's easy to make mistakes in even the easiest calculations in this chapter, unless you pay close attention to the units. Densities of gases are typically given in units of g/L, whereas, you may remember that densities of solids and liquids are often given in units of g/mL. The units of temperature are commonly given in °C. Gas law calculations must have all temperatures converted to Kelvin. The value of the ideal gas law constant, R, is different with different units, so make sure you are using the correct value of R.

Solutions to Blue-Numbered Questions for Review and Thought for Chapter 10

Topical Questions

The Atmosphere

8. *Answer/Explanation:* Convert values in Table 10.1 to parts per million (ppm) and parts per billion (ppb). To accomplish these conversions, we need to describe a relationship between percent and ppm and ppb:

$$\text{Percent gas in air} = \frac{\text{L of gas}}{100 \text{ L of air}} \qquad \text{Parts per million gas in air} = \frac{\text{L of gas}}{1,000,000 \text{ L of air}}$$

So, 1% = 10,000 ppm. Multiply the number in units of percent by 10,000 to get ppm.

$$\text{Parts per billion gas in air} = \frac{\text{L of gas}}{1,000,000,000 \text{ L of air}}$$

So, 1ppm = 1,000 ppb. Multiply the number in units of ppm by 1,000 to get ppm.

Molecule	ppm	ppb	
N_2	780,840	780,840,000	↑
O_2	209,480	209,480,000	
Ar	9,340	9,340,000	
CO_2	330	330,000	
Ne	18.2	18,200	> 1 ppm
H_2	10.	10,000	
He	5.2	5,200	
CH_4	2	2,000	↓
Kr	1	1,000	↑
CO	0.1	100	between
Xe	0.08	80	1ppm
O_3	0.02	20	and
NH_3	0.01	10	1ppb
NO_2	0.001	1	↓
SO_2	0.0002	0.2	< 1ppb

✓ *Reasonable Answer Check:* The numbers are different only by the appropriate factor of 10.

10. *Answer:* **1.5×10^8 metric tons added, 2×10^6 metric tons total**

Strategy and Explanation: Given the mass of a sample of coal, the percentage of sulfur in the coal, the weight fraction of SO_2 in the atmosphere, and (given in Question 9) the mass of the atmosphere, determine the mass of SO_2 added to the atmosphere, and the total amount of SO_2 in the atmosphere.

To answer the first Question, use metric conversions and the mass fraction as a conversion factor to determine the mass of sulfur. Use mole and molar mass conversion factors to determine the mass of SO_2, assuming that all the sulfur in the coal is converted to SO_2 and released into the atmosphere. Convert the mass back to metric tons. To answer the second Question, use the mass of the atmosphere (given in Question 9) and the mass fraction of SO_2 in terms of metric tons as a conversion factor to determine the total mass of SO_2.

First find the mass of S:

$$3.1 \times 10^9 \text{ metric tons coal} \times \frac{1000 \text{ kg coal}}{1 \text{ metric ton coal}} \times \frac{2.5 \text{ kg S}}{100 \text{ kg coal}} \times \frac{1000 \text{ g S}}{1 \text{ kg S}} = 7.8 \times 10^{13} \text{ g S}$$

Then, find the mass of SO_2:

$$7.8 \times 10^{13} \text{ g S} \times \frac{1 \text{ mol S}}{32.066 \text{ g S}} \times \frac{1 \text{ mol } SO_2}{1 \text{ mol S}} \times \frac{64.0638 \text{ g } SO_2}{1 \text{ mol } SO_2} = 1.5 \times 10^{14} \text{ g } SO_2$$

Then, convert the mass of SO_2 back to metric tons:

$$1.5 \times 10^{14} \text{ g } SO_2 \times \frac{1 \text{ kg } SO_2}{1000 \text{ g } SO_2} \times \frac{1 \text{ metric ton } SO_2}{1000 \text{ kg } SO_2}$$

$$= 1.5 \times 10^8 \text{ metric tons } SO_2 \text{ was added to the atmosphere in 1980}$$

Get the total mass of SO_2 currently in the atmosphere

$$5.3 \times 10^{15} \text{ metric tons air} \times \frac{0.4 \text{ metric tons } SO_2}{1,000,000,000 \text{ metric tons air}} = 2 \times 10^6 \text{ metric tons } SO_2$$

✓ *Reasonable Answer Check:* It is clear that some of the SO_2 presumably released into the atmosphere in 1980 is no longer there, since the total mass of SO_2 is less than what was introduced that year. SO_2 is a reactive gas, getting oxidized to SO_3 in the presence of air and then producing sulfuric acid when reacting with rainwater. This removes the sulfur from the air.

Properties of Gases

11. *Answer:* **(a) 0.947 atm (b) 950. mm Hg (c) 542 torr (d) 98.7 kPa (e) 6.91 atm**

Strategy and Explanation: Convert a series of pressure quantities into other pressure units.

Use Table 10.2 to design conversion factors to achieve the conversions.

(a)
$$720. \text{ mmHg} \times \frac{1 \text{ atm}}{760. \text{ mmHg}} = 0.947 \text{ atm}$$

(b)
$$1.25 \text{ atm} \times \frac{760 \text{ mmHg}}{1 \text{ atm}} = 950. \text{ mmHg}$$

(c)
$$542 \text{ mmHg} \times \frac{760 \text{ torr}}{760 \text{ mmHg}} = 542 \text{ torr}$$

(d)
$$740. \text{ mmHg} \times \frac{1 \text{ atm}}{760. \text{ mmHg}} \times \frac{101.325 \text{ kPa}}{1 \text{ atm}} = 98.7 \text{ kPa}$$

(e)
$$700. \text{ kPa} \times \frac{1 \text{ atm}}{101.325 \text{ kPa}} = 6.91 \text{ atm}$$

✓ *Reasonable Answer Check:* The unit of atm represents a lot more pressure than the units of kPa, torr, and mm Hg, so it makes sense that the numbers of atmospheres are always much smaller than the pressure expressed in these other units. The unit of kPa represents more pressure than the units of torr and mm Hg, so it makes sense that the numbers of kPa are always smaller than the pressure expressed in the other units of torr and mm Hg. Torr and mm Hg are the same size, so their quantities should be identical.

13. *Answer:* **14 m**

Strategy and Explanation: Given the density of mercury, the density of an oil used to construct a barometer, and the atmospheric pressure, determine the height in meters of the oil column in the oil barometer.

Use Table 10.2 to convert the pressure into mm Hg. Pressure on the liquid pushes the liquid up the barometer until its mass exerts the same force per unit area as the air pressure: $P_{liquid} = Force/Area = mg^2/Area$. Since g is a constant, the force per unit area of the liquid counteracting the air pressure is proportional to just the mass per unit area. Relate the mass per unit area of mercury to the mass per unit area of the oil for 1 atm pressure. Relate the mass of each liquid to its respective density and volume. Relate the volume of each liquid to the dimensions of its respective barometers, including the height. Relate the height of mercury in the mercury barometer to the height of the oil in the oil barometer.

According to Table 10.2, at 1.0 atm the height of a column of mercury in a mercury barometer is 760 mm. Let m = mass of the liquid, A = cylindrical area of the barometer's column, d = density of the liquid, V = volume of the liquid, and h = the height of the liquid in the barometer. Now, equate the mass per unit area of each barometer and derives an equation relating their heights:

$$m_{Hg}/A_{Hg} = m_{oil}/A_{oil}$$

$$d_{Hg}V_{Hg}/A_{Hg} = d_{oil}V_{oil}/A_{oil}$$

$$V = (Area) \times (height) = Ah$$

$$d_{Hg}A_{Hg}h_{Hg}/A_{Hg} = d_{oil}A_{oil}h_{oil}/A_{oil}$$

$$d_{Hg}h_{Hg} = d_{oil}h_{oil}$$

$$h_{oil} = \frac{d_{Hg}h_{Hg}}{d_{oil}} = \frac{\left(13.596 \text{ g/cm}^3\right)\left(760 \text{ mm}\right)}{\left(0.75 \text{ g/cm}^3\right)} \times \frac{1 \text{ m}}{1000 \text{ mm}} = 14 \text{ m}$$

✓ *Reasonable Answer Check:* A larger mass of oil will be needed to counterbalance the atmospheric pressure because it is less dense than mercury. A higher column of oil makes sense.

Kinetic-Molecular Theory

15. *Answer:* **See I-V below; assumption III and IV become false at high P or low T; assumption I becomes false at high P; assumption II is most nearly correct.**

Strategy and Explanation: The five basic concepts of kinetic-molecular theory are given in Section 10.3:

I.   A gas is composed of molecules whose size is much smaller than the distances between them.

II. Gas molecules move randomly at various speeds and in every possible direction.

III. Except when molecules collide, forces of attraction and repulsion between them are negligible.

IV. When collisions occur, they are elastic.

V.   The average kinetic energy of gas molecules is proportional to the absolute temperature.

A discussion of real gas behavior is provided in Section 10.8. Of these five basic assumptions, the ones that become false at very high pressures or very low temperatures are assumptions III and IV. Slow molecules crowded together are much more likely to interact even when they are not colliding. Collisions may not be elastic under these circumstances, because colliding molecules might stick together due to large enough interactive forces between them. Assumption I may become false at very high pressures, because the molecules are crowded together reducing the distance between them.

The assumption that seems most likely to always be most nearly correct is assumption II. As long as a substance is a gas, it retains the capacity to disperse to uniformly fill any container it is introduced into, though the rate of dispersal may vary.

17. *Answer:* $CH_2Cl_2 < Kr < N_2 < CH_4$

Strategy and Explanation: Given the formulas of various atoms and molecules and their common temperature, put their gases in order of increasing average molecular speed.

Kinetic energy is proportional to temperature. With all of the samples at the same temperature, their average kinetic energies are the same. Kinetic energy is related to mass and velocity. $E_{kin} = \frac{1}{2}mv^2$. Velocity is related to kinetic energy and mass: $v^2 = \frac{2E}{m}$. Therefore, molecules with smaller mass have the faster molecular speed. To rank the molecules with increasing speed, rank them from the largest molar mass to the smallest.

Estimate the molar masses: Kr molar mass 83.8 g/mol, CH_4 molar mass = 16.0 g/mol, N_2 molar mass is 28.0 g/mol, CH_2Cl_2 molar mass = 84.9 g/mol.

slowest speed: $CH_2Cl_2 < Kr < N_2 < CH_4$ fastest speed

✓ *Reasonable Answer Check:* It makes sense that lightweight things go faster.

19. *Answer:* **Ne**

Strategy and Explanation: Given the formulas of various gases introduced into one end of a tube, determine which gas will reach the end of the tube first.

Velocity is related to kinetic energy and mass: $v^2 = \frac{2E}{m}$. Therefore, molecules with smaller mass have the faster molecular speed. The molecule with the fastest speed will travel a fixed distance the quickest. To determine which molecule arrives first, rank them from the largest molar mass to the smallest.

Estimate the molar masses: Ar molar mass = 40 g/mol, Ne molar mass = 20 g/mol, Kr molar mass = 84 g/mol, Xe molar mass = 131 g/mol. Ne will arrive first.

✓ *Reasonable Answer Check:* The lightest weight things go faster, so they will arrive sooner.

Gas Behavior and the Ideal Gas Law

20. *Answer:* 4.2×10^{-5} **mol**

Strategy and Explanation: Given the volume of a sample of air at STP and the volume fraction of CO in ppm, determine the moles of CO.

Use the volume fraction in terms of liters as a conversion factor to determine the liters of CO. Use the molar volume of a gas at STP as a conversion factor to determine the moles of CO.

$$1.0 \text{ L air} \times \frac{950 \text{ L CO}}{1,000,000 \text{ L air}} \times \frac{1 \text{ mol CO}}{22.414 \text{ L CO}} = 4.2 \times 10^{-5} \text{ mol CO}$$

✓ *Reasonable Answer Check:* Air has a very small proportion of CO. This sample has a small amount of CO.

22. *Answer:* **62.5 mm Hg**

Strategy and Explanation: Given the volume and pressure of a sample of gas in one flask and the volume of a flask it is transferred to at the same temperature, determine the new pressure.

Use Boyle's law to relate volume to pressure. $P_1V_1 = P_2V_2$ (unchanging T and n)

$$P_2 = \frac{P_1V_1}{V_2} = \frac{(100. \text{ mm Hg}) \times (125 \text{ mL})}{(200. \text{ mL})} = 62.5 \text{ mm Hg}$$

✓ *Reasonable Answer Check:* The larger volume should have a smaller pressure.

25. *Answer:* **26.5 mL**

Strategy and Explanation: Given the original volume and temperature of a sample of gas in a syringe (presumably at atmospheric pressure) and the new temperature of the sample, determine the new volume (presumably still at atmospheric pressure).

Convert the temperatures to Kelvin. Use Charles' law to relate volume to absolute temperature.

$$\frac{V_1}{T_1} = \frac{V_2}{T_2} \qquad \text{(P and n constant)}$$

$$T_1 = 20. \,°C + 273.15 = 293 \text{ K} \qquad T_2 = 37 \,°C + 273.15 = 310. \text{ K}$$

$$V_2 = V_1 \times \frac{T_2}{T_1} = (25.0 \, \text{mL}) \times \frac{(310. \, \text{K})}{(293 \, \text{K})} = 26.5 \, \text{mL}$$

✓ *Reasonable Answer Check:* Gas at higher temperature should have a larger volume.

27. *Answer:* **– 96 °C**

Strategy and Explanation: Use Charles' law to relate volume to absolute temperature, as described in the solution to Question 25. $T_1 = 80. \, °C + 273.15 = 353 \, K$

$$T_2 = T_1 \times \frac{V_2}{V_1} = (353 \, \text{K}) \times \frac{(1.25 \, \text{L})}{(2.50 \, \text{L})} = 177 \, \text{K}$$

$$177 \, \text{K} - 273.15 = -96 \, °C$$

✓ *Reasonable Answer Check:* Gas at smaller volume should have a lower temperature.

29. *Answer:* **501 mL**

Strategy and Explanation: Given the original pressure, volume, and temperature of gas and the new pressure and volume, determine the new volume exerted by the gas in the tire.

Convert the temperature to Kelvin.

$$T_1 = 22 \, °C + 273.15 = 295 \, K \qquad\qquad T_2 = 42 \, °C + 273.15 = 315 \, K$$

Use the combined gas law to relate volume to pressure and absolute temperature.

$$\frac{P_1 V_1}{T_1} = \frac{P_2 V_2}{T_2} \qquad\qquad \text{(n constant)}$$

$$V_2 = V_1 \times \frac{T_2}{T_1} \times \frac{P_1}{P_2} = (754 \, \text{mL}) \times \frac{(315 \, \text{K})}{(295 \, \text{K})} \times \frac{(165 \, \text{mmHg})}{(265 \, \text{mmHg})} = 501 \, \text{mL}$$

✓ *Reasonable Answer Check:* The temperature fraction: $\frac{(315 \, \text{K})}{(295 \, \text{K})}$ is larger than one, consistent with increasing

the volume due to the increased temperature. The pressure fraction: $\frac{(165 \, \text{mmHg})}{(265 \, \text{mmHg})}$ is smaller than one,

consistent with decreasing the volume due to an increased pressure. Clearly these two effects counteract each other, but this pressure change affects the volume more than the temperature change does.

31. *Answer:* **0.507 atm**

Strategy and Explanation: Given the mass, identity, temperature, and volume of a gas sample, determine the pressure of the sample.

Use the ideal gas law: $PV = nRT \qquad R = 0.08206 \, \dfrac{\text{L} \cdot \text{atm}}{\text{mol} \cdot \text{K}}$

The units of R remind us to determine the moles of gas (using mass and molar mass), to convert the temperature to Kelvin, and to convert the volume to liters.

$$1.55 \, \text{g Xe} \times \frac{1 \, \text{mol Xe}}{131.29 \, \text{g Xe}} = 0.0118 \, \text{mol Xe}$$

$$T = 20. \, °C + 273.15 = 293 \, K$$

$$V = 560. \, \text{mL} \times \frac{1 \, \text{L}}{1000 \, \text{mL}} = 0.560 \, \text{L}$$

$$P = \frac{nRT}{V} = \frac{(0.0118 \, \text{mol}) \times \left(0.08206 \, \dfrac{\text{L} \cdot \text{atm}}{\text{mol} \cdot \text{K}} \right) \times (293 \, \text{K})}{(0.560 \, \text{L})} = 0.507 \, \text{atm}$$

✓ *Reasonable Answer Check:* The small fraction of a mole makes sense with the small mass. The units in the pressure calculation cancel properly to give atm.

33. *Answer:* **Largest number in sample (d); smallest number in sample (c)**

Strategy and Explanation: Given a set of gas samples, determine which has the largest number of molecules and which has the smallest number of molecules.

Some of the samples are at STP, so use the molar volume of a gas at STP to determine the number of molecules (in units of moles).

$$\text{1 mol of \textbf{any} gas occupies 22.414 L.} \quad 1.0 \text{ L} \times \frac{1 \text{ mol gas}}{22.414 \text{ L}} = 0.045 \text{ mol gas}$$

In the other cases, use the ideal gas law as described in the solution to Question 31. Once all the moles are calculated, identify which sample has the most moles and which sample has the least moles.

(a) 0.045 mol H_2 at STP

(b) 0.045 mol N_2 at STP

(c) $T = 27\ °C + 273.15 = 300.\ K$ $P = 760.\ \text{mmHg} \times \dfrac{1 \text{ atm}}{760.\ \text{mmHg}} = 1.00 \text{ atm}$

$$n = \frac{PV}{RT} = \frac{(1.0 \text{ atm}) \times (1.00 \text{ L})}{\left(0.08206\ \dfrac{\text{L} \cdot \text{atm}}{\text{mol} \cdot \text{K}}\right) \times (300.\ \text{K})} = 0.0406 \text{ mol}$$

(d) $T = 0.\ °C + 273.15 = 273\ K$ $P = 800.\ \text{mmHg} \times \dfrac{1 \text{ atm}}{760.\ \text{mmHg}} = 1.05 \text{ atm}$

$$n = \frac{PV}{RT} = \frac{(1.05 \text{ atm}) \times (1.0 \text{ L})}{\left(0.08206\ \dfrac{\text{L} \cdot \text{atm}}{\text{mol} \cdot \text{K}}\right) \times (273\ \text{K})} = 0.047 \text{ mol}$$

Of these samples, sample (d) has the most molecules and sample (c) has the smallest number of molecules.

✓ *Reasonable Answer Check:* To keep a 1.0-L gas sample at standard temperature and still have larger than standard pressure suggests that there must be more molecules hitting the walls than a sample at STP. To have a 1.0-L gas sample at higher than standard temperature and still stay at standard pressure suggests that there must be fewer molecules hitting the walls harder, than a sample at STP.

Quantities of Gases in Chemical Reactions

34. *Answer:* **6.0 L H_2**

Strategy and Explanation: Given the balanced chemical equation for a chemical reaction, the mass of one reactant, and excess other reactant, determine the volume of the product produced at a specified pressure and temperature.

Convert from grams to moles. Use the stoichiometric relationship from the balanced equation to determine moles of product. Use the ideal gas law to determine the pressure of the product.

$$6.5 \text{ g Al} \times \frac{1 \text{ mol Al}}{26.9815 \text{ g Al}} \times \frac{2 \text{ mol } H_2}{2 \text{ mol Al}} = 0.24 \text{ mol } H_2$$

$$T = 22\ °C + 273.15 = 295\ K \qquad P = 742 \text{ mmHg} \times \frac{1 \text{ atm}}{760 \text{ mmHg}} = 0.976 \text{ atm}$$

$$P = \frac{n_{H_2} RT}{V} = \frac{(0.24 \text{ mol } H_2) \times \left(0.08206\ \dfrac{\text{L} \cdot \text{atm}}{\text{mol} \cdot \text{K}}\right) \times (295\ \text{K})}{(0.976 \text{ atm})} = 6.0 \text{ L } H_2$$

✓ *Reasonable Answer Check:* All units cancel properly in the calculation of atmosphere. A quarter mole of gas occupies a volume is a little larger than one quarter of the molar volume of a gas at STP, which make sense since the sample's temperature is larger than the STP temperature.

36. *Answer:* **1.9 L CO_2; approximately half the typical volume of two loaves of bread**

Strategy and Explanation: Given the mass of sucrose, the formula and the product of a reaction, determine the maximum volume of CO_2 produced at STP. Compare that volume with the typical volume of two loaves of bread. Let's use French bread.

Balance the equation. Determine the moles of sucrose from the molar mass, then use the stoichiometric relationships given in the balanced equation to determine the moles of CO_2 produced. Last, use the molar volume of a gas at STP to determine the volume of CO_2. Estimate the total volume of two loaves of French bread assuming they are cylinders. Compare the two volumes.

Balance the equation:

$$C_{12}H_{22}O_{11}(s) + 12\ O_2(g) \longrightarrow 12\ CO_2\ (g) + 11\ H_2O\ (\ell)$$

$$2.4\ g\ C_{12}H_{22}O_{11} \times \frac{1\ mol\ C_{12}H_{22}O_{11}}{342.2956\ g\ C_{12}H_{22}O_{11}} \times \frac{12\ mol\ CO_2}{1\ mol\ C_{12}H_{22}O_{11}} \times \frac{22.414\ L\ CO_2\ at\ STP}{1\ mol\ CO_2} = 1.9\ L\ CO_2$$

Assume one loaf of French bread is a cylinder, 3.0 inches in diameter and 18 inches long.

$$r = 1.5\ in$$
$$A = \pi r^2 = \pi(1.5\ in)^2 = 7.1\ in^2$$
$$V = A\ell = (7.1\ in^2) \times (18\ in) = 130\ in^3$$

$$130\ in^3 \times \left(\frac{2.54\ cm}{1\ in}\right)^3 \times \frac{1\ mL}{1\ cm^3} \times \frac{1\ L}{1000\ mL} = 2.1\ L$$

Two loaves would have twice this volume: $2 \times (2.1\ L) = 4.2\ L$. The CO_2 bubbles produced in the bread are nearly half of its volume.

✓ *Reasonable Answer Check:* Slicing open French bread we see that it has a vast "honeycomb" of bubble-shaped spaces in it. It makes sense that approximately half the loaf's volume can be associated with the CO_2 bubbles formed by the yeast when the bread was rising.

38. *Answer:* **10.4 L O_2; 10.4 L H_2O**

Strategy and Explanation: Given the balanced chemical equation for a chemical reaction and the volume of one reactant at a specified pressure and temperature, determine the volume of the other reactant at a specified pressure and temperature that will cause complete reaction, and the volume of one of the products at a specified pressure and temperature that will be produced.

Avogadro's law allows us to interpret a balanced equation with gas reactants and products in terms of gas volumes, as long as their temperatures and pressures are the same. Use the stoichiometry to relate liters of $SiH_4(g)$ that react with liters of $O_2(g)$ and liters of $H_2O(g)$.

The balanced equation tells us that one volume of $SiH_4(g)$ reacts with two volumes of $O_2(g)$ to make two volumes of $H_2O(g)$, since all the volumes are measured at the same temperature and pressure.

$$5.2\ L\ H_2(g) \times \frac{2\ L\ O_2\ (g)}{1\ L\ SiH_4\ (g)} = 10.4\ L\ O_2\ (g)$$

$$5.2\ L\ H_2(g) \times \frac{2\ L\ H_2O(g)}{1\ L\ SiH_4\ (g)} = 10.4\ L\ H_2O(g)$$

Notice on sig figs: Multiplying a number by an exact whole number, n, is like adding that number to itself n times. $2 \times (5.2) = 5.2\ L + 5.2\ L = 10.4\ L$, that is why we use the addition rule for assessing the significant figures, here.

✓ *Reasonable Answer Check:* Twice as many O_2 molecules are needed compared to the number of SiH_4 molecules, forming twice as many H_2O molecules. So, it makes sense that both the volume of O_2 and the volume of H_2O are twice the volume of SiH_4.

40. *Answer:* **21 mm Hg**

Strategy and Explanation: Given the balanced chemical equation for a chemical reaction, the mass of one reactant, and excess other reactant, determine the pressure of the product produced at a specified volume and temperature.

Convert from grams to moles. Use the stoichiometric relationship from the balanced equation to determine moles of product. Use the ideal gas law to determine the pressure of the product.

$$0.050 \text{ g B}_4\text{H}_{10} \times \frac{1 \text{ mol B}_4\text{H}_{10}}{53.32 \text{ g B}_4\text{H}_{10}} \times \frac{10 \text{ mol H}_2\text{O}}{2 \text{ mol B}_4\text{H}_{10}} = 0.0047 \text{ mol H}_2\text{O}$$

$$T = 30. \text{ °C} + 273.15 = 303 \text{ K}$$

$$P = \frac{n_{\text{H}_2\text{O}}RT}{V} = \frac{(0.0047 \text{ mol H}_2\text{O}) \times \left(0.08206 \dfrac{\text{L} \cdot \text{atm}}{\text{mol} \cdot \text{K}}\right) \times (303 \text{ K})}{(4.25 \text{ L})} = 0.027 \text{ atm}$$

$$0.027 \text{ atm} \times \frac{760 \text{ mm Hg}}{1 \text{ atm}} = 21 \text{ mm Hg}$$

✓ *Reasonable Answer Check:* The relative quantities of B_4H_{10} and H_2O seem sensible. All units cancel properly in the calculation of pressure. This is a relatively low pressure for water but the sample is also small.

42. *Answer:* **1.44 g Ni(CO)$_4$**

Strategy and Explanation: Given the description of a chemical reaction and the volume of one reactant at a specified pressure and temperature, determine the mass of the product that can be produced.

Balance the equation. Use the ideal gas law to determine the moles of the reactant. Use the stoichiometric relationship from the balanced equation to determine moles of product. Then, convert from moles to grams.

$$\text{Ni(s)} + 4 \text{ CO(g)} \longrightarrow \text{Ni(CO)}_4$$

$$T = 25.0 \text{ °C} + 273.15 = 298.2 \text{ K} \qquad 418 \text{ mm Hg} \times \frac{1 \text{ atm}}{760 \text{ mm Hg}} = 0.550 \text{ atm}$$

$$n_{\text{CO}} = \frac{PV}{RT} = \frac{(0.550 \text{ atm}) \times (1.50 \text{ L})}{\left(0.08206 \dfrac{\text{L} \cdot \text{atm}}{\text{mol} \cdot \text{K}}\right) \times (298.2 \text{ K})} = 0.0337 \text{ mol CO}$$

$$0.0337 \text{ mol CO} \times \frac{1 \text{ mol Ni(CO)}_4}{4 \text{ mol CO}} \times \frac{170.7 \text{ g Ni(CO)}_4}{1 \text{ mol Ni(CO)}_4} = 1.44 \text{ g Ni(CO)}_4$$

✓ *Reasonable Answer Check:* The units all cancel properly in the calculation of moles of $Ni(CO)_4$. A reasonable mass of product is formed considering the molar quantities and the molar mass.

Gas Density and Molar Mass

44. *Answer:* **130. g/mol**

Strategy and Explanation: Given the density of a gas at STP, determine the molar mass of the gas.

Density is related to mass and volume: $d = \dfrac{m}{V}$. Because we know the molar volume of a gas at STP (22.414 liters per mol), we can use this equation to get the molar mass (M = grams per mole).

$$d \times V = m$$
$$d \times (\text{molar volume}) = (\text{molar mass})$$

$$5.79 \frac{\text{g}}{\text{L}} \times 22.414 \frac{\text{L}}{\text{mol}} = 130. \frac{\text{g}}{\text{mol}}$$

45. *Answer:* **2.7×10^3 mL**

Strategy and Explanation: Given the mass of a gas, the identity of the gas, its temperature, and its pressure, determine the volume occupied by the gas.

Convert from grams to moles. Use the ideal gas law to determine the volume of the gas.

$$4.4 \text{ g CO}_2 \times \frac{1 \text{ mol CO}_2}{44.0095 \text{ g CO}_2} = 0.10 \text{ mol CO}_2$$

$$T = 27 \text{ °C} + 273.15 = 300. \text{ K} \qquad 730. \text{ mm Hg} \times \frac{1 \text{ atm}}{760. \text{ mm Hg}} = 0.961 \text{ atm}$$

$$V = \frac{nRT}{P} = \frac{(0.10 \text{ mol}) \times \left(0.08206 \dfrac{L \cdot atm}{mol \cdot K}\right) \times (300. \text{ K})}{(0.961 \text{ atm})} = 2.7 \text{ L}$$

$$2.7 \text{ L} \times \frac{1000 \text{ mL}}{1 \text{ L}} = 2.7 \times 10^3 \text{ mL}$$

✓ *Reasonable Answer Check:* The molar volume of a gas under these conditions is going to be similar to that at STP (22.4 L/mol), because T is slightly higher than standard T and P is slightly lower than standard P. When using one tenth of a mole of CO_2, one would expect approximately one tenth of the molar volume, roughly 2.2 liters. This answer seems right, though it is unclear why it needed to be given in milliliters.

47. *Answer:* **P_{He} is 7.000 times greater than P_{N_2}**

Strategy and Explanation: Given the formulas of the compounds composing two gas samples with equal density, volume and temperature, determine the relationship between the pressures of the two gas samples.

We have insufficient information to take a straightforward approach to this task, so find a way to relate pressure to molar mass when the density, volume and temperature are equal: Density is mass per unit volume. Density can be calculated using the molar mass (M, the grams per mole, or m/n) and the molar volume (the volume per mole, or V/n).

$$d = \frac{m}{V} = \frac{m/n}{V/n} = \frac{M}{V/n} \qquad \text{so,} \qquad \frac{V}{n} = \frac{M}{d}$$

Rearranging the ideal gas law:

$$P\frac{V}{n} = RT$$

Substituting the first equation into the second, gives:

$$P \times \frac{M}{d} = RT \qquad \text{or} \qquad PM = dRT$$

Therefore, at constant d and T:

$$P_2 M_2 = M_1 P_1 \qquad \text{or} \qquad \frac{P_2}{P_1} = \frac{M_1}{M_2}$$

$$\frac{P_{He}}{P_{N_2}} = \frac{M_{N_2}}{M_{He}} = \frac{(28.0134 \text{ g/mol})}{(4.0026 \text{ g/mol})} = 7.000$$

The pressure of helium is seven times greater than the pressure of nitrogen.

✓ *Reasonable Answer Check:* A much larger number of atoms would need to be in the helium sample for it to have the same density as the nitrogen sample. Because pressure is proportional to number of particles, it makes sense that the pressure in the helium sample is much larger than in the nitrogen sample.

48. *Answer:* **3.7×10^{-4} g/L**

Strategy and Explanation: Given the molar mass of a gas and its pressure and temperature, determine the density of the gas.

We have insufficient information to take a straightforward approach to this task, so let's look at what we know: Density is mass per unit volume. If we can calculate the molar volume (the volume per mole) and divide that into the molar mass (M, the grams per mole), we can get the density. Use the ideal gas law to determine the molar volume of the gas.

$$\text{molar volume} = \frac{V}{n} = \frac{RT}{P} \qquad\qquad d = \frac{m}{V} = \frac{M}{\left(\dfrac{RT}{P}\right)} = \frac{MP}{RT}$$

$$T = -23 \text{ °C} + 273.15 = 250. \text{ K}$$

$$0.20 \text{ mmHg} \times \frac{1 \text{ atm}}{760 \text{ mmHg}} = 2.6 \times 10^{-4} \text{ atm}$$

$$d = \frac{MP}{RT} = \frac{\left(29.0 \dfrac{g}{mol}\right) \times (2.6 \times 10^{-4} \text{ atm})}{\left(0.08206 \dfrac{L \cdot atm}{mol \cdot K}\right) \times (250. \text{ K})} = 3.7 \times 10^{-4} \frac{g}{L}$$

✓ *Reasonable Answer Check:* The low density makes sense at the very low pressure.

Partial Pressures of Gases

49. *Answer:* **4.51 atm total**

Strategy and Explanation: Given the masses of gases in a mixture at a specified volume and temperature, determine the total pressure.

Convert from grams to moles. Use the ideal gas law to determine the partial pressure of each gas. Use Dalton's law of partial pressures to determine the total pressure. Dalton's law and its applications are described in Section 10.7, page 361. Dalton's law of partial pressures states that the total pressure (P) exerted by a mixture of gases is the sum of their partial pressures (p_1, p_2, p_3, etc.), if the volume (V) and temperature (T) are constant.

$$P_{tot} = P_1 + P_2 + P_3 + ... \qquad \text{(V and T constant)}$$

$$1.50 \text{ g H}_2 \times \frac{1 \text{ mol H}_2}{2.0158 \text{ g H}_2} = 0.744 \text{ mol H}_2$$

$$T = 25 \text{ °C} + 273.15 = 298 \text{ K}$$

$$P_{H_2} = \frac{n_{H_2} RT}{V} = \frac{(0.744 \text{ mol H}_2) \times \left(0.08206 \frac{L \cdot atm}{mol \cdot K}\right) \times (298 \text{ K})}{(5.00 \text{ L})} = 3.64 \text{ atm H}_2$$

$$5.00 \text{ g N}_2 \times \frac{1 \text{ mol N}_2}{28.0134 \text{ g N}_2} = 0.178 \text{ mol N}_2$$

$$P_{N_2} = \frac{n_{N_2} RT}{V} = \frac{(0.178 \text{ mol H}_2) \times \left(0.08206 \frac{L \cdot atm}{mol \cdot K}\right) \times (298 \text{ K})}{(5.00 \text{ L})} = 0.873 \text{ atm N}_2$$

Using Dalton's law: $P_{tot} = P_{H_2} + P_{N_2} = 3.64 \text{ atm H}_2 + 0.873 \text{ atm N}_2 = 4.51 \text{ atm total}$

✓ *Reasonable Answer Check:* The relative pressure of H_2 and N_2 makes sense from the relative number of moles. All un-needed units cancel properly in the calculation of atmosphere.

51. *Answer:* **(a) 154 mm Hg (b) X_{N_2}=0.777, X_{O_2} = 0.208, X_{Ar} = 0.0093, X_{CO_2} = 0.0003, X_{H_2O} = 0.0053**
(c) 77.7% N_2, 20.8% O_2, 0.93% Ar, 0.03% CO_2, 0.54% H_2O; slight difference, since this sample is wet

Strategy and Explanation: Given the partial pressure of several gases in a sample of the atmosphere, and the total pressure of the sample, determine the partial pressure of O_2, the mole fraction of each gas, and the percent by volume. Compare the percentages to Table 10.1.

Dalton's law and its applications are described in Section 10.7, page 361. Dalton's law of partial pressures states that the total pressure (P) exerted by a mixture of gases is the sum of their partial pressures (p_1, p_2, p_3, etc.), if the volume (V) and temperature (T) are constant.

$$P_{tot} = P_1 + P_2 + P_3 + ... \qquad \text{(V and T constant)}$$

Because $p_i = X_i P_{tot}$, we can use the total pressure and the partial pressure of a component to determine its mole fraction:

$$X_i = \frac{P_i}{P_{tot}}$$

According to Avogadro's law, moles and gas volumes are proportional, so the mole fraction is equal to the volume fraction. To get percent, multiply the volume fraction by 100%.

(a) $$P_{tot} = P_{N_2} + P_{O_2} + P_{Ar} + P_{CO_2} + P_{H_2O}$$

$$p_{O_2} = P_{tot} - P_{N_2} - P_{Ar} - P_{CO_2} - P_{H_2O}$$

$$p_{O_2} = (740. \text{ mm Hg}) - (575 \text{ mm Hg}) - (6.9 \text{ mm Hg}) - (0.2 \text{ mm Hg}) - (4.0 \text{ mm Hg}) = 154 \text{ mm Hg}$$

(b) $X_{N_2} = \dfrac{P_{N_2}}{P_{tot}} = \dfrac{575 \text{ mmHg}}{740. \text{ mmHg}} = 0.777$ $X_{O_2} = \dfrac{P_{O_2}}{P_{tot}} = \dfrac{154 \text{ mmHg}}{740. \text{ mmHg}} = 0.208$

$X_{Ar} = \dfrac{P_{Ar}}{P_{tot}} = \dfrac{6.9 \text{ mmHg}}{740. \text{ mmHg}} = 0.0093$ $X_{CO_2} = \dfrac{P_{CO_2}}{P_{tot}} = \dfrac{0.2 \text{ mmHg}}{740. \text{ mmHg}} = 0.0003$

$$X_{H_2O} = \dfrac{P_{H_2O}}{P_{tot}} = \dfrac{4.0 \text{ mmHg}}{740. \text{ mmHg}} = 0.0054$$

(c)

$$\% \ N_2 = X_{N_2} \times 100\% = 0.777 \times 100\% = 77.7\% \ N_2$$

$$\% \ O_2 = X_{O_2} \times 100\% = 0.208 \times 100\% = 20.8\% \ O_2$$

$$\% \ Ar = X_{Ar} \times 100\% = 0.0093 \times 100\% = 0.93\% \ Ar$$

$$\% \ CO_2 = X_{CO_2} \times 100\% = 0.0003 \times 100\% = 0.03\% \ CO_2$$

$$\% \ H_2O = X_{H_2O} \times 100\% = 0.0054 \times 100\% = 0.54\% \ H_2O$$

The Table 10.1 figures are slightly different. This sample is wet, whereas the proportions given in Table 10.1 are for dry air.

✓ *Reasonable Answer Check:* The percentages are very close to those provided in the table, and the variations are explainable. The sum of the mole fractions is 1, and the sum of the percentages is 100%.

53. *Answer:* **(a) 1.98 atm (b) P_{O_2} = 0.438 atm, P_{N_2} = 0.182 atm, P_{Ar} = 1.36 atm**

Strategy and Explanation: Given three containers of gases, with known volume and pressure, determine the total pressure and partial pressures of the gases when the three containers are opened to each other.

Use Boyle's law to determine how the pressure of each gas changes with the increase in total volume and then use Dalton's law to determine the total pressures after mixing. Use the definition of partial pressure to show that the pressures calculated are the partial pressure.

The gas in one chamber is allowed to diffuse into all three chambers, so its final volume increases to the total of the three volumes: $V_{tot} = 3.00 \text{ L} + 2.00 + 5.00 \text{ L} = 10.00 \text{ L}$

Boyle's law: $P_{f,gas} = \dfrac{P_{i,gas} V_{i,gas}}{V_{f,gas}}$ For O_2, $P_{f,O_2} = \dfrac{P_i V_i}{V_f} = \dfrac{(1.46 \text{ atm}) \times (3.00 \text{ L})}{(10.00 \text{ L})} = 0.438 \text{ atm}$

For N_2, $P_{f,N_2} = \dfrac{(0.908 \text{ atm}) \times (2.00 \text{ L})}{(10.00 \text{ L})} = 0.182 \text{ atm}$ For Ar, $P_{f,Ar} = \dfrac{(2.71 \text{ atm}) \times (5.00 \text{ L})}{(10.00 \text{ L})} = 1.36 \text{ atm}$

(a) $P_{f, tot} = P_{f,O_2} + P_{f,N_2} + P_{f,Ar} = 0.438 \text{ atm} + 0.182 \text{ atm} + 1.36 \text{ atm} = 1.98 \text{ atm}$

(b) Partial pressure is the pressure each gas would cause if it were alone in the container.
$P_{f,O_2} = P_{O_2} = 0.438 \text{ atm}, \ P_{f,N_2} = P_{N_2} = 0.182 \text{ atm}, \ P_{f,Ar} = P_{Ar} = 1.36 \text{ atm}$

✓ *Reasonable Answer Check:* All the individual final pressures are less than the initial pressures, which makes sense because the volume is larger. The total pressure is a weighted average of the three pressures, and is influenced most by the gas present in largest quantity (Ar).

The Behavior of Real Gases

56. *Answer:* **18 mL $H_2O(\ell)$; 22.4 L $H_2O(g)$; No we cannot achieve a pressure of 1 atm at this temperature, because the vapor pressure of water at 0 °C to be 4.6 mm Hg, at pressures higher than this, the water would liquefy.**

Strategy and Explanation: Use standard conversion factors:

$$1 \text{ mol } H_2O \times \dfrac{18.02 \text{ g } H_2O}{1 \text{ mol } H_2O} \times \dfrac{1 \text{ mL } H_2O}{1.0 \text{ g } H_2O} = 18 \text{ mL } H_2O \text{ liquid}$$

1 mol H_2O at STP occupies 22.4 L.

Table 10.4 gives the vapor pressure of water at 0 °C to be 4.6 mm Hg. At pressures higher than this, water would liquefy. We cannot achieve 1 atm pressure of water vapor at this low temperature, so we cannot achieve the standard state condition for water vapor.

58. *Answer/Explanation:* The behavior of real gases is discussed in Section 10.8. At low temperatures, the molecules are moving relatively slowly; however, when the pressure is very low, they are still quite far apart. As external pressures increases, the gas volume decreases, the slow molecules are squeezed closer together, and the attractions among the molecules get stronger. Figure 10.15 shows that a gas molecule strikes the walls of the container with less force due to the attractive forces between it and its neighbors. This makes the mathematical product PV smaller than the mathematical product nRT.

Greenhouse Gases and Global Warming

60. *Answer/Explanation:* Greenhouse effect is the trapping of heat radiation by atmospheric gases. Global warming is the increase of the average global temperature. Global warming is caused by an increase in the amount of greenhouse gases in the atmosphere.

62. *Answer/Explanation:* CO_2 gets into the atmosphere by animal respiration, by burning fossil fuels and other plant materials, and by the decomposition of organic matter. CO_2 gets removed from the atmosphere by plants during photosynthesis, when it is dissolved in rainwater, and when it is incorporated into carbonate and bicarbonate compounds in the oceans. Currently, atmospheric CO_2 production exceeds CO_2 removal.

General Questions

64. *Answer:* (a) Before: P_{H_2} = 3.7 atm; P_{Cl_2} = 4.9 atm (b) Before: P_{tot} = 8.6 atm (c) After: P_{tot} = 8.6 atm
(d) Cl_2; 0.5 mol remain (e) P_{HCl} = 7.4 atm; P_{Cl_2} = 1.2 atm (f) P = 8.9 atm

Strategy and Explanation: Given the description of a reaction, the masses of two gaseous reactants, and the volume and temperature of the mixture, determine the partial pressure of the reactants, the total pressure due to the gases in the flask before and after the reaction, the excess reactant, the number of moles of it left over, the partial pressures of the gases in the flask, and the total pressure after the temperature is increased to a given value.

Balance the equation. Find moles of each reactant and use the ideal gas law to determine the pressures. Use Dalton's law for total pressures. Find limiting reactant and excess reactant as described in Chapter 4. Use the combined gas law to determine the pressure at a different temperature.

$$H_2(g) + Cl_2(g) \longrightarrow 2HCl(g)$$

(a) $\quad 3.0 \text{ g } H_2 \times \dfrac{1 \text{ mol } H_2}{2.0158 \text{ g } H_2} = 1.5 \text{ mol } H_2 \qquad\qquad 140. \text{ g } Cl_2 \times \dfrac{1 \text{ mol } Cl_2}{70.906 \text{ g } Cl_2} = 1.97 \text{ mol } Cl_2$

$$T = 28 °C + 273.15 = 301 \text{ K}$$

$$P_{H_2} = \frac{nRT}{V} = \frac{(1.5 \text{ mol } H_2) \times \left(0.08206 \frac{L \cdot atm}{K \cdot mol}\right) \times (301 \text{ K})}{10. \text{ L}} = 3.7 \text{ atm}$$

$$P_{Cl_2} = \frac{nRT}{V} = \frac{(1.97 \text{ mol } Cl_2) \times \left(0.08206 \frac{L \cdot atm}{K \cdot mol}\right) \times (301 \text{ K})}{10. \text{ L}} = 4.9 \text{ atm}$$

(b) $\qquad\qquad P_{tot, \text{ before}} = P_{H_2} + P_{Cl_2} = (3.7 \text{ atm } H_2) + (4.9 \text{ atm } Cl_2) = 8.6 \text{ atm total}$

(c) The number of moles of gas reactants is equal to the number of moles of gaseous products, so the total pressure will not change. $P_{tot, \text{ after}}$ = 8.6 atm total

(d) The balanced chemical equation shows equal molar quantities of each reactant reacting, so the H_2 is the limiting reactant and Cl_2 is the excess reactant, and 1.5 mol of each reactant react. Subtracting the moles of Cl_2 that react from the moles of Cl_2 describes how many moles of Cl_2 remain.

$$2.0 \text{ mol } Cl_2 \text{ initial} - 1.5 \text{ mol } Cl_2 \text{ react} = 0.5 \text{ mol } Cl_2 \text{ remain}$$

(e) When 1.5 mol of each reactant react, 3.0 mol of HCl are formed.

$$P_{HCl} = \frac{nRT}{V} = \frac{(3.0 \text{ mol HCl}) \times \left(0.08206 \frac{L \cdot atm}{K \cdot mol}\right) \times (301 \text{ K})}{10. \text{ L}} = 7.4 \text{ atm}$$

$$P_{Cl_2} = PP_{tot,\ after} - P_{HCl} = (8.6 \text{ atm total}) - (7.4 \text{ atm HCl}) = 1.2 \text{ atm } Cl_2$$

(f) $T_1 = 301$ K, $T_2 = 40.\ °C + 273.15 = 313$ K

$$P_2 = P_1 \times \frac{V_1}{V_2} \times \frac{T_2}{T_1} = P_1 \times \frac{T_2}{T_1} \text{ at constant volume}$$

$$P_2 = P_1 \times \frac{T_2}{T_1} = (8.6 \text{ atm}) \times \frac{(313 \text{ K})}{(301 \text{ K})} = 8.9 \text{ atm}$$

✓ *Reasonable Answer Check:* The partial pressure of Cl_2 calculated in (e) can also be found using the ideal gas law:

$$P_{Cl2} = \frac{nRT}{V} = \frac{(0.5 \text{ mol } Cl_2) \times \left(0.08206 \frac{L \cdot atm}{K \cdot mol}\right) \times (301 \text{ K})}{10. \text{ L}} = 1. \text{ atm}$$

We could construct mole fractions of the products and multiply them by the total pressure to determine the partial pressures after the reaction was completed:

$$X_{HCl} = \frac{3.0 \text{ mol HCl}}{3.0 \text{ mol HCl} + 0.50 \text{ mol } Cl_2} = 0.86$$

$$P_{HCl} = X_{HCl}P_{tot} = (0.86) \times (8.6 \text{ atm}) = 7.4 \text{ atm}$$

$$X_{Cl_2} = \frac{0.50 \text{ mol } Cl_2}{3.0 \text{ mol HCl} + 0.5 \text{ mol } Cl_2} = 0.14$$

$$P_{Cl_2} = X_{Cl_2}P_{tot} = (0.14) \times (8.6 \text{ atm}) = 1.2 \text{ atm}$$

$$P_{tot,\ after} = P_{HCl} + P_{Cl_2} = 7.4 \text{ atm} + 1.2 \text{ atm} = 8.6 \text{ atm total}$$

These answers are the same as calculated above. The results are self-consistent and reasonable size (i.e., smaller number of moles produce smaller partial pressures and smaller mole fractions).

Applying Concepts

67. *Answer:* **See graph below**

Strategy and Explanation: The molar masses of C_2H_6 and F_2 are 30 g/mol and 38 g/mol. That means the average speed of F_2 is somewhat less than that of C_2H_6. The total pressure of the gases is 720. mm Hg, and the partial pressure of F_2 is 540. mm Hg. Dalton's law tells us that the sum of the partial pressures is the total pressure, and so the partial pressure of C_2H_6 is 180. mm Hg, or one third that of the F_2 molecules. The partial pressure is directly proportional to the mole fraction of the molecules, in the container, so there should be one third as many C_2H_6 molecules as F_2 molecules. So, the graph of number of molecules verses molecular speed will have the F_2 curve peaking at a slightly smaller speed value and the C_2H_6 curve will have one third of the vertical rise.

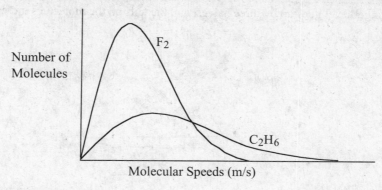

68. *Answer:* **See drawings below**

Strategy and Explanation: The speed of the molecules is related to the temperature. In particular, $T \propto mv^2$. So, when the temperature changes by a certain factor, the speed of the molecules (v) changes by the square root of that same factor. We'll indicate that fact by making the tails of the arrows proportionately shorter. The pressure describes how close together the molecules are. When the pressure is changed by a certain factor, the molecules will be represented as that much closer together. The volume will change the space occupied by the gas, and that will be indicated in the movement of the syringe plunger.

For reference, the initial state looks like this:

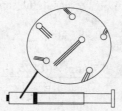

(a) When the temperature is decreased to one half of its original value, Charles' law tells us that the volume decreases by half. The molecules are just as far apart as they were (since the pressure is the same), but the plunger level goes down by half, and the tails on the molecules decrease in length by about 0.7 times.

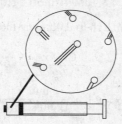

(b) When the pressure decreased to one half of its original value, Boyle's law tells us that the volume increases by half. The tails on the molecules are the same length, but the molecules are twice as far apart as they were (so we put half as many in this view), and the plunger level moves up by half.

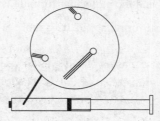

(c) When the temperature is tripled and the pressure is doubled, the combined gas law tells us that the volume will have a net increase by 3/2 (three times larger due to the temperature change and half as large due to the pressure change). The molecules are two times closer together than they were (so we'll add twice as many), the plunger level goes up by three halves, and the tails on the molecules increase in length by about 1.7 times.

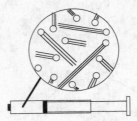

70. *Answer:* **Box (b)**

Strategy and Explanation: The initial volume is 1.8 L and the final volume is 0.9 L. This 2:1 ratio in the gas volumes means for every two molecules of gas reactants there must be one molecule of gas products. The reactant count is six, so the box that has three product molecules fits these observations. The correct box is Box (b):

$$6\ AB_2(g) \longrightarrow 3\ A_2B_4(g)$$

72. *Answer:* **(a) 64.1 g/mol (b) empirical: CHF, molecular: $C_2H_2F_2$ (c) see structures below**

Strategy and Explanation:

(a) Use the ideal gas law to determine the moles of gas in the sample. $T = 50.0\ °C + 273.15 = 323.2\ K$

$$750.\ mmHg \times \frac{1\ atm}{760.\ mmHg} = 0.987\ atm \qquad 125\ mL \times \frac{1\ L}{1000\ mL} = 0.125\ L$$

$$n_{C_xH_yF_z} = \frac{PV}{RT} = \frac{(0.987\ atm) \times (0.125\ L)}{\left(0.08206\ \dfrac{L \cdot atm}{mol \cdot K}\right) \times (323.2\ K)} = 4.65 \times 10^{-3}\ mol\ C_xH_yF_z$$

Divide the grams of gas in the sample by the calculated moles of gas in the sample to get the molar mass.

$$Molar\ Mass = \frac{0.298\ g\ C_xH_yF_z}{4.65 \times 10^{-3}\ mol\,C_xH_yF_z} = 64.1\ g/mol\ C_xH_yF_z$$

(b) Use the methods described in Section 3.10 to determine the empirical formula.

$$37.5\ g\ C \times \frac{1\ mol\ C}{12.0107\ g\ C} = 3.12\ mol\ C \qquad 3.15\ g\ H \times \frac{1\ mol\ H}{1.0079\ g\ H} = 3.13\ mol\ H$$

$$59.3\ g\ F \times \frac{1\ mol\ F}{18.9984\ g\ F} = 3.12\ mol\ F$$

Set up a mole ratio and simplify:

$$3.12\ mol\ C : 3.13\ mol\ H : 3.12\ mol\ F$$

$$1\ C : 1\ H : 1\ F$$

The empirical formula is CHF and the molecular formula is $(CHF)_n$. So, $n = x = y = z$

The molar mass of the empirical formula = 32.02 g/mol

$$n = \frac{63\ g/mol}{32.02\ g/mol} = 2.0$$

Therefore, the molecular formula is $C_2H_2F_2$.

(c) $C_2H_2F_2$ can have *cis-* and *trans-*isomers if the F atoms are on different C atoms. (described in Chapter 8)

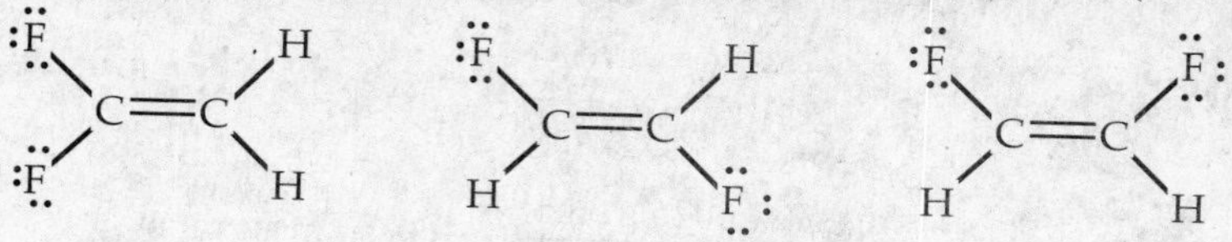

More Challenging Questions

73. *Answer:* **0.88 atm**

Strategy and Explanation: Given the names and masses of three gases and the temperature and volume of the vessel they occupy, determine the total pressure in the vessel.

Calculate the moles of each substance, using the molar masses. Add the moles of each gas to determine the total moles of gas in the vessel. Use the volume and temperature in the ideal gas law to determine the total pressure.

Carbon dioxide is CO_2 and methane is CH_4. $V_{tot} = 8.7$ L

Find moles of CO_2, CH_4, and Ar in the sample:

$$3.44 \text{ g CO}_2 \times \frac{1 \text{ mol CO}_2}{44.009 \text{ g CO}_2} = 0.0782 \text{ mol CO}_2$$

$$1.88 \text{ g CH}_4 \times \frac{1 \text{ mol CH}_4}{16.0423 \text{ g CH}_4} = 0.117 \text{ mol CH}_4$$

$$4.28 \text{ g Ar} \times \frac{1 \text{ mol Ar}}{39.948 \text{ g Ar}} = 0.107 \text{ mol Ar}$$

$$n_{tot} = 0.0781 \text{ mol CO}_2 + 0.117 \text{ mol CH}_4 + 0.107 \text{ mol Ar} = 0.302 \text{ mol tot}$$

$$T = 37 \,°C + 273.15 = 310. \text{ K}$$

$$P = \frac{n_{tot}RT}{V_{tot}} = \frac{(0.302 \text{ mol}) \times \left(0.08206 \frac{\text{L} \cdot \text{atm}}{\text{mol} \cdot \text{K}}\right) \times (310. \text{ K})}{(8.7 \text{ L})} = 0.88 \text{ atm}$$

✓ *Reasonable Answer Check:* The percentages are very close to those provided in the table, and the variations are explainable. The sum of the mole fractions is 1, and the sum of the percentages is 100%.

75. *Answer:* $\dfrac{m_{Ne}}{m_{Ar}} = 0.2$

Strategy and Explanation: Given two known gaseous substances in separate balloons at the same temperature and pressure with one volume designated as double the other, determine the mass ratio of the two substances.

Use the ideal gas law and the relationship between mass and moles to determine the mass ratio in terms of volume and molar mass: $M = m/n$

$$PV = nRT = \left(\frac{m}{M}\right)RT$$

Rearranging to solve for m gives:

$$m = \frac{MPV}{RT}$$

At equal T and P:

$$\frac{m_2}{m_1} = \frac{M_2 V_2}{M_1 V_1}$$

$$\frac{m_{Ne}}{m_{Ar}} = \frac{M_{Ne} V_{Ne}}{M_{Ar} V_{Ar}}$$

The volume of the neon balloon is twice that of the argon balloon.

$$V_{Ne} = 2 V_{Ar}$$

$$\frac{V_{Ne}}{V_{Ar}} = 2$$

$$\frac{m_{Ne}}{m_{Ar}} = \frac{4.0026 \text{ g/mol}}{39.948 \text{ g/mol}} (2) = 0.2 = \text{mass ratio}$$

The mass of the neon in the orange balloon is 0.2 times the mass of the argon in the blue balloon.

✓ *Reasonable Answer Check:* Even though the neon balloon is twice the volume, the heavier argon atoms result in a more massive argon sample.

77. *Answer:* (a) **29.1 mol CO_2; 14.6 mol N_2; 2.43 mol O_2** (b) **1.1×10^3 L** (c) **p_{CO_2}= 0.631 atm;**

p_{N_2}= 0.317 atm; p_{O_2} = 0.0527 atm

Strategy and Explanation: Given the mass of a reactant and the balanced chemical equation for the explosive decomposition of the reactant, determine the total moles of gases produced, the volume occupied by the product gases at a given temperature and pressure, and the partial pressures of each gas.

Use molar mass to determine the moles of reactant, and use equation stoichiometry to calculate the moles of each *gas phase* reactant. (Notice: reactants that are not gas phase will not contribute this sum). Add the moles of each gas to determine the total moles of gas in the vessel. Use the pressure and temperature in the ideal gas law to determine the total volume. Calculate mole fraction with the following equation: $X_i = \dfrac{n_i}{n_{tot}}$ then partial pressure is calculated with $p_i = X_i P_{tot}$.

Calculate the moles of reactant, using the molar masses.

$$1.00 \text{ kg } C_3H_5(NO_3)_3 \times \frac{1000 \text{ g}}{1 \text{ kg}} \times \frac{1 \text{ mol } C_3H_5(NO_3)_3}{103.0765 \text{ g } C_3H_5(NO_3)_3} = 9.70 \text{ mol } C_3H_5(NO_3)_3$$

(a) Find moles of CO_2, N_2, and O_2 in the sample: (Water is a liquid.)

$$9.70 \text{ mol } C_3H_5(NO_3)_3 \times \frac{12 \text{ mol } CO_2}{4 \text{ mol } C_3H_5(NO_3)_3} = 29.1 \text{ mol } CO_2$$

$$9.70 \text{ mol } C_3H_5(NO_3)_3 \times \frac{6 \text{ mol } N_2}{4 \text{ mol } C_3H_5(NO_3)_3} = 14.6 \text{ mol } N_2$$

$$9.70 \text{ mol } C_3H_5(NO_3)_3 \times \frac{1 \text{ mol } O_2}{4 \text{ mol } C_3H_5(NO_3)_3} = 2.43 \text{ mol } O_2$$

(b) $$n_{tot} = 29.1 \text{ mol } CO_2 + 14.6 \text{ mol } N_2 + 2.43 \text{ mol } O_2 = 46.1 \text{ mol tot}$$

$$T = 25 \,°C + 273.15 = 298 \text{ K}$$

$$V_{tot} = \frac{n_{tot}RT}{P} = \frac{(46.1 \text{ mol}) \times \left(0.08206 \dfrac{L \cdot atm}{mol \cdot K}\right) \times (298 \text{ K})}{(1.0 \text{ atm})} = 1.1 \times 10^3 \text{ L}$$

(c) $$X_{CO2} = \frac{n_{CO2}}{n_{tot}} = \frac{29.1 \text{ mole } CO_2}{46.1 \text{ mole tot}} = 0.631 \qquad p_{CO_2} = (0.631)(1.00 \text{ atm}) = 0.631 \text{ atm}$$

$$X_{N2} = \frac{n_{N2}}{n_{tot}} = \frac{14.6 \text{ mole } N_2}{46.1 \text{ mole tot}} = 0.317 \qquad p_{N_2} = (0.317)(1.00 \text{ atm}) = 0.317 \text{ atm}$$

$$X_{O2} = \frac{n_{O2}}{n_{tot}} = \frac{2.43 \text{ mole } O_2}{46.1 \text{ mole tot}} = 0.0527 \qquad p_{O_2} = (0.0527)(1.00 \text{ atm}) = 0.0527 \text{ atm}$$

✓ *Reasonable Answer Check:* The decomposition of TNT produces a large volume of gas product, which explains one reason why this decomposition is explosive. The sum of the mole fractions is 1, and the sum of the partial pressures is the total pressure.

78. *Answer:* **458 torr**

Strategy and Explanation: Given the initial pressure of a gas, the relative amount of it that undergoes reaction, and the stoichiometric relationship between the reactants and products, determine the new pressure in the container.

Determine the pressure of the gas reactant that reacts. Use the equation stoichiometry interpreted in units of pressure for the gases to determine pressure of the gas products. Then add this number to the pressure of the unreacted reactant gas to calculate the final pressure.

Half of the reactant reacts, so the pressure of the reactant that undergoes reaction is:

$$0.5 \times (550. \text{ torr}) = 275 \text{ torr}$$

$$275 \text{ torr reactant} \times \frac{2 \text{ torr product}}{3 \text{ torr reactant}} = 183 \text{ torr product}$$

The unreacted reactant has a pressure of 550. torr – 275 torr = 275 torr

The total pressure after this reaction is complete = 275 torr + 183 torr = 458 torr

✓ *Reasonable Answer Check:* The reaction reduces the number of gasses, so it makes sense that the pressure after some of the reactant has reacted is less.

79. *Answer:* **4.5 mm³**

Strategy and Explanation: Given the volume and pressure of a gas bubble at the bottom of a lake, determine the volume of the same bubble at the surface of the lake with given pressure.

Use Boyle's law to determine how the volume of gas changes with the decrease in pressure:

$$V_{f,bubble} = \frac{P_{i,bubble} V_{i,bubble}}{P_{f,bubble}} = \frac{\left(4.4 \text{ atm} \times \dfrac{760 \text{ torr}}{1 \text{ atm}}\right) \times \left(1 \text{ mm}^3\right)}{(740. \text{ torr})} = 4.5 \text{ mm}^3$$

✓ *Reasonable Answer Check:* At the lower pressure, it makes sense that the bubble will increase in volume.

81. *Answer:* **(a) More significant, because of more collisions   (b) More significant, because of more collisions   (c) Less significant, because the molecules will move faster**

Strategy and Explanation:

(a) When the gas is compressed to a smaller volume, the molecules will be closer together and they will collide with each other more often. The temperature is fixed, so the molecules will hit each other at the same average speed; however, with more collisions the interactions between the molecules will be more significant.

(b) When more molecules of the same gas are added to the container, the molecules will be closer together and they will collide with each other more often. The temperature is fixed, so the molecules will hit each other at the same average speed; however, with more collisions the interactions between the molecules will be more significant.

(c) When the temperature is increased, the average kinetic energy of the molecules is increased and the molecules will move faster. That means they will collide with each other more often. Because the speed has increased the time these molecules spend in proximity to each other will decrease, so, compared to the situation in (a) and (b), the effect of intermolecular interactions will less significant.

Chapter 11: Liquids, Solids, and Materials

Introduction

This chapter should be thought of as a continuation of Chapter 10. It describes the other two typical states of matter: liquids and solids, and the physical processes of converting from one state to another. Physical properties (rather than chemical one) will be important in this chapter. It is important to learn the different types of non-chemical interactions to be able to understand why substances will be found in each of the physical states.

As in Chapter 10, the calculations in this chapter are straightforward, but you must be careful with units. The conceptual information is more challenging and represents key elements of many uses of physical chemistry.

Solutions to Blue-Numbered Questions
for Review and Thought for Chapter 11

Topical Questions

The Liquid State

10. *Answer:* **Reduce the pressure**

    Strategy and Explanation: A liquid can be converted to a vapor without changing the temperature by reducing the pressure above it. This can be accomplished by putting the liquid in a container whose pressure can be altered using a vacuum pump. Pumping the gases out of the container reduces the atmospheric pressure acting on the liquid allowing it to reach the boiling point at a much lower temperature. Incidentally, a liquid in an open container will eventually evaporate at temperatures much lower than its boiling point because it absorbs energy from the surroundings and its vapor pressure never reaches the equilibrium vapor pressure, therefore, evaporation continues.

13. *Answer:* $\mathbf{1.5 \times 10^6 \ kJ}$

    Strategy and Explanation: Given the molar enthalpy of vaporization of a compound, determine the heat energy required to vaporize a given mass of a compound.

    Convert the mass to moles, and use the molar enthalpy of vaporization to get total heat energy.

    $$1.0 \text{ metric ton } NH_3 \times \frac{10^3 \text{ kg}}{1 \text{ metric ton}} \times \frac{1000 \text{ g}}{1 \text{ kg}} \times \frac{1 \text{ mol } NH_3}{17.0304 \text{ g } NH_3} \times \frac{25.1 \text{ kJ}}{1 \text{ mol } NH_3} = 1.5 \times 10^6 \text{ kJ}$$

    ✓ *Reasonable Answer Check:* Units cancel appropriately, and it makes sense that a large amount of ammonia needs a large amount of heat energy.

14. *Answer:* **181 kJ**

    Strategy and Explanation: Given the molar enthalpy of vaporization of a compound, determine the heat energy required to vaporize a given mass of a compound.

    Convert the mass to moles, and use the molar enthalpy of vaporization to get total heat energy.

    $$1.00 \text{ kg } CCl_3F \times \frac{1000 \text{ g}}{1 \text{ kg}} \times \frac{1 \text{ mol } CCl_3F}{137.368 \text{ g } CCl_3F} \times \frac{24.8 \text{ kJ}}{1 \text{ mol } CCl_3F} = 181 \text{ kJ}$$

    ✓ *Reasonable Answer Check:* Units cancel appropriately, and it makes sense that a quantity of about seven moles needs about seven times the molar enthalpy.

15. *Answer:* **233 kJ**

Strategy and Explanation: Given the molar enthalpy of vaporization of a compound and the density of the liquid, determine the heat energy required to vaporize a given volume of the liquid.

Convert the volume to moles, then use the molar enthalpy of vaporization to get total heat energy.

$$250. \text{ mL CH}_3\text{OH} \times \frac{0.787 \text{ g CH}_3\text{OH}}{1 \text{ mL CH}_3\text{OH}} \times \frac{1 \text{ mol CH}_3\text{OH}}{32.0417 \text{ g CH}_3\text{OH}} \times \frac{38.0 \text{ kJ}}{1 \text{ mol CH}_3\text{OH}} = 233 \text{ kJ}$$

✓ *Reasonable Answer Check:* Units cancel appropriately, and it makes sense that a quantity of about six moles needs about six times the molar enthalpy.

17. *Answer:* **2.00 kJ for Hg sample; 1.13 kJ for water sample; even though the heat of vaporization for water is greater than for Hg, the Hg sample takes more energy because of its greater mass.**

Strategy and Explanation: Given the enthalpy of vaporization of two compounds and the density of their liquids, determine the heat energy required to vaporize a given volume of each liquid and compare them.

Convert to grams and use the heat of vaporization for the mercury. Convert the volume of water to moles, then use the molar heat of vaporization to get total heat energy for the water.

$$0.500 \text{ mL Hg} \times \frac{13.6 \text{ g Hg}}{1 \text{ mL Hg}} \times \frac{294 \text{ J}}{1 \text{ g Hg}} \times \frac{1 \text{ kJ}}{1000 \text{ J}} = 2.00 \text{ kJ} \text{ to vaporize Hg sample}$$

$$0.500 \text{ mL H}_2\text{O} \times \frac{1.00 \text{ g H}_2\text{O}}{1 \text{ mL H}_2\text{O}} \times \frac{1 \text{ mol H}_2\text{O}}{18.0152 \text{ g H}_2\text{O}} \times \frac{40.7 \text{ kJ}}{1 \text{ mol H}_2\text{O}} = 1.13 \text{ kJ} \text{ to vaporize water sample}$$

It takes more energy to vaporize the Hg sample than the water sample. Even though the heat of vaporization for water is greater than for Hg, the Hg sample takes more energy because of its greater mass.

✓ *Reasonable Answer Check:* Units cancel appropriately. Equal volumes of water and Hg have very different masses, due to the large difference in their densities. So, while water has a larger heat of vaporization $(2.26 \times 10^3 \text{ J/g})$, the number of grams of water in the sample volume is quite a bit smaller.

19. *Answer/Explanation:* NH_3 has a relatively large boiling point because the molecules interact using relatively strong hydrogen bonding intermolecular forces. The increase in the boiling points of the series PH_3, AsH_3, and SbH_3 is related to the increasing London dispersion intermolecular forces experienced due to the larger central atom in the molecule and the greater polarizability of the valence electrons. (size: $P < As < Sb$)

Vapor Pressure

21. *Answer/Explanation:* Methanol molecules are capable of hydrogen bonding, whereas formaldehyde molecules use dipole-dipole forces to interact. Molecules experiencing stronger intermolecular forces (such as methanol here) will have higher boiling points and lower vapor pressures compared to molecules experiencing weaker intermolecular forces (such as formaldehyde here).

23. *Answer:* **0.21 atm, approximately 57 °C**

Strategy and Explanation: Given the altitude on a mountain above sea level, a simple relationship between altitude and pressure changes, and Figure 11.5, determine the atmospheric pressure on the mountain and the boiling point of water at that altitude.

Use the altitude/pressure relationship to get the atmospheric pressure on the mountain, then look up the boiling temperature on the graph.

$$22834 \text{ ft} \times \frac{3.5 \text{ mbar decrease}}{100 \text{ ft}} \times \frac{1 \text{ bar}}{1000 \text{ mbar}} \times \frac{10^5 \text{ Pa}}{1 \text{ bar}} \times \frac{1 \text{ kPa}}{1000 \text{ Pa}} \times \frac{1 \text{ atm}}{101.325 \text{ kPa}} = 0.79 \text{ atm decrease}$$

$$1.00 \text{ atm} - 0.79 \text{ atm} = 0.21 \text{ atm}$$

$$0.21 \text{ atm} \times \frac{760 \text{ mm Hg}}{1 \text{ atm}} = 160 \text{ mm Hg}$$

Looking at Figure 11.5, this corresponds to a boiling temperature between 40 °C and 60 °C, about 57 °C.

✓ *Reasonable Answer Check:* It makes sense that a lower pressure causes a liquid to have a lower boiling T.

25. *Answer:* **1600 mm Hg**

Strategy and Explanation: Given the enthalpy of vaporization, and the vapor pressure at a given temperature, determine the vapor pressure at another given temperature.

Use the two-set Clausius-Claypeyron equation:

$$\ln\left(\frac{P_2}{P_1}\right) = \frac{-\Delta H_{vap}}{R}\left(\frac{1}{T_2} - \frac{1}{T_1}\right)$$

R is given in Table 10.4. $T_1 = 90°C + 273.15 = 363.15$ K $\cong 360$ K *(round to tens place)*

$$T_2 = 130°C + 273.15 = 403.15 \text{ K} \cong 4.0 \times 10^2 \text{ K}$$

$$\ln\left(\frac{P_2}{370 \text{ mmHg}}\right) = \frac{-\left(44.0 \text{ kJ / mol}\right)}{\left(8.314 \frac{J}{K \cdot mol}\right)\left(\frac{1 \text{ kJ}}{1000 \text{ J}}\right)}\left(\frac{1}{400 \text{ K}} - \frac{1}{360 \text{ K}}\right)$$

$$\ln\left(\frac{P_2}{370 \text{ mmHg}}\right) = \frac{-\left(44.0\right)}{\left(0.008314\right)}\left(0.0025 - 0.0028\right) = \frac{-\left(44.0\right)}{\left(0.008314\right)}\left(-0.0003\right)$$

$$\ln\left(\frac{P_2}{370 \text{ mmHg}}\right) = 1.5$$

$$P_2 = \left(370 \text{ mmHg}\right) e^{1.5} = 1600 \text{ mm Hg}$$

✓ *Reasonable Answer Check:* Since the temperature increases, the vapor pressure should be larger.

27. *Answer:* **70. kJ/mol**

Strategy and Explanation: Given two temperatures that represent the temperature changes required to double the vapor pressure. Determine the enthalpy of vaporization.

Use the two-set Clausius-Claypeyron equation (given in the solution to Question 25).

R is given in Table 10.4. The pressure doubles, so the pressure ratio is: $\frac{P_2}{P_1} = 2$

$$T_1 = 70.0°C + 273.15 = 343.2 \text{ K}$$

$$T_2 = 80.0°C + 273.15 = 353.2 \text{ K}$$

$$\ln\left(2\right) = \frac{-\Delta H_{vap}}{\left(8.314 \frac{J}{K \cdot mol}\right)\left(\frac{1 \text{ kJ}}{1000 \text{ J}}\right)}\left(\frac{1}{353.2 \text{ K}} - \frac{1}{343.2 \text{ K}}\right)$$

$$\ln\left(2\right) = \frac{-\Delta H_{vap}}{\left(0.008314 \frac{kJ}{mol}\right)}\left(0.002914 - 0.002832\right) = \frac{-\Delta H_{vap}}{\left(0.008314 \frac{kJ}{mol}\right)}\left(-0.000083\right)$$

$$0.693 = -\Delta H_{vap}\left(-0.0099 \frac{mol}{kJ}\right)$$

$$70. \frac{kJ}{mol} = \Delta H_{vap}$$

✓ *Reasonable Answer Check:* The enthalpy of vaporization must be positive, since energy is required to take a liquid into the gas state. The size is similar to the enthalpy of vaporizations given in earlier Questions.

Phase Changes: Solids, Liquids, and Gases

28. *Answer/Explanation:* A high melting point and a high heat of fusion tell us that a large amount of energy is required to melt a solid. That is the case when the interparticle interactions between the particles in the solid are very strong, such as in solids composed of ions.

30. *Answer/Explanation:* A higher heat of fusion occurs when the intermolecular forces are stronger. The intermolecular forces between the molecules of H_2O in the solid (hydrogen bonding) are stronger than the intermolecular forces between the molecules of H_2S (dipole-dipole).

31. *Answer:* **27 kJ**

Strategy and Explanation: Given the molar enthalpies of fusion and vaporization of a compound, determine the heat energy required to raise a given number of moles of solid to the melting point, melt it, raise the liquid to the boiling point and boil it.

Find the mass from the moles. Use the molar enthalpies to get total heat energy for both of the phase transitions. Rearrange Equation 6.2 and use Table 6.1 to get the heat energy needed to change the temperature of the solid and liquid water (as described in Chapter 6, Section 6.3). Add together the heat energy for each stage to get the total quantity of heat energy.

$$0.50 \text{ mol } H_2O \times \frac{18.0152 \text{ g } H_2O}{1 \text{ mol } H_2O} = 9.0 \text{ g } H_2O$$

Total heat energy is the sum of the heat energies to warm the ice to the melting point (0 °C), the heat energy required to melt the ice, the heat energy to warm the water to the boiling point (100. °C), and the heat energy required to vaporize the water.

$$q_{tot} = q_{heat\ ice} + q_{melt} + q_{heat\ water} + q_{boil}$$

$$q_{tot} = (c_{ice} \times m \times \Delta T) + (n\Delta H_{fus}) + (c_{liquid} \times m \times \Delta T) + (n\Delta H_{vap})$$

$$q_{tot} = \left(\frac{2.06 \text{ J}}{g \, °C}\right) \times \left(\frac{1 \text{ kJ}}{1000 \text{ J}}\right) \times (9.0 \text{ g}) \times [0 \, °C - (-5 \, °C)] \ + \ (0.50 \text{ mol}) \times \left(\frac{6.020 \text{ kJ}}{mol}\right)$$

$$+ \left(\frac{4.184 \text{ J}}{g \, °C}\right) \times \left(\frac{1 \text{ kJ}}{1000 \text{ J}}\right) \times (9.0 \text{ g}) \times [100. \, °C - (0 \, °C)] \ + \ (0.50 \text{ mol}) \times \left(\frac{40.07 \text{ kJ}}{mol}\right)$$

$$q_{tot} = 0.09 \text{ kJ} + 3.0 \text{ kJ} + 3.8 \text{ kJ} + 20. \text{ kJ} = 27 \text{ kJ}$$

✓ *Reasonable Answer Check:* The relative size of the four terms seems sensible, comparing the sizes of the heat capacities and the heats of the phase transitions.

33. *Answer:* **51.9 g CCl_2F_2**

Strategy and Explanation: Given the heat of vaporization of a compound, the heat of fusion for ice, and the specific heat capacity of water, determine what mass of the liquid compound must evaporate to lower the temperature of a sample of water to the freezing point and freeze it.

Find the mass from the moles of water. Use the heat for changing the temperature of the liquid water (as described in Chapter 6, Section 6.3). Determine the molar heat of crystallization from the molar heat of fusion and use it to get total heat energy for the phase transition. Add together the heat energy for each stage to get the total quantity of heat energy required. Use the heat of vaporization of the compound to determine the mass.

$$2.00 \text{ mol } H_2O \times \frac{18.0152 \text{ g } H_2O}{1 \text{ mol } H_2O} = 36.0 \text{ g } H_2O$$

Total heat energy for freezing the water is the sum of the heat energy to cool the ice to the freezing point (0 °C) and the heat energy required to freeze it.

$$q_{tot\ for\ water} = q_{cool\ ice} + q_{freeze}$$

$$q_{tot\ for\ water} = (c_{water} \times m \times \Delta T) + (n\Delta H_{cryst})$$

$$\Delta H_{cryst} = -\Delta H_{fus} = -6.02 \text{ kJ/mol}$$

$$q_{\text{tot for water}} = \left(\frac{4.184 \text{ J}}{\text{g} \,^\circ\text{C}}\right) \times \left(\frac{1 \text{ kJ}}{1000 \text{ J}}\right) \times (36.0 \text{ g}) \times [0 \,^\circ\text{C} - (-20. \,^\circ\text{C})] \;\; + \;\; (2.00 \text{ mol}) \times \left(\frac{-6.02 \text{ kJ}}{\text{mol}}\right)$$

$$q_{\text{tot for water}} = -3.0 \text{ kJ} + (-12.0 \text{ kJ}) = -15.0 \text{ kJ}$$

15.0 kJ must be removed to freeze the water, so determine how many grams of CCl_2F_2 must evaporate to use up the 15.0 kJ of thermal energy.

$$15.0 \text{ kJ} \times \frac{1000 \text{ J}}{1 \text{ kJ}} \times \frac{1 \text{ g } CCl_2F_2}{289 \text{ J}} = 51.9 \text{ g } CCl_2F_2 \text{ must evaporate}$$

✓ *Reasonable Answer Check:* The relative size of the two terms seems sensible, comparing the sizes of the heat capacity and the heat energy of the phase transition. The calculated mass of CCl_2F_2 would definitely fit inside a typical freezer compressor.

35. *Answer/Explanation:* A higher melting point is a result of stronger interparticle forces. Coulomb's law describes the attraction between charged particles; the ion-ion coulombic interaction is stronger with smaller distance between the charges because the charges are more localized and closer together. According to the periodic trends described in Section 7.10, Li^+ and F^- are smaller than Cs^+ and I^-. Therefore, LiF experiences higher coulombic interactions than those in solid CsI, and LiF has a higher melting point.

37. *Answer:* **Highest melting point is (a) SiC. Lowest melting point is (d) $CH_3CH_2CH_2CH_3$.**

Strategy and Explanation: The highest melting point is a result of strongest interparticle forces. Covalent interactions are stronger forces in solids than intermolecular forces between separate molecules. The nonmetal-nonmetal compound, SiC, would be held together by an extended covalent network. Rb is a metal from Group 1A, and alkali metals have relatively low melting points. The other two interact by London forces, so the highest melting point is (a) SiC.

The lowest melting point is a result of the weakest intermolecular forces. We need to compare the intermolecular forces in the two molecules, I_2 (molar mass 254 g/mol) and $CH_3CH_2CH_2CH_3$ (molar mass 54 g/mol). The larger I_2 molecules can experience more significant London forces than the smaller $CH_3CH_2CH_2CH_3$ molecules, so (d) $CH_3CH_2CH_2CH_3$ has the lowest melting point.

40. *Answer:* **(a) gas phase (b) liquid phase (c) solid phase**

Strategy and Explanation:

(a) The point at $-70\,^\circ$C and 1 atm is just below and to the right of the solid/gas equilibrium line, so the phase present will be the gas phase.

(b) The point at $-40\,^\circ$C and 15.5 atm is to the right of the solid/liquid equilibrium line and above the liquid/gas equilibrium line, so the phase present will be the liquid phase.

(c) The point at $-80\,^\circ$C and 4.7 atm is below and to the left of the solid/liquid equilibrium line and above the solid/gas equilibrium line, so the phase present will be the solid phase.

Types of Solids

42. *Answer:* **(a) molecular solid (b) metallic solid (c) network solid (d) ionic solid**

Strategy and Explanation:

(a) P_4O_{10} is a molecule, so it forms a molecular solid.

(b) Brass is composed of various metals, so it is a metallic solid.

(c) Graphite is an extended array of covalently bonded carbon, so it is a network solid.

(d) $(NH_4)_3PO_4$ is composed of common ions, NH_4^+ and PO_4^{3-}, so it is an ionic solid.

44. *Answer:* **(a) amorphous solid (b) molecular solid (c) ionic solid (d) metallic solid; see explanations below.**

Strategy and Explanation: Pure substances have fixed melting temperatures, whereas mixtures and amorphous solids have ill-defined melting points. Network and ionic solids have much higher melting points, because of stronger interparticle bonds. Solids that conduct are made from metals. Liquids that conduct could be metals or ions.

(a) A soft, slippery solid that has no definite melting point is probably an amorphous solid.

(b) Violet crystals with moderate (not high) melting point that do not conduct in either phase probably represent a molecular solid.

(c) Hard colorless crystal with a high melting point and liquid that conducts is probably an ionic solid.

(d) A hard solid that melts at a high temperature and conducts in both solid and liquid states is a metallic solid.

45. *Answer:* **(a) molecular solid (b) ionic solid (c) metallic solid or network solid (d) amorphous solid**

Strategy and Explanation: Use the characterization conclusions as described in the solution to Question 44:

(a) A solid that melts below 100 °C and is insoluble in water is probably a nonpolar molecular solid.

(b) An ionic solid will conduct electricity only when melted.

(c) A solid that is insoluble in water and conducts electricity is probably a metallic solid, though it might be a network solid like graphite.

(d) A noncrystalline solid that has a wide melting point range is an amorphous solid.

Crystalline Solids

47. *Answer/Explanation:* See Figure 11.21 and its description.

49. *Answer:* **220 pm**

Strategy and Explanation: Given the length of the edge of a face-centered cubic unit cell of xenon, determine the radius of the xenon atoms.

Use the figure and method similar to those described in Problem-Solving Example 11.7. The face diagonal distance is the hypotenuse of an equilateral triangle. Diagonal distance = $\sqrt{2} \times$ (edge). We can see from the figure that the diagonal distance represents four times the radius of the atom.

Use these two relationships to find the radius of the atom.

$$\text{Diagonal distance} = \sqrt{2} \times (\text{edge length}) = \sqrt{2} \times (620 \text{ pm}) = 880 \text{ pm}$$

$$\text{Radius} = \frac{\text{Diagonal distance}}{4} = \frac{880 \text{ pm}}{4} = 220 \text{ pm}$$

✓ *Reasonable Answer Check:* The geometric relationships are logical. We expect the radius to be less than the edge length. The van der Waals radius for Xenon is 216 pm.

52. *Answer:* **body diagonal length = 696 pm; side length = 401 pm.**

Strategy and Explanation: Given the radii of cesium and chloride ions, determine the length of the body diagonal and the length of the side of the unit cell.

Use Figure 11.23 showing the unit cell. The body diagonal runs through one whole Cs^+ ion and from the center to the edge of two Cl^- ions. The body diagonal is the hypotenuse of a triangle with sides represented by the face diagonal and the edge, so:

$$\left(\text{Body diagonal length}\right)^2 = (\text{edge length})^2 + \left(\text{face diagonal length}\right)^2$$

The face diagonal is the hypotenuse of an equilateral triangle.

$$\text{Face diagonal length} = \sqrt{2} \times (\text{edge length})$$

Use these three relationships to find the length of the body diagonal and the length of the side of the unit cell.

$$\text{Diagonal distance} = 2 \times \text{Radius } Cs^+ + 2 \times \text{Radius } Cs^+$$

$$= 2 \times 167 \text{ pm} + 2 \times 181 \text{ pm} = 696. \text{ pm}$$

$$\left(\text{Body diagonal length}\right)^2 = (\text{edge length})^2 + \left[\sqrt{2} \times (\text{edge length})\right]^2 = 3 \times (\text{edge length})^2$$

$$\text{(edge length)}^2 = \frac{\left(\text{Body diagonal length}\right)^2}{3} = \frac{\left(696 \text{ pm}\right)^2}{3} = 1.61 \times 10^3 \text{ pm}^2$$

$$\text{edge length} = \sqrt{1.61 \times 10^3 \text{ pm}^3} = 401 \text{ pm}$$

54. *Answer:* **(a) 152 pm (b) r_{I^-} = 212 pm, r_{Li^+} = 88.0 pm (c) see discussion below**

Strategy and Explanation: Given the lengths of the sides of the unit cells for solid lithium metal and solid lithium iodide, the type of structure, and some assumptions of which atoms are touching, determine the radius of Li atom and the radii of Li^+ and I^-.

Use the geometrical relationships described in the solution to Question 52.

(a) Looking at Figure 11.21 at the body-centered cubic, and assuming that the atoms touch along the body diagonal, we get: Body diagonal length = $4r_{Li}$

Given the edge length of the Li unit cell, 351 pm, calculate the body diagonal length:

$$\text{Body diagonal length} = \sqrt{3 \times \text{(edge length)}^2} = \sqrt{3} \times 351 \text{ pm} = 608 \text{ pm}$$

Use body diagonal length to get radius of Li atom:

$$r_{Li} = \frac{\text{Body diagonal length}}{4} = \frac{608 \text{ pm}}{4} = 152 \text{ pm}$$

(b) Looking at the cubic crystal lattice structure shown for NaCl in Figure 11.24, and assuming that the I^- ions touch each other along the face diagonal and the I^- ions touch the Na^+ ion along the edge:

$$\text{Face diagonal length} = 4 \, r_{I^-} \qquad\qquad \text{Edge length} = 2 \, r_{Li^+} + 2 \, r_{I^-}$$

Given the edge length of the LiI unit cell, 600. pm, calculate the face diagonal length:

(For sig. figs, assume that this length is as precise as the length in (a), ± 1 pm.)

$$\text{Face diagonal length} = \sqrt{2} \times \text{(edge length)} = \sqrt{2} \times 600. \text{ pm} = 849 \text{ pm}$$

Use face diagonal length to get radius of I^- ion:

$$r_{I^-} = \frac{\text{Body diagonal length}}{4} = \frac{849 \text{ pm}}{4} = 212 \text{ pm}$$

Use the edge length of the LiI unit cell radius of I^- ion to get the radius of Li^+ ion:

$$2 \, r_{Li^+} = \text{Edge length} - 2 \, r_{I^-} = 600. \text{ pm} - 2 \times 212 \text{ pm} = 176 \text{ pm}$$

$$r_{Li^+} = 88.0 \text{ pm}$$

(c) It is reasonable that the Li atom is larger than the Li^+ cation. Figure 7.21 gives exactly the same value for the radius of the Li atom (152 pm). It gives a slightly larger value for Li^+ ion (90 pm) and slightly larger value for I^- (206 pm). The assumption that I^- anions touch each other is reasonable, since the very tiny Li^+ ions cannot span the entire gap between the I^- ions in the unit cell.

✓ *Reasonable Answer Check:* The geometric relationships are logical. The relative sizes of the atom and ion radii, the body diagonal length, and the edge length are all reasonable. The sum of the Figure 7.21 radii to back-calculate the edge length (2 × 90 pm + 2 × 206 pm) gives 592 pm, a little less than the actual measured length, suggesting that the ions do not touch. Similarly, the face diagonal length (4 × 206 pm) gives 824 pm, a little less than the actual measured length.

Network Solids

55. *Answer/Explanation:* Carbon atoms in diamond are sp^3 hybridized and are tetrahedrally bonded to four other carbon atoms. Carbon atoms in pure graphite are sp^2 hybridized and bonded with a triangular planar shape to other carbon atoms. These bonds are partially double bonded so they are shorter than the single bonds in diamond. However, the planar sheets of sp^2 hybridized carbon atoms are only weakly attracted by intermolecular forces to adjacent layers, so these interplanar distances in graphite are much longer than the C–C single bonds in the diamond. The net result is that graphite is less dense than diamond.

57. *Answer/Explanation:* Diamond is an electrical insulator because all the electrons are in single bonds, which are shared between two specific atoms and cannot move around. However, graphite is a good conductor of electricity because its electrons are delocalized in conjugated double bonds, which allow the electrons to move easily through the graphite sheets.

Metals, Semiconductors, and Insulators

59. *Answer/Explanation:* In a conductor, the valence band is only partially filled, whereas, in an insulator, the valence band is completely full, the conduction band is empty, and there is a wide energy gap between the two. In a semiconductor, the gap between the valence band and the conduction band is very small so that electrons are easily excited into the conduction band.

61. *Answer/Explanation:* Substance (c), Ag, has the greatest electrical conductivity because it is a metal. Substance (d), P_4, has the smallest electrical conductivity because it is a nonmetal. (The other two are metalloids.)

General Questions

64. *Answer:* **780 kJ to heat the liquid, 1.6×10^4 kJ total to reach the final state and temperature.**

Strategy and Explanation: Given the normal boiling point, molar heat of vaporization, and the specific heat capacities of the gas and liquid states of a compound, determine the heat energy evolved when a given mass of the substance is cooled from an initial to a final temperature.

Use techniques similar to those described in the solution to Question 31.

$$10. \text{ kg NH}_3 \times \frac{1000 \text{ g NH}_3}{1 \text{ kg NH}_3} = 1.0 \times 10^4 \text{ g NH}_3$$

$$10. \text{ kg NH}_3 \times \frac{1000 \text{ g NH}_3}{1 \text{ kg NH}_3} \times \frac{1 \text{ mol NH}_3}{17.0304 \text{ g NH}_3} = 5.9 \times 10^2 \text{ mol NH}_3$$

A change in temperature in Kelvin degrees is the same as the change in temperature in Celsius degrees. So, use $c_{gas} = 2.2$ J/g °C and $c_{liquid} = 4.7$ J/g °C

Find the heat energy absorbed when heating the liquid to the boiling point (– 33.4 °C)

$$q_{\text{heating liquid}} = c_{\text{liquid}} \times m \times \Delta T$$

$$q_{\text{heating liquid}} = \left(\frac{4.7 \text{ J}}{\text{g °C}}\right) \times (1.0 \times 10^4 \text{ g NH}_3) \times [-33.4 \text{ °C} - (-50.0 \text{ °C})] \times \left(\frac{1 \text{ kJ}}{1000 \text{ J}}\right) = 780 \text{ kJ}$$

The total heat energy absorbed to get to 0.0 °C is the sum of the heat energy absorbed heating the liquid, the heat energy absorbed when vaporizing the liquid to a gas, and the heat energy absorbed when heating the gas to the final temperature (0.0 °C). $q_{tot} = q_{\text{heating liquid}} + (n \times \Delta H_{vap}) + (c_{gas} \times m \times \Delta T)$

$$q_{tot} = 780 \text{ kJ} + 5.9 \times 10^2 \text{ mol} \times \left(\frac{23.5 \text{ kJ}}{\text{mol}}\right) + \left(\frac{2.2 \text{ J}}{\text{g °C}}\right) \times (1.0 \times 10^4 \text{ g NH}_3) \times [0.0 \text{ °C} - (-33.4 \text{ °C})] \times \left(\frac{1 \text{ kJ}}{1000 \text{ J}}\right)$$

$$q_{tot} = 780 \text{ kJ} + (14000 \text{ kJ}) + (730 \text{ kJ}) = 16000 \text{ kJ} = 1.6 \times 10^4 \text{ kJ}$$

✓ *Reasonable Answer Check:* The relative size of the three terms seems sensible, comparing the sizes of the heat capacities and the heat of vaporization.

66. *Answer:* **(a) dipole-dipole forces and London forces (b) $CH_4 < NH_3 < SO_2 < H_2O$**

Strategy and Explanation:

(a) The molecular geometry of the SO_2 molecule is determined in the solution to Question 11(c) in Chapter 9:

SO_2 (18 e⁻)

The type is AX_2E_1, so the electron-pair geometry is triangular planar and the molecular geometry is angular (120°).

The asymmetric shape of the SO_2 molecule means that it is polar. Therefore, the strongest intermolecular forces between molecules in solid and liquid SO_2 are dipole-dipole forces. All molecules experience London forces, so that force is part of the attractions in the liquid and solid states of SO_2, also.

(b) The normal boiling point is smaller when the intermolecular attractions are weaker. The normal boiling point is larger when the intermolecular attractions are stronger.

$$CH_4 \ (-161.5 \ °C) < NH_3 \ (-33.4 \ °C) < SO_2 \ (-10 \ °C) < H_2O \ (100 \ °C)$$

✓ *Reasonable Answer Check:* Ordering the molecules by boiling points indicates the surprising result that the intermolecular attractions in SO_2 are stronger than the attractions in NH_3. This is probably a result of variable size. The molar mass of SO_2 is 64 g/mol, but the molar mass of NH_3 is only 17 g/mol. SO_2 has three atoms from period two and NH_3 has only one atom from period two. So it makes sense SO_2 molecules may have significant enough London forces to be more attractive to each other than are tiny, polar molecules of NH_3 undergoing hydrogen bonding.

Applying Concepts

67. *Answer:* (a) 80 mm Hg (b) 18 °C (c) 640 mm Hg (d) diethyl ether and ethanol (e) diethyl ether would evaporate; ethanol and water would remain liquids (f) water

Strategy and Explanation:

(a) To find the equilibrium vapor pressure for ethyl alcohol at room temperature, look at Figure 11.5 and find the pressure where the ethyl alcohol curve passes a temperature of 25 °C. This is about 80 mm Hg.

(b) To find the temperature when the equilibrium vapor pressure for diethyl ether is 400 mm Hg, look at Figure 11.5 and find the temperature where the diethyl ether curve passes a pressure of 400 mm Hg. This is about 18 °C.

(c) Find the equilibrium vapor pressure for water at 95 °. To do this, look at Figure 11.5 and find the pressure where the water curve passes a temperature of 95 °C. This is about 640 mm Hg.

(d) To find out which substances will be gases under specific condition of 200 mm Hg and 60 °C, look at Figure 11.5 and find the pressure where each curve passes a temperature of 60 °C. If their vapor pressure is above 200 mm Hg, they will be gases. Here, both diethyl ether and ethanol are gases.

(e) To find out which substance will immediately evaporate in your hand, look at Figure 11.5 and find the pressure where each curve passes body temperature of 37 °C. If their pressure is close to 760 mm Hg they will readily evaporate. Here, diethyl ether will evaporate readily. The ethanol and water would evaporate more slowly.

(f) Water has the strongest intermolecular attractions as empirically indicated by having a lower vapor pressure than the other two liquids at any common temperature.

69. *Answer/Explanation:* The butane in the lighter is under great enough pressure that the vapor pressure of butane at room temperature is less than the pressure inside the lighter. Hence, it exists as a liquid.

70. *Answer:* **Diagram 1 and region C; diagram 2 and line E, diagram 3 and region B, diagram 4 and line F, diagram 5 and line G, diagram 6 and point H, diagram 7 and region A**

Strategy and Explanation: Refer to Section 11.3, if you need to be reminded about the different parts of a phase diagram.

Nanoscale diagram 1 looks like the substance atoms are all in the gas phase. That is region C on the phase diagram.

Nanoscale diagram 2 looks like some of the substance atoms are in the solid phase (atoms piled up in a regular array) and some of the substance atoms are in the gas phase. That is a point on the line described by E on the phase diagram.

Nanoscale diagram 3 looks like the substance atoms are all in the liquid phase. That is region B on the phase diagram.

Nanoscale diagram 4 looks like some of the substance atoms are in the solid phase and some of the substance atoms are in the liquid phase. That is a point on the line described by F on the phase diagram.

Nanoscale diagram 5 looks like some of the substance atoms are in the liquid phase and some of the substance atoms are in the gas phase. That is a point on the line described by G on the phase diagram.

Nanoscale diagram 6 looks like some of the substance atoms are in the liquid phase, some of the substance atoms are in the liquid phase, and some of the substance atoms are in the gas phase. That is the point described by H on the phase diagram.

Nanoscale diagram 7 looks like the substance atoms are all in the solid phase. That is region A on the phase diagram.

More Challenging Questions

72. *Answer/Explanation:* Vapor-phase water condenses on contact with cool skin. The condensation of steam into water is exothermic. That energy is also absorbed by the skin causing more burning, along with the burn resulting from the heat energy given off as the temperature drops (common in each case).

75. *Answer:* **(a) condensation then freezing (b) triple point (c) melting point curve**

Strategy and Explanation: Phase diagrams are described in Section 11.3.

(a) The substance's initial state (F) is gaseous (at low-enough pressure and high-enough temperature). The temperature remains fixed as the transition toward point G, but the pressure is increased. At a sufficiently high pressure (point E), the gas undergoes **condensation** to a liquid. Just before arriving at point G, the liquid undergoes a change to solid by **freezing**.

(b) The point (A) where the vapor pressure curve meets the point curve and the sublimation point curve, is called the **triple point**.

(c) The AB line is a segment of the **melting point curve**, and indicates how the melting temperature change over a range of pressures, from the triple point, point A, to atmospheric pressure (1 atm), at point B.

77. *Answer:* **(a) 560 mm Hg (b) benzene (c) 73°C (d) methyl ethyl ether is 7°C; carbon disulfide is 47°C; benzene is 81°C (these are all approximate, due to the size of the graph provided)**

Strategy and Explanation: Answers to (a), (c) and (d) are obtained from the graph, by locating the given information on the appropriate axis and following that horizontal or vertical to the point on the appropriate curve.

(a) Finding 0°C on the horizontal axis and tracing a vertical line to the methyl ethyl ether curve shows that the vapor pressure of methyl ethyl ether at 0°C is approximately 560 mm Hg.

(b) A substance will remain in the liquid state at higher temperatures if its molecules experience higher intermolecular forces; hence, its vapor pressure will be lower. Therefore, of the three liquids compared here, molecules in benzene experience the greatest intermolecular forces.

(c) Finding 600 mm Hg on the vertical axis and tracing a horizontal line to the benzene curve gives a temperature of approximately 73°C.

(d) Normal boiling point is the boiling temperature at 1 atm (760 mm Hg). Finding 760 mm Hg on the vertical axis and tracing horizontal lines to each curve gives the following normal boiling points: for methyl ethyl ether, approximately 7°C; for carbon disulfide, approximately 47°C, for benzene, approximately 81°C.

79. *Answer:* **22.2°C**

Strategy and Explanation: Given the enthalpy of vaporization for water, determine the boiling point at a specific vapor pressure.

Use the two-set Clausius-Claypeyron equation (given in the solution to Question 25) and the normal boiling point of water (See Table 11.2). R is given in Table 10.4. $P_1 = 24$ mm Hg

$$T_2 = 100.00°C + 273.15 = 373.15 \text{ K}$$
$$P_2 = 1 \text{ atm} = 760.0 \text{ mm Hg}$$

$$\ln\left(\frac{760.0 \text{ mm Hg}}{24 \text{ mm Hg}}\right) = \frac{-(40.7 \text{ kJ/mol})}{\left(8.314 \frac{J}{K \cdot mol}\right)\left(\frac{1 \text{ kJ}}{1000 \text{ J}}\right)}\left(\frac{1}{373.15 \text{ K}} - \frac{1}{T_1}\right)$$

$$\ln(31.67) = 3.455 = \frac{-(40.7)}{(0.008314)}\left(0.0026799 - \frac{1}{T_1}\right)$$

$$-\frac{(3.4\underline{5}5)(0.008314)}{(40.7)} = 0.0026799 - \frac{1}{T_1}$$

$$-0.000706 = 0.0026799 - \frac{1}{T_1}$$

$$\frac{1}{T_1} = 0.0026799 + 0.000706 = 0.00339$$

$$T_1 = 295.3 \text{ K}$$

$$T_1 = 295.3 \text{ K} - 273.15 = 22.2°C$$

✓ *Reasonable Answer Check:* Since the pressure is lower than the pressure for the normal boiling temperature, the boiling point must be lower than the normal boiling point.

80. *Answer:* **0.533 g/cm³**

Strategy and Explanation: Given the lengths of the edge of the unit cells for solid lithium metal, the type of structure, and the temperature, determine the density of the metal at this temperature.

Find the edge length in units of centimeters so that the density will come out in g/cm³. Determine the volume of the unit cell, and the mass of the atoms in that cell. Then, divide mass by volume to get density.

$$351 \text{ pm} \times \frac{10^{-12} \text{ m}}{1 \text{ pm}} \times \frac{1 \text{ cm}}{10^{-2}} = 3.51 \times 10^{-8} \text{ cm}$$

$$V = (\text{edge}) = (3.81 \times 10^{-8} \text{ cm})^3 = 4.32 \times 10^{-23} \text{ cm}^3$$

In Section 11.6 we learn that a body-centered cubic unit cell has 2 atoms in it.

$$2 \text{ atoms Li} \times \frac{1 \text{ mole Li}}{6.022 \times 10^{23} \text{ atoms Li}} \times \frac{6.941 \text{ g Li}}{1 \text{ mole Li}} = 2.305 \times 10^{-23} \text{ g Li}$$

$$\text{Density} = \frac{2.305 \times 10^{-23} \text{ g}}{4.32 \times 10^{-23} \text{ cm}^3} = 0.533 \text{ g/cm}^3$$

✓ *Reasonable Answer Check:* Lithium has a small molar mass. Table 1.1 shows various metals (not Li) and the metals with small molar masses have smaller densities.

Chapter 12: Chemical Kinetics: Rates of Reactions

Introduction

Try to get used to thinking about kinetics at a particulate level by visualizing changing concentrations, orientations, surface area, and speed of reactants and the effect on the rate.

If you are studying Chapters 12 and 13 together, be careful not to confuse kinetics and equilibrium by (1) confusing k, the rate constant for a kinetic process, and K, the equilibrium constant for an equilibrium process, (2) thinking a small value for k indicates that the reaction will go slowly, and (3) using coefficients from balanced chemical equations for the powers in a rate law.

The effect of temperature on the rate of a reaction is seen in cooking and in the difficulty in starting a car engine on a cold winter morning.

A dust explosion is a good example of the effect of surface area on the rate of a reaction. There are some interesting television shows on TV that routinely feature videos of explosions, such as MythBusters and Time Warp on the Discovery Channel. Slow-motion clips of such explosions are often easy to find on the internet.

Solutions to Blue-Numbered Questions
for Review and Thought for Chapter 12

Topical Questions

Reaction Rate

7. *Answer/Explanation:* The smaller the grains of sugar, the faster the sugar will dissolve, because the surface contact with the water is essential to the dissolving process. The most finely granulated sugar is (d) powdered sugar. The second most finely granulated sugar is (c) granulated sugar. Granulated sugar compressed into cubes, or (b) sugar cubes, would be somewhat slower than granulated sugar to dissolve. The slowest of the four to dissolve would be large sugar crystals in (a) rock candy sugar.

9. *Answer:* **(a) 0.23 mol/L·h   (b) 0.20 mol/L·h   (c) 0.161 mol/L·h   (d) 0.12 mol/L·h   (e) 0.090 mol/L·h (f) 0.066 mol/L·h**

Strategy and Explanation: Given data of concentrations at various times during the course of a chemical reaction, find the rate of change for each time interval.

To calculate the average rate for each time interval, use the method described in Section 12.1:

(a) From 0 to 0.50 h: $\text{Rate}_{0-0.5} = \dfrac{0.849\ \text{mol}/\text{L} - 0.733\ \text{mol}/\text{L}}{0.50\ \text{h} - 0.00\ \text{h}} = \dfrac{0.116\ \text{mol}/\text{L}}{0.50\ \text{h}} = 0.23\ \dfrac{\text{mol}}{\text{L}\cdot\text{h}}$

(b) From 0.5 to 1.00 h: $\text{Rate}_{0.5-1.0} = \dfrac{0.733\ \text{mol}/\text{L} - 0.633\ \text{mol}/\text{L}}{1.00\ \text{h} - 0.50\ \text{h}} = \dfrac{0.100\ \text{mol}/\text{L}}{0.50\ \text{h}} = 0.20\ \dfrac{\text{mol}}{\text{L}\cdot\text{h}}$

(c) From 1.00 to 2.00 h: $\text{Rate}_{1.0-2.0} = \dfrac{0.633\ \text{mol}/\text{L} - 0.472\ \text{mol}/\text{L}}{2.00\ \text{h} - 1.00\ \text{h}} = \dfrac{0.161\ \text{mol}/\text{L}}{1.00\ \text{h}} = 0.161\ \dfrac{\text{mol}}{\text{L}\cdot\text{h}}$

(d) From 2.00 to 3.00 h: $\text{Rate}_{2.0-3.0} = \dfrac{0.472\ \text{mol}/\text{L} - 0.352\ \text{mol}/\text{L}}{3.00\ \text{h} - 2.00\ \text{h}} = \dfrac{0.120\ \text{mol}/\text{L}}{1.00\ \text{h}} = 0.120\ \dfrac{\text{mol}}{\text{L}\cdot\text{h}}$

(e) From 3.00 to 4.00 h: $\text{Rate}_{3.0-4.0} = \dfrac{0.352\ \text{mol}/\text{L} - 0.262\ \text{mol}/\text{L}}{4.00\ \text{h} - 3.00\ \text{h}} = \dfrac{0.090\ \text{mol}/\text{L}}{1.00\ \text{h}} = 0.090\ \dfrac{\text{mol}}{\text{L}\cdot\text{h}}$

(f)   From 4.00 to 5.00 h: $\text{Rate}_{3.0\text{-}4.0} = \dfrac{0.262 \text{ mol/L} - 0.196 \text{ mol/L}}{5.00 \text{ h} - 4.00 \text{ h}} = \dfrac{0.066 \text{ mol/L}}{1.00 \text{ h}} = 0.066 \dfrac{\text{mol}}{\text{L} \cdot \text{h}}$

✓ *Reasonable Answer Check:* According to the discussion of rate dependence on concentration in Section 12.2, it makes sense that the rate decreases as the reactant concentration decreases.

11. *Answer:* **(a) The average rate is directly proportional to average $[N_2O_5]$ for each pair. (b) k = 0.29 h^{-1}**

Strategy and Explanation: Given data of concentrations at various times during the course of a chemical reaction and the average rate calculated over different time intervals, show that the reaction obeys a specific given rate law, evaluate the rate constant by averaging.

The average rates over each time interval were calculated in the solution to Question 10. Find the average concentration over that interval. If the rate is proportional to the concentration, then a graph of concentration vs. rate should be linear.

(a)   From 0 to 0.50 h, the rate was 0.23 mol/L·h.

$$\text{Average concentration}_{0\text{-}0.5} = \frac{0.849 \text{ mol/L} + 0.733 \text{ mol/L}}{2} = 0.791 \frac{\text{mol}}{\text{L}}$$

From 0.5 to 1.00 h, the rate was 0.20 mol/L·h.

$$\text{Average concentration}_{0.5\text{-}1.0} = \frac{0.733 \text{ mol/L} + 0.633 \text{ mol/L}}{2} = 0.683 \frac{\text{mol}}{\text{L}}$$

From 1.00 to 2.00 h, the rate was 0.161 mol/L·h

$$\text{Average concentration}_{1.0\text{-}2.0} = \frac{0.633 \text{ mol/L} + 0.472 \text{ mol/L}}{2} = 0.553 \frac{\text{mol}}{\text{L}}$$

From 2.00 to 3.00 h, the rate was 0.120 mol/L·h

$$\text{Average concentration}_{2.0\text{-}3.0} = \frac{0.472 \text{ mol/L} + 0.352 \text{ mol/L}}{2} = 0.412 \frac{\text{mol}}{\text{L}}$$

From 3.00 to 4.00 h, the rate was 0.090 mol/L·h.

$$\text{Average concentration}_{3.0\text{-}4.0} = \frac{0.352 \text{ mol/L} + 0.262 \text{ mol/L}}{2} = 0.307 \frac{\text{mol}}{\text{L}}$$

From 4.00 to 5.00 h, the rate was 0.066 mol/L·h.

$$\text{Average concentration}_{4.0\text{-}5.0} = \frac{0.262 \text{ mol/L} + 0.196 \text{ mol/L}}{2} = 0.229 \frac{\text{mol}}{\text{L}}$$

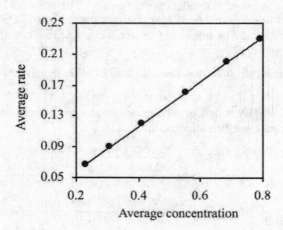

The linear relationship between rate and concentration shows the data satisfies the equation described in the Question:

$$\text{Rate} = k \, [N_2O_5]$$

(b)   From 0 to 0.50 h, the rate is 0.23 mol/L·h and the average concentration is 0.791 mol/L.

$$k = \frac{Rate}{[N_2O_5]} = \frac{0.23\frac{mol}{L \cdot h}}{0.791\frac{mol}{L}} = 0.29h^{-1}$$

Similarly, for each of the other time increments:

Average rate, mol/L·h	Average concentration, mol/L	$k\ (h^{-1})$
0.23	0.791	0.29
0.20	0.683	0.29
0.161	0.553	0.291
0.120	0.412	0.291
0.090	0.307	0.29
0.066	0.229	0.29

The average value of k is 0.29 h^{-1}.

✓ *Reasonable Answer Check:* According to the discussion of rate dependence on concentration in Section 12.2, it makes sense that the rate decreases as the reactant concentration decreases. It also makes sense that the value of k is the same each time it was calculated.

13. *Answer:* **See graph and explanations below**

Strategy and Explanation: Qualitatively, the concentration of the reactant will drop nonlinearly (first-order reaction) as the concentrations of the reactants increase. The rate of increase of NO concentration will mirror the rate of decrease of NO_2, and will be twice that of O_2, due to their stoichiometric ratios being 2 : 2 : 1. At late times, the concentrations of all species will level off to an unchanging value. (The dashed line on the graph is used to find the initial rate in (a).)

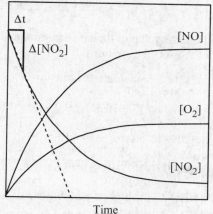

(a) As shown on the graph, the initial rate can be determined by taking a small time interval (Δt) near the initial time where the curve is still well approximated by a straight line (represented by the dashed line) and finding the slope of that line.

$$Rate = \frac{-\Delta[NO_2]}{\Delta t} = -\text{ slope of the straight line}$$

(b) As shown on the graph at very long times, the final rate will be zero because the concentration stops changing and the numerator of the rate expression is zero. That will happen for one of two reasons, either all the reactants will have been used up or the reaction will reach an equilibrium state.

Effect of Concentration on Reaction Rates

15. *Answer:* **(a) rate = k[NO$_2$]^2 (b) rate is quartered (c) rate is unchanged**

Strategy and Explanation:

(a) The reaction is second-order in NO_2 and zeroth-order in CO, so the rate law has powers of 2 and zero on the concentrations of NO_2 and CO, respectively.

$$\text{rate} = k[NO_2]^2[CO]^0$$

Because any number raised to the zero power results in 1, that says the rate is unaffected by the concentration of CO.

$$\text{rate} = k[NO_2]^2$$

(b) Knowing the rate law, the rate increase can be determined by creating a ratio to provide a simple relationship between the rate changes and the concentration changes.

Make a ratio of the rate law of the reaction under two different conditions: the initial condition (designated by subscript "1", $\text{rate}_1 = k[NO_2]_1^2$) compared to the final condition (designated by subscript "2", $\text{rate} = k[NO_2]_2^2$). A ratio gives the rate change factor:

$$\text{rate change factor} = \frac{\text{rate}_2}{\text{rate}_1} = \frac{k[NO_2]_2^2}{k[NO_2]_1^2} = \left(\frac{[NO_2]_2}{[NO_2]_1}\right)^2$$

When the concentration of NO_2 is halved, the concentration ratio is $\frac{[NO_2]_2}{[NO_2]_1} = \frac{1}{2}$, so the rate change factor

$= \left(\frac{1}{2}\right)^2 = \frac{1}{4}$. The rate will be quartered.

(c) Because the rate law does not include concentration dependence for CO, the rate will be unchanged when the CO concentration is doubled.

17. *Answer:* **(a) (i) 9.0×10^{-4} M/h, (ii) 1.8×10^{-3} M/h, (iii) 3.6×10^{-3} M/h; (b)-(d) see explanations below**

Strategy and Explanation:

(a) Given the rate law of a reaction, the value of the rate constant, and several concentrations, determine the instantaneous rates at each of those concentrations.

Plug the concentration and the rate constant into the rate law: $\text{rate} = k[Pt(NH_3)_2Cl_2]$

(i) $\text{rate}_{0.010} = (0.090 \text{ h}^{-1})(0.010 \text{ M}) = 9.0 \times 10^{-4}$ M/h

(ii) $\text{rate}_{0.020} = (0.090 \text{ h}^{-1})(0.020 \text{ M}) = 1.8 \times 10^{-3}$ M/h

(iii) $\text{rate}_{0.040} = (0.090 \text{ h}^{-1})(0.040 \text{ M}) = 3.6 \times 10^{-3}$ M/h

(b) If initial concentration of $Pt(NH_3)_2Cl_2$ is high, the initial rate of disappearance of $Pt(NH_3)_2Cl_2$ will be high. If initial concentration of $Pt(NH_3)_2Cl_2$ is low, the initial rate of disappearance of $Pt(NH_3)_2Cl_2$ will be small. The rate of disappearance of $Pt(NH_3)_2Cl_2$ is directly proportional to the concentration of $Pt(NH_3)_2Cl_2$.

(c) The rate law shows direct proportionality between rate and the concentration of $Pt(NH_3)_2Cl_2$.

(d) When the initial concentration of $Pt(NH_3)_2Cl_2$ is high, the rate of appearance of Cl^- will be high. When initial concentration is low, rate of appearance Cl^- will be small. The rate of appearance of Cl^- is directly proportional to the concentration of $Pt(NH_3)_2Cl_2$.

✓ *Reasonable Answer Check:* The rate is faster when the concentration is higher in (a).

19. *Answer:* **(a) rate = k [I][II]   (b) k = 1.04 L/mol·s**

Strategy and Explanation: Given the initial concentrations and initial rates of a reaction for several different experimental conditions for the same chemical reaction, determine the rate law and the rate constant for the reaction.

In Section 12.2, the method of finding the rate law from initial rates is described for getting the orders. However, comparing pairs of experiments where only one of the concentrations is different and relating that to the changes in the rate does not give consistent results. So, here we will derive a linear equation relating the concentrations and the rates and graph the results. Once the orders are determined, plug the data into the rate law to determine the value of k.

Because I and II are reactants and are both varied in the different experiments, we will seek a rate law looks like this: rate = $k[I]^i[II]^j$ where k, i, and j are currently unknown.

(a) Comparing experiments with [I] constant, the rate law can be simplified to:

$$rate = k'[II]^j \qquad\qquad where\ k' = k[I]^i$$

If we take the log of both sides of the equation, we can derive a linear relationship:

$$log(rate) = log(k') + log([II]^j)$$

$$log(rate) = log(k') + j\ log[II]$$

$$log(rate) = j\ log[II] + log(k')$$

Comparing to the equation of a line: $y = m\ x + b$, we see that if we plot

log(rate) against log[II], the slope of the line will be the reaction order, j.

Similarly comparing experiments with [II] constant, the rate law can be simplified to:

$$rate = k''[II]^I \qquad\qquad where\ k'' = k[II]^j$$

This gives a similar linear relationship: $log(rate) = i\ log[I] + log(k'')$

Comparing to the equation of a line: $y = m\ x + b$, we see that if we plot log(rate) against log[I], the slope of the line will be the reaction order, i.

Four data sets have constant [I] and four data sets have constant [II].

log[II] with constant [I]	log rate with constant [I]	log[I] with constant [II]	log rate with constant [II]
-4.745	-8.509	-4.783	-8.824
-4.453	-8.201	-4.606	-8.569
-4.149	-7.951	-4.304	-8.345
-3.975	-7.752	-3.827	-7.752

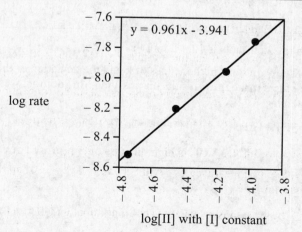

So, the reaction order of reactant II (j) is one.

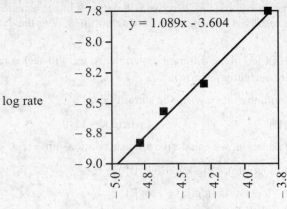

log[I] with [II] constant

So, the reaction order of reactant I (i) is also one. So, rate = k[I][II]

(b) Solve the rate law for k:

$$k = \frac{\text{rate}}{[I][II]}$$

Plug in each experiment's data. Here is an example of the first experiment's calculation:

$$k_1 = \frac{1.50 \times 10^{-9}\ \text{mol L}^{-1}\text{s}^{-1}}{(1.65 \times 10^{-5}\ \text{mol}/\text{L})(10.6 \times 10^{-5}\ \text{mol}/\text{L})} = 0.858\ \text{L mol}^{-1}\text{s}^{-1}$$

$[I] \times 10^5$ (mol/L)	$[I] \times 10^5$ (mol/L)	Initial rate $\times 10^9$ (mol L^{-1}s^{-1})	Rate constant (L mol^{-1}s^{-1})
1.65	10.6	1.50	0.858
14.9	10.6	17.7	1.12
14.9	7.10	11.2	1.06
14.9	3.52	6.30	1.20
14.9	1.76	3.10	1.18
4.97	10.6	4.52	0.858
2.48	10.6	2.70	1.03

The average of these seven rate constants is 1.04 L mol^{-1}s^{-1}

✓ *Reasonable Answer Check:* The variations in the values of k and the deviation of the slopes from the exact integer values expected for the orders make these results unsatisfying. In addition, we can try to get the value of k from the y-intercepts of the graphs:

First Graph: b = log(k') = log(k[I]^i)

$$k = \frac{10^b}{[I]^1} = \frac{10^{-3.941}}{\left(14.9 \times 10^{-5}\right)^1} = 0.769\ \text{L mol}^{-1}\text{s}^{-1}$$

Second Graph: b = log(k'') = log(k[II]^j)

$$k = \frac{10^b}{[II]^j} = \frac{10^{-3.604}}{\left(10.6 \times 10^{-5}\right)^1} = 2.35\ \text{L mol}^{-1}\text{s}^{-1}$$

These are not very close to each other either. These wide variations might suggest a systematic error in the collection of the data under various conditions.

20. *Answer:* **(a) The order with respect to NO is one and with respect to H_2 is one. (b) The reaction is second-order. (c) rate = k[NO][H_2] (d) k = 2.4×10^2 L mol^{-1}s^{-1} (e) initial rate = 1.5×10^{-2} mol L^{-1}s^{-1}**

Strategy and Explanation: Follow the method described in the answers to Questions 17 and 19.

The reactants are NO and H_2. Data are available for the changes in each of these reactants' concentrations, so the rate law looks like this:

$$\text{rate} = k[NO]^i[H_2]^j \quad \text{where k, i and j are currently unknown.}$$

(a) Looking at Experiments 1 and 2, the initial concentration of NO triples, the initial concentration of H_2 stays constant, and the initial rate changes by a factor of three. The rate change is the same as the concentration change, which suggests that the rate is proportional to the concentration of NO, and the order with respect to NO is one.

Looking at Experiments 2 and 3, the initial concentration of NO stays constant, the initial concentration of H_2 goes up by a factor of four ($10.0 \times 10^{-3}/2.50 \times 10^{-3}$), and the initial rate changes by a factor of four ($3.6 \times 10^{-2}/9.0 \times 10^{-3}$). The rate change is the same as the concentration change, which suggests that the rate is proportional to the concentration of H_2, and the order with respect to H_2 is also one.

(b) The overall order is the sum of each reactant order. Here, the reaction is second-order.

(c) The rate law now looks like this: rate = k[NO]^1[H_2]^1

(d) Solve the rate law for k: $k = \dfrac{\text{rate}}{[NO][H_2]}$

Plug in each experiment's data. Here is an example of the calculation for Experiment 1:

$$k_1 = \frac{3.0 \times 10^{-3} \text{ mol L}^{-1}\text{s}^{-1}}{(5.00 \times 10^{-3} \text{ mol / L})(2.50 \times 10^{-3} \text{ mol / L})} = 2.4 \times 10^2 \text{ L mol}^{-1}\text{s}^{-1}$$

[NO] (mol/L)	[H_2] (mol/L)	Initial rate (mol L^{-1}s^{-1})	Rate constant (L mol^{-1}s^{-1})
5.00×10^{-3}	2.50×10^{-3}	3.0×10^{-3}	2.4×10^2
15.0×10^{-3}	2.50×10^{-3}	9.0×10^{-3}	2.4×10^2
15.0×10^{-3}	10.0×10^{-3}	3.6×10^{-2}	2.4×10^2

The average of these three rate constants is 2.4×10^2 L mol^{-1}s^{-1}.

(e)   rate = k[NO]^1[H_2]^1 = $(2.4 \times 10^2$ L mol^{-1}s$^{-1})(8.0 \times 10^{-3}$mol L$^{-1})^1(8.0 \times 10^{-3}$ mol L$^{-1})^1$

$$= 1.5 \times 10^{-2} \text{ mol L}^{-1}\text{s}^{-1}$$

Rate Law and Order of Reaction

21. *Answer:* **(a) First-order in A, third-order in B, and fourth-order overall (b) First-order in A, first-order in B, and second-order overall (c) First-order in A, zero-order in B, and first-order overall (d) Third-order in A, first-order in B, and fourth-order overall**

Strategy and Explanation: Reaction order is the power to which the concentration of a component is raised. The overall order is the sum of all the individual orders.

(a) In the rate law: Rate = k[A][B]^3 the order with respect to A is one and the order with respect to B is three. The overall order is (1 + 3 =) four.

(b) In the rate law: Rate = k[A][B] the order with respect to A is one and the order with respect to B is one. The overall order is (1 + 1 =) two.

(c) In the rate law: Rate = k[A] the order with respect to A is one and the order with respect to B is zero. The overall order is (1 + 0 =) one.

(d) In the rate law: Rate = k[A]^3[B] the order with respect to A is three and the order with respect to B is one. The overall order is (3 + 1 =) four.

23. *Answer:* **(a) rate = k[CH₃COCH₃][H₃O⁺]; first order in both H₃O⁺ and CH₃COCH₃, and zero order in Br₂ (b) 4 × 10⁻³ L mol⁻¹s⁻¹ (c) 2 × 10⁻⁵ mol L⁻¹s⁻¹**

Strategy and Explanation: Given the initial concentrations and initial rates of a reaction for several different experimental conditions for the same chemical reaction, determine the rate law, the reaction orders, the rate constant and the rate of the reaction given specific concentrations.

In Section 12.2, the method of finding the rate law from initial rates is described for getting the orders. Then compare pairs of experiments where only one of the concentrations is different and relate that to the changes in the rate. Once the orders are determined, plug the data into the rate law to determine the value of k. Use stoichiometric relationships between reactants and products, as described in Section 12.1, to relate the rates of reactants and products.

(a) The reactants are CH_3COCH_3, Br_2 and water. The reaction is catalyzed by acid, so the rate is affected by the concentration of H_3O^+. Data are available for the changes in the first two reactants' concentrations and the catalyst concentration, so we will assume the quantity of water is relatively unchanged in this reaction. Hence, the rate law looks like this:

$$\text{rate} = k[CH_3COCH_3]^i[Br_2]^j[H_3O^+]^h \quad \text{where k, i, j and h are currently unknown.}$$

Looking at Experiments 1 and 2, the initial concentration of CH_3COCH_3 stays constant, the initial concentration of Br_2 doubles, the initial concentration of H_3O^+ stays constant, and the initial rate does not change. The rate is independent of the concentration of Br_2, and the order with respect to Br_2 is zero.

Looking at Experiments 1 and 3, the initial concentration of CH_3COCH_3 stays constant, the initial concentration of Br_2 stays constant, the initial concentration of H_3O^+ doubles, and the initial rate changes by a factor of approximately two $(12.0 \times 10^{-5}/5.7 \times 10^{-5} = 2.1)$. The rate change is the same as the concentration change, which suggests that the rate is proportional to the concentration of H_3O^+, and the order with respect to H_3O^+ is one.

Looking at Experiments 1 and 5, the initial concentration of CH_3COCH_3 increases by a factor of 1.3 $(0.40/0.30)$, the initial concentration of Br_2 stays constant, the initial concentration of H_3O^+ stays constant, and the initial rate changes by a factor of 1.3 $(7.6 \times 10^{-5}/5.7 \times 10^{-5})$. The rate change is the same as the concentration change, which suggests that the rate is proportional to the concentration of CH_3COCH_3, and the order with respect to CH_3COCH_3 is one.

The rate law now looks like this: $\text{rate} = k[CH_3COCH_3]^1[H_3O^+]^1$

(b) Solve the rate law for k: $$k = \frac{\text{rate}}{[CH_3COCH_3][H_3O^+]}$$

Plug in each experiment's data. Here is an example of the calculation for Experiment 1:

$$k_1 = \frac{5.7 \times 10^{-5} \text{ mol L}^{-1}\text{s}^{-1}}{(0.30 \text{ mol / L})(0.05 \text{ mol / L})} = 4 \times 10^{-3} \text{ L mol}^{-1}\text{s}^{-1}$$

[CH₃COCH₃] (mol/L)	[H₃O⁺] (mol/L)	Initial rate (mol L⁻¹s⁻¹)	Rate constant (L mol⁻¹s⁻¹)
0.30	0.05	5.7 × 10⁻⁵	4 × 10⁻³
0.30	0.10	12.0 × 10⁻⁵	4.0 × 10⁻³
0.40	0.20	31.0 × 10⁻⁵	3.9 × 10⁻³
0.40	0.05	7.6 × 10⁻⁵	4 × 10⁻³

The first two experiments have the same rate (represented in the first row of the chart above). The average of these five rate constants is 4×10^{-3} L mol⁻¹s⁻¹.

(c) $\text{rate} = k[CH_3COCH_3]^1[H_3O^+]^1 = (4 \times 10^{-3} \text{ L mol}^{-1}\text{s}^{-1})(0.10 \text{ mol/L})^1(0.050 \text{ mol/L})^1 = 2 \times 10^{-5} \text{ mol L}^{-1}\text{s}^{-1}$

25. *Answer:* **(a) 0.16 mol/L (b) 90. s (c) 120 s**

Strategy and Explanation: Given the order of a reaction, the initial concentration, and the half-life, determine the concentration at a new time and the time it takes to get to a new concentration.

Use the first-order half-life to get the value of k (Equation 12.7). Then use the integrated rate law to find concentration and time.

$$k = \frac{\ln 2}{t_{1/2}} = \frac{\ln 2}{30. \text{ s}} = 2.3 \times 10^{-2} \text{ s}^{-1}$$

(a)   *For enough sig figs, assume the time data is ± 1 s.*

$$\ln[A]_t = -kt + \ln[A]_0 = -(2.3 \times 10^{-2} \text{ s}^{-1})(60. \text{ s}) + \ln(0.64 \text{ mol/L})$$

$$\ln[A]_t = -1.8$$

$$[A]_t = e^{-1.8} = 0.16 \text{ mol/L}$$

(b) $$[A]_t = \frac{1}{8} \times [A]_0 = \frac{1}{8} \times (0.64 \text{ mol/L}) = 0.080 \text{ mol/L}$$

$$kt = \ln[A]_0 - \ln[A]_t = \ln(0.64 \text{ mol/L}) - \ln(0.080 \text{ mol/L}) = 2.1$$

$$t = \frac{2.1}{k} = \frac{2.1}{2.3 \times 10^{-2} \text{ s}^{-1}} = 90. \text{ s}$$

(c) $$[A]_t = 0.040 \text{ mol/L}$$

$$kt = \ln[A]_0 - \ln[A]_t = \ln(0.64 \text{ mol/L}) - \ln(0.040 \text{ mol/L}) = 2.8$$

$$t = \frac{2.8}{k} = \frac{2.8}{2.3 \times 10^{-2} \text{ s}^{-1}} = 120 \text{ s}$$

✓ *Reasonable Answer Check:* The concentration after some time has passed is always smaller than the initial concentration. It also makes sense that the longer the time, the smaller the concentration.

27. *Answer:* **t = 5.49 × 10⁴ s**

Strategy and Explanation: Use the method described in the solution to Question 25.

$$k = \frac{\ln 2}{t_{1/2}} = \frac{\ln 2}{1.47 \times 10^4 \text{ s}} = 4.72 \times 10^{-5} \text{ s}^{-1}$$

$$[A]_0 = \frac{1.6 \times 10^{-3} \text{ mol}}{2.0 \text{ L}} = 8.0 \times 10^{-4} \text{ mol/L}$$

$$[A]_t = \frac{1.2 \times 10^{-4} \text{ mol}}{2.0 \text{ L}} = 6.0 \times 10^{-5} \text{ mol/L}$$

$$kt = \ln[A]_0 - \ln[A]_t = \ln(8.0 \times 10^{-4} \text{ mol/L}) - \ln(6.0 \times 10^{-5} \text{ mol/L}) = 2.59$$

$$t = \frac{2.59}{k} = \frac{2.59}{4.72 \times 10^{-5} \text{ s}^{-1}} = 5.49 \times 10^4 \text{ s}$$

28. *Answer:* **810 s**

Strategy and Explanation: Use the method described in the solution to Question 27.

$$k = \frac{\ln 2}{t_{1/2}} = \frac{\ln 2}{2.3 \times 10^2 \text{ s}} = 3.0 \times 10^{-3} \text{ s}^{-1}$$

$$kt = \ln[A]_0 - \ln[A]_t = \ln(4.32 \times 10^{-2} \text{ mol/L}) - \ln(3.75 \times 10^{-3} \text{ mol/L}) = 2.444$$

$$t = \frac{2.444}{k} = \frac{2.444}{3.0 \times 10^{-3} \text{ s}^{-1}} = 810 \text{ s}$$

A Nanoscale View: Elementary Reactions

29. *Answer:* **Reaction (d) is unimolecular and elementary; reaction (b) is bimolecular and elementary; reactions (a) and (c) are not elementary.**

Strategy and Explanation: A reaction with exactly one reactant is unimolecular and elementary. A reaction with exactly two reactants, either two of the same reactants or one of each of two different reactants, is bimolecular and elementary. A reaction that has more than two reactants, in any combination, is not elementary. We will assume that these reactions fit one of these three descriptions.

(a) The reaction has three reactants (one CH_4 and two O_2 molecules), so it is not elementary.

(b) The reaction has two reactants (one O_3 molecule and one O atom), so it is bimolecular and elementary.

(c) The reaction has three reactants (one Mg atom and two H_2O molecules), so it is not elementary.

(d) The reaction has one reactant (one O_3 molecule), so it is unimolecular and elementary.

31. *Answer:* **Reaction NO with O_3; NO is asymmetrical and Cl is symmetrical.**

Strategy and Explanation: The reaction of NO with O_3 will have a more important steric factor than the reaction of Cl with O_3, because NO is an asymmetrical molecule and Cl is a symmetrical atom. All collisions with a symmetrical atom could be effective, if they have enough energy. Collisions with the "wrong end" of the asymmetric molecule might be ineffective just because of which atoms came in contact during the collision.

Temperature and Reaction Rate

33. *Answer:* $E_a = 19$ **kJ/mol;** $\dfrac{rate_1}{rate_2} = 1.8$

Strategy and Explanation: Given activation energy of a reaction find the ratio of the rates at two different Ts.

A ratio of the rates when only the temperature changes must be equal to a ratio of the rate constants. Convert the temperatures to Kelvin, then use the equation derived in Problem-Solving Example 12.9 to calculate the rate constant ratio.

$$E_a = 19 \text{ kJ/mol (from Problem-Solving Example 12.8).}$$

$$T_1 = 50.\,°C + 273.15 = 323 \text{ K}, \qquad\qquad T_2 = 25\,°C + 273.15 = 298 \text{ K}$$

$$\frac{rate_1}{rate_2} = \frac{k_1}{k_2} = e^{\frac{E_a}{R}\left(\frac{1}{T_2} - \frac{1}{T_1}\right)} = e^{\frac{(19 \text{ kJ/mol})}{(0.008314 \text{ kJ/mol·K})}\left(\frac{1}{(298 \text{ K})} - \frac{1}{(323 \text{ K})}\right)} = e^{0.6} = 1.8$$

✓ *Reasonable Answer Check:* The rate increases at a higher temperature.

35. *Answer:* **(a) 22.2 kJ/mol, 6.66×10^7 L^{2}mol^{-2}s^{-1} (b) 8.39×10^4 L^{2}mol^{-2}s^{-1}**

Strategy and Explanation: Given the values of the rate constant and several temperatures, determine the activation energy, the frequency factor and the rate constant at a new temperature.

As described in Section 12.5, we can make a linear graph by taking the logarithm of the rate constant and the reciprocal of the absolute temperature. The slope is related to the activation energy and the y-intercept is related to the frequency factor. The Arrhenius equation can then be used to find the rate constant at a new temperature.

(a) Notice, according to the table heading, that the k values all have a multiplier of 10^{-5}:

$\dfrac{1}{T}$ (K^{-1})	ln(k)
0.002393	11.626
0.002080	12.468
0.001923	12.889
0.001579	13.752
0.001500	13.955
0.001408	14.292
0.001355	14.433

Now make a graph of 1/T versus ln k:

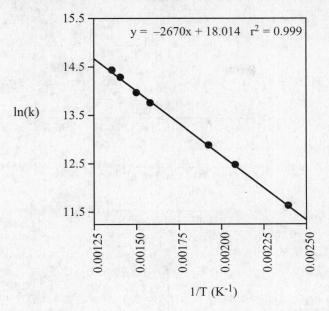

$y = -2670x + 18.014$ $r^2 = 0.999$

ln(k)

1/T (K^{-1})

The slope, $m = -\dfrac{E_a}{R}$. $E_a = -mR = -(-2670)(0.008314) = 22.2$ kJ/mol

The y-intercept, $b = \ln A$. $A = e^b = e^{18.014} = 6.66 \times 10^7$ L^{2}mol^{-2}s^{-1}

(b) Use the Arrhenius equation to estimate the rate constant. $k = A\,e^{-E_a/RT}$

$$k = (6.66 \times 10^7 \text{ L}^2\text{mol}^{-2}\text{s}^{-1})\,e^{-(22.2\,\text{kJ}/\text{mol})/(0.008314\,\text{kJ}/\text{mol·K})(400.0\,\text{K})} = 8.39 \times 10^4 \text{ L}^2\text{mol}^{-2}\text{s}^{-1}$$

37. *Answer:* **(a) 8×10^{-4} mol L^{-1}s^{-1}** **(b) 3×10^1 mol L^{-1}s^{-1}**

Strategy and Explanation: Given the activation energy, the concentration of the reactant, the frequency factor, and two temperatures, determine the rates of the reaction at those two temperatures.

The units of the frequency factor are also the units of the rate constant. Use that information to determine the order of the reaction. Write the rate law for the reaction in terms of k and concentration. Use the Arrhenius equation to find the value of k, then plug it and the concentration into the rate law.

The reaction is first-order since the units of A and therefore the units of k are time^{-1}. (Refer to Table 12.2)

$$\text{Rate} = k[CH_3CH_2I]^1 \qquad\qquad k = A\,e^{-E_a/RT}$$

An example of the calculation is here for answer (a): Convert temperatures to Kelvin.

$$400.\,°C + 273.15 = 673 \text{ K}$$

$$k = (1.2 \times 10^{14}\,\text{s}^{-1}) \times e^{-(221\,\text{kJ}/\text{mol})/(0.008314\,\text{kJ}/\text{mol·K})(673\,\text{K})} = (1.2 \times 10^{14}\,\text{s}^{-1}) \times e^{-39.5} = 8 \times 10^{-4}\,\text{s}^{-1}$$

$$\text{Rate} = (8 \times 10^{-4}\,\text{s}^{-1}) \times (0.012\,\text{mol/L})^1 = 1 \times 10^{-5}\,\text{mol L}^{-1}\text{s}^{-1}$$

	T (K)	k (s^{-1})	Rate (mol L^{-1}s^{-1})
(a)	673	8×10^{-4}	1×10^{-5} *(one sig fig)*
(b)	1073	3×10^1	$25 = 3 \times 10^1$ *(one sig fig)*

✓ *Reasonable Answer Check:* It makes sense that the rate of the reaction increases with increasing temperature.

39. *Answer:* **3×10^2 kJ/mol**

Strategy and Explanation: Given two different temperatures and the rate constants at those temperatures, determine the activation energy of a reaction.

Convert the temperatures to Kelvin, then use the equation derived in Problem-Solving Example 12.9 to calculate the activation energy:

$$\ln\left(\frac{k_1}{k_2}\right) = \frac{E_a}{R}\left(\frac{1}{T_2} - \frac{1}{T_1}\right)$$

25. °C + 273.15 = 298 K   55. °C + 273.15 = 328 K

$$\ln\left(\frac{3.46 \times 10^{-5}\ s^{-1}}{1.5 \times 10^{-3}\ s^{-1}}\right) = \frac{E_a}{R}\left(\frac{1}{328\ K} - \frac{1}{298\ K}\right)$$

$$\ln\left(2.3 \times 10^{-2}\right) = \frac{E_a}{(0.008134\,kJ\,/\,mol \cdot K)}\left(0.00305\ K^{-1} - 0.00336\ K^{-1}\right)$$

$$-3.77 = \frac{E_a}{(0.008134\,kJ\,/\,mol \cdot K)}\left(0.00031\ K^{-1}\right)$$

$$-3.77 = E_a\,(-.037\ mol/kJ)$$

$$E_a = 3 \times 10^2\ kJ/mol$$

✓ *Reasonable Answer Check:* The calculated E_a has a reasonable size and sign, similar to those seen in previous Questions. The result of the subtraction limits the number of significant figures we can report, suggesting that the temperature data should be kept more precisely.

Rate Laws for Elementary Reactions

41. *Answer:* **Exothermic**

Strategy and Explanation: Given the activation energy for the forward reaction of an elementary reaction and the activation energy for the reverse of the same reaction, determine if the forward reaction is exothermic or endothermic.

The difference between the forward and reverse activation energies is the ΔE for the reaction. If ΔE is positive, the forward reaction is endothermic, and if it is negative, the forward reaction is exothermic.

$$\Delta E = E_{a,forward} - E_{a,reverse} = 32\ kJ/mol - 58\ kJ/mol = -26\ kJ/mol$$

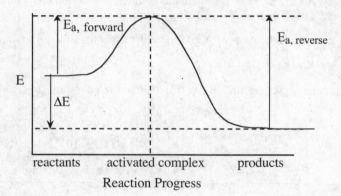

The reaction is exothermic.

✓ *Reasonable Answer Check:* The reverse reaction requires a higher activation energy than the forward reaction, suggesting that the reactants are higher in energy than the reactants. It makes sense that the reaction is exothermic.

43. *Answer:* **See graphs below**

Strategy and Explanation: Draw these diagrams using the information described in the solution to Question 41, and in Section 12.4. ΔH and ΔE are identical for these diagrams.

(a) $E_{a,reverse} = E_{a,forward} - \Delta H = (75\ kJ\ mol^{-1}) - (-145\ kJ\ mol^{-1}) = 220.\ kJ\ mol^{-1}$

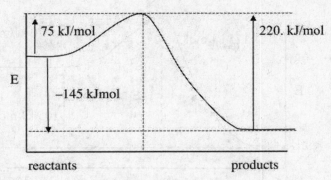

(b) $E_{a,reverse} = E_{a,forward} - \Delta H = (65\ kJ\ mol^{-1}) - (-70.\ kJ\ mol^{-1}) = 135\ kJ\ mol^{-1}$

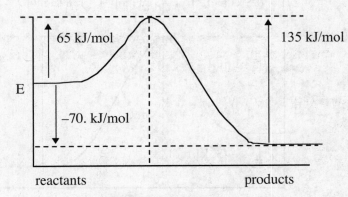

(c) $E_{a,reverse} = E_{a,forward} - \Delta H = (85\ kJ\ mol^{-1}) - (+70.\ kJ\ mol^{-1}) = 15\ kJ\ mol^{-1}$

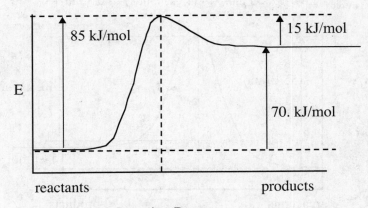

44. *Answer:* **see diagrams below**

Strategy and Explanation: Draw these diagrams using the information described in the solution to Question 43 and in Section 12.4. ΔH and ΔE are identical for these diagrams.

(a) $E_{a,reverse} = E_{a,forward} - \Delta H = (175 \text{ kJ mol}^{-1}) - (105 \text{ kJ mol}^{-1}) = 70. \text{ kJ mol}^{-1}$

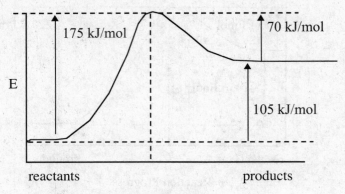

(b) $E_{a,reverse} = E_{a,forward} - \Delta H = (95 \text{ kJ mol}^{-1}) - (-43 \text{ kJ mol}^{-1}) = 138 \text{ kJ mol}^{-1}$

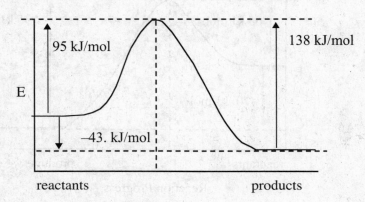

(c) $E_{a,reverse} = E_{a,forward} - \Delta H = (55 \text{ kJ mol}^{-1}) - (15 \text{ kJ mol}^{-1}) = 40. \text{ kJ mol}^{-1}$

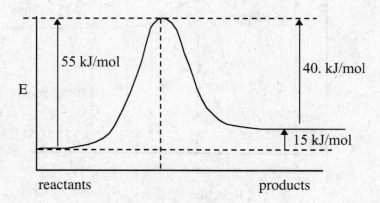

45. *Answer:* **(a) Reaction (b) (b) Reaction (c)**

Strategy and Explanation: Assuming everything about the forward reactions in this Question is identical except the activation energy, the reaction with the largest $E_{a,forward}$ will be the slowest, and that with the smallest $E_{a,forward}$ will be the fastest.

(a) Of the three, the smallest $E_{a,forward}$ is 65 kJ mol^{-1}. That is reaction (b).

(b) Of the three, the largest $E_{a,forward}$ is 85 kJ mol^{-1}. That is reaction (c).

47. *Answer:* **(a) rate = k[NO][NO$_3$] (b) rate = k[O][O$_3$] (c) rate = [(CH$_3$)$_3$CBr] (d) rate = k[HI]^2**

Strategy and Explanation: Given the chemical equations of elementary reactions write their rate laws.

As described in Section 12.6, the stoichiometric coefficient of a reactant in an elementary equation is the reaction order for that reactant.

(a)	$NO + NO_3 \longrightarrow$ products	rate = k[NO]^1[NO$_3$]^1
(b)	$O + O_3 \longrightarrow$ products	rate = k[O]^1[O$_3$]^1
(c)	$(CH_3)_3CBr \longrightarrow$ products	rate = [(CH$_3$)$_3$CBr]^1
(d)	$2\ HI \longrightarrow$ products	rate = k[HI]^2

Reaction Mechanisms

49. *Answer:* **(a) see equations below (b) the first step is rate-determining**

Strategy and Explanation:

(a) As we did in Chapter 6 with Hess's law Questions, cancel anything that ends up on both sides of the equation and add up the rest.

$$NO_2 + F_2 \longrightarrow FNO_2 + \cancel{F}$$
$$+ NO_2 + \cancel{F} \longrightarrow FNO_2$$
$$\overline{2\ NO_2 + F_2 \longrightarrow 2\ FNO_2}$$

(b) The rate law looks like this: rate = k[NO$_2$][F$_2$]

That matches the stoichiometry of the first reaction, so it must be the rate-determining step.

✓ *Reasonable Answer Check:* It makes sense that the first reaction would be the rate limiting step, since a fluorine free radical is going to be extremely reactive.

51. *Answer:* **(a) CH$_3$COOCH$_3$ + H$_2$O $\rightleftharpoons$ CH$_3$COOH + CH$_3$OH (b) catalyst: H$_3$O$^+$ (c) intermediates: H$_3$C(OH)OCH$_3^+$, H$_3$C(H$_2$O)(OH)OCH$_3^+$, H$_3$C(OH)$_2$OHCH$_3^+$, and H$_2$O**

Strategy and Explanation:

(a) As we did in Chapter 6 with Hess's law Questions, cancel anything that ends up on both sides of the equation and add up the rest.

$$CH_3COOCH_3 + \cancel{H_3O^+} \rightleftharpoons \cancel{CH_3C(OH)OCH_3^+} + \cancel{H_2O}$$
$$+ \cancel{CH_3C(OH)OCH_3^+} + \cancel{H_2O} \rightleftharpoons \cancel{CH_3C(H_2O)(OH)OCH_3^+}$$
$$+ \cancel{CH_3C(H_2O)(OH)OCH_3^+} \rightleftharpoons \cancel{CH_3C(OH)_2OHCH_3^+}$$
$$+ \cancel{CH_3C(OH)_2OHCH_3^+} + H_2O \rightleftharpoons CH_3COOH + CH_3OH + \cancel{H_3O^+}$$
$$\overline{CH_3COOCH_3 + H_2O \rightleftharpoons CH_3COOH + CH_3OH}$$

(b) A catalyst shows up in a mechanism as a reactant in an early step and then again as a product in a later step. There is a catalyst in this reaction. It is H$_3$O$^+$, introduced in reaction 1 and reproduced in step 4.

(c) An intermediate appears in a mechanism as a product in an early step and then again as a reactant in a later step. Three intermediates are formed in this reaction: H$_3$C(OH)OCH$_3^+$, H$_3$C(H$_2$O)(OH)OCH$_3^+$, and H$_3$C(OH)$_2$OHCH$_3^+$. A molecule of H$_2$O is also created then destroyed in this reaction, but another H$_2$O molecule is consumed, so water may be technically considered to be an intermediate, as well as a reactant.

53. *Answer:* **Only mechanism (a) is compatible with the observed rate law.**

    Strategy and Explanation: Follow the method described in Question 51. The observed rate law is given:

    $$\text{rate} = k[(CH_3)_3CBr]$$

    (a) The rate law for this mechanism comes from the rate of the slow first step:

    rate = $k[(CH_3)_3CBr]$, so it is compatible with the observed rate law.

    (b) The single step mechanism suggests an elementary bimolecular reaction with a rate law of that looks like this: rate = $k[(CH_3)_3CBr][OH^-]$ It is incompatible with the observed rate law.

    So, only mechanism (a) is compatible.

Catalysts and Reaction Rate

54. *Answer:* **(a) is true.**

    Strategy and Explanation:

    (a) The concentration of a homogeneous catalyst may appear in the rate law.

    (b) A catalyst may be consumed in a reaction, but must always be recreated.

    (c) A homogeneous catalyst is in the same phase as the reactants, but a heterogeneous catalyst is always in a different phase.

    (d) A catalyst can change the mechanism of a reaction and allow different intermediates to be produced, but it cannot change the products formed. A chemical that changes the products is not called a catalyst. Notice that if a multiple set of reactions occurs simultaneously with the same reactants, a catalyst that speeds only one of these reactions would help produce one product in favor of some others. In such case, the statement might be considered true, but it is likely not something students studying this chapter will think of.

56. *Answer:* **(a) and (c); (a) homogeneous (c) heterogeneous**

    Strategy and Explanation: A catalyst may appear in the rate law, though it will not appear in the overall equation representing the reaction. Any rate law with chemicals that are not reactants or products, involve catalysts. If the catalyst has the same phase as the reactants, it is homogeneous. If the catalyst has a different phase from the reactants, it is heterogeneous.

    (a) This aqueous reaction involves a homogeneous catalyst, $H_3O^+(aq)$.

    (b) This reaction does not appear to involve a catalyst. Both of the concentrations in the rate law are of reactants.

    (c) This gas phase reaction involves a heterogeneous catalyst, Pt(s).

    (d) This reaction does not appear to involve a catalyst. Both of the concentrations in the rate law are of reactants.

Enzymes: Biological Catalysts

58. *Answer/Explanation:*

    Enzyme: A protein that catalyzes a biological reaction.

    Substrate: The reactant of a reaction catalyzed by an enzyme.

    Active site: That part of an enzyme where the substrate binds to the enzyme in preparation for conversion into products.

    Proteins: Large polypeptide molecules that serve as structural and functional molecules in living organisms.

    Enzyme-substrate complex: A loose affiliation, usually involving intermolecular forces, between the enzyme and the substrate.

    Induced fit: When the natural shape of either the enzyme or the substrate changes slightly in the formation of the enzyme-substrate complex.

Globular proteins: Proteins whose structure allows them to fold into tight, compact spherical shapes, in contrast to fibrous proteins which are long and slender.

59. *Answer:* **The rate is 30. times faster.**

Strategy and Explanation: We will use information described in Section 12.2, and methods similar to those in Questions 33 - 38.

The reaction rate is proportional to the concentration in a first-order reaction, rate = k[E]. That means the rate increase is related to the ratios:

$$\frac{rate_2}{rate_1} = \frac{[E]_2}{[E]_1} = \frac{4.5 \times 10^{-6} \text{ M}}{1.5 \times 10^{-6} \text{ M}} = 30.$$

The reaction rate is increased by a factor of 30. times.

Catalysts in Industry

62. *Answer/Explanation:*

(a) The honeycomb arrangement of the ceramic support is mostly likely to maximize the surface area and increase contact with the heterogeneous catalyst.

(b) The catalysis happens only at the surface of the metal. So, using expensive platinum in the form of strips or rods would be less efficient and more costly.

General Questions

63. *Answer:* **see diagram and energy relationships below**

Strategy and Explanation: An exothermic reaction has the products at a lower-energy state than the reactants.

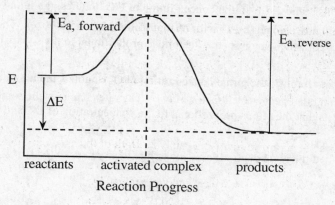

$$\Delta E = E_{a,forward} - E_{a,reverse}$$

65. *Answer:* **(a) true (b) false. "The reaction rate decreases as a first-order reaction proceeds at a constant temperature." (c) true (d) false. "As a second-order reaction proceeds at a constant temperature, the rate constant does not change." (Other corrections for (b) and (d) are also possible.)**

Strategy and Explanation:

(a) True: It is possible to change the rate constant for a reaction by changing the temperature. The Arrehenius equation in Section 12.5 shows the temperature dependence of the rate constant, k. Unless E_a is zero, k would change when T changes.

(b) False. A related true statement is: The reaction rate decreases as a first-order reaction proceeds at a constant temperature.

(c) True: The rate constant for a reaction is independent of reaction concentrations. The rate law describes the rate constant as the proportionality constant between the concentrations and the rate.

(d) False. A related true statement is: As a second-order reaction proceeds at a constant temperature, the rate constant does not change.

66. *Answer:* **rate = k[H$_2$][NO]^2**

Strategy and Explanation: We will use information described in Section 12.2, and methods similar to those in Questions 33 - 38.

The reactants are H$_2$ and NO, so the rate law looks like this:

$$\text{rate} = k[H_2]^i[NO]^j \quad \text{where k, i and j are currently unknown.}$$

When the initial concentration of H$_2$ is halved (presuming that the initial concentration of NO stays constant), the initial rate is halved. The rate change is the same as the concentration change, which suggests that the rate is proportional to the concentration of H$_2$, and the order with respect to H$_2$ is one.

When the initial concentration of NO increases by a factor of 3 (presuming that the initial concentration of H$_2$ stays constant), the initial rate changes by a factor of 9. The rate change is the square of the concentration change, which suggests that the rate is proportional to the square of the concentration of NO, and the order with respect to NO is two.

The rate equation, also called the rate law, now looks like this: $\text{rate} = k[H_2]^1[NO]^2 = k[H_2][NO]^2$

67. *Answer:* **(a) NO is second order, O$_2$ is first order. (b) rate = k[NO]^2[O$_2$] (c) 25 L^{2}mol^{-2}s^{-1} (d) 7.8 × 10^{-4} mol L^{-1}s^{-1} (e) rate for NO: 2.0 × 10^{-4} mol L^{-1}s^{-1}; rate for NO$_2$: 2.0 × 10^{-4} mol L^{-1}s^{-1}**

Strategy and Explanation: Follow the method described in the solution to Question 20.

The reactants are NO and O$_2$. Data are available for the changes in each of these reactants' concentrations, so the rate law looks like this:

$$\text{rate} = k[NO]^i[O_2]^j \quad \text{where k, i and j are currently unknown.}$$

(a) Looking at Experiments 1 and 2, the initial concentration of NO doubles, the initial concentration of O$_2$ stays constant, and the initial rate changes by a factor of four ($1.0 \times 10^{-4}/2.5 \times 10^{-5}$). The rate change is the square of the concentration change, which suggests that the rate is proportional to the square of the concentration of NO, and the order with respect to NO is two.

Looking at Experiments 1 and 3, the initial concentration of O$_2$ doubles, the initial concentration of NO stays constant, and the initial rate changes by a factor of two. The rate change is the same as the concentration change, which suggests that the rate is proportional to the concentration of O$_2$, and the order with respect to O$_2$ is one.

(b) The rate law (also called the rate equation) now looks like this: $\text{rate} = k[NO]^2[O_2]^1$

(c) Solve the rate law for k:

$$k = \frac{\text{rate}}{[NO]^2[O_2]}$$

Plug in each experiment's data. Here is an example of the calculation for Experiment 1:

$$k_1 = \frac{2.5 \times 10^{-5} \text{ mol L}^{-1}\text{s}^{-1}}{(0.010 \text{ mol}/\text{L})^2(0.010 \text{ mol}/\text{L})} = 25 \text{ L}^2 \text{ mol}^{-2}\text{s}^{-1}$$

[NO] (mol/L)	[O$_2$] (mol/L)	Initial rate (mol L^{-1}s^{-1})	Rate constant (L^{2}mol^{-2}s^{-1})
0.010	0.010	2.5 × 10^{-5}	25
0.020	0.010	1.0 × 10^{-4}	25
0.010	0.020	5.0 × 10^{-5}	25

The average of these three rate constants is 25 L^{2}mol^{-2}s^{-1}.

(d) rate = k[NO]^2[O$_2$]^1 = (25 L^{2}mol^{-2}s^{-1})(0.025 mol/L)^2(0.050 mol/L)^1= 7.8 × 10^{-4}mol L^{-1}s^{-1}

(e) The relative rates depend on the stoichiometric coefficients (Equation 12.3). Here, the stoichiometric relationship is $2NO : 1O_2 : 2NO_2$

$$-\frac{1}{2}\left(\frac{\Delta[NO]}{\Delta t}\right) = -\frac{1}{1}\left(\frac{\Delta[O_2]}{\Delta t}\right) = \frac{1}{2}\left(\frac{\Delta[NO_2]}{\Delta t}\right)$$

$$-\frac{\Delta[NO]}{\Delta t} = 2\left(-\frac{\Delta[O_2]}{\Delta t}\right) = 2\left(1.0\times10^{-4}\ mol\,L^{-1}s^{-1}\right) = 2.0\times10^{-4}\ mol\,L^{-1}s^{-1}$$

$$\frac{\Delta[NO_2]}{\Delta t} = 2\left(-\frac{\Delta[O_2]}{\Delta t}\right) = 2\left(1.0\times10^{-4}\ mol\,L^{-1}s^{-1}\right) = 2.0\times10^{-4}\ mol\,L^{-1}s^{-1}$$

Applying Concepts

69. *Answer:* **Curve A represents $[H_2O(g)]$ increase with time, Curve B represents $[O_2(g)]$ increase with time, and Curve C represents $[H_2O_2(g)]$ decrease with time.**

Strategy and Explanation: Cuve A represents the increase in the concentration of the product $H_2O(g)$ with time. Curve B represents the increase in the concentration of product O_2 with time. The O_2 curve is half as steep as the H_2O line because the stoichiometric relationship between them is 1:2. Curve C represents the decrease in the concentration of reactant $H_2O_2(g)$ with time.

71. *Answer:* **Snapshot (b)**

Strategy and Explanation: Products form more quickly at higher temperatures, so (b) the snapshot with fewer HI molecules is the one corresponding to a lower temperature.

73. *Answer:* **rate = k[A]^3[B][C]^2**

Strategy and Explanation: We will use the corrected information described above, the techniques described in Section 12.2, and methods similar to those in the solutions to Questions 23-24.

The reactants are A, B and C, so the rate law looks like this:

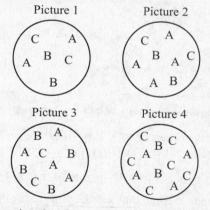

rate = k[A]^i[B]^j[C]^h where k, i, j, and h are currently unknown.

Between pictures 1 and 2, the initial concentration of A doubles (2 A → 4 A), and the initial concentrations of B and C stay constant (2 of each), the initial rate changes by a factor of (12/1.5 =) eight. The rate change is the cube of the concentration change ($8 = 2^3$), which suggests that the order with respect to A is three.

Between pictures 2 and 3, the initial concentration of B doubles (2 B → 4 B), and the initial concentrations of A and C stay constant (4 A and 2 C), the initial rate changes by a factor of (23/12 =) two. The rate change is equal to the concentration change, which suggests that the order with respect to B is one.

Between pictures 2 and 4, the initial concentration of C triples (2 C → 6 C), and the initial concentrations of A and B stay constant (4A and 2B), the initial rate changes by a factor of nine. The rate change is the square of the concentration change($9 = 3^2$), which suggests that the the order with respect to C is two.

The rate equation, also called the rate law, now looks like this: rate = k[A]^3[B][C]^2

74. *Answer:* **E_a is very, very small – approximately zero.**

Strategy and Explanation: The Arrhenius equation described in Section 12.5 describes the temperature dependence of the rate constant.

$$k = A e^{-E_a/RT}$$

If a radioactive decay reaction is independent of temperature, that says the activation energy E_a is indistinguishable from zero, and $k = A e^{-0/RT} = Ae^0 = A(1) = A$, with no functional dependence on T.

More Challenging Questions

76. *Answer:* **(a) rate $= 2.4 \times 10^{-7}$ mol L^{-1}s^{-1} + k[BSC][F$^-$] (b) 0.3 L mol^{-1}s^{-1}**

Strategy and Explanation: Given the initial concentrations and initial rates of a reaction at several different experimental conditions, determine the rate law and rate constant for the reaction.

In Section 12.2, the method of finding the rate law from initial rates is described for getting the orders. However, before we compare the experimental rates, we need to subtract the residual rate. Then try comparing pairs of experiments where only one of the concentrations is different and relating that to the changes in the rate. If the result of the pair-wise comparison is ambiguous, make a linear graph. Once the orders are determined, plug the data into the rate law to determine the value of k.

Benzenesulfonyl chloride (abbreviated BSC) has a known effect on the rate (first-order), but it has a constant concentration in this experiment, 2×10^{-4} M. As a result, we can't say how it is involved in the constant term representing the residual rate, so we will seek a rate law looks like this:

$$\text{rate} = 2.4 \times 10^{-7} \text{mol L}^{-1}\text{s}^{-1} + k[BSC][F^-]^i \quad \text{where k and i are currently unknown.}$$

To study how [F$^-$] affects the rate of the reaction, subtract the residual rate from the observed rate.

Experiment	[F$^-$] $\times 10^2$ (mol/L)	Initial rate $\times 10^7$ (mol L^{-1}s^{-1})	Adjusted initial rate $\times 10^7$ (mol L^{-1}s^{-1})
1	0	2.4	2.4 − 2.4 = 0.0
2	0.5	5.4	5.4 − 2.4 = 3.0
3	1.0	7.9	7.9 − 2.4 = 5.5
4	2.0	13.9	13.9 − 2.4 = 11.5
5	3.0	20.2	20.2 − 2.4 = 17.8
6	4.0	25.2	25.2 − 2.4 = 22.8
7	5.0	32.0	32.0 − 2.4 = 29.6

(a) The adjusted rate law has the functional form: adjusted rate $= k[BSC][F^-]^i$

To get the order of F$^-$, compare experiments 2 to 3, 3 to 4, and 4 to 6. They each show a concentration increase of a factor of two. The ratio of the adjusted initial rate in each of these instances is:

$$2 \text{ to } 3: \frac{5.5}{3.0} = 1.8, \quad 3 \text{ to } 4: \frac{11.5}{5.5} = 2.1, \quad 4 \text{ to } 6: \frac{22.8}{11.5} = 2.0$$

It looks like the adjusted rate change is approximately the same as the concentration change, suggesting that the adjusted rate is proportional to the concentration of fluoride ions. The relationship is only approximate, though, so let's prepare a graph that uses all of the data. By taking the log of the rate law, the order becomes the slope of a linear graph.

$$\log(\text{adjusted rate}) = \log(k[BSC]) + \log([F^-]^i)$$
$$\log(\text{adjusted rate}) = i \log[F^-] + \log(k[BSC])$$

log [F$^-$]	log(adjusted initial rate) (mol L^{-1}s^{-1})	log [F$^-$]	log(adjusted initial rate) (mol L^{-1}s^{-1})
−2.3	−6.52	−1.52	−5.750
−2.00	−6.26	−1.40	−5.642
−1.70	−5.939	−1.30	−5.529

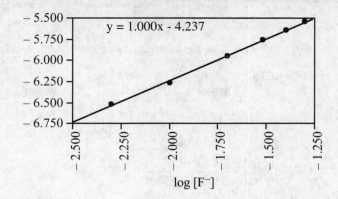

This confirms that the order of the adjusted rate law with respect to fluoride (i) is one, and the most complete rate law we can write for the reaction looks like this:

$$\text{rate} = 2.4 \times 10^{-7} \text{mol L}^{-1}\text{s}^{-1} + k[\text{BSC}][\text{F}^-]$$

(b) Solve the adjusted rate law for k: $k = \dfrac{\text{adjusted rate}}{[\text{BSC}][\text{F}^-]}$

Plug in each set of data. Here is a sample of a calculation for the first experiment with nonzero [F⁻]:

$$k_2 = \frac{3.0 \times 10^{-7} \text{ mol L}^{-1}\text{s}^{-1}}{(2 \times 10^{-4} \text{ M})(0.5 \times 10^{-2} \text{ mol/L})} = 0.3 \text{ M}^{-1}\text{s}^{-1}$$

Experiment	$[\text{F}^-] \times 10^2$ (mol/L)	Adjusted initial rate $\times$ 10^7 (mol L^{-1}s^{-1})	Adjusted rate constant (M^{-1}s^{-1})
2	0.5	3.0	0.3
3	1.0	5.5	0.3
4	2.0	11.5	0.3
5	3.0	17.8	0.3
6	4.0	22.8	0.3
7	5.0	29.6	0.3

The average of these seven rate constants is 0.3 L mol⁻¹s⁻¹

✓ *Reasonable Answer Check:* It is satisfying to get the same rate constant for each data set.

78. *Answer:* **(a) 3×10^2 kJ (b) 2×10^3 s**

Strategy and Explanation: Given two different temperatures and the rate constants at those temperatures, determine the activation energy of a reaction.

(a) Convert the temperatures to Kelvin, then use the equation derived in Problem-Solving Example 12.9 to calculate the activation energy:

$$\ln\!\left(\frac{k_1}{k_2}\right) = \frac{E_a}{R}\left(\frac{1}{T_2} - \frac{1}{T_1}\right)$$

470. °C + 273.15 = 743 K   510. °C + 273.15 = 783 K

$$\ln\!\left(\frac{1.10 \times 10^{-4} \text{ s}^{-1}}{1.02 \times 10^{-3} \text{ s}^{-1}}\right) = \frac{E_a}{R}\left(\frac{1}{783 \text{ K}} - \frac{1}{743 \text{ K}}\right)$$

$$\ln(0.108) = \frac{E_a}{(0.008134 \text{kJ/mol} \cdot \text{K})}\left(0.00128 \text{ K}^{-1} - 0.00135 \text{ K}^{-1}\right)$$

$$-2.227 = \frac{E_a}{(0.008134 \text{kJ/mol} \cdot \text{K})}\left(0.00007 \text{ K}^{-1}\right) \quad \textit{(subtraction leaves only one sig fig)}$$

$$-2.227 = E_a \, (-.008 \text{ mol/kJ})$$

$$269 \text{ kJ} = E_a$$

$$3 \times 10^2 \text{ kJ} = E_a \quad \textit{(with one sig fig)}$$

(b) Use the same equation to determine the rate constant at 500. °C.

$$500. \text{ °C} + 273.15 = 773 \text{ K}$$

$$\ln\left(\frac{1.10 \times 10^{-4} \text{ s}^{-1}}{k_2}\right) = \frac{E_a}{R}\left(\frac{1}{773 \text{ K}} - \frac{1}{743 \text{ K}}\right)$$

$$\ln\left(\frac{1.10 \times 10^{-4} \text{ s}^{-1}}{k_2}\right) = \frac{(3 \times 10^2 \text{ kJ})}{(0.008134 \text{ kJ / mol} \cdot \text{K})}\left(0.00129 \text{ K}^{-1} - 0.00135 \text{ K}^{-1}\right) \quad \textit{(with 1 sig fig in } E_a)$$

$$\ln\left(\frac{1.10 \times 10^{-4} \text{ s}^{-1}}{k_2}\right) = -2 \quad \textit{(with one sig fig in } E_a \textit{ and in subtraction)}$$

$$\frac{1.10 \times 10^{-4} \text{ s}^{-1}}{e^{-2}} = k_2 \quad \textit{(with one sig fig in exponent of e)}$$

$$8 \times 10^{-4} \text{ s}^{-1} = k_2 \quad \textit{(round to one sig fig)}$$

The rate constants are first-order units, so use first order integrated rate law to determine the time:

$$kt = \ln[A]_0 - \ln[A]_t = \ln(0.10) - \ln(0.023) = 1.47$$

$$t = \frac{1.47}{k} = \frac{1.47}{8 \times 10^{-4} \text{ s}^{-1}} \quad \textit{(with one sig fig in k)}$$

$$t = 2 \times 10^3 \text{ s} \quad \textit{(rounded to one sig fig)}$$

80. *Answer:* **See mechanisms below (There can be other correct answers.)**

Strategy and Explanation: You may come up with correct answers to this Question that are NOT the same as the answers given here. In all cases, a proposed mechanism must have three things: First, it should be composed of first or second order elementary reactions. Second, it must recreate the given overall net reaction, including the elimination of all intermediates and the recreation of all the catalysts. Third, it must recreate the observed rate law. If the rate law is simple first or second order, then make the first step the slow step, and use the chemicals in the rate law as reactants. If the rate law is more complicated, start with a first step that is fast, and include some of the chemicals from the rate law in that reaction, then use the product of that reaction as a reactant in the next reaction. In each case, once you have written the mechanism, confirm that its mechanism matches the observed (given) mechanism.

(a) $CH_3CO_2CH_3 + H_3O^+ \longrightarrow CH_3COHOCH_3 + H_2O$ slow

$CH_3COHOCH_3 + H_2O \longrightarrow CH_3COH(OH_2)OCH_3$ fast

$CH_3COH(OH_2)OCH_3 \longrightarrow CH_3C(OH)_2 + CH_3OH$ fast

$CH_3C(OH)_2 + H_2O \longrightarrow CH_3COOH + H_3O^+$ fast

Check this mechanism: Each reaction is unimolecular or bimolecular Check the overall reaction:

$CH_3CO_2CH_3 + \cancel{H_3O^+} \longrightarrow CH_3COHOCH_3 + \cancel{H_2O}$

$CH_3COHOCH_3 + \cancel{H_2O} \longrightarrow CH_3COH(OH_2)OCH_3$

$CH_3COH(OH_2)OCH_3 \longrightarrow CH_3C(OH)_2 + CH_3OH$

$CH_3C(OH)_2 + H_2O \longrightarrow CH_3COOH + \cancel{H_3O^+}$

$\overline{CH_3CO_2CH_3 + H_2O \longrightarrow CH_3COOH + CH_3OH}$ This is the net equation.

The rate law of the slow first elementary reaction is the rate law for the whole mechanism. As described in Section 12.6, the stoichiometric coefficient of a reactant in an elementary reaction is the reaction order for that reactant.

$$CH_3CO_2CH_3 + H_3O^+ \longrightarrow \text{products} \qquad\qquad \text{rate} = k_1[CH_3CO_2CH_3][H_3O^+]$$

This matches the reported rate law for the reaction. Therefore, the proposed mechanism is plausible.

(b)
$$H_2 + I_2 \longrightarrow 2\,HI \qquad\qquad\qquad\qquad \text{slow}$$

Check this mechanism: The reaction is bimolecular and represents the overall reaction. As described in (a), check the mechanism.

$$H_2 + I_2 \longrightarrow \text{products} \qquad\qquad \text{rate} = k[H_2][I_2]$$

This matches the reported rate law for the reaction. Therefore, the proposed mechanism is plausible.

(c) The presence of I_2 and Pt in the rate law suggests that they are catalyst in this reaction:

$$H_2 + Pt(s) \longrightarrow PtH_2 \qquad\qquad\qquad \text{fast}$$

$$PtH_2 + I_2 \longrightarrow PtH_2I_2 \qquad\qquad\qquad \text{slow}$$

$$PtH_2I_2 + O_2 \longrightarrow PtI_2O + H_2O \qquad\qquad \text{fast}$$

$$PtI_2O + H_2 \longrightarrow Pt(s) + I_2 + H_2O \qquad\quad \text{fast}$$

Check this mechanism: Each reaction is unimolecular or bimolecular. Overall:

$$H_2 + Pt \longrightarrow PtH_2$$
$$PtH_2 + I_2 \longrightarrow PtH_2I_2$$
$$PtH_2I_2 + O_2 \longrightarrow PtI_2O + H_2O$$
$$PtI_2O + H_2 \longrightarrow Pt + I_2 + H_2O$$

$$2\,H_2 + O_2 \longrightarrow 2\,H_2O \qquad\qquad \text{This is the overall equation.}$$

Determine the rate law for this mechanism. as described in the solution to Question 80.

The slow step is step 2, so the rate of the reaction is equal to the rate of this rate-determining step. Because this reaction is elementary, its rate law is related to the stoichiometry of its reactants:

$$\text{rate of reaction} = \text{rate}_{\text{reaction 2}} = k_2[PtH_2][I_2]$$

PtH_2 is an intermediate, so we need to eliminate it from the proposed mechanism. Look for all the ways this intermediate is created and destroyed during the mechanism. It is only created in the forward reaction of step 1. It is destroyed in the reverse reaction of step 1 and also in the reaction of step 2. Because all three of these reactions are elementary, use their reactants' stoichiometric coefficients to write rate laws for each of these three reactions:

$$\text{rate}_{\text{forward reaction 1}} = k_1[H_2](\text{area of Pt surface})$$

$$\text{rate}_{\text{reverse reaction 1}} = k_{-1}[PtH_2]^1$$

$$\text{rate}_{\text{reaction 2}} = k_2[PtH_2][I_2]$$

The rate of PtH_2 creation is equal to the rate of its destruction once the steady state condition is reached:

$$\text{rate}_{\text{forward reaction 1}} = \text{rate}_{\text{reverse reaction 1}} + \text{rate}_{\text{reaction 2}}$$

$$k_1[H_2](\text{area of Pt surface}) = k_{-1}[PtH_2]^1 + k_2[PtH_2][I_2]$$

Because the rate of step 2 is presumed to be much smaller than the rate of step 1, the second term is presumed to be negligibly small compared to the first term:

$$\text{rate}_{\text{forward reaction 1}} \cong \text{rate}_{\text{reverse reaction 1}}$$

$$k_1[H_2](\text{area of Pt surface}) = k_{-1}[PtH_2]^1$$

$$[PtH_2] = \frac{k_1}{k_{-1}}[H_2](\text{area of Pt surface})$$

Substitute the equality just derived into the rate law: $\text{rate} = k[PtH_2][I_2]$

$$\text{rate} = k_2\frac{k_1}{k_{-1}}[H_2](\text{area of Pt surface})[I_2]$$

We will define the new group of rate constants by one variable, $k' = k_2\dfrac{k_1}{k_{-1}}$.

$$\text{rate} = k'[H_2][I_2](\text{area of Pt surface})$$

This matches the reported rate law for the reaction. Therefore, the proposed mechanism is a plausible one.

81. *Answer:* **Estimate 402×10^{-21} J/molecule (similar to 435×10^{-21} J/molecule)**

Strategy and Explanation: Given the description of a partial reaction and bond enthalpies, estimate the activation energy for the whole reaction and compare to the activation energy for the whole reaction given elsewhere.

The "rotation reaction" looks like this: $CH_3CH=CHCH_3 \longrightarrow CH_3CH-CHCH_3$

Use Table 8.2 to find the bond enthalpy for the single and double bonds between carbon atoms. The difference between them gives an estimate of the enthalpy required to break the π-bond in kJ/mol. Active Figure 12.7 gives energies in J/molecule, so convert the estimated energy to J/molecule.

$$D_{C=C} - D_{C-C} = 598 \text{ kJ/mol} - 356 \text{ kJ/mol} = 242 \text{ kJ/mol}$$

$$\frac{242 \text{ kJ}}{1 \text{ mole}} \times \frac{1 \text{ mol}}{6.022 \times 10^{23} \text{ molecules}} \times \frac{1000 \text{ J}}{1 \text{ kJ}} = 402 \times 10^{-21} \text{ J / molecule}$$

In Active Figure 12.7, the activation energy is given 435×10^{-21} J/molecule. It is comparable in the first significant figure to the estimated energy for breaking a π-bond.

✓ *Reasonable Answer Check:* The calculation of the estimated energy of breaking a π-bond uses bond enthalpies, which are derived from the average of the enthalpies for several molecules containing these kinds of bonds.

Chapter 13: Chemical Equilibrium

Introduction

Even when a reaction appears to stop, reactants are constantly converting to products to the same extent that products are converting to reactants. The macroscopic image of a stationary condition is still microscopically dynamic. Molecules with a tendency to react won't just stop reacting. What makes a reaction appear to stop is that the reverse reaction is happening at the same rate as the forward reaction.

Try to avoid confusing equilibrium constant expressions with kinetic rate laws from Chapter 12. Kinetic rate laws are determined experimentally and not by the coefficients of the balanced chemical equation. Equilibrium constant expressions on the other hand do have a relationship to the coefficients of the balanced chemical equations. Try not to mix up the definitions of k and K.

Solutions to Blue-Numbered Questions for Review and Thought for Chapter 13

Topical Questions

Characteristics of Chemical Equilibrium

7. *Answer/Explanation:* There are many answers to this Question. Prepare a sample of N_2O_4 in which the N atoms are the heavier isotopes ^{15}N. Introduce the heavy isotope of N_2O_4 into an equilibrium mixture of N_2O_4 and NO_2. Dynamic equilibria representing the decomposition of the dimer $N_2O_4(g)$ produces NO_2 (g) and $^{15}NO_2$ (g), which will occasionally recombine into the mixed dimer, $O_2{}^{15}N–NO_2$ (g)

$$O_2N–NO_2 \text{ (g)} \rightleftharpoons 2\ NO_2 \text{ (g)}$$
$$O_2{}^{15}N–{}^{15}NO_2 \text{ (g)} \rightleftharpoons 2\ {}^{15}NO_2 \text{ (g)}$$
$$O_2{}^{15}N–NO_2 \text{ (g)} \rightleftharpoons {}^{15}NO_2 \text{ (g)} + NO_2 \text{ (g)}$$

Use spectroscopic methods, such as infrared spectroscopy to observe the distribution of the radioisotope among the reactants and products.

The Equilibrium Constant

9. *Answer:* **see drawings below**

Strategy and Explanation: Draw reactant concentration vs. time as a downward curve to level off at the equilibrium concentration. Draw product concentration vs. time curve as an upward curve to level off at the equilibrium concentration. The slope of [NO] increase is steeper than slope of $[N_2]$ decrease, and the equilibrium concentration of NO is equal to half of the equilibrium concentration of N_2 or O_2.

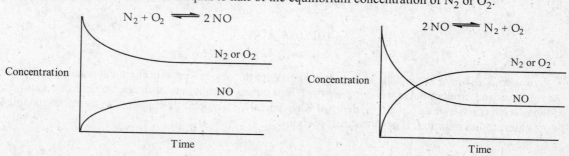

11. *Answer:* (a) $K_c = \dfrac{[H_2O]^2[O_2]}{[H_2O_2]^2}$ (b) $K_c = \dfrac{[PCl_5]}{[PCl_3][Cl_2]}$ (c) $K_c = [CO]^2$ (d) $K_c = \dfrac{[H_2S]}{[H_2]}$

Strategy and Explanation: For the expression of the equilibrium constant, use this form: $K_c = \dfrac{[\text{products}]}{[\text{reactants}]}$; if a

stoichiometric coefficient precedes a species, that number is used as the mathematical power of the concentration in the expression. Remember that the equilibrium state is not affected by the relative quantity of any solids or liquids, so those materials do not show up in the equilibrium expression. Hence, in particular, the answer for (c) does not have $SiO_2(s)$, $C(s)$, or $SiC(s)$ in the expression, and the answer for (d) does not have $S_8(s)$ in the expression.

13. *Answer:* **Equation (e)**

Strategy and Explanation: The two equations are related, so we will use the information in Section 13.2 to identify how their equilibrium constants are related.

Multiplying the first equation by a constant factor of 2 and then reversing that equation gives the second equation. The first change means we need to raise the K from the first reaction to the power of 2. The second change means we need to take the reciprocal of the K after the first change. Therefore the second equation's equilibrium constant is related to the first equation's equilibrium constant as represented in (e), $K_{c_2} = 1/K_{c_1}^2$.

15. *Answer:* (a) $K_P = \dfrac{P_{H_2O}^2 P_{O_2}}{P_{H_2O_2}^2}$ (b) $K_P = \dfrac{P_{PCl_5}}{P_{PCl_3} P_{Cl_2}}$ (c) $K_P = P_{CO}^2$ (d) $K_P = \dfrac{P_{H_2S}}{P_{H_2}}$

Strategy and Explanation: For the expression of the equilibrium constant, use this form: $K_P = \dfrac{P_{\text{products}}}{P_{\text{reactants}}}$; if a

stoichiometric coefficient precedes a species, that number is used as the mathematical power of the pressure in the expression. Remember that the equilibrium is not affected by the quantity of any solids or liquids, so those materials do not show up in the equilibrium expression. Hence, in particular, the answer for (c) does not have $SiO_2(s)$, $C(s)$, or $SiC(s)$ in the expression, and the answer for (d) does not have $S_8(s)$ in the expression.

Determining Equilibrium Constants

17. *Answer:* **(a) 1.0 (b) 1.0 (c) 1**

Strategy and Explanation: Given the equation describing the interconversion of two isomers and the concentrations of the gaseous reactant and product at equilibrium, find the value of K_c.

Write the K_c expression for the reaction and plug in the known values.

$$A(g) \rightleftharpoons B(g)$$

$$K_c = \dfrac{[B]}{[A]}$$

(a) $K_c = \dfrac{0.74 \text{ mol}/L}{0.74 \text{ mol}/L} = 1.0$

(b) $K_c = \dfrac{2.0 \text{ mol}/L}{2.0 \text{ mol}/L} = 1.0$

(c) $K_c = \dfrac{0.01 \text{ mol}/L}{0.01 \text{ mol}/L} = 1$

✓ *Reasonable Answer Check:* The change in temperature does not affect this chemical reaction, suggesting that the reactants and products have the same energy. That means that the rate of the forward reaction and the rate of the reverse reaction are affected in identical ways when the temperature changes. It seems reasonable for two isomers to have similar, if not identical, energies.

19. *Answer:* $K_p = 1.6$

Strategy and Explanation: Given an equation for a reaction and the moles of the gaseous reactants and products at equilibrium in a known volume, find the value of K_p.

Write the K_c expression for the reaction, calculate the concentrations of the gases, and plug them into the expression to get the value of K_c. Use the relationship between K_p and K_c (Equation 13.5) to get K_p:

$$K_p = K_c(RT)^{\Delta n}$$

$$H_2(g) + CO_2(g) \rightleftharpoons H_2O(g) + CO(g)$$

$$K_c = \frac{[H_2O][CO]}{[H_2][CO_2]} = \frac{\left(\dfrac{0.11 \text{ mol}}{1.0 \text{ L}}\right) \times \left(\dfrac{0.11 \text{ mol}}{1.0 \text{ L}}\right)}{\left(\dfrac{0.087 \text{ mol}}{1.0 \text{ L}}\right) \times \left(\dfrac{0.087 \text{ mol}}{1.0 \text{ L}}\right)} = 1.6$$

$$T = 986\ {}^\circ C + 273.15 = 1259 \text{ K}$$

$$\Delta n = 1 \text{ mol } H_2O(g) + 1 \text{ mol } CO(g) - 1 \text{ mol } H_2(g) - 1 \text{ mol } CO_2(g) = 0$$

$$K_p = K_c(RT)^{\Delta n}$$

$$K_p = (1.6) \times \left\{\left(0.08206\ \frac{L \cdot atm}{mol \cdot K}\right) \times (1259 \text{ K})\right\}^0 = 1.6$$

✓ *Reasonable Answer Check:* The slightly larger moles of products suggest this reaction is slightly product-favored, so it makes sense that the K_c is larger than 1. The equal moles of gas-phase reactants and gas-phase products, which is responsible for the power in the $(RT)^0$ factor, makes K_p the same as K_c.

21. *Answer:* $K_c = 75$

Strategy and Explanation: Use a method similar to that described in the first part of the solution to Question 19.

$$CO(g) + Cl_2(g) \rightleftharpoons COCl_2(g) \qquad\qquad K_c = \frac{[COCl_2]}{[CO][Cl_2]}$$

We will assume the volume is measured with a precision of ± 1 L.

$$K_c = \frac{\left(\dfrac{9.00 \text{ mol}}{50.\text{ L}}\right)}{\left(\dfrac{3.00 \text{ mol}}{50.\text{ L}}\right) \times \left(\dfrac{2.00 \text{ mol}}{50.\text{ L}}\right)} = 75$$

The larger moles of products suggest this reaction is slightly product-favored, so it makes sense that the K_c is larger than 1.

The Meaning of the Equilibrium Constant

23. *Answer:* **Reactions (b) and (c) are product-favored. Most reactant-favored (a), then (b), then (c).**

Strategy and Explanation: Given a table of equations with values of K_c and K_p, order the members of a set or equations from most reactant-favored to most product-favored.

Look up the given chemical equation or a related chemical equation, determine the size of its K_c, using techniques described at the end of Section 13.2, as needed. K_c values larger than 1 are product-favored. The smaller K_c is, the more reactant-favored the reaction is. The larger K_c is, the more product-favored the reaction is. Order the equations from smallest K_c to largest K_c.

(a) $2 NH_3(g) \rightleftharpoons N_2(g) + 3 H_2(g)$

     is the reverse of the third reaction in Table 13.1, so $K_{c,(a)} = 1/K_c = 1/(3.5 \times 10^8) = 2.9 \times 10^{-9}$.

     K_c is smaller than 1, so the reaction is not product-favored.

(b) $NH_4^+(aq) + OH^-(aq) \rightleftharpoons NH_3(aq) + H_2O(\ell)$

     is the reverse of the twelfth reaction in Table 13.1, so $K_{c,(b)} = 1/K_c = 1/(1.8 \times 10^{-5}) = 5.6 \times 10^4$.

     K_c is larger than 1, so the reaction is product-favored.

(c) $2 NO(g) \rightleftharpoons N_2(g) + O_2(g)$ is the reverse of the fourth reaction in Table 13.1, so

     $K_{c,(c)} = 1/K_c = 1/(4.5 \times 10^{-31}) = 2.2 \times 10^{30}$. K_c is larger than 1, so the reaction is product-favored.

Therefore, the order is (a) 2.9×10^{-9}, then (b) 5.6×10^4, then (c) 2.2×10^{30}.

✓ *Reasonable Answer Check:* Reversing a reaction takes what were the products and makes them the reactants and vice versa, so the reverse of a product-favored reaction will be a reactant-favored reaction and vice versa. The values of K_c make sense. Comparing the values of K to determine which is the most reactant-favored and product-favored among reaction with different stoichiometric relationships is not always legitimate. Here, while the denominators all have two concentration values, we find four concentration values in the numerator of (a), only two in (c), and only one in (b). These differences will affect the size of K_c dramatically. In general, one should compare the values of K_c only among reactions with the same stoichiometric relationships (i.e., with the same number of reactants and products.)

Using Equilibrium Constants

25. *Answer:* (a) see table below   (b) $\dfrac{0.100 + x}{0.100 - x} = 2.5$, $x = 0.043$ (c) [2-methylpropane] = 0.024 M,

     [butane] = 0.010 M

Strategy and Explanation: Given an equation for a reaction, the initial concentration or moles of the gaseous reactant in a known volume, and the value of the equilibrium constant, determine the change in the concentrations and the equilibrium concentrations of the reactants and products.

Follow the procedure given in Section 13.5. When necessary, find Q to determine direction of the reaction. Then write the equation and construct a reaction table. Describe the stoichiometric changes in terms of one variable, x, the change in the concentration of butane. Write the K_c expression for the reaction in terms of x. Solve that equation to get x, and use it to calculate the concentrations of the gases at equilibrium.

$$butane\ (g) \rightleftharpoons 2\text{-methylpropane}\ (g)$$

(a)   (conc. butane) = (conc. 2-methylpropane) = 0.100 M

$$Q_c = \frac{(conc.\ 2\text{-methylpropane})}{(conc.\ butane)} = \frac{0.100\ M}{0.100\ M} = 1.00 < 2.5$$

$Q_c < K_c$, so reaction goes from reactants toward products:

	butane (g) $\rightleftharpoons$	2-methylpropane (g)
initial conc. (M)	0.100	0.100
change as reaction occurs (M)	− x	+ x
equilibrium conc. (M)	0.100 − x	0.100 + x

(b)
$$K_c = \frac{[2\text{-methylpropane}]}{[\text{butane}]} = 2.5$$

At equilibrium
$$\frac{0.100 + x}{0.100 - x} = 2.5$$

$$0.100 + x = 2.5(0.100 - x) = 0.25 - 2.5x$$

$$x + 2.5x = 0.25 - 0.100$$

$$3.5x = 0.15$$

$$x = 0.043$$

(c) _____ (conc. butane) = 0.017 mol butane/0.50 L = 0.034 M

	butane (g) ⇌	2-methylpropane (g)
initial conc. (M)	0.034	0
change as reaction occurs (M)	− x	+ x
equilibrium conc. (M)	0.034 − x	x

At equilibrium
$$\frac{x}{0.034 - x} = 2.5$$

$$x = 2.5(0.034 - x) = 0.085 - 2.5x$$

$$x + 2.5x = 0.085$$

$$3.5x = 0.085$$

$$x = 0.024 \text{ M} = [2\text{-methylpropane}]$$

$$[\text{butane}] = 0.034 \text{ M} - 0.024 \text{ M} = 0.010 \text{ M}$$

✓ *Reasonable Answer Check:* Substituting the equilibrium concentrations into the equilibrium expression will reproduce K_c:

Part (b), [butane] = 0.100 − 0.043 = 0.057 M, [2-methylpropane] = 0.100 + 0.043 = 0.143 M. $K_c = \frac{0.143}{0.057} = 2.5$.

(c) $K_c = \frac{0.024}{0.010} = 2.4$. These are both right, within the uncertainty of the data.

27. *Answer:* **[HI] = 4 × 10⁻² M, [I₂]= 5 × 10⁻³ M, [H₂] = 5 × 10⁻³ M**

Strategy and Explanation: Follow the procedure given in Section 13.3. Write the equation and a construct a reaction table. Describe the stoichiometric changes in terms of one variable, x. Write the K_c expression for the reaction in terms of x. Solve that equation to get x, and use it to calculate the concentrations of the gases at equilibrium.

$$(\text{conc. HI}) = 0.05 \text{ mol}/1.0 \text{ L} = 0.05 \text{ M}$$

	H₂ (g)	+ I₂ (g) ⇌	2 HI (g)
initial conc. (M)	0	0	0.05
change as reaction occurs (M)	+ x	+ x	− 2x
equilibrium conc. (M)	x	x	0.05− 2x

At equilibrium
$$K_c = \frac{[\text{HI}]^2}{[\text{H}_2][\text{I}_2]} = \frac{(0.05 - 2x)^2}{(x)(x)} = 76$$

Take the square root of each side:
$$\frac{(0.05 - 2x)}{x} = 8.7$$

$$0.05 - 2x = 8.7x$$

$$0.05 = 8.7x + 2x = 10.7x$$

$$x = 5 \times 10^{-3} \text{ M} = [H_2] = [I_2]$$

$$[HI] = 0.05 - 2x = 0.05 \text{ M} - 2 \times (5 \times 10^{-3} \text{ M}) = 4 \times 10^{-2} \text{ M}$$

✓ *Reasonable Answer Check:* The larger concentration of product is consistent with a K greater than 1. Substituting the equilibrium concentrations into the equilibrium expression will reproduce K_c within the

precision of the data: $K_c = \dfrac{(4 \times 10^{-2})^2}{(5 \times 10^{-3}) \times (5 \times 10^{-3})} = 6 \times 10^1.$

29. *Answer:* **(a) 1.94 mol (b) 1.92 mol (c) 1.98 mol**

Strategy and Explanation: Use a method similar to that described in the solution to Question 25.

(a)   (conc. I_2) = 1.00 mol/10.00 L = 0.100 M

(conc. H_2) = 3.00 mol/10.00 L = 0.300 M

	H_2 (g)	+ I_2 (g)	⇌	2 HI (g)
initial conc. (M)	0.300	0.100		0
change as reaction occurs (M)	– x	– x		+ 2x
equilibrium conc. (M)	0.300 – x	0.100 – x		2x

At equilibrium $K_c = \dfrac{[HI]^2}{[H_2][I_2]} = \dfrac{(2x)^2}{(0.300 - x)(0.100 - x)} = 50.0$

Set up to use the quadratic equation (see Appendix A, Section A.7, page A.12):

$$50.0\{(0.300)(0.100) - (0.300 + 0.100)x + x^2\} = 4x^2$$

$$1.50 - 20.0x + 50.0x^2 = 4x^2$$

$$1.50 - 20.0x + 46.0x^2 = 0$$

Plug into the quadratic equation:

$$x = \frac{-b \pm \sqrt{b^2 - 4ac}}{2a} = \frac{20.0 \pm \sqrt{(-20.0)^2 - 4(46.0)(1.50)}}{2(46.0)}$$

The two roots are:

$$x = \frac{20.0 - 11.1}{2(46.0)} = \frac{8.9}{92.0} = 0.097$$

$$x = \frac{20.0 + 11.1}{2(46.0)} = \frac{31.1}{92.0} = 0.338$$

Notice: Both of these roots are mathematically sound solutions for the quadratic equation, but the second of these roots must be discarded because it is too large and produces a nonsensical result. In particular, here, the equilibrium concentration of I_2 (0.100 – x) must be a positive number, but using the second root produces a negative answer, 0.100 – 0.338 = – 0.238.

$$x = 0.097 \text{ M}$$

$$2x = 2(0.097 \text{ M}) = 0.194 \text{ M} = [HI]$$

$$10.00 \text{ L} \times \frac{0.194 \text{ mol}}{\text{L}} = 1.94 \text{ mol}$$

$$[H_2] = 0.300 - x = 0.300 \text{ M} - (0.097 \text{ M}) = 0.203 \text{ M}$$

$$[I_2] = 0.100 - x = 0.100 \text{ M} - (0.097 \text{ M}) = 0.003 \text{ M}$$

✓ *Reasonable Answer Check:* (a) The larger concentration of product is consistent with a K greater than 1. Substituting the equilibrium concentrations into the equilibrium expression will reproduce K_c within the

precision of the data: $K_c = \dfrac{(0.194)^2}{(0.203) \times (0.003)} = 6 \times 10^1 \cong 50.0$ within the uncertainty of 1 significant figure.

(b) (conc. I_2) = 1.00 mol/5.00 L = 0.200 M   (conc. H_2) = 3.00 mol/5.00 L = 0.600 M

	H_2 (g)	+ I_2 (g)	$\rightleftharpoons$ 2 HI (g)
initial conc. (M)	0.600	0.200	0
change as reaction occurs (M)	– x	– x	+ 2x
equilibrium conc. (M)	0.600 – x	0.200 – x	2x

At equilibrium $K_c = \dfrac{[HI]^2}{[H_2][I_2]} = \dfrac{(2x)^2}{(0.600-x)(0.200-x)} = 50.0$

Set up to use the quadratic equation (Appendix A, Section A.7, page A.12):

$$50.0\{(0.600)(0.200) - (0.600 + 0.200)x + x^2\} = 4x^2$$
$$6.00 - 40.0x + 46.0x^2 = 0$$

Plug into the quadratic equation:

$$x = \frac{-b \pm \sqrt{b^2 - 4ac}}{2a} = \frac{40.0 \pm \sqrt{(-40.0)^2 - 4(46.0)(6.00)}}{2(46.0)} = \frac{17.7}{92.0} = 0.192$$

$$2x = 0.384 \text{ M} = [HI]$$

$$5.00 \text{ L} \times \frac{0.384 \text{ mol}}{L} = 1.92 \text{ mol}$$

✓ *Reasonable Answer Check:* (b) The larger concentration of product is consistent with a K greater than 1. Substituting the equilibrium concentrations into the equilibrium expression will reproduce K_c within the precision of the data:

$$[H_2] = 0.600 - x = 0.600 \text{ M} - (0.192 \text{ M}) = 0.408 \text{ M}$$
$$[I_2] = 0.200 - x = 0.200 \text{ M} - (0.192 \text{ M}) = 0.008 \text{ M}$$

$K_c = \dfrac{(0.384)^2}{(0.408) \times (0.008)} = 5 \times 10^1 \cong 50.0$ with only one significant figure. The answers for (a) and (b) should

be the same, because the volume dependence cancels due to the fact that the number of products and reactants are the same. Within the cited uncertainty (a) and (b) give the same results.

(c)   (conc. I_2) = 1.00 mol/10.00 L = 0.100 M

(conc. H_2) = (3.00 mol + 3.00 mol)/10.00 L = 0.600 M

	H_2 (g)	+ I_2 (g)	$\rightleftharpoons$ 2 HI (g)
initial conc. (M)	0.600	0.100	0
change as reaction occurs (M)	– x	– x	+ 2x
equilibrium conc. (M)	0.600 – x	0.100 – x	2x

At equilibrium $K_c = \dfrac{[HI]^2}{[H_2][I_2]} = \dfrac{(2x)^2}{(0.600-x)(0.100-x)} = 50.0$

Set up to use the quadratic equation (Appendix A, Section A.7, page A.12):

$$50.0\{(0.600)(0.100) - (0.600 + 0.100)x + x^2\} = 4x^2$$
$$3.00 - 35.0x + 46.0x^2 = 0$$

Plug into the quadratic equation:

$$x = \frac{-b \pm \sqrt{b^2 - 4ac}}{2a} = \frac{35.0 \pm \sqrt{(-35.0)^2 - 4(46.0)(3.00)}}{2(46.0)} = \frac{9.1}{92.0} = 0.099$$

$$2x = 0.198 \text{ M} = [HI]$$

$$10.00 \text{ L} \times \frac{0.198 \text{ mol}}{L} = 1.98 \text{ mol}$$

✓ *Reasonable Answer Check:* (c) The larger concentration of product is consistent with a K greater than 1. Substituting the equilibrium concentrations into the equilibrium expression will reproduce K_c within the precision of the data:

$$[H_2] = 0.300 - x = 0.600\ M - (0.099\ M) = 0.501\ M$$

$$[I_2] = 0.100 - x = 0.100\ M - (0.099\ M) = 0.001\ M$$

$$K_c = \frac{(0.198)^2}{(0.501) \times (0.001)} = 7 \times 10^1 \cong 50.0.$$ With only one significant figure round off errors have started to dramatically affect the smallest number, $[I_2]$. These results still look reasonable.

31. *Answer:* **(a) No (b) proceeds toward products**

Strategy and Explanation: Given an equation for a reaction, the initial moles of the gaseous reactant in a known volume, and the value of the equilibrium constant, determine if the system is at equilibrium and, if it is not, determine which direction it must proceed to reach equilibrium.

Follow the procedure given in Section 13.5. Calculate Q and compare it with K. If $Q = K$, then the system is at equilibrium. If $Q < K$, the reaction proceeds toward products to reach equilibrium; if $Q > K$, the reaction proceeds toward reactants to reach equilibrium. $K_c = 3.58 \times 10^{-3}$

$$(\text{conc. } SO_3) = 0.15\ mol/10.0\ L = 0.015\ M$$

$$(\text{conc. } SO_2) = 0.015\ mol/10.0\ L = 0.0015\ M$$

$$(\text{conc. } O_2) = 0.0075\ mol/10.0\ L = 0.00075\ M$$

$$Q_c = \frac{(\text{conc. } SO_2)^2 (\text{conc. } O_2)}{(\text{conc. } SO_3)^2} = \frac{(0.0015)^2 (0.00075)}{(0.015)^2} = 7.5 \times 10^{-6}$$

(a) $Q_c \neq K_c$, so the reaction is not at equilibrium.

(b) $Q_c < K_c$, so the reaction must proceeds toward products to reach equilibrium.

✓ *Reasonable Answer Check:* Considering the powers of 10, the size of Q makes sense

$$\frac{(10^{-3})^2 (10^{-4})}{(10^{-2})^2} = \frac{(10^{-6})(10^{-4})}{(10^{-4})} = 10^{-6}$$

33. *Answer:* **(a) No (b) proceed toward reactants (c) $[N_2] = [O_2] = 0.002\ M;\ [NO] = 0.061\ M$**

Strategy and Explanation: Given an equation for a reaction, the initial moles of the gaseous reactant in a known volume, and the value of the equilibrium constant, determine if the system is at equilibrium and, if it is not, determine which direction it must proceed to reach equilibrium, and calculate the equilibrium concentrations.

Start with the procedure given in Question 31. Calculate Q and compare it to K to see if the system is at equilibrium, and, as needed, which direction the reaction proceeds in order to reach equilibrium. If the system is not already at equilibrium, use a method similar to that described in the solution to Question 27 to find the equilibrium concentrations. $K_c = 1.7 \times 10^{-3}$

$$(\text{conc. } NO) = 0.015\ mol/10.0\ L = 0.0015\ M$$

$(\text{conc. } N_2) = 0.25\ mol/10.0\ L = 0.025\ M$ $\qquad\qquad$ $(\text{conc. } O_2) = 0.25\ mol/10.0\ L = 0.025\ M$

$$Q_c = \frac{(\text{conc. } NO)^2}{(\text{conc. } N_2)(\text{conc. } O_2)} = \frac{(0.0015)^2}{(0.025)(0.025)} = 3.6 \times 10^{-3}$$

(a) $Q_c \neq K_c$, so the reaction is not at equilibrium.

(b) $Q_c > K_c$, so the reaction must proceeds toward reactants to reach equilibrium.

(c)

	$N_2(g)$	$+\ O_2(g) \rightleftharpoons$	$2\ NO(g)$
initial conc. (M)	0.025	0.025	0.0015
change as reaction occurs (M)	$+ x$	$+ x$	$- 2x$
equilibrium conc. (M)	$0.025 - x$	$0.025 - x$	$0.015 + 2x$

At equilibrium

$$K_c = \frac{[NO]^2}{[N_2][O_2]} = \frac{(0.0015 - 2x)^2}{(0.025 + x)(0.025 + x)} = 1.7 \times 10^{-3}$$

Take the square root of each side:

$$\frac{(0.0015 - 2x)}{(0.025 + x)} = \sqrt{1.7 \times 10^{-3}} = 0.041$$

Solve for x:

$$0.0015 - 2x = 0.041(0.025 + x)$$

$$0.0015 - 2x = 0.001025 + 0.041x$$

$$0.0015 - 0.001025 = (2 + 0.041)x$$

$$x = 2.3 \times 10^{-4} \text{ M}$$

$$[N_2] = [O_2] = 0.025 - x = 0.025 - 2.3 \times 10^{-4} \text{ M} = 0.025 \text{ M}$$

$$[NO] = 0.0015 - 2x = 0.0015 \text{ M} - 2 \times (2.3 \times 10^{-4} \text{ M}) = 0.0010 \text{ M}$$

✓ *Reasonable Answer Check:* The smaller concentrations of products are consistent with a Q > K and a K less than 1.

Shifting a Chemical Equilibrium: Le Chatlier's Principle

35. *Answer:* **Choice (b); heat energy is absorbed, and increasing the temperature drives the reaction towards products.**

Strategy and Explanation: As described in Section 13.6, the positive ΔH indicates that heat energy is a reactant in this reaction. Increasing the temperature will drive the reaction toward products and away from N_2O_4. That means the N_2O_4 concentration will decrease.

37. *Answer:* **see chart below**

Strategy and Explanation: Use the methods described in Section 13.6 to answer this Question. Changing concentrations or pressures of reactants and products or changing the available energy in the system will take the system out of equilibrium. The response to that change will be a shift away from the increase. Changing the amounts of solids or liquids will not affect the equilibrium position. Only temperature can affect the value of the equilibrium constant. Because this reaction is exothermic, energy is released.

$$H_2(g) + Br_2(g) \rightleftharpoons 2 \, HBr(g) \qquad\qquad \Delta H = -103.7 \text{ kJ}$$

Change	$[Br_2]$	$[HBr]$	K_c	K_p
Some H_2 *(a reactant)* is added to the container	decrease	increase	no change	no change
The temperature of the gases in the container is increased. *(increase in energy, energy released in the reaction)*	increase	decrease	decrease	decrease
The pressure of HBr *(a product)* is increased.	increase	increase (since extra is added)	no change	no change

39. *Answer:* **(a) no shift (b) left (c) left**

Strategy and Explanation: Use the methods described in Section 13.6 to answer this Question. Changing concentrations or pressures of reactants and products or changing the available energy in the system will take the system out of equilibrium. The response to that change will be to shift away from an increase (or shift towards a decrease). Changing the amounts of solids or liquids will not affect the equilibrium position. Because the reaction is endothermic, energy is absorbed.

$$2 \, NOBr(g) \rightleftharpoons 2 \, NO(g) + Br_2(\ell) \qquad\qquad \Delta H = 16.1 \text{ kJ}$$

(a) Equilibrium is not affected by the addition of a pure liquid, so adding more liquid Br_2 will cause no change.

(b) Removing reactant, NOBr, shifts the reaction toward the reactants (left) to increase the [NOBr].

(c) Decreasing the temperature removes energy. The reaction shifts toward the reactants (left), to evolve more energy.

41. *Answer:* **(a) decrease (b) see sketch below**

Strategy and Explanation:

(a) Use the methods described in Question 39 and Section 13.6. $PbCl_2(s) \rightleftharpoons Pb^{2+}(aq) + 2\ Cl^-(aq)$

Adding the soluble ionic compound, NaCl, produces a solution with more of the product, Cl^-. The reaction shifts toward the reactants (left) and decreases the $[Pb^{2+}]$.

(b) Each time NaCl is added, the chloride concentration rises, then the reaction shifts toward reactants producing more solid and dropping the concentrations of both reactants until a new equilibrium condition is achieved. The following qualitative sketch (similar to that of Figure 13.6) illustrates these changes.

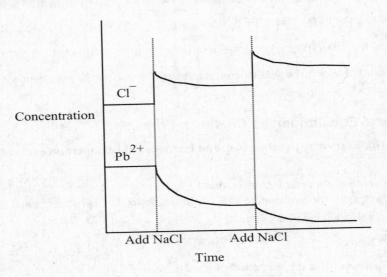

Equilibrium at the Nanoscale

43. Answer: **(a) energy effect (b) entropy effect (c) neither**

Strategy and Explanation: We will use methods described in Section 13.7 to answer this Question.

The entropy effect is related to randomness and probability. The more random a system is, the more probable it is. In many of the chemical reactions we study, we can get a qualitative idea of the entropy change by looking at the physical phases of the reactants and products. In general, gases have more entropy than liquids, which have more entropy than solids. Aqueous substances have more entropy than solids, but less than gases.

The energy effect is related to the relative stability of the reactants and products. If the products are more stable than the reactants (exothermic, ΔH = negative), then the products are favored energetically. If the reactants are more stable than the products (endothermic, ΔH = positive), then the reactants are favored energetically.

(a) 4 mol gas $\rightleftharpoons$ 2 mol gas   Entropy effect favors reactants, not products.

Reaction is exothermic, so energy effect favors products.

Reaction (a) (in the forward direction) is favored only by the energy effect.

(b) 1 mol gas $\rightleftharpoons$ 2 mol gas   Entropy effect favors products.

Reaction is endothermic, so energy effect favors reactants, not products.

Reaction (b) (in the forward direction) is favored only by the entropy effect.

(c) 4 mol gas $\rightleftharpoons$ 2 mol gas   Entropy effect favors reactants, not products.

Reaction is endothermic, so energy effect favors reactants, not products.

Reaction (c) (in the forward direction) is not favored by either effect.

Controlling Chemical Reactions: The Haber-Bosch Process

45. *Answer:* **(a) Step 1: –296.830 kJ/mol, Step 2: –197.78 kJ/mol, Step 3: –132.44 kJ/mol (b) all three steps are exothermic (c) None have entropy increase, Steps 2 and 3 have entropy decrease, the entropy in Step 1 stays about the same. (d) All three steps have products favored more at low temperatures.**

Strategy and Explanation:

(a) Follow the method described in the solution to Question 57 in Chapter 6 and Section 6.10.

Look up the ΔH_f° values in Appendix J.

Step 1: $\Delta H^\circ = (1 \text{ mol}) \times \Delta H_f^\circ(SO_2) - (1 \text{ mol}) \times \Delta H_f^\circ(S) - (1 \text{ mol}) \times \Delta H_f^\circ(O_2)$

$\Delta H^\circ = (1 \text{ mol}) \times (-296.830 \text{ kJ/mol}) - (1 \text{ mol}) \times (0 \text{ kJ/mol}) - (1 \text{ mol}) \times (0 \text{ kJ/mol}) = -296.830 \text{ kJ/mol}$

Step 2: $\Delta H^\circ = (2 \text{ mol}) \times \Delta H_f^\circ(SO_3) - (2 \text{ mol}) \times \Delta H_f^\circ(SO_2) - (1 \text{ mol}) \times \Delta H_f^\circ(O_2)$

$\Delta H^\circ = (2 \text{ mol}) \times (-395.72 \text{ kJ/mol}) - (2 \text{ mol}) \times (-296.830 \text{ kJ/mol}) - (1 \text{ mol}) \times (0 \text{ kJ/mol}) = -197.78 \text{ kJ/mol}$

Step 3: $\Delta H^\circ = (1 \text{ mol}) \times \Delta H_f^\circ(H_2SO_4(\ell)) - (1 \text{ mol}) \times \Delta H_f^\circ(SO_3) - (1 \text{ mol}) \times \Delta H_f^\circ(H_2O(\ell))$

$\Delta H^\circ = (1 \text{ mol}) \times (-813.989 \text{ kJ/mol}) - (1 \text{ mol}) \times (-395.72 \text{ kJ/mol})$

$- (1 \text{ mol}) \times (-285.830 \text{ kJ/mol}) = -132.44 \text{ kJ/mol}$

(b) All three reactions are exothermic (ΔH = negative)

(c) Follow the method described in the solution to Question 43.

Step 1:	1 mol gas ⇌ 1 mol gas	No large entropy change.
Step 2:	3 mol gas ⇌ 2 mol gas	Entropy decreases.
Step 3:	1 mol gas ⇌ 0 mol gas	Entropy decreases.

(d) Heat is produced in all three steps. A low temperature in each step would serve to reduce the energy and drive the reaction toward products. All three steps would favor products more at lower temperatures.

General Questions

48. *Answer:* **(a)**

Species	Br$_2$	Cl$_2$	F$_2$	H$_2$	N$_2$	O$_2$
[E] (mol/L)	0.28	0.057	1.44	1.76×10^{-5}	4×10^{-14}	4.0×10^{-6}

(b) see explanation below.

Strategy and Explanation: Use a method similar to that described in the answer to Question 25.

(a)   (conc. E$_2$) = 1.00 mol E$_2$/1.0 L = 1.0 M

	E$_2$ (g) ⇌	2 E (g)
initial conc. (M)	1.0	0
change as reaction occurs (M)	– x	+ 2x
equilibrium conc. (M)	1.0 – x	2x

At equilibrium $K_c = \dfrac{[E]^2}{[E_2]} = \dfrac{(2x)^2}{1.0 - x}$

Set up to use the quadratic equation (Appendix A, Section A.7, page A.12):

$$(2x)^2 = K_c(1.0 - x) = K_c - K_c x$$

$$4x^2 + K_c x - K_c = 0$$

Plug into the quadratic equation:

$$x = \frac{-b \pm \sqrt{b^2 - 4ac}}{2a} = \frac{-K_c \pm \sqrt{K_c^2 - 4(K_c)(-K_c)}}{2(4)}$$

$$[E] = 2x$$

Here is an example of the calculation for the first species: $E = Br$, $K_c = 8.9 \times 10^{-2}$.

$$x = \frac{-(8.9 \times 10^{-2}) - \sqrt{(8.9 \times 10^{-2})^2 - 4(8.9 \times 10^{-2})(-8.9 \times 10^{-2})}}{2(4)} = \frac{1.1}{8} = 0.14$$

$$[Br] = 2 \times (0.14\ M) = 0.28\ M$$

Similar calculations give these results:

E species	K_c	x (M)	[E] (M)
Br	8.9×10^{-2}	0.14	0.28
Cl	3.4×10^{-3}	0.029	0.057
F	7.4	0.72	1.44
H	3.1×10^{-10}	8.8×10^{-6}	1.76×10^{-5}
N	1×10^{-27}	2×10^{-14}	4×10^{-14}
O	1.6×10^{-11}	2.0×10^{-6}	4.0×10^{-6}

✓ *Reasonable Answer Check:* [E] is small when K is small. [E] is large when K is large. The equilibrium

concentrations combine to reproduce $K_c = \dfrac{(2x)^2}{1.0 - x}$.

(b) At this temperature, the lowest bond enthalpy is predicted from the reaction that gives the most products, so F_2 is predicted to have the lowest bond dissociation enthalpy.

Comparing these results to the bond enthalpy values in Table 8.2, the product production decreases as the bond enthalpy increases: 156kJ F_2, 193kJ Br_2, 242kJ Cl_2, 436kJ H_2, 498kJ O_2, 946kJ N_2.

Lewis structures of F_2, Br_2, Cl_2, H_2 have a single bonds and more products in these calculations than O_2 with double bond and N_2 with a triple bond.

49. *Answer:* **(a) 0.67 (b) same result, see explanation below**

Strategy and Explanation:

(a) Following the procedure given in Section 13.3, write the equation and construct a reaction table. No products are initially present. *(Assume percent is known within ± 1%.)*

(conc. ^{14}C-*cis*-dichloroethene) = 100.% and [^{14}C-*trans*-dichloroethene] = 40.% at equilibrium

	^{14}C-*cis*-dichloroethene ⇌	^{14}C-*trans*-dichloroethene
initial conc. (%)	100.	0
change as reaction occurs (%)	_____	_____
equilibrium conc. (%)		40.

Set up the equilibrium changes using x and the stoichiometry of the equation

	^{14}C-*cis*-dichloroethene ⇌	^{14}C-*trans*-dichloroethene
initial conc. (%)	100.	0
change as reaction occurs (%)	– x	+ x
equilibrium conc. (%)	100. – x	40. = x

Using the equilibrium concentration of *trans*-dichloroethene gives x = 40.%.

$$[^{14}C\text{-}cis\text{-dichloroethene}] = 100.\% - 40.\% = 60.\%$$

$$K_c = \frac{[^{14}C\text{-}trans\text{-dichloroethane}]}{[^{14}C\text{-}cis\text{-dichloroethane}]} = \frac{x}{100. - x} = \frac{40.\%}{60.\%} = 0.67$$

(b) If we started with 100% ^{14}C-*trans*-dichloroethene, the reaction would have run backwards producing the ^{14}C-*cis*-isomer and depleting the ^{14}C-*trans*-isomer until the concentration reached exactly the same equilibrium ratio as determined above.

51. *Answer:* **Cases a and b will have decreased [HI] at equilibrium; cases c and d will have increased [HI]**

Strategy and Explanation: Use some of the methods described in the answer to Question 25.

$$2\ HI(g) \rightleftharpoons H_2(g) + I_2(g) \qquad\qquad K_c = \frac{[H_2][I_2]}{[HI]^2} = 0.0200$$

Construct Q_c:
$$Q_c = \frac{(\text{conc. } H_2)(\text{conc. } I_2)}{(\text{conc. } HI)^2}$$

Then compare to K_c.

Case a:
$$Q_c = \frac{\left(\dfrac{0.10\ \text{mol } H_2}{10.00\ \text{L}}\right)\left(\dfrac{0.10\ \text{mol } I_2}{10.00\ \text{L}}\right)}{\left(\dfrac{1.0\ \text{mol } HI}{10.00\ \text{L}}\right)^2} = 0.010 < 0.0200$$

Reaction goes toward products, and the concentration of HI decreases.

Case b:
$$Q_c = \frac{\left(\dfrac{1.0\ \text{mol } H_2}{10.00\ \text{L}}\right)\left(\dfrac{1.0\ \text{mol } I_2}{10.00\ \text{L}}\right)}{\left(\dfrac{10.\ \text{mol } HI}{10.00\ \text{L}}\right)^2} = 0.010 < 0.0200$$

Reaction goes toward products, and the concentration of HI decreases.

Case c:
$$Q_c = \frac{\left(\dfrac{10.\ \text{mol } H_2}{10.00\ \text{L}}\right)\left(\dfrac{1.0\ \text{mol } I_2}{10.00\ \text{L}}\right)}{\left(\dfrac{10.\ \text{mol } HI}{10.00\ \text{L}}\right)^2} = 0.10 > 0.0200$$

Reaction goes toward reactants, and the concentration of HI increases.

Case d:
$$Q_c = \frac{\left(\dfrac{0.381\ \text{mol } H_2}{10.00\ \text{L}}\right)\left(\dfrac{1.75\ \text{mol } I_2}{10.00\ \text{L}}\right)}{\left(\dfrac{5.62\ \text{mol } HI}{10.00\ \text{L}}\right)^2} = 0.0211 < 0.0200$$

Reaction goes toward reactants a very small amount, and the concentration of HI increases a very small amount.

In conclusion, the HI concentration increases in case c and very slightly in case d. The HI concentration decreases in cases a and b.

Applying Concepts

53. *Answer:* **Yes, the system is at equilibrium; no further experiments would be needed.**

Strategy and Explanation: For the system to be at equilibrium, it must contain some quantity of both reactants and products, the concentrations of all the reactants and products must be constant over time, and the same relative concentrations are established regardless of which direction the equilibrium was approached. The 42% *cis* concentration was achieved in two different experiments; therefore, the system is at equilibrium. No further experiments are needed.

55. *Answer:* **(a) Reaction (iii) (b) Reaction (i)**

Strategy and Explanation: As described near the end of Section 13.6, an endothermic reaction will have an increased K_c at higher temperatures, and an exothermic reaction will have a decrease in K_c at higher temperatures. In addition, the larger the change, the more exothermic or endothermic the reaction is, respectively.

(a) When R = CH_3, the values of K_c for the *cis-trans* isomerism decrease at higher temperatures. This is the only data set that shows this trend, so (iii) the *cis-trans* isomerism for R = CH_3 is the most exothermic.

(b) When R = F and R = Cl, the values of K_c for the *cis-trans* isomerism increase at higher temperatures. That means these are both endothermic *cis-trans* isomerism reactions. The R = F isomerism has a larger change, so (i) the *cis-trans* isomerism for R = F is the most endothermic.

57. *Answer/Explanation:* In the warmer sample, the molecules would be moving faster, and more NO_2 molecules would be seen. In the cooler sample, the molecules would be moving slower, and fewer NO_2 molecules would be seen. In both samples, the molecules are moving very fast. The average speed of gas molecules is commonly hundreds of miles per hour. In both samples, I would see some N_2O_4 molecules decomposing and some NO_2 molecules reacting with each other, at equal rates.

59. *Answer:* **Diagrams (b), (c), and (d)**

Strategy and Explanation: The equilibrium constant expression for this reaction is: $K_c = \dfrac{[AB]^2}{[A_2][B_2]}$.

We need to find diagrams that fit this range: $10^2 > = \dfrac{[AB]^2}{[A_2][B_2]} > 0.1$

Diagram (a) has only reactants, A_2 and B_2. No AB molecules are found in the mixture. That means the value of $K_c = 0$. Diagram (e) has only product molecules, AB. No A_2 or B_2 are found in the mixture. That means the value of $K_c = \infty$. Neither of these K_c values is within the stated range.

In the other three diagrams, we notice that the number of A_2 molecules and the number of B_2 molecules are equal.

$$K = \dfrac{[AB]^2}{[A_2][B_2]} = \dfrac{[(\text{Number AB}) \times N_0 / V]^2}{(\text{Number } A_2) \times N_0 / V \times (\text{Number } B_2) \times N_0 / V}$$

$$= \dfrac{(\text{Number AB})^2}{(\text{Number } A_2) \times (\text{Number } B_2)} = \dfrac{(\text{Number AB})^2}{(\text{Number } A_2)^2} = \left(\dfrac{\text{Number AB}}{\text{Number } A_2}\right)^2$$

That means we can simplify the search range by taking the square root of the range variables and compare them to the ratio of the numbers:

$$\sqrt{10^2} = 10 > = \dfrac{\text{Number of AB}}{\text{Number of } A_2} > = \sqrt{0.1} = 0.3$$

Diagram (b) has 4 AB and 2 A_2 = 2 B_2 : $\dfrac{\text{Number of AB}}{\text{Number of } A_2} = \dfrac{4}{2} = 2$

Diagram (c) has 6 AB and 1 A_2 = 1 B_2 : $\dfrac{\text{Number of AB}}{\text{Number of } A_2} = \dfrac{6}{1} = 6$

Diagram (d) has 2 AB and 3 A_2 = 3 B_2 : $\dfrac{\text{Number of AB}}{\text{Number of } A_2} = \dfrac{2}{3} = 0.667$

Therefore, all three of these remaining diagrams, (b), (c), and (d), represent equilibrium mixtures where $10^2 > K_c > 0.1$.

61. *Answer:* **See diagram below (many other answers are possible)**

Strategy and Explanation: The nanoscale-level diagram will have CO, H_2O, CO_2, and H_2 gas molecules in relative proportions dictated by the relationship described by the equilibrium constant expression.

$$K_c = 4.00 = \dfrac{[CO_2][H_2]}{[CO][H_2O]}$$

Assume the container has a volume of V and contains one CO molecule and one H_2O molecule. We can use the equilibrium expression to calculate how many product molecules to put in the container at equilibrium with these reactants.

The concentrations of each of the reactants is: $[CO] = [H_2O] = 1$ molecule $/ V = 1/V$. If we let x represent the number of CO_2 and H_2 molecule, their concentrations are: $[CO_2] = [H_2] = x$ molecules $/ V$. Substitute the concentrations into the equilibrium expression and solve for x:

$$4.00 = \frac{(\text{Number of } CO_2 / V)(\text{Number of } H_2 / V)}{(\text{Number of } CO / V)(\text{Number of } H_2O / V)} = \frac{(x/V)^2}{(1/V)^2}$$

$$4.00 = x^2$$

$$2.00 = x = \text{number of } CO_2 \text{ molecules} = \text{number of } H_2 \text{ molecules}$$

To represent an equilibrium state, we need to put two CO_2 molecules and two H_2 molecules in the box with one CO molecule and one H_2O molecule.

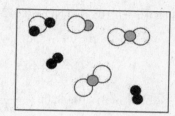

Many other answers are possible as long as K_c is satisfied, such as: four CO_2 molecules and one H_2 molecules in the box with one CO molecule and one H_2O molecule $\frac{(4/V)(1/V)}{(1/V)(1/V)} = 4$, or: four CO_2 molecules and two H_2 molecules in the box with one CO molecule and two H_2O molecule $\frac{(4/V)(2/V)}{(1/V)(2/V)} = 4$, etc.

✓ *Reasonable Answer Check:* The equilibrium constant can be reproduced by plugging in the concentrations of the reactants and products: $\frac{(2/V)(2/V)}{(1/V)(1/V)} = 4$

63. *Answer/Explanation:* Dynamic equilibria with small values of K introduce a small amount of D^+ ions in place of H^+ ions in the place of the acidic hydrogen.

$$H_2O(\ell) \rightleftharpoons H^+(aq) + OH^-(aq)$$

$$D_2O(\ell) \rightleftharpoons D^+(aq) + OH^-(aq)$$

$$C_6H_5COOH (s) \rightleftharpoons C_6H_5COOH (aq) \rightleftharpoons H^+(aq) + C_6H_5COO^-(aq)$$

$$C_6H_5COOD (s) \rightleftharpoons C_6H_5COOD (aq) \rightleftharpoons D^+(aq) + C_6H_5COO^-(aq)$$

$$D^+(aq) + C_6H_5COO^-(aq) \rightleftharpoons C_6H_5COOD (aq) \rightleftharpoons C_6H_5COOD (s)$$

More Challenging Questions

65. *Answer:* (a) $K_c = 0.0168$ (b) $P_{tot} = 0.661$ atm   (c) $K_p = 0.514$

Strategy and Explanation:

(a) Convert grams to moles, then use a method similar to that described in the first part of the answer to Question 19.

$$6.44 \text{ g NOBr} \times \frac{1 \text{ mol NOBr}}{109.9 \text{ g NOBr}} = 0.0586 \text{ mol NOBr}$$

$$3.15 \text{ g NO} \times \frac{1 \text{ mol NO}}{30.01 \text{ g NO}} = 0.105 \text{ mol NO}$$

$$8.38 \text{ g Br}_2 \times \frac{1 \text{ mol Br}_2}{159.8 \text{ g Br}_2} = 0.0524 \text{ mol Br}_2$$

$$2 \text{ NOBr(g)} \rightleftharpoons 2 \text{ NO(g)} + \text{Br}_2\text{(g)}$$

$$K_c = \frac{[\text{NO}]^2[\text{Br}_2]}{[\text{NOBr}]^2} = \frac{\left(\dfrac{0.105 \text{ mol}}{10.00 \text{ L}}\right)^2\left(\dfrac{0.0524 \text{ mol}}{10.00 \text{ L}}\right)}{\left(\dfrac{0.0586 \text{ mol}}{10.00 \text{ L}}\right)^2} = 0.0168$$

(b) Use a combination of Dalton's law of partial pressures (Section 10.7, page 361) and the ideal gas law, PV = nRT (Section 10.4, page 354), to calculate the total pressure.

$$n_{tot} = 0.0586 \text{ mol NOBr} + 0.105 \text{ mol NO} + 0.0524 \text{ mol Br}_2 = 0.216 \text{ mol total}$$

$$P_{tot}V = n_{tot}RT$$

$$T = 100.\ °C + 273.15 = 373 \text{ K}$$

$$P_{tot} = \frac{n_{tot}RT}{V} = \frac{(0.261 \text{ mol}) \times \left(0.08206 \dfrac{\text{mol} \cdot \text{L}}{\text{K} \cdot \text{mol}}\right) \times (373 \text{ K})}{(10.00 \text{ L})} = 0.661 \text{ atm}$$

(c) Use a method similar to that described in the answer to Question 19.

$$\Delta n = 2 \text{ mol NO(g)} + 1 \text{ mol Br}_2\text{(g)} - 2 \text{ mol H}_2\text{(g)} = 1$$

$$K_p = K_c(RT)^{\Delta n}$$

$$K_p = (0.0168) \times \left\{\left(0.08206 \frac{\text{L} \cdot \text{atm}}{\text{mol} \cdot \text{K}}\right) \times \left(373 \text{ K}\right)\right\}^1 = 0.514$$

This $(RT)^1$ factor makes K_p larger than K_c.

67. *Answer:* (a) **$[\text{PCl}_5]$ = 0.005 M, $[\text{PCl}_3]$ = $[\text{Cl}_2]$ = 0.14 M   (b) $[\text{Cl}_2]$ = 0.14 M, $[\text{PCl}_5]$ = 0.01 M, $[\text{PCl}_3]$ = 0.29 M**

Strategy and Explanation: Use a method similar to that described in the answer to Question 25.

(a) (conc. PCl_5) = 0.75 mol PCl_5/5.00 L = 0.15 M

	PCl_5 (g) $\rightleftharpoons$	PCl_3 (g)	+ Cl_2 (g)
initial conc. (M)	0.15	0	0
change as reaction occurs (M)	− x	+ x	+ x
equilibrium conc. (M)	0.15 − x	x	x

At equilibrium
$$K_c = \frac{[\text{PCl}_3][\text{Cl}_2]}{[\text{PCl}_5]} = \frac{(x)(x)}{(0.15 - x)} = 3.30$$

Set up to use the quadratic equation (Appendix A, Section A.7, page A.12):

$$x^2 = 3.30(0.15 - x) = 0.50 - 3.30x$$
$$x^2 + 3.30x - 0.5 = 0$$

Plug into the quadratic equation:

$$x = \frac{-b \pm \sqrt{b^2 - 4ac}}{2a} = \frac{-3.30 \pm \sqrt{(3.30)^2 - 4(1)(-0.15)}}{2(1)} = \frac{0.29}{2} = 0.14$$

$$x = [\text{PCl}_3] = [\text{Cl}_2] = 0.14 \text{ M}$$

$$[\text{PCl}_5] = 0.15 - x = 0.15 \text{ M} - 0.14 \text{ M} = 0.01 \text{ M}$$

The reaction condition in (a) causes more than 90% formation of products, so let's take the reaction completely to the product's side, then bringing it back to equilibrium. This is a common tactic when reactants form products in the reaction with a K larger than 1. Use Chapter 4 stoichiometric "limiting reactant" procedure to accomplish this (Section 4.5).

	PCl$_5$ (g)	$\rightleftharpoons$	PCl$_3$ (g)	+ Cl$_2$ (g)
initial conc. (M)	0.15		0	0
final conc. (M)	0		0.15	0.15
change as reaction occurs (M)	+ x		− x	− x
equilibrium conc. (M)	x		0.15 − x	0.15 − x

At equilibrium

$$K_c = \frac{\left[PCl_3\right]\left[Cl_2\right]}{\left[PCl_5\right]} = \frac{(0.15-x)(0.15-x)}{(x)} = 3.30$$

Set up to use the quadratic equation (Appendix A, Section A.7, page A.12):

$$(0.15)^2 - 2(0.15)x + x^2 = 3.30x$$

$$x^2 - 3.60x - 0.023 = 0$$

Plug into the quadratic equation:

$$x = \frac{-b \pm \sqrt{b^2 - 4ac}}{2a} = \frac{3.60 \pm \sqrt{(-3.60)^2 - 4(1)(0.023)}}{2(1)} = \frac{0.01}{2} = 0.005$$

$$[PCl_5] = 0.005 \text{ M}$$

$$x = [PCl_3] = [Cl_2] = 0.15 - 0.005 \text{ M} = 0.14 \text{ M}$$

This way we get slightly more reliable information about the smallest concentration, [PCl$_5$].

(b)

$$(\text{conc. } PCl_5) = 0.75 \text{ mol } PCl_5/5.00 \text{ L} = 0.15 \text{ M}$$

$$(\text{conc. } PCl_3) = 0.75 \text{ mol } PCl_5/5.00 \text{ L} = 0.15 \text{ M}$$

	PCl$_5$ (g)	$\rightleftharpoons$	PCl$_3$ (g)	+ Cl$_2$ (g)
initial conc. (M)	0.15		0.15	0
change as reaction occurs (M)	− x		+ x	+ x
equilibrium conc. (M)	0.15 − x		0.15 + x	x

At equilibrium

$$K_c = \frac{\left[PCl_3\right]\left[Cl_2\right]}{\left[PCl_5\right]} = \frac{(0.15+x)(x)}{(0.15-x)} = 3.30$$

Set up to use the quadratic equation (Appendix A, Section A.7, page A.12):

$$0.15x + x^2 = 3.30(0.15 - x) = 0.50 - 3.30x$$

$$x^2 + 3.45x - 0.50 = 0$$

Plug into the quadratic equation:

$$x = \frac{-b \pm \sqrt{b^2 - 4ac}}{2a} = \frac{-3.45 \pm \sqrt{(3.45)^2 - 4(1)(0.50)}}{2(1)} = \frac{0.28}{2} = 0.14$$

$$x = [Cl_2] = 0.14 \text{ M}$$

$$[PCl_5] = 0.15 - x = 0.15 \text{ M} - 0.14 \text{ M} = 0.01 \text{ M}$$

$$[PCl_3] = 0.15 + x = 0.15 \text{ M} + 0.14 \text{ M} = 0.29 \text{ M}$$

The reaction condition in (b) also causes more than 90% formation of products, but taking the reaction completely to the products side, as we did in (a), produces the same result, within the given significant figures, so it is not shown here.

✓ *Reasonable Answer Check:* Substituting the equilibrium concentrations into the equilibrium expression will reproduce K$_c$:

$$K_{c,(a)} = \frac{(0.14)(0.14)}{(0.005)} = 4 \qquad\qquad K_{c,(a)} = \frac{(0.29)(0.14)}{(0.01)} = 4$$

Both are approximately equal to 3.30, within the uncertainty of the results (±1).

69. *Answer:* **(a) (i) right (ii) right (iii) left (b) (i) left (ii) no shift (iii) right (c) (i) right (ii) right (iii) no shift (d) (i) right (ii) right (iii) left**

Strategy and Explanation: Use methods similar to those in the solution to Question 39 and Section 13.6.

(a) $C(s) + H_2O(g) \rightleftharpoons CO(g) + 2 H_2(g)$ $\Delta H = 131.3$ kJ

  (i) Increasing the temperature adds more energy. The reaction shifts toward the products (right), to absorb the energy.

  (ii) Decreasing the pressure affects the products (3 mol of gas) more than the reactants (1 mol gas). That results in a shift of the reaction toward the products (right), to increase the pressure.

  (iii) Increasing the amount of CO(g) increases the concentration of a product. The reaction shifts toward the reactants (left), to decrease the [CO].

(b) $3 Fe(s) + 4 H_2O(g) \rightleftharpoons Fe_3O_4(s) + 4 H_2(g) + 149.9$ kJ

  (i) Increasing the temperature adds energy. The reaction shifts toward the reactants (left), to absorb the energy.

  (ii) Decreasing the pressure affects the products and the reactants equally (4 mol gas on each side). That results in no shift at all.

  (iii) Increasing the amount of $H_2O(g)$ increases the concentration of a reactant. The reaction shifts toward the products (right), to decrease the $[H_2O]$.

(c) $C(s) + CO_2(g) \rightleftharpoons 2 CO(g)$ $\Delta H = + 172.5$ kJ

  (i) Increasing the temperature adds more energy. The reaction shifts toward the products (right), to absorb the energy.

  (ii) Decreasing the pressure affects the products (2 mol of gas) more than the reactants (1 mol gas). That results in a shift of the reaction toward the products (right), to increase the pressure.

  (iii) The equilibrium is not affected by the addition of any solid, so adding more C(s) will cause no shift.

(d) $N_2O_4(g) + 54.8$ kJ $\rightleftharpoons 2 NO_2(g)$

  (i) Increasing the temperature adds more energy. The reaction shifts toward the products (right), to absorb the energy.

  (ii) Decreasing the pressure affects the products (2 mol of gas) more than the reactants (1 mol gas). That results in a shift of the reaction toward the products (right), to increase the pressure.

  (iii) Increasing the amount of $NO_2(g)$ increases the concentration of a product. The reaction shifts toward the reactants (left), to decrease the $[NO_2]$.

71. *Answer:* **(a) $H_2(g) + Br_2(g) \rightleftharpoons 2 HBr(g)$ (b) –103 kJ (c) energy effect (d) to the left (e) no effect (f) The reactants must achieve activation energy.**

Strategy and Explanation:

(a) $H_2(g) + Br_2(g) \rightleftharpoons 2 HBr(g)$

(b) $\Delta H \cong (1 \text{ mol}) \times D_{H-H} + (1 \text{ mol}) \times D_{Br-Br} - (2 \text{ mol}) \times D_{H-Br}$

   $= (1 \text{ mol}) \times (436 \text{ kJ/mol}) + (1 \text{ mol}) \times (193 \text{ kJ/mol}) - (2 \text{ mol}) \times (366 \text{ kJ/mol}) = -103$ kJ

(c) The energy effect drives the reaction toward products (exothermic reaction, means the products are more stable). The entropy effect is minor, since the number and size of the products and reactants are about the same. Entropy would favor products only slightly because the HBr molecule includes a mixture of both H and Br, and is therefore more random and entropically favorable.

(d) An increase of temperature will drive the exothermic reaction back to the left, toward products.

(e) An increase in the pressure will not affect the equilibrium state, since the same number of gas molecules are found on each side of the equation. However, if the pressure change was significant enough to liquefy the bromine, then a shift toward reactants would be seen, as the increase of pressure would be relieved by the smaller number of molecules in the gas phase.

(f) Establishing equilibrium requires that the reactants have at least enough energy to overcome the activation barrier. At temperatures lower than 200°, the reactants did not have enough energy for activation.

Chapter 14: The Chemistry of Solutes and Solutions

Introduction

The rule-of-thumb "like dissolves like" is most helpful when trying to decide whether a solute will dissolve in a solvent. However, you need to know the ionic and covalent character of substances to use this rule. You may wish to briefly review ionic compounds and covalent molecules from Chapter 3.

This chapter reviews most of the concentration units used in the first half of the book and describes one new one, molality. Please note the distinction—though the words look quite similar, molarity and molality are calculated differently and used differently.

Everyday examples and important applications, such as melting ice on snowy roads, ice skating, water purification, blood cells, pickles, etc., should help you see how colligative properties are central to life.

Solutions to Blue-Numbered Questions
for Review and Thought for Chapter 14

Topical Questions

How Substances Dissolve

18. *Answer/Explanation:* If the solid interacts with the solvent using similar or stronger intermolecular forces, it will dissolve readily. If the solute interacts with the solvent using different intermolecular forces than those experienced in the solvent, it will be almost insoluble. For example, consider dissolving an ionic solid in water and oil. The interactions between the ions in the solid and water are very strong, since ions would be attracted to the highly polar water molecule; hence the solid would have a relatively high solubility. However, the ions in the solid interact with each other much more strongly than the London dispersion forces experienced between the nonpolar hydrocarbons in the oil; hence, the solid would have a low solubility.

19. *Answer/Explanation:* The dissolving process was endothermic, so the temperature dropped as more solute was added. The solubility of the solid at the lower temperature is lower, so some of the solid did not dissolve. As the solution warmed up, however, the solubility increased again. What remained of the solid dissolved. The solution was saturated at the lower temperature, but is no longer saturated at the current temperature.

21. *Answer/Explanation:* When an organic acid has a large (non-polar) piece, it interacts primarily using London dispersion intermolecular forces. Since water interacts via hydrogen bonding intermolecular forces, it would rather interact with itself than with the acid. Hence, the solubility of the large organic acids drops, and some are completely insoluble.

23. *Answer/Explanation:* The positive side (H side) of the very polar water molecule interacts with the negative ions. The negative side of the very polar water molecule (O side) interacts with the positive ions.

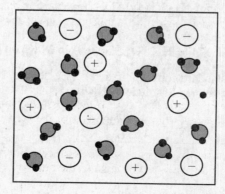

25. *Answer:* 1×10^{-3} **M** N_2

Strategy and Explanation: Given the partial pressure of a gas, the temperature, and the Henry's law constant, determine the concentration of the gas dissolved in the water.

Find the partial pressure of the gas in units of mm Hg, then use Henry's law described in Section 14.5: $s_g = k_H P_g$ Where s_g is the solubility of the gas, k_H is the Henry's law constant, and P_g is the partial pressure of a gas.

78% by volume of gaseous N_2 in air is directly proportional to the percentage by mole (Avogadro's law, Section 10.4). The mole percentage is directly proportional to the pressure (Section 10.7), so we get 0.78 atm N_2 in every 1.00 atm air.

$$2.5 \text{ atm air} \times \frac{0.78 \text{ atm } N_2}{1 \text{ atm air}} \times \frac{760 \text{ mmHg } N_2}{1 \text{ atm } N_2} = 1.5 \times 10^3 \text{ mmHg } N_2$$

$$s_g = (8 \times 10^{-7} \text{ mol L}^{-1}\text{mm Hg}^{-1})(1.5 \times 10^3 \text{ mm Hg}) = 1 \times 10^{-3} \text{ mol L}^{-1} = 1 \times 10^{-3} \text{ M}$$

✓ *Reasonable Answer Check:* Always double check to be sure that the gas does not react with the water, since Henry's law does not apply to those gases.

Concentration Units

27. *Answer:* **90. g ethanol**

Strategy and Explanation: Use standard conversion factors (see Section 14.6 for examples).

$$750 \text{ mL solution} \times \frac{1.00 \text{ g solution}}{1 \text{ mL solution}} \times \frac{12 \text{ g ethanol}}{100 \text{ g solution}} = 90. \text{ g ethanol}$$

29. *Answer:* 1.6×10^{-6} **g lead**

Strategy and Explanation: Use standard conversion factors (see Section 14.6 for examples). *(Assume that "1 gallon" of paint is measured with at least two significant figures.)*

$$1.0 \text{ cm}^2 \times \left(\frac{1 \text{ in}}{2.54 \text{ cm}}\right)^2 \times \left(\frac{1 \text{ ft}}{12 \text{ in}}\right)^2 \times \frac{1 \text{ gal paint}}{500. \text{ ft}^2} \times \frac{8.0 \text{ lb paint}}{1 \text{ gal paint}} \times \frac{453.6 \text{ g paint}}{1 \text{ lb paint}} \times \frac{200. \text{ g Pb}}{10^6 \text{ g paint}} = 1.6 \times 10^{-6} \text{ g Pb}$$

31. *Answer:* **(a) 160. g NH_4Cl (b) 83.9 g KCl (c) 7.46 g Na_2SO_4**

Strategy and Explanation: Use standard conversion factors (see Section 14.6 for examples).

(a) $\quad 750. \text{ mL} \times \dfrac{1 \text{ L}}{1000 \text{ mL}} \times \dfrac{4.00 \text{ mol } NH_4Cl}{1 \text{ L}} \times \dfrac{53.491 \text{ g } NH_4Cl}{1 \text{ mol } NH_4Cl} = 160. \text{ g } NH_4Cl$

(b) $\quad 1.50 \text{ L} \times \dfrac{0.750 \text{ mol KCl}}{1 \text{ L}} \times \dfrac{74.5513 \text{ g KCl}}{1 \text{ mol KCl}} = 83.9 \text{ g KCl}$

(c) $\quad 150. \text{ mL} \times \dfrac{1 \text{ L}}{1000 \text{ mL}} \times \dfrac{0.350 \text{ mol } Na_2SO_4}{1 \text{ L}} \times \dfrac{142.0422 \text{ g } Na_2SO_4}{1 \text{ mol } Na_2SO_4} = 7.46 \text{ g } Na_2SO_4$

32. *Answer:* **(a) 0.762 M KCl   (b) 0.174 M K_2CrO_4 (c) 0.0126 M $KMnO_4$ (d) 0.167 M $C_6H_{12}O_6$**

Strategy and Explanation: Use standard conversion factors (see Section 14.6 for examples). Calculate quantity of solute in moles then divide by volume of solution in liters.

(a) $\quad \dfrac{14.2 \text{ g KCl} \times \dfrac{1 \text{ mol KCl}}{74.5513 \text{ g KCl}}}{250. \text{ mL} \times \dfrac{1 \text{ L}}{1000 \text{ mL}}} = 0.762 \text{ M KCl}$

The calculation in (a) can be written more concisely as:

$$\frac{14.2 \text{ g KCl}}{250. \text{ mL}} \times \frac{1 \text{ mol KCl}}{74.5513 \text{ g KCl}} \times \frac{1000 \text{ mL}}{1 \text{ L}} = 0.762 \text{ M KCl}$$

(b) $$\frac{5.08 \text{ g K}_2\text{CrO}_4}{150. \text{ mL}} \times \frac{1 \text{ mol K}_2\text{CrO}_4}{194.1903 \text{ g K}_2\text{CrO}_4} \times \frac{1000 \text{ mL}}{1 \text{ L}} = 0.174 \text{ M K}_2\text{CrO}_4$$

(c) $$\frac{0.799 \text{ g KMnO}_4}{400. \text{ mL}} \times \frac{1 \text{ mol KMnO}_4}{158.0339 \text{ g KMnO}_4} \times \frac{1000 \text{ mL}}{1 \text{ L}} = 0.0126 \text{ M KMnO}_4$$

(d) $$\frac{15.0 \text{ g C}_6\text{H}_{12}\text{O}_6}{500. \text{ mL}} \times \frac{1 \text{ mol C}_6\text{H}_{12}\text{O}_6}{180.1554 \text{ g C}_6\text{H}_{12}\text{O}_6} \times \frac{1000 \text{ mL}}{1 \text{ L}} = 0.167 \text{ M C}_6\text{H}_{12}\text{O}_6$$

33. *Answer:* **(a) 0.180 M MgNH$_4$PO$_4$ (b) 0.683 M NaCH$_3$COO (c) 0.0260 M CaC$_2$O$_4$ (d) 0.0416 M (NH$_4$)$_2$SO$_4$**

Strategy and Explanation: Use standard conversion factors (See Section 14.6 for examples). Calculate quantity of solute in moles then divide by volume of solution in liters (See the solution to Question 32).

(a) $$\frac{6.18 \text{ g MgNH}_4\text{PO}_4}{250. \text{ mL}} \times \frac{1 \text{ mol MgNH}_4\text{PO}_4}{137.3147 \text{ g MgNH}_4\text{PO}_4} \times \frac{1000 \text{ mL}}{1 \text{ L}} = 0.180 \text{ M MgNH}_4\text{PO}_4$$

(b) $$\frac{16.8 \text{ g NaCH}_3\text{COO}}{300. \text{ mL}} \times \frac{1 \text{ mol NaCH}_3\text{COO}}{82.0337 \text{ g NaCH}_3\text{COO}} \times \frac{1000 \text{ mL}}{1 \text{ L}} = 0.683 \text{ M NaCH}_3\text{COO}$$

(c) $$\frac{2.50 \text{ g CaC}_2\text{O}_4}{750. \text{ mL}} \times \frac{1 \text{ mol CaC}_2\text{O}_4}{128.097 \text{ g CaC}_2\text{O}_4} \times \frac{1000 \text{ mL}}{1 \text{ L}} = 0.0260 \text{ M CaC}_2\text{O}_4$$

(d) $$\frac{2.20 \text{ g (NH}_4)_2\text{SO}_4}{400. \text{ mL}} \times \frac{1 \text{ mol (NH}_4)_2\text{SO}_4}{132.139 \text{ g (NH}_4)_2\text{SO}_4} \times \frac{1000 \text{ mL}}{1 \text{ L}} = 0.0416 \text{ M (NH}_4)_2\text{SO}_4$$

35. *Answer:* **96% H$_2$SO$_4$**

Strategy and Explanation: Use standard conversion factors (see Section 14.6 for examples).

$$\frac{18 \text{ mol H}_2\text{SO}_4}{1 \text{ L solution}} \times \frac{1 \text{ L solution}}{1000 \text{ cm}^3 \text{ solution}} \times \frac{1 \text{ cm}^3 \text{ solution}}{1.84 \text{ g solution}} \times \frac{98.078 \text{ g H}_2\text{SO}_4}{1 \text{ mol H}_2\text{SO}_4} \times 100 \% = 96\% \text{ H}_2\text{SO}_4$$

37. *Answer:* **59 g C$_2$H$_4$(OH)$_2$**

Strategy and Explanation: Use standard conversion factors (see Section 14.6 for examples)

$$950. \text{ g H}_2\text{O} \times \frac{1 \text{ kg H}_2\text{O}}{1000 \text{ g H}_2\text{O}} \times \frac{1.0 \text{ mol C}_2\text{H}_4(\text{OH})_2}{1 \text{ kg H}_2\text{O}} \times \frac{62.07 \text{ g C}_2\text{H}_4(\text{OH})_2}{1 \text{ mol C}_2\text{H}_4(\text{OH})_2} = 59 \text{ g C}_2\text{H}_4(\text{OH})_2$$

Colligative Properties

42. *Answer:* **boiling point of (a) < boiling point of (d) < boiling point of (b) < boiling point of (c)**

Strategy and Explanation: In Section 14.7, we learn that the freezing point decreases as the molality of solute particles increases. As described near the end of Section 14.7, page 527, solutes that ionize provide a larger number of particles in the solution than solutes that do not ionize. So, we must calculate the concentration of particles in each solution and order the solutions from the smallest concentration to the largest concentration.

(a) Each methanol (a covalent molecule) provides one particle. = 0.10 mol/kg particles

(b) Each KCl ionizes into K$^+$ and Cl$^-$, two particles.

$$0.10 \text{ mol/kg KCl} \times (2 \text{ particles/mol KCl}) = 0.20 \text{ mol/kg particles}$$

(c) Each BaCl$_2$ ionizes into Ba^{2+} and 2 Cl$^-$, three particles.

$$0.080 \text{ mol/kg BaCl}_2 \times (3 \text{ particles/mol BaCl}_2) = 0.24 \text{ mol/kg particles}$$

(d) Each Na_2SO_4 ionizes into 2 Na^+ and SO_4^{2-}, three particles.

$$0.040 \text{ mol/kg } Na_2SO_4 \times (\text{ 3 particles/mol } Na_2SO_4) = 0.12 \text{ mol/kg particles}$$

Order of decreasing freezing points: (a) < (d) < (b) < (c)

44. *Answer:* $T_f = -1.65 \,°C, \; T_b = 100.46 \,°C$

Strategy and Explanation: Use methods described in Section 14.7.

Calculate the molality of the solute. $\qquad m_{solute} = \dfrac{\text{mol solute}}{\text{kg solvent}}$

$$T_b(\text{solution}) = T_b(\text{solvent}) + \Delta T_b \qquad\qquad \Delta T_b = K_b m_{solute}$$

Use molality to get ΔT_b. Then add ΔT_b to the normal boiling point to get the boiling point of the solution. Use a similar method for the freezing point calculation:

$$T_f(\text{solution}) = T_f(\text{solvent}) - \Delta T_f \qquad\qquad \Delta T_f = K_f m_{solute}$$

Molal boiling-point-elevation and molal freezing-point-lowering constants for water are given in Problem-Solving Example 14.11.

$$m_{solute} = \frac{4.00 \text{ g } CO(NH_2)_2}{75.0 \text{ g water}} \times \frac{1 \text{ mol } CO(NH_2)_2}{60.06 \text{ g } CO(NH_2)_2} \times \frac{1000 \text{ g water}}{1 \text{ kg water}} = 0.888 \text{ mol / kg}$$

$$\Delta T_b = (0.52 \,°C \text{ kg mol}^{-1}) \times (0.888 \text{ mol/kg}) = 0.46 \,°C$$

$$T_b(\text{solution}) = 100.00 \,°C + 0.46 \,°C = 100.46 \,°C$$

$$\Delta T_f = (1.86 \,°C \text{ kg mol}^{-1}) \times (0.888 \text{ mol/kg}) = 1.65 \,°C$$

$$T_f(\text{solution}) = 0.00 \,°C - 1.65 \,°C = -1.65 \,°C$$

46. *Answer:* $X_{H_2O} = 0.79999; \; 712 \text{ g sucrose}$

Strategy and Explanation: Given the vapor pressure of the solvent at a given temperature, the vapor pressure of a solution at the same temperature and the mass of solvent, determine the mole fraction of the solvent and the mass of solute in the solution. Use Raoult's law, described in Section 14.7 and in the solution to Question 45.

$$X_{H_2O} = \frac{P_{H_2O}}{P^0_{H_2O}} = \frac{119.55 \text{ mmHg}}{149.44 \text{ mmHg}} = 0.79999$$

$$150. \text{ g } H_2O \times \frac{1 \text{ mol } H_2O}{18.02 \text{ g } H_2O} = 8.32 \text{ mol } H_2O$$

$$X_{H_2O} = \frac{n_{H_2O}}{n_{H_2O} + n_{sucrose}}$$

Solve for $n_{sucrose}$:

$$X_{H_2O}(n_{H_2O} + n_{sucrose}) = X_{H_2O}n_{H_2O} + X_{H_2O}n_{sucrose} = n_{H_2O}$$

$$X_{H_2O}n_{sucrose} = n_{H_2O} - X_{H_2O}n_{H_2O}$$

$$n_{sucrose} = \frac{(1 - X_{H_2O})n_{H_2O}}{X_{H_2O}} = \frac{(1 - 0.79999)(8.32 \text{ mol})}{0.79999} = \frac{(0.20001)(8.32 \text{ mol})}{0.79999} = 2.08 \text{ mol}$$

$$2.08 \text{ mol } C_{12}H_{22}O_{11} \times \frac{342.297 \text{ g } C_{12}H_{22}O_{11}}{1 \text{ mol } C_{12}H_{22}O_{11}} = 712 \text{ g } C_{12}H_{22}O_{11}$$

✓ *Reasonable Answer Check:* The mole fraction of sucrose = 2.08 mol/(8.32 mol + 2.08 mol) = 0.200. The sum of all the mole fractions (0.79999 + 0.200) is one.

48. *Answer:* **190 g/mol**

Strategy and Explanation: Adapt the method described in the solution to Question 44 to find moles of solute. Divide mass by moles to get molar mass.

$$\Delta T_b = 0.65 \ ^\circ C \qquad\qquad \Delta T_b = K_b m_{unknown}$$

$$m_{unknown} = \frac{\Delta T_b}{K_b} = \frac{0.65 \ ^\circ C}{2.53 \ ^\circ C \ kg \ mol^{-1}} = 0.26 \ mol / kg$$

$$100. \ g \ benzene \times \frac{1 \ kg \ benzene}{1000 \ g \ benzene} \times \frac{0.26 \ mol \ unknown}{1 \ kg \ benzene} = 0.026 \ mol \ unknown$$

$$Molar \ mass = \frac{5.0 \ g}{0.026 \ mol} = 190 \ g / mol$$

50. *Answer:* **3.6×10^2 g/mol; $C_{20}H_{16}Fe_2$**

Strategy and Explanation: Adapt the method described in the solution to Question 43 to find moles of solute, then divide mass by moles for molar mass. Use methods described in the solution to Question 49 in Chapter 3 to find molecular formula.

$$\Delta T_b = T_b(solution) - T_b(solvent) = 80.26 \ ^\circ C - 80.10 \ ^\circ C = 0.16 \ ^\circ C$$

$$\Delta T_b = K_b m_{C_{10}H_8Fe}$$

$$m_{C_{10}H_8Fe} = \frac{\Delta T_b}{K_b} = \frac{0.16 \ ^\circ C}{2.53 \ ^\circ C \ kg \ mol^{-1}} = 0.063 \ mol / kg$$

$$11.12 \ g \ benzene \times \frac{1 \ kg \ benzene}{1000 \ g \ benzene} \times \frac{0.063 \ mol \ C_{10}H_8Fe}{1 \ kg \ benzene} = 0.00070 \ mol \ C_{10}H_8Fe$$

$$Molar \ mass \ of \ the \ compound = \frac{0.255 \ g}{0.00070 \ mol} = 3.6 \times 10^2 \ g / mol$$

The molecular formula is a multiple of the empirical formula: $(C_{10}H_8Fe)_n$

The molar mass of the empirical formula $C_{10}H_8Fe$ is 184.01 g/mol

$$n = \frac{3.6 \times 10^2 \ g / mol}{184.01 \ g / mol} = 2.0 \cong 2$$

The molecular formula is: $C_{20}H_{16}Fe_2$

53. *Answer:* **(a) 2.5 kg $C_2H_6O_2$ (b) 104.2 °C**

Strategy and Explanation: Adapt the method described in the solution to Question 44:

(a)
$$\Delta T_f = T_f(solvent) - T_f(solution) = 0.0 \ ^\circ C - (-15.0 \ ^\circ C) = 15.0 \ ^\circ C$$

$$m_{C_2H_6O_2} = \frac{\Delta T_f}{K_f} = \frac{15.0 \ ^\circ C}{1.86 \ ^\circ C \ kg \ mol^{-1}} = 8.06 \ mol / kg$$

$$5.0 \ kg \ water \times \frac{8.06 \ mol \ C_2H_6O_2}{1 \ kg \ water} \times \frac{62.07 \ g \ C_2H_6O_2}{1 \ mol \ C_2H_6O_2} \times \frac{1 \ kg \ C_2H_6O_2}{1000 \ g \ C_2H_6O_2} = 2.5 \ kg \ C_2H_6O_2$$

You must add 2.5 kg of ethylene glycol to 5.0 kg of water for this much freezing protection.

(b)
$$\Delta T_b = (0.52 \text{ °C kg mol}^{-1}) \times (8.06 \text{ mol/kg}) = 4.2 \text{ °C}$$

$$T_b(\text{solution}) = 100.00 \text{ °C} + 4.2 \text{ °C} = 104.2 \text{ °C}$$

General Questions

57. *Answer/Explanation:* Water in the cells of the wood leaked out, since the osmotic pressure inside the cells was less than that of the seawater in which the wood was sitting. See the discussion about hypertonic solutions and cells in Section 14.7.

59. *Answer:* **28% NH$_3$**

Strategy and Explanation: Use standard conversion factors (see Section 14.6 for examples).

$$\frac{14.8 \text{ mol NH}_3}{1 \text{ L solution}} \times \frac{1 \text{ L solution}}{1000 \text{ cm}^3 \text{ solution}} \times \frac{1 \text{ cm}^3 \text{ solution}}{0.90 \text{ g solution}} \times \frac{17.03 \text{ g NH}_3}{1 \text{ mol NH}_3} \times 100\% = 28\% \text{ NH}_3$$

Applying Concepts

63. *Answer:* **(a)**

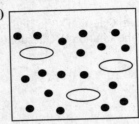

(b)

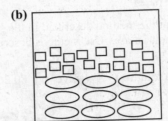

Strategy and Explanation:

(a) Sugar and water interact with the same hydrogen bonding attractive forces, so they will commingle.

(b) Carbon tetrachloride and sugar interact with very different interactive forces, so they will remain separate phases.

65. *Answer:* **(a) unsaturated (b) supersaturated (c) supersaturated (d) unsaturated**

Strategy and Explanation: Figure 14.11 gives curves showing solubility versus temperature. If the concentration is on the curve at the given temperature, the solution is saturated. If the concentration is above the curve at the given temperature, the solution is supersaturated. If the concentration is below the curve at the given temperature, the solution is unsaturated.

(a) 40 g NH$_4$Cl/100 g H$_2$O at 80 °C: This point is below the curve, so the solution is unsaturated.

(b) 100 g LiCl/100 g H$_2$O at 30 °C: This point is above the curve, so the solution is supersaturated.

(c) 120 g NaNO$_3$/100 g H$_2$O at 40 °C: This point is above the curve, so the solution is supersaturated.

(d) 25 g Li$_2$SO$_4$/100 g H$_2$O at 50 °C: This point is below the curve, so the solution is unsaturated.

67. *Answer:* **see chart below**

Strategy and Explanation: Mass fraction is calculated by dividing the mass of the substance by the total mass of the solution. Calculate weight percent by multiplying mass fraction by 100%. Calculate ppm, using the equality 1% = 10,000 ppm.

Compound	Mass of compound	Mass of water	Mass fraction	Weight percent	ppm of solute
Lye	**75.0 g**	125 g	0.375	**37.5%**	**3.75×10^5**
Glycerol	33 g	200. g	**0.14**	**14%**	**1.4×10^5**
Acetylene	0.0015 g	**2×10^2 g**	**0.000009**	0.0009%	**9**

69. *Answer/Explanation:*

(a) Seawater contains more dissolved solutes than fresh water. The presence of a solute lowers the freezing point. That means a lower temperature is required to freeze the seawater than to freeze fresh water.

(b) Salt added to a mixture of ice and water will lower the freezing point of the water. If the ice cream is mixed at a lower temperature, its temperature will drop faster; hence, it will freeze faster.

More Challenging Questions

71. *Answer:* **(a) Empirical formula is $C_{18}H_{24}Cr$. (b) Molecular formula is $C_{18}H_{24}Cr$.**

Strategy and Explanation: This is a combination of the problem-solving from several different chapters.

(a) From Chapter 3: 100.00 g compound – 73.94 g C – 8.27 g H = 17.79 g Cr

$$73.94 \text{ g C} \times \frac{1 \text{ mol C}}{12.0107 \text{ g C}} = 6.156 \text{ mol C}$$

$$8.27 \text{ g H} \times \frac{1 \text{ mol H}}{1.0079 \text{ g H}} = 8.21 \text{ mol H}$$

$$17.79 \text{ g Cr} \times \frac{1 \text{ mol Cr}}{51.996 \text{ g Cr}} = 0.342 \text{ mol Cr}$$

Mole Ratio: 6.156 mol C : 8.21 mol H : 0.342 mol Cr

Simplify the ratio: 18 C : 24 H : 1 Cr

Empirical formula is $C_{18}H_{24}Cr$.

(b) Adapt the solution to Question 54.   T = 25 °C + 273.15 = 298 K

$$c = \frac{\Pi}{RT} = \frac{3.17 \text{ mmHg}\left(\dfrac{1 \text{ atm}}{760 \text{ mmHg}}\right)}{\left(0.08206 \dfrac{\text{L} \cdot \text{atm}}{\text{mol} \cdot \text{K}}\right)(298 \text{ K})} = 1.71 \times 10^{-4} \text{ mol} / \text{L}$$

Assuming that the addition of solute does not change the volume of the solution, so the volume of the solvent is equal to the volume of the solution:

$$100. \text{ mL chloroform} \times \frac{1 \text{ L chloroform}}{1000 \text{ mL chloroform}} \times \frac{1 \text{ L solution}}{1 \text{ L chloroform}}$$

$$\times \frac{1.71 \times 10^{-4} \text{ mol solute}}{1 \text{ L solution}} = 1.71 \times 10^{-5} \text{ mol solute}$$

$$\text{Molar mass of the compound} = \frac{5.00 \text{ mg}}{1.71 \times 10^{-5} \text{ mol}} \times \frac{1 \text{ g}}{1000 \text{ mg}} = 292 \text{ g} / \text{mol}$$

The molecular formula is a multiple of the empirical formula: $(C_{18}H_{24}Cr)_n$

The molar mass of the empirical formula is 292.37 g/mol, so the molecular formula is $C_{18}H_{24}Cr$.

73. *Answer:* **(a) No (b) 108.9 °C**

Strategy and Explanation: Adapt the method described in the answers to Question 44:

(a) Calculate the molality from the proof: $\dfrac{100 \text{ proof ethanol}}{2} = 50\%$ by volume ethanol

$$\frac{50.0 \text{ mL C}_2\text{H}_5\text{OH}}{50.0 \text{ mL water}} \times \frac{0.789 \text{ g C}_2\text{H}_5\text{OH}}{1 \text{ mL C}_2\text{H}_5\text{OH}} \times \frac{1 \text{ mL H}_2\text{O}}{1.00 \text{ g H}_2\text{O}} \times \frac{1000 \text{ g}}{1 \text{ kg}}$$

$$\times \frac{1 \text{ mol C}_2\text{H}_5\text{OH}}{46.068 \text{ g C}_2\text{H}_5\text{OH}} = \frac{17.13 \text{ mol C}_2\text{H}_5\text{OH}}{\text{kg H}_2\text{O}}$$

$$\Delta T_f = m_{C_2H_6O_2} K_f = (17.13 \text{ mol/kg})(1.86 \text{ °C kg mol}^{-1}) = 31.9 \text{ °C}$$

$$T_f(50\% \text{ solution}) = T_f(H_2O) - \Delta T_f = 0.0 \text{ °C} - (31.9 \text{ °C}) = -31.9 \text{ °C}$$

So, the vodka will not freeze at a temperature of −15°C.

(b) $\Delta T_b = m_{C_2H_6O_2} K_b = (17.13 \text{ mol/kg})(0.52 \text{ °C kg mol}^{-1}) = 8.9 \text{ °C}$

$$T_b(50\% \text{ solution}) = T_b(H_2O) + \Delta T_b = 100.00 \text{ °C} + 8.9 \text{ °C} = 108.9 \text{ °C}$$

(Notice: This answer assumes that the dissolved alcohol will not evaporate from the solution as the temperature is raised, which is probably not true. If the alcohol does evaporate, that will lower the concentration of the solute, and the boiling temperature will be closer to that of pure water than that calculated here.)

75. *Answer:* **28 m**

Strategy and Explanation: Given the concentration of the non-electrolyte solute present in tree sap, determine the height of the sap in a tree.

Use method described in Section 14.7. The osmotic pressure equation for a nonelectrolyte (with i = 1) is: $\Pi = cRT$ Π is the osmotic pressure, c is the concentration of the solute in the solution, R is the familiar gas constant with liter and atmosphere units, 0.08206 L·atm/mol·K, and T is the absolute temperature in kelvin units. For c, use the difference in the concentration of the sap inside and outside of the tree, then compare the measure of liquid height by relating their densities (see Table 1.1 on page 8 of the textbook for these densities). Assume the temperature is 25.00°C and the density of a dilute aqueous solution is the same as water.

$$\Pi = (0.13 \text{ M} - 0.020 \text{ M}) \times \left(0.08206 \frac{\text{L} \cdot \text{atm}}{\text{mol} \cdot \text{K}}\right) \times (298.15 \text{ K}) \times \left(\frac{760 \text{ mmHg}}{1 \text{ atm}}\right) = 2.0 \times 10^3 \text{ mmHg } \textit{(two sig figs)}$$

Since pressure is proportional to density of the liquid, 1 mm Hg is directly related to the density of mercury with the same proportionality constant as 1 mm sap is related to the density of sap.

$$2.0 \times 10^3 \text{ mmHg} \times \frac{13.55 \text{ g/mL}}{1 \text{ mmHg}} \times \frac{1 \text{ mm sap}}{0.998 \text{ g/mL}} \times \frac{1 \text{ m}}{1000 \text{ mm}} = 28 \text{ m sap}$$

76. *Answer:* **0.30 mol/L**

Strategy and Explanation: Use method described in Section 14.7. The osmotic pressure equation for is: $\Pi = cRTi$ Π is the osmotic pressure, c is the concentration of the solute in the solution, R is the familiar gas constant with liter and atmosphere units, 0.08206 L·atm/mol·K, and T is the absolute temperature in kelvin units, and i is the number of particles per formula unit of solute.

Glucose is a nonelectrolye, so I

$$T = 37 \text{ °C} + 273.15 = 310. \text{ K}$$

$$c = \frac{\Pi}{RTi} = \frac{7.7 \text{ atm}}{\left(0.08206 \dfrac{\text{L} \cdot \text{atm}}{\text{mol} \cdot \text{K}}\right)(310. \text{ K})(1.00)} = 0.30 \text{ mol/L}$$

78. *Answer:* **(a) 6300 ppm, 6,300,000 ppb (b) 0.040 M (c) 4.99 × 10⁵ bottles**

Strategy and Explanation: A sample of exactly 100 g of solution contains 0.63 g SnF_2 and 99.37 g water.

(a) The units ppm and ppb are discussed in the solution to Question 8 in Chapter 10. In the solution to Question 67 is given 1% = 10,000 ppm.

$$0.63 \% \text{ SnF}_2 \times \frac{10,000 \text{ ppm}}{1 \%} = 6300 \text{ ppm SnF}_2$$

$$6300 \text{ ppm SnF}_2 \times \frac{1000 \text{ ppb SnF}_2}{1 \text{ ppm SnF}_2} = 6,300,000 \text{ ppb SnF}_2$$

(b) $\dfrac{0.63 \text{ g SnF}_2}{10^2 \text{ g solution}} \times \dfrac{0.998 \text{ g solution}}{1 \text{ mL}} \times \dfrac{1 \text{ mol SnF}_2}{156.707 \text{ g SnF}_2} \times \dfrac{1000 \text{ mL}}{1 \text{ L}} = 0.040 \text{ M SnF}_2$

(c) In one metric ton, there are 10^6 grams.

$$10^6 \text{ g SnO}_2 \times \dfrac{1 \text{ mol SnO}_2}{150.709 \text{ g SnO}_2} \times \dfrac{1 \text{ mol Sn}}{1 \text{ mol SnO}_2} \times \dfrac{118.710 \text{ g Sn}}{1 \text{ mol Sn}} = 7.88 \times 10^5 \text{ g Sn theoretical}$$

$$7.88 \times 10^5 \text{ g Sn theoretical} \times \dfrac{80 \text{ g Sn actual}}{100 \text{ g Sn theoretical}} = 6.30 \times 10^5 \text{ g Sn actual}$$

$$6.30 \times 10^5 \text{ g Sn} \times \dfrac{1 \text{ mol Sn}}{118.710 \text{ g Sn}} \times \dfrac{1 \text{ mol SnF}_2}{1 \text{ mol Sn}} \times \dfrac{156.707 \text{ g SnF}_2}{1 \text{ mol SnF}_2} = 8.32 \times 10^5 \text{ g SnF}_2 \text{ theoretical}$$

$$8.32 \times 10^5 \text{ g SnF}_2 \text{ theoretical} \times \dfrac{94 \text{ g SnF}_2 \text{ actual}}{100 \text{ g SnF}_2 \text{ theoretical}} = 7.82 \times 10^5 \text{ g SnF}_2 \text{ actual}$$

Use the answer from (b), in units of molarity to determine volume.

$$7.82 \times 10^5 \text{ g SnF}_2 \times \dfrac{1 \text{ mol SnF}_2}{156.707 \text{ g SnF}_2} \times \dfrac{1 \text{ L}}{0.040 \text{ mol SnF}_2} \times \dfrac{1000 \text{ mL}}{1 \text{ L}} \times \dfrac{1 \text{ bottle}}{250. \text{ mL}} = 4.99 \times 10^5 \text{ bottles}$$

Chapter 15: Acids and Bases

Introduction

Teaching for Conceptual Understanding

Chapter 15 is very important because the concepts of acids and bases apply to all science, engineering and pre-professional health career majors as well as many real-world experiences.

You may want to review the basic concepts of acids and bases that were introduced in Chapter 5. They are critical in being successful with the more quantitative aspects of acid-base strength and pH.

Use your instincts when studying acid strength. Most people would agree that hydrochloric acid should not be substituted for vinegar in a salad dressing. We know that hydrochloric acid is a harsher or more dangerous acid. In this chapter, you will learn more about why.

Keep in mind that this chapter is simply building on the basic ideas of equilibrium introduced in Chapter 13. Try not to be confused by the variety of subscripts on the equilibrium constants, e.g., K_w, K_a, and K_b. They are all still equilibrium constants. The subscripts (w, a, b, etc.) simply designate specific chemical reactions.

Understanding and being able to use Table 15.2 (Ionization Constants for Some Acids and Their Conjugate Bases) is paramount to understanding and predicting chemical reactivity of acids and bases. Take the time to study this table in detail and practice in gleaning information from it.

Solutions to Blue-Numbered Questions for Review and Thought for Chapter 15

Topical Questions

The Brønsted-Lowry Concept of Acids and Bases

8. *Answer:* **see equations below**

   Strategy and Explanation: An $H^+(aq)$ ion is transferred from the acid to the water molecule to make $H_3O^+(aq)$ and the conjugate base of the acid.

   (a) $HCO_3^-(aq) + H_2O(\ell) \rightleftharpoons H_3O^+(aq) + CO_3^{2-}(aq)$

   (b) $HCl(aq) + H_2O(\ell) \rightleftharpoons H_3O^+(aq) + Cl^-(aq)$

   (c) $CH_3COOH(aq) + H_2O(\ell) \rightleftharpoons H_3O^+(aq) + CH_3COO^-(aq)$

   (d) $HCN(aq) + H_2O(\ell) \rightleftharpoons H_3O^+(aq) + CN^-(aq)$

   Notice: There is variable extent to which these reactions will proceed toward products, depending on their strength.

10. *Answer:* **see equations below**

    Strategy and Explanation: Use the method shown in the solution to Question 8.

    (a) $HIO(aq) + H_2O(\ell) \rightleftharpoons H_3O^+(aq) + IO^-(aq)$

    (b) $CH_3(CH_2)_4COOH(aq) + H_2O(\ell) \rightleftharpoons H_3O^+(aq) + CH_3(CH_2)_4COO^-(aq)$

(c) $HOOCCOOH(aq) + H_2O(\ell) \rightleftharpoons H_3O^+(aq) + HOOCCOO^-(aq)$

$HOOCCOO^-(aq) + H_2O(\ell) \rightleftharpoons H_3O^+(aq) + {}^-OOCCOO^-(aq)$

(d) $CH_3NH_3^+(aq) + H_2O(\ell) \rightleftharpoons H_3O^+(aq) + CH_3NH_2(aq)$

Notice: There is variable extent to which these reactions will proceed toward products, depending on their strength.

12. *Answer:* **see equations below**

Strategy and Explanation: An $H^+(aq)$ ion is transferred from the water molecule to the base to make the conjugate acid of the base and $OH^-(aq)$.

(a) $HSO_4^-(aq) + H_2O(\ell) \rightleftharpoons H_2SO_4(aq) + OH^-(aq)$

(b) $CH_3NH_2(aq) + H_2O(\ell) \rightleftharpoons CH_3NH_3^+(aq) + OH^-(aq)$

(c) $I^-(aq) + H_2O(\ell) \rightleftharpoons HI(aq) + OH^-(aq)$

(d) $H_2PO_4^-(aq) + H_2O(\ell) \rightleftharpoons H_3PO_4(aq) + OH^-(aq)$

Notice: There is variable extent to which these reactions will proceed toward products, depending on their strength. In particular, the two bases in (a) and (c) I^- are very weak bases. The reaction that is actually observed between HSO_4^- and water is shown in the answer to Question 7(c). I^- ion is not observed to react with water.

16. *Answer:* **(a) conjugate base, I^-, iodide ion   (b) conjugate acid, HNO_3, nitric acid   (c) conjugate acid, HCO_3^-, hydrogen carbonate ion (d) conjugate base, HCO_3^-, hydrogen carbonate ion   (e) conjugate acid of H_2SO_4, sulfuric acid, and conjugate base, SO_4^{2-}, sulfate ion (f) conjugate acid, HSO_3^-, hydrogen sulfite ion**

Strategy and Explanation: Conjugate acid-base pairs differ by one H^+ ion.

(a) HI is a Brønsted-Lowry acid. Its conjugate base is I^-, iodide ion.

(b) NO_3^- is a Brønsted-Lowry base. Its conjugate acid is HNO_3, nitric acid.

(c) CO_3^{2-} is a Brønsted-Lowry base. Its conjugate acid is HCO_3^-, hydrogen carbonate ion.

(d) H_2CO_3 is a Brønsted-Lowry acid. Its conjugate base is HCO_3^-, hydrogen carbonate ion.

(e) HSO_4^- as a Brønsted-Lowry base, has a conjugate acid of H_2SO_4, sulfuric acid. HSO_4^- as a Brønsted-Lowry acid, has a conjugate base of SO_4^{2-}, sulfate ion.

(f) SO_3^{2-} is a Brønsted-Lowry base. Its conjugate is HSO_3^-, hydrogen sulfite ion.

18. *Answer:* **pairs (b), (c), and (d)**

Strategy and Explanation: Conjugate acid-base pairs differ by only one H^+ ion.

(a) NH_2^- is the conjugate base of NH_3, not NH_4^+. NH_3 is the conjugate base of NH_4^+.

(b) NH_3 is the conjugate acid of NH_2^-.

(c) H_3O^+ is the conjugate acid of H_2O.

(d) OH^- is the conjugate acid of O^{2-}.

(e) H_3O^+ is the conjugate acid of H_2O, not OH^-. H_2O is the conjugate acid of OH^-.

To conclude: Proper conjugate acid-base pairs are described in (b), (c), and (d).

20. *Answer:* **see identifications below**

Strategy and Explanation: The conjugate base of an acid has one less H^+ than its acid partner; the conjugate acid of a base has one more H^+ than its base partner.

(a) $HS^-(aq)$ $+$ $H_2O(\ell)$ $\rightleftharpoons$ $H_2S(aq)$ $+$ $OH^-(aq)$

    reactant base   **reactant acid**   **conj. acid**   **conj. base**
                                                                                       of HS^-   **of H_2O**

      H_2O/OH^- and H_2S/HS^- are the two acid-base conjugate pairs.

(b) $S^{2-}(aq)$ $+$ $NH_4^+(aq)$ $\rightleftharpoons$ $NH_3(aq)$ $+$ $HS^-(aq)$

    reactant base   **reactant acid**   **conj. base**   **conj. acid**
                                                                                       of NH_4^+   **of S^{2-}**

      NH_4^+/NH_3 and H_2S/HS^- are the two acid-base conjugate pairs.

(c) $HCO_3^-(aq)$ $+$ $HSO_4^-(aq)$ $\rightleftharpoons$ $H_2CO_3(aq)$ $+$ $SO_4^{2-}(aq)$

    reactant base   **reactant acid**   **conj. acid**   **conj. base**
                                                                                       of HCO_3^-   **of HSO_4^-**

      HSO_4^-/SO_4^{2-} and H_2CO_3/HCO_3^- are the two acid-base conjugate pairs.

(d) $NH_3(aq)$ $+$ $NH_2^-(aq)$ $\rightleftharpoons$ $NH_2^-(aq)$ $+$ $NH_3(aq)$

    reactant acid   **reactant base**   **conj. base**   **conj. acid**
                                                                                       of NH_3   **of NH_2^-**

      NH_3/NH_2^- and NH_3/NH_2^- are the two acid-base conjugate pairs.

22. *Answer:* **see equations below**

Strategy and Explanation: For acids, sequentially transfer H^+ from the acid to water. For bases, sequentially transfer H^+ from water to the base.

(a) CO_3^{2-} is the anion of a diprotic acid, H_2CO_3, so write two protonation equations.

    $CO_3^{2-}(aq) + H_2O(\ell) \rightleftharpoons HCO_3^-(aq) + OH^-(aq)$

    $HCO_3^-(aq) + H_2O(\ell) \rightleftharpoons H_2CO_3(aq) + OH^-(aq)$

(b) $H_3AsO_4(aq)$ is a triprotic acid, write three deprotonation equations.

    $H_3AsO_4(aq) + H_2O(\ell) \rightleftharpoons H_3O^+(aq) + H_2AsO_4^-(aq)$

    $H_2AsO_4^-(aq) + H_2O(\ell) \rightleftharpoons H_3O^+(aq) + HAsO_4^{2-}(aq)$

    $HAsO_4^{2-}(aq) + H_2O(\ell) \rightleftharpoons H_3O^+(aq) + AsO_4^{3-}(aq)$

(c) $NH_2CH_3COO^-(aq)$ can be protonated in two places, at the N atom and at the O^- atom, so write two protonation equations.

    $NH_2CH_2COO^-(aq) + H_2O(\ell) \rightleftharpoons {}^+NH_3CH_2COO^-(aq) + OH^-(aq)$

    ${}^+NH_3CH_2COO^-(aq) + H_2O(\ell) \rightleftharpoons {}^+NH_3CH_2COOH + OH^-(aq)$

pH Calculations

USE THE FOLLOWING INFORMATION FOR QUESTIONS 24-30. The following is a summary of relationships between H_3O^+ concentration, OH^- concentration, pH, and pOH.

$$pH = -\log[H_3O^+] \qquad pOH = -\log[OH^-] \qquad 14 = pH + pOH$$

To get $[H_3O^+]$ or $[OH^-]$ concentrations from pH or pOH, use the following relationships:

$$[H_3O^+] = 10^{-pH} \quad \text{and} \quad [OH^-] = 10^{-pOH}$$

When pH < 7 and pOH > 7, the solution is acidic; when pH = 7 = pOH, the solution is neutral; when pH < 7 and pOH > 7, the solution is acidic.

24. *Answer:* $\mathbf{3 \times 10^{-11}}$ **M; basic**

Strategy and Explanation: Use the appropriate relationship from among those provided in the box above. When pH = 10.5, $[H_3O^+] = 10^{-10.5} = 3 \times 10^{-11}$ M. The solution is basic.

The high pH gives an $[H_3O^+]$ that is relatively small ($< 10^{-7}$). $Mg(OH)_2$, found in some antacids, is an ionic compound that puts hydroxide ions in the saturated solution.

26. *Answer:* **pH = 12.40, pOH = 1.60**

Strategy and Explanation: Use the appropriate relationship from among those provided in the box before the solution to Question 24.

NaOH is a soluble hydroxide and a strong base. It completely ionizes to form Na^+ and OH^-:

$$[OH^-] = 0.025 \text{ M}, \qquad pOH = -\log[OH^-] = 1.60 \qquad pH = 14.00 - pOH = 14.00 - 1.60 = 12.40$$

28. *Answer:* $\mathbf{5 \times 10^{-2}}$ **M; 2 g HCl**

Strategy and Explanation: Use the appropriate relationship from among those provided in the box before the solution to Question 24 and Chapter 5.   When pH = 1.3, $[H_3O^+] = 10^{-1.3} = 5 \times 10^{-2}$ M.

HCl is a strong acid. One mole of HCl completely ionizes to one mole of H_3O^+ and one mole Cl^- ions.

$$1000. \text{ mL solution} \times \frac{1 \text{ L}}{1000 \text{ mL}} \times \frac{5 \times 10^{-2} \text{ mol } H_3O^+}{1 \text{ L solution}} \times \frac{1 \text{ mol HCl}}{1 \text{ mol } H_3O^+} \times \frac{36.4609 \text{ g HCl}}{1 \text{ mol HCl}} = 2 \text{ g HCl}$$

30. *Answer:*

	pH	$[H_3O^+]$	$[OH^-]$	acidic or basic
(a)	**6.21**	6.1×10^{-7}	**1.6×10^{-8}**	**acidic**
(b)	**5.34**	**4.5×10^{-6}**	2.2×10^{-9}	**acidic**
(c)	4.67	**2.1×10^{-5}**	**4.7×10^{-10}**	**acidic**
(d)	**1.60**	2.5×10^{-2}	**4.0×10^{-13}**	**acidic**
(e)	9.12	**7.6×10^{-10}**	**1.3×10^{-5}**	**basic**

Strategy and Explanation: Use the appropriate relationship from among those provided in the box before the solution to Question 24.

(a) $[H_3O^+] = 6.1 \times 10^{-7}$ M $pH = -\log[H_3O^+] = 6.21$ acidic solution (pH $< 10^{-7}$)

   $pOH = 14 - pH = 7.79$ $[OH^-] = 10^{-pOH} = 1.6 \times 10^{-8}$ M

(b) $[OH^-] = 2.2 \times 10^{-9}$ M $pOH = -\log[OH^-] = 8.66$

   $pH = 14 - pOH = 5.34$ $[H_3O^+] = 10^{-pH} = 4.5 \times 10^{-6}$ M   acidic solution (pH $< 10^{-7}$)

(c) pH = 4.67 $[H_3O^+] = 10^{-pH} = 2.1 \times 10^{-5}$ M   acidic solution (pH < 10^{-7})

   pOH = 14 − pH = 9.33 $[OH^-] = 10^{-pOH} = 4.7 \times 10^{-10}$ M

(d) $[H_3O^+] = 2.5 \times 10^{-2}$ M   pH = −log$[H_3O^+]$ = 1.60   acidic solution (pH < 10^{-7})

   pOH = 14 − pH = 12.40 $[OH^-] = 10^{-pOH} = 4.0 \times 10^{-13}$ M

(e) pH = 9.12 $[H_3O^+] = 10^{-pH} = 7.6 \times 10^{-10}$ M   basic solution (pH > 10^{-7})

   pOH = 14 − pH = 4.88 $[OH^-] = 10^{-pOH} = 1.3 \times 10^{-5}$ M

Acid-Base Strengths

32. *Answer:* **see chemical equations and equilibrium expressions below**

Strategy and Explanation: In acid ionization reactions, the acid donates H^+ to a water molecule to make H_3O^+ and the conjugate base. In base ionization reactions, the base receives H^+ from a water molecule to make OH^- and the conjugate acid.

(a) $F^-(aq) + H_2O(\ell) \rightleftharpoons HF(aq) + OH^-(aq)$ $K = \dfrac{[HF][OH^-]}{[F^-]}$

(b) $NH_3(aq) + H_2O(\ell) \rightleftharpoons NH_4^+(aq) + OH^-(aq)$ $K = \dfrac{[NH_4^+][OH^-]}{[NH_3]}$

(c) $H_2CO_3(aq) + H_2O(\ell) \rightleftharpoons HCO_3^-(aq) + H_3O^+(aq)$ $K = \dfrac{[HCO_3^-][H_3O^+]}{[H_2CO_3]}$

(d) $H_3PO_4(aq) + H_2O(\ell) \rightleftharpoons H_2PO_4^-(aq) + H_3O^+(aq)$ $K = \dfrac{[H_2PO_4^-][H_3O^+]}{[H_3PO_4]}$

(e) $CH_3COO^-(aq) + H_2O(\ell) \rightleftharpoons CH_3COOH(aq) + OH^-(aq)$ $K = \dfrac{[CH_3COOH][OH^-]}{[CH_3COO^-]}$

(f) $S^{2-}(aq) + H_2O(\ell) \rightleftharpoons HS^-(aq) + OH^-(aq)$ $K = \dfrac{[HS^-][OH^-]}{[S^{2-}]}$

34. *Answer:* **(a) 0.10 M NH_3 (b) 0.10 M K_2S (c) 0.10 M $NaCH_3COO$ (d) 0.10 M KCN**

Strategy and Explanation: In all these comparisons, the concentrations are the same, so we can use the ionization constants from Table 15.2 to compare their strengths. The larger the K_b, the more basic the solution. Ionic compounds provide cations or anions that may be bases, so watch for them.

(a) NH_3 is a base. ($K_b = 1.8 \times 10^{-5}$) NaF(s) $\longrightarrow$ Na$^+$(aq) + F$^-$(aq), and F$^-$ is a base. ($K_b = 1.4 \times 10^{-11}$)

   Given equal concentrations, a solution of NH_3 is more basic than a solution of NaF.

(b) $K_2S(s) \longrightarrow 2 K^+(aq) + S^{2-}(aq)$, and S^{2-} is a base. ($K_b = 1 \times 10^5$)

            $K_3PO_4(s) \longrightarrow 3 K^+(aq) + PO_4^{3-}(aq)$, and PO_4^{3-} is a base. ($K_b = 2.8 \times 10^{-2}$)

   Given equal concentrations, a solution of K_2S is more basic than a solution of K_3PO_4.

(c) $NaNO_3(s) \longrightarrow Na^+(aq) + NO_3^-(aq)$, NO_3^- is base. ($K_b = 5 \times 10^{-16}$)

            $NaCH_3COO(s) \longrightarrow Na^+(aq) + CH_3COO^-(aq)$, and CH_3COO^- is a base. ($K_b = 5.6 \times 10^{-10}$)

   Given equal concentrations, a solution of CH_3COONa is more basic than a solution of $NaNO_3$.

(d) KCN(s) $\longrightarrow$ K$^+$(aq) + CN$^-$(aq), and CN$^-$ is a base. ($K_b = 2.5 \times 10^{-5}$) NH_3 is a base. ($K_b = 1.8 \times 10^{-5}$)

   Given equal concentrations, a solution of KCN is more basic than a solution of NH_3.

Using K_a and K_b

36. *Answer:* $\mathbf{1.6 \times 10^{-5}}$

Strategy and Explanation: Follow the procedure described in Section 15.7, adapting the methods described in the solution to Question 25 in Chapter 13. Then use the information described before Question 23, as needed.

We do not know the formula of butyric acid, so we will assume it is monoprotic and give it the symbol HBu. The equation for the equilibrium and the equilibrium expression are:

$$HBu(aq) + H_2O(\ell) \rightleftharpoons H_3O^+(aq) + Bu^-(aq) \qquad K_a = \frac{[H_3O^+][Bu^-]}{[HBu]}$$

As the reactants decompose, the concentrations of the products increase stoichiometrically, until they reach equilibrium concentrations.

At equilibrium, pH = 3.21, so $[H_3O^+] = 10^{-pH} = 10^{-3.21} = 6.2 \times 10^{-4}$. Since the H_3O^+ ions are produced from the decomposition of HBu, an equal quantity of Bu^- is formed, and the concentrations change:

	HBu(aq)	H_3O^+(aq)	Bu^-(aq)
initial conc. (M)	0.015	$1.0 \times 10^{-7*}$	0
change as reaction occurs (M)	-6.2×10^{-4}	$+6.2 \times 10^{-4}$	$+6.2 \times 10^{-4}$
equilibrium conc. (M)	$0.015 - 6.2 \times 10^{-4}$	6.2×10^{-4}	6.2×10^{-4}

* from the dissociation of pure water. This number is small compared to the acid added.

$$K_a = \frac{(6.2 \times 10^{-4})(6.2 \times 10^{-4})}{(0.025 - 6.2 \times 10^{-4})} = 1.6 \times 10^{-5}$$

38. *Answer:* $\mathbf{[H_3O^+] = 1.3 \times 10^{-5}; [A^-] = 1.3 \times 10^{-5}; [HA] = 0.040}$

Strategy and Explanation: Follow the procedure described in Section 15.7, adapting the methods described in the solution to Question 25 in Chapter 13. Then use the information described before Question 23, as needed. The equation for the equilibrium and the equilibrium expression are:

$$HA(aq) + H_2O(\ell) \rightleftharpoons H_3O^+(aq) + A^-(aq) \qquad K_a = \frac{[H_3O^+][A^-]}{[HA]}$$

As the reactants decompose, the concentrations of the products increase stoichiometrically, until they reach equilibrium concentrations.

	HA(aq)	H_3O^+(aq)	A^-(aq)
initial conc. (M)	0.040	$1.0 \times 10^{-7*}$	0
change as reaction occurs (M)	$-x$	$+x$	$+x$
equilibrium conc. (M)	$0.040 - x$	x	x

* from dissociation of pure water. Assume this concentration will be small compared to the acid added.

At equilibrium $\qquad K_a = \frac{(x)(x)}{(0.040 - x)} = 4.0 \times 10^{-9}$

Assume x is very small and: $0.040 - x \cong 0.040$.

$$4.0 \times 10^{-9} = \frac{x^2}{(0.040)}$$

$$x^2 = (4.0 \times 10^{-9})(0.040)$$

$$x = 1.3 \times 10^{-5} \text{ M} = [H_3O^+] = [A^-]$$

$$[HA] = 0.040 \text{ M} - x = 0.040 \text{ M} - 1.3 \times 10^{-5} \text{ M} = 0.040 \text{ M (as assumed)}$$

40. *Answer:* **1.4 × 10⁻⁵**

Strategy and Explanation: Use the method described in the solution to Question 36. The equation for the equilibrium and the equilibrium expression are:

$$CH_3CH_2COOH(aq) + H_2O(\ell) \rightleftharpoons H_3O^+(aq) + CH_3CH_2COO^-(aq) \qquad K_a = \frac{[H_3O^+][CH_3CH_2COO^-]}{[CH_3CH_2COOH]}$$

At equilibrium, pH = 2.93, so $[H_3O^+] = 10^{-pH} = 10^{-2.93} = 1.2 \times 10^{-3}$ M. Since the H_3O^+ ions are produced from the decomposition of C_6H_5COOH, an equal quantity of $C_6H_5COO^-$ is formed, and the concentrations change in the following way:

	$C_6H_5COOH(aq)$	$H_3O^+(aq)$	$C_6H_5COO^-(aq)$
initial conc. (M)	0.15	$1.0 \times 10^{-7*}$	0
change as reaction occurs (M)	-1.2×10^{-3}	$+1.2 \times 10^{-3}$	$+1.2 \times 10^{-3}$
equilibrium conc. (M)	$0.015 - 1.2 \times 10^{-3}$	1.2×10^{-3}	1.2×10^{-3}

* from dissociation of pure water. Assume this concentration will be small compared to the acid added.

$$K_a = \frac{(1.2 \times 10^{-3})(1.2 \times 10^{-3})}{(0.015 - 1.2 \times 10^{-3})} = 1.4 \times 10^{-5}$$

42. *Answer:* **8.85**

Strategy and Explanation: Follow the procedure described in Section 15.7, adapting the methods described in the solution to Question 25 in Chapter 13. Then use the information described in the box before Question 23, as needed. The equation for the equilibrium and the equilibrium expression are:

$$C_6H_5NH_2(aq) + H_2O(\ell) \rightleftharpoons C_6H_5NH_3^+(aq) + OH^-(aq) \qquad K_b = \frac{[C_6H_5NH_3^+][OH^-]}{[C_6H_5NH_2]}$$

As the reactants decompose, the concentrations of the products increase stoichiometrically, until they reach equilibrium concentrations.

	$C_6H_5NH_2(aq)$	$C_6H_5NH_3^+(aq)$	$OH^-(aq)$
initial conc. (M)	0.12	0	$1.0 \times 10^{-7*}$
change as reaction occurs (M)	$-x$	$+x$	$+x$
equilibrium conc. (M)	$0.12 - x$	x	x

* from dissociation of pure water. Assume this concentration will be small compared to the base added.

At equilibrium $\qquad K_b = \frac{(x)(x)}{(0.12 - x)} = 4.2 \times 10^{-10}$

Assume x is very small and: $0.12 - x \cong 0.12$.

$$x^2 = (4.2 \times 10^{-10})(0.12)$$

$$x = 7.1 \times 10^{-6} \text{ M} = [OH^-]$$

$$pOH = -\log[OH^-] = -\log(7.1 \times 10^{-6}) = 5.15$$

$$pH = 14.00 - pOH = 14.00 - 5.15 = 8.85$$

43. *Answer:* **(a) $C_{10}H_{15}NH_2(aq) + H_2O(\ell) \rightleftharpoons C_{10}H_{15}NH_3^+(aq) + OH^-(aq)$ (b) 10.47**

Strategy and Explanation:

(a) Amantadine reacts with water, undergoing hydrolysis to form a basic solution. The equation for the equilibrium is:

$$C_{10}H_{15}NH_2(aq) + H_2O(\ell) \rightleftharpoons C_{10}H_{15}NH_3^+(aq) + OH^-(aq)$$

(b) The K_a given is for the dissociation of $C_{10}H_{15}NH^+(aq)$. We need to find the K_b from the K_a:

$$K_b = \frac{K_w}{K_a} = \frac{1.00 \times 10^{-14}}{7.9 \times 10^{-11}} = 1.3 \times 10^{-4}$$

The reactants decompose, the concentrations of the products increase stoichiometrically, until they reach equilibrium concentrations.

	$C_{10}H_{15}N(aq)$	$C_{10}H_{15}NH^+(aq)$	$OH^-(aq)$
initial conc. (M)	0.0010	0	$1.0 \times 10^{-7*}$
change as reaction occurs (M)	$-x$	$+x$	$+x$
equilibrium conc. (M)	$0.0010 - x$	x	x

* from dissociation of pure water. Assume this concentration will be small compared to the base added.

At equilibrium $K_b = \dfrac{[C_{10}H_{15}NH_3^+][OH^-]}{[C_{10}H_{15}NH_2]} = \dfrac{(x)(x)}{(0.0010 - x)} = 1.3 \times 10^{-4}$

$$x^2 = (1.3 \times 10^{-4})(0.0010 - x)$$

$$x^2 + 1.3 \times 10^{-4}x - 1.5 \times 10^{-7} = 0$$

Use the quadratic equation: (see Appendix A, Section A.7, page A.12)

$$x = 3.0 \times 10^{-4} \, M = [\,OH^-\,]$$

$$pOH = -\log[OH^-] = -\log(3.0 \times 10^{-4}) = 3.53$$

$$pH = 14.00 - pOH = 14.00 - 3.53 = 10.47$$

44. *Answer:* **3.28**

Strategy and Explanation: First use methods from Chapters 3 and 4 to find the initial concentration of the $C_3H_6O_3$. Then follow the procedure described in Section 15.7, adapting the methods described in the solution to Question 25 in Chapter 13. Then use the information described in the box before the solution to Question 23, as needed.

$$\frac{56 \text{ mg } C_3H_6O_3}{250 \text{ mL soln}} \times \frac{1 \text{ g}}{1000 \text{ mg}} \times \frac{1 \text{ mol } C_3H_6O_3}{90.08 \text{ g } C_3H_6O_3} \times \frac{1000 \text{ mL}}{1 \text{ L}} = 0.0025 \text{ M}$$

According to the structure shown in Section 15.2, lactic acid is a monoprotic acid, so we'll write the formula as: $HC_3H_5O_3$. The equation for the equilibrium and the equilibrium expression are:

$$HC_3H_5O_3(aq) + H_2O(\ell) \rightleftharpoons H_3O^+(aq) + C_3H_5O_3^-(aq) \qquad K_a = \frac{[H_3O^+][C_3H_5O_3^-]}{[HC_3H_5O_3]}$$

As the reactants decompose, the concentrations of the products increase stoichiometrically, until they reach equilibrium concentrations.

	$HC_3H_5O_3(aq)$	$H_3O^+(aq)$	$C_3H_5O_3^-(aq)$
initial conc. (M)	0.0025	$1.0 \times 10^{-7*}$	0
change as reaction occurs (M)	$-x$	$+x$	$+x$
equilibrium conc. (M)	$0.0025 - x$	x	x

* from dissociation of pure water. Assume this concentration will be small compared to the acid added.

At equilibrium $K_a = \dfrac{(x)(x)}{(0.0025 - x)} = 1.4 \times 10^{-4}$

$$x^2 = (1.4 \times 10^{-4})(0.0025 - x)$$

$$x^2 + 1.4 \times 10^{-4}x - 3.5 \times 10^{-7} = 0$$

Use the quadratic equation: (see Appendix A, Section A.7, page A.12)

$$x = 5.2 \times 10^{-4} \text{ M} = [H_3O^+]$$

$$pH = -\log[H_3O^+] = -\log(5.2 \times 10^{-4}) = 3.28$$

Acid-Base Reactions

46. *Answer:* **(a) CN⁻; product-favored (b) HS⁻; reactant-favored (c) H₂(g); product-favored**

Strategy and Explanation: Use the methods described in the answers to Questions 16, 20, and 32. Compare the reactant acid to the product acid and identify which is stronger and which is weaker. Do the same with the bases. Equilibrium favors the weaker species in the reaction.

(a) **CN⁻(aq)**   + HSO_4^-(aq)   ⇌   HCN(aq)   + SO_4^{2-}(aq)

  stronger base   stronger acid   weaker acid   weaker base

  The reaction is product-favored.

(b) H_2S(aq)   + $H_2O(\ell)$ ⇌ H_3O^+(aq)   +   **HS⁻(aq)**

  weaker acid   weaker base   stronger acid   stronger base

  The reaction is reactant-favored.

(c) H⁻(aq)   + $H_2O(\ell)$ ⇌   OH⁻(aq)   +   **H₂(g)**

  stronger base   stronger acid   weaker base   weaker acid

  The reaction is product-favored.

48. *Answer:* **(a) less than 7 (b) greater than 7 (c) equal to 7; see explanations below**

Strategy and Explanation: Adapt the method described in the answers to Questions 33-34.

(a) $AlCl_3$(s) ⟶ Al^{3+}(aq) + 3 Cl⁻(aq)   The Cl⁻ ions do not affect the pH of a water solution, because HCl is a strong acid. However, the hydrated aqueous aluminum ion is formed:

$$Al^{3+} + 6\ H_2O(aq) \longrightarrow [Al(H_2O)_6]^{3+}$$

It is a weak acid, so we predict pH less than 7.

(b) Na_2S(s) ⟶ 2 Na⁺(aq) + S^{2-}(aq)   The Na⁺ ions do not affect the pH of a water solution. S^{2-} is a strong base, so we predict pH greater than 7.

(c) $NaNO_3$(s) ⟶ Na⁺(aq) + NO_3^-(aq)   Neither of these ions affects the pH of a water solution, because NaOH is a strong base and HNO_3 is a strong acid, so we predict pH equal to 7.

50. *Answer:* **(a) greater than 7 (b) greater than 7 (c) greater than 7; see explanations below**

Strategy and Explanation: Adapt the method described in the solution to Question 48.

(a) Na_2HPO_4(s) ⟶ 2 Na⁺(aq) + HPO_4^{2-}(aq)   The Na⁺ ions do not affect the pH of a water solution. The HPO_4^{2-} ion is a weak acid and a weak base, so we must compare the size of K_a and K_b:

HPO_4^{2-} (aq) + H_2O(aq) ⇌ H_3O^+(aq) + PO_4^{3-}(aq) $K_a = 3.6 \times 10^{-13}$

HPO_4^{2-} (aq) + H_2O(aq) ⇌ $H_2PO_4^-$(aq) + OH⁻(aq) $K_b = 1.7 \times 10^{-7}$

$H_2PO_4^-$ is a stronger base than it is an acid, so we predict pH greater than 7.

(b) $(NH_4)_2S$(s) ⟶ 2 NH_4^+(aq) + S^{2-}(aq)   The NH_4^+ ion is a weak acid, but the S^{2-} ion is a strong base, so we predict pH greater than 7.

(c) KCH_3COO(s) ⟶ K⁺(aq) + CH_3COO^-(aq)   The K⁺ ions do not affect the pH of a water solution. CH_3COO^- is a weak base, so we predict pH greater than 7.

Lewis Acids and Bases

52. *Answer:* **Lewis acid: molecule (b); Lewis bases: molecules (a), (b), and (c)**

Strategy and Explanation: The Lewis model focuses on what electron pairs are doing. The substance capable of donating the electron pair to form a new bond is called a Lewis base. The substance capable of accepting an electron pair is a Lewis acid.

(a) O^{2-} has a lone pair of electrons that can form a new bond, so O^{2-} can be a Lewis base. It cannot accept any more electrons, so it is not a Lewis acid.

$$:\overset{..}{\underset{..}{O}}:$$

(b) CO_2 has a lone pair of electrons on the O atoms that can form a new bond, so CO_2 can be a Lewis base. Its central C atom can interact with lone pairs on other Lewis bases, so CO_2 can also be a Lewis acid.

$$:\overset{..}{O}=C=\overset{..}{O}:$$

(c) H^- has a lone pair of electrons that can form a new bond, so H^- can be a Lewis base. It cannot accept any more electrons, so it is not a Lewis acid.

$$H:$$

54. *Answer:* **(a) Lewis acid: I_2; Lewis base: I^- (b) Lewis acid: BF_3; Lewis base: SO_2 (c) Lewis acid: Au^+; Lewis base: CN^- (d) Lewis acid: CO_2; Lewis base: H_2O**

Strategy and Explanation: Identify which reactant is donating electrons and which is accepting them.

(a) The curved arrow shows how the lone pair on I^- becomes a new bond on the right I of I_2.

Therefore, I_2 is the Lewis acid and I^- is the Lewis base.

(b) The curved arrow shows how the lone pair on S becomes a new bond between the S atom and B atom.

Therefore, BF_3 is the Lewis acid and SO_2 is the Lewis base.

(c) Au^+ metal ion has no bonds to start with and only two low-energy valence electrons. Electrons on CN^- make new bonds with Au^+, so Au^+ is Lewis acid and CN^- is Lewis base.

(d) A similar reaction is shown at the end of Section 15.9 on page 569. The curved arrow below shows how the lone pair on O becomes a new bond between C atom and O atom. (Notice that after the initial Lewis acid-base reaction, one of the H atoms migrates to a nearby O atom to balance the positive and negative charges.)

Therefore, CO_2 is the Lewis acid and H_2O is the Lewis base.

General Questions

56. *Answer:* **(a) CH$_3$COOH is a weak acid. (b) Na$_2$O, contains O^{2-},a strong base. (c) H$_2$SO$_4$ is a strong acid. (d) NH$_3$ is a weak base. (e) Ba(OH)$_2$, contains OH$^-$,a strong base. (f) H$_2$PO$_4^-$ is amphiprotic.**

Strategy and Explanation: The term "amphiprotic" describes a species that can function both as a Bronsted-Lowry base and as a Bronsted-Lowry acid. In general, any species with an H can donate H$^+$ ions, though sometimes that will not happen in the common water solvent. In addition, any species with one or more lone pairs can be a base, since the electrons could make a bond to a proton. However, because this Question asks for a judgment of "weak" and "strong," we will restrict our designations to the reactions of these species in water and use Table 15.2 to assist us. Notice that Na$_2$O, contains O^{2-}.

58. *Answer:* **(a) less than 7 (b) equal to 7 (c) greater than 7**

Strategy and Explanation: If equal molar amounts of an acid and a base are combined, the stronger of them will dictate the pH of the solution.

(a) A weak base and a strong acid will have an acidic pH, less than 7.

(b) A strong base and a strong acid will have a neutral pH, equal to 7.

(c) A strong base and a weak acid will have a basic pH, greater than 7.

60. *Answer:* **2.85**

Strategy and Explanation: Follow the methods described in the answers to Questions 35-40.

$$\frac{5.0 \text{ mg C}_6\text{H}_8\text{O}_6}{1 \text{ mL soln}} \times \frac{1 \text{ g}}{1000 \text{ mg}} \times \frac{1 \text{ mol C}_6\text{H}_8\text{O}_6}{176 \text{ g C}_6\text{H}_8\text{O}_6} \times \frac{1000 \text{ mL}}{1 \text{ L}} = 0.028 \text{ M}$$

Ascorbic acid, C$_6$H$_8$O$_6$, is diprotic acid, so give it a formula of H$_2$A. The equation for the equilibrium and the equilibrium expression are:

$$H_2A(aq) + H_2O(\ell) \rightleftharpoons H_3O^+(aq) + HA^-(aq) \qquad K_a = \frac{[H_3O^+][HA^-]}{[H_2A]}$$

As the reactants decompose, the concentrations of the products increase stoichiometrically, until they reach equilibrium concentrations.

	H$_2$A(aq)	H$_3$O$^+$(aq)	HA$^-$(aq)
initial conc. (M)	0.028	$1.0 \times 10^{-7*}$	0
change as reaction occurs (M)	$-x$	$+x$	$+x$
equilibrium conc. (M)	$0.028 - x$	x	x

* from dissociation of pure water. Assume this concentration will be small compared to the acid added.

At equilibrium $\qquad K_a = \dfrac{(x)(\hat{x})}{(0.028 - x)} = 7.9 \times 10^{-5}$

$$x^2 = (7.9 \times 10^{-5})(0.028 - x)$$

If we assume that x is small, and $0.028 - x \cong 0.028$

$$x^2 = (7.9 \times 10^{-5})(0.028)$$

$$x = 1.5 \times 10^{-3} \text{ M} = [H_3O^+]$$

$$pH = - \log[H_3O^+] = -\log(1.5 \times 10^{-3}) = 2.82$$

If we use the quadratic equation: (see Appendix A, Section A.7, page A.12)

$$x^2 + 7.9 \times 10^{-5}x - 2.2 \times 10^{-6} = 0$$

$$x = 1.4 \times 10^{-3} \text{ M} = [H_3O^+]$$

$$pH = - \log[H_3O^+] = -\log(1.4 \times 10^{-3}) = 2.85$$

62. *Answer:* **9.73**

Strategy and Explanation: Use the method similar to that given in the solution to Question 42. The equation for the equilibrium and the equilibrium expression are:

$$ClO^-(aq) + H_2O(\ell) \rightleftharpoons HClO(aq) + OH^-(aq) \qquad K_b = \frac{[HClO][OH^-]}{[ClO^-]}$$

As the reactants decompose, the concentrations of the products increase stoichiometrically, until they reach equilibrium concentrations.

	$ClO^-(aq)$	$HClO(aq)$	$OH^-(aq)$
initial conc. (M)	0.010	0	$1.0 \times 10^{-7*}$
change as reaction occurs (M)	$-x$	$+x$	$+x$
equilibrium conc. (M)	$0.010 - x$	x	x

* from dissociation of pure water. Assume this concentration will be small compared to the base added.

At equilibrium $\qquad\qquad K_b = \dfrac{(x)(x)}{(0.010 - x)} = 2.9 \times 10^{-7}$

Assume x is very small and: $0.010 - x \cong 0.010$.

$$x^2 = (2.9 \times 10^{-7})(0.010)$$

$$x = 5.4 \times 10^{-5} \text{ M} = [OH^-]$$

$$pOH = -\log[OH^-] = -\log(5.4 \times 10^{-5}) = 4.27$$

$$pH = 14.00 - pOH = 14.00 - 4.27 = 9.73$$

Applying Concepts

65. *Answer/Explanation:* Conjugates in an acid-base pair must differ by only one H^+ ion. The acids and bases were identified correctly, but the conjugates were not. HCO_3^- is the conjugate of H_2CO_3 and HSO_4^- is the conjugate of SO_4^{2-}.

66. *Answer/Explanation:* The other two acid-base theories are described by the Lewis theory in the following ways:

Arrhenius theory: Electron pairs on the solvent water molecules (Lewis base) form a bond with the hydrogen ion (Lewis acid) producing aqueous H^+ ions. Electron pairs on the OH^- ions (Lewis base) form a bond with the hydrogen ion (Lewis acid) in the solvent water molecule, producing aqueous OH^- ions.

Bronsted-Lowry Theory: The H^+ ion from the Bronsted-Lowry acid is bonded to a Bronsted-Lowry base using an electron pair on the base. The electron-pair acceptor, the H^+ ion, is the Lewis acid and the electron-pair donor is the Lewis base.

68. *Answer:* **(a) 2.01 (b) 23 times**

Strategy and Explanation: Adapt the method shown in Question 38.

(a) The equation for the equilibrium and the equilibrium expression are:

$$CCl_3COOH(aq) + H_2O(\ell) \rightleftharpoons H_3O^+(aq) + CCl_3COO^-(aq) \qquad K_a = \frac{[H_3O^+][CCl_3COO^-]}{[CCl_3COOH]}$$

The concentrations of products increase stoichiometrically, until equilibrium is reached.

	$CCl_3COOH(aq)$	$H_3O^+(aq)$	$CCl_3COO^-(aq)$
initial conc. (M)	0.010	$1.0 \times 10^{-7*}$	0
change as reaction occurs (M)	$-x$	$+x$	$+x$
equilibrium conc. (M)	$0.010 - x$	x	x

* from dissociation of pure water. Assume this concentration will be small compared to the acid added.

At equilibrium $K_a = \dfrac{(x)(x)}{(0.010 - x)} = 3.0 \times 10^{-1}$

$$x^2 = (3.0 \times 10^{-1})(0.010 - x)$$

$$x^2 + 3.0 \times 10^{-1}x - 3.0 \times 10^{-5} = 0$$

Use the quadratic equation: (see Appendix A, Section A.7, page A.12)

$$x = 9.7 \times 10^{-3}\ M = [H_3O^+]$$

$$pH = -\log[H_3O^+] = -\log(9.7 \times 10^{-3}) = 2.01$$

(b) The equation for the equilibrium and the equilibrium expression are:

$$CH_3COOH(aq) + H_2O(\ell) \rightleftharpoons H_3O^+(aq) + CH_3COO^-(aq) \qquad K_a = \dfrac{[H_3O^+][CH_3COO^-]}{[CH_3COOH]}$$

The concentrations of products increase stoichiometrically, until equilibrium is reached.

	$CH_3COOH(aq)$	$H_3O^+(aq)$	$CH_3COO^-(aq)$
initial conc. (M)	0.010	$1.0 \times 10^{-7*}$	0
change as reaction occurs (M)	$-x$	$+x$	$+x$
equilibrium conc. (M)	$0.010 - x$	x	x

* from dissociation of pure water. Assume this concentration will be small compared to the acid added.

At equilibrium $K_a = \dfrac{(x)(x)}{(0.010 - x)} = 1.8 \times 10^{-5}$

Assume x is very small and: $0.20 - x \cong 0.20$.

$$1.8 \times 10^{-5} = \dfrac{x^2}{(0.010)}$$

$$x^2 = (1.8 \times 10^{-5})(0.010)$$

$$x = 4.2 \times 10^{-4}\ M = [H_3O^+]$$

To determine the number of times more hydronium ions in the trichloroacetic acid versus the acetic acid, divide the hydronium ion concentration in the trichloroacetic acid solution by that in the acetic acid solution.

$$\dfrac{9.7 \times 10^{-3}\ M}{4.2 \times 10^{-4}\ M} = 23 \text{ times more } H_3O^+ \text{ in } CCl_3COOH(aq)$$

More Challenging Questions

70. *Answer:* **0.76 L; probably not**

Strategy and Explanation: This problem can be solved by methods described in Chapter 5 and 10. Use the chemical reaction in Section 5.2 on page 150. The product-favored reaction goes essentially to completion in the presence of sufficient acid.

$$2.5 \text{ g NaHCO}_3 \times \dfrac{1 \text{ mol NaHCO}_3}{84.0066 \text{ g NaHCO}_3} \times \dfrac{1 \text{ mol CO}_2(g)}{1 \text{ mol NaHCO}_3} = 0.030 \text{ mol CO}_2(g)$$

$$V_{CO_2} = \dfrac{n_{CO_2}RT}{P} = \dfrac{(0.030 \text{ mol}) \times \left(0.08206 \dfrac{L \cdot atm}{mol \cdot K}\right) \times (37 + 273)K}{1 \text{ atm}} = 0.76 \text{ L}$$

This, by itself, is an insufficient volume of CO_2 to rupture a 1-L stomach, though it would be enough to be uncomfortable. To be able to predict a rupture, we'd need to know how much volume the food occupied and whether this added CO_2 would exceed the capacity of his stomach. We don't know what pressure of gas the stomach at maximum volume can take, nor do we know if some other involuntary action such as burping or vomiting would occur to prevent the excess gas from causing damage. One would need to be a gastroenterologist, to accurately answer this part of the Question.

71. *Answer:* **Yes, pH increases**

Strategy and Explanation: When the can is initially pressurized with CO_2, the following chemical equilibrium expression is shifted toward the formation of products. The beverage becomes acidic.

$$CO_2(aq) + 2\,H_2O(\ell) \rightleftharpoons H_3O^+(aq) + HCO_3^-(aq)$$

When the can is opened and warmed, carbon dioxide escapes from the carbonated beverage. The loss of $CO_2(aq)$ shifts the above equilibrium to the left. Hydronium ion concentration is decreased and the solution pH increases (i.e. becomes more basic).

73. *Answer/Explanation:* Br has a higher electronegativity than H. The bromine withdraws electron density from the nitrogen atom, reducing the nitrogen's ability to bind the positively charged proton. The K_b value is smaller for $BrNH_2$ than for NH_3. $ClNH_2$ is a weaker base than $BrNH_2$ because Cl is more electronegative than Br.

75. *Answer:* **HM > HQ > HZ ; HZ, $K_{a,HZ} = 1 \times 10^{-5}$; HQ, $K_{a,HQ} = 1 \times 10^{-3}$; HM, $K_{a,HM} = 1 \times 10^{-1}$ or larger**

Strategy and Explanation: Stronger acids result in weaker conjugate bases. The stronger base has the larger pH. So, the salt solution with the highest pH contains the anion of the weakest acid.

Smallest pH: NaM > NaQ > NaZ   :Largest pH

Strongest: HM > HQ > HZ   :Weakest

The equation for the equilibrium and the equilibrium expression are:

$$A^-(aq) + H_2O(\ell) \rightleftharpoons HA(aq) + OH^-(aq) \qquad K_b = \frac{[HA][OH^-]}{[A^-]}$$

The initial $[OH^-]$ of neutral water is 1×10^{-7} M. The initial concentration of the base is 0.1 M. The concentrations of products increase stoichiometrically, until they reach equilibrium concentrations.

In general:	A^-(aq)	HA(aq)	OH^-(aq)
initial conc. (M)	0.1	0	$1 \times 10^{-7*}$
change as reaction occurs (M)	$-x$	$+x$	$+x$
equilibrium conc. (M)	$0.1 - x$	x	$10^{-7} + x$

* from the dissociation of pure water.

At equilibrium $K_b = \dfrac{(x)(1 \times 10^{-7} + x)}{(0.1 - x)}$

$$1 \times 10^{-7} + x = [OH^-] = 10^{-pOH} = 10^{-(14.00 - pH)} = 10^{(pH - 14.00)}$$

$$x = 10^{(pH - 14.00)} - 1 \times 10^{-7}$$

For NaZ, A^- above is Z^- and pH = 9.0, so

$$x = 10^{(9.0 - 14.00)} - 1 \times 10^{-7} = 1 \times 10^{-5} - 1 \times 10^{-7} = 1 \times 10^{-5}, \text{ and}$$

$$K_b = \frac{(1 \times 10^{-5})(1 \times 10^{-7} + 1 \times 10^{-5})}{(0.1 - 1 \times 10^{-5})} = 1 \times 10^{-9}$$

For NaQ, A^- above is Q^- and pH = 8.0, so

$$x = 10^{(8.0 - 14.00)} - 1 \times 10^{-7} = 1 \times 10^{-6} - 1 \times 10^{-7} = 1 \times 10^{-6}, \text{ and}$$

$$K_b = \frac{(1 \times 10^{-6})(1 \times 10^{-7} + 1 \times 10^{-6})}{(0.1 - 1 \times 10^{-6})} = 1 \times 10^{-11}$$

For NaM, A^- above is M^- and pH = 7.0, so $x = 10^{(7.0 - 14.00)} - 1 \times 10^{-7}$

$x = 1 \times 10^{-7} - 1 \times 10^{-7} = 0$, suggesting that the reaction with water does not go toward products at all. That means K_b of M^- may be so small that the pH of the solution is accounted for exclusively by the ionization of water. If indeed the base does provide all the OH^- ions to make the solution's pH: $[HM] = [OH^-] = 10^{-7.0} = 1 \times 10^{-7}$:

$$K_b = \frac{(1 \times 10^{-7})(1 \times 10^{-7})}{(0.1)} = 1 \times 10^{-13}$$

A relationship between K_a and K_b is described at the end of Section 15.7 on page 561.

For HZ, $K_a = K_w/K_b = (1.0 \times 10^{-14})/(1 \times 10^{-9}) = 1 \times 10^{-5}$

For HQ, $K_a = K_w/K_b = (1.0 \times 10^{-14})/(1 \times 10^{-11}) = 1 \times 10^{-3}$

For HM, $K_a = K_w/K_b = (1.0 \times 10^{-14})/(1 \times 10^{-13}) = 1 \times 10^{-1}$ or larger

77. *Answer:* **(a) weak (b) weak (c) conjugate acid-base pair (d) 6.26**

Strategy and Explanation:

(a) The acid ionization constant for hydrogen is very small. H_2O_2 is a weak acid.

(b) K_b for OOH^- is larger than K_a, but it is still small. OOH^- is a weak base.

(c) H_2O_2 and OOH^- are a conjugate acid and base.

(d) Hydrogen peroxide will ionize in solution according to the following equation with the following equilibrium expression:

$$HOOH(aq) + H_2O(\ell) \rightleftharpoons H_3O^+(aq) + HOO^-(aq) \qquad K_b = \frac{[H_3O^+][HOO^-]}{[HOOH]}$$

	HOOH	H_3O^+(aq)	HOO^-(aq)
initial conc. (M)	0.100	$1.0 \times 10^{-7}*$	0
change as reaction occurs (M)	$-x$	$+x$	$+x$
equilibrium conc. (M)	$0.100 - x$	$1.0 \times 10^{-7} + x$	x

* from the dissociation of pure water.

$$K_b = \frac{(1.0 \times 10^{-7} + x)(x)}{(0.100 - x)} = 2.5 \times 10^{-12}$$

$$(1.0 \times 10^{-7} + x)x = (2.5 \times 10^{-12})(0.100 - x)$$

If we assume that x is small, and $0.100 - x \cong 0.100$

$$(1.0 \times 10^{-7} + x)x = (2.5 \times 10^{-12})(0.100)$$

$$x^2 + x(1.0 \times 10^{-7}) - 2.5 \times 10^{-13} = 0$$

Use the quadratic equation: (see Appendix A, Section A.7, page A.12)

$$x = 4.5 \times 10^{-7}$$

$$[H_3O^+] = 1.0 \times 10^{-7} + 4.5 \times 10^{-7} = 5.5 \times 10^{-7} \text{ M}$$

$$pH = -\log[H_3O^+] = -\log(5.5 \times 10^{-7}) = 6.26$$

79. *Answer:* **0.1% solution is 9% dissociation; saturated solution is 2% dissociation.**

Strategy and Explanation: Determination of the percent dissociation for adipic acid requires the calculation of the initial concentration of the acid and the concentration of H_3O^+ created. The concentration of adipic acid is determined from the mass of the acid per 100 mL of solution.

Assume the 0.1% solution is a mass percent. At low concentrations, the density of a solution can be assumed equal to the density of water. Convert to molarity as follows:

$$\frac{0.1 \text{ g } C_5H_9O_2COOH}{100 \text{ g solution}} \times \frac{0.998 \text{ g solution}}{1 \text{ mL}} \times \frac{1000 \text{ mL}}{L} \times \frac{1 \text{ mol } C_5H_9O_2COOH}{146.1408 \text{ g } C_5H_9O_2COOH} = 7 \times 10^{-3} \text{ M}$$

Determine the concentration of H_3O^+ from pH = 3.2.

$$[H_3O^+] = 10^{-pH} = 10^{-3.2} = 6 \times 10^{-4}$$

The percent ionization of the 0.1% adipic acid solution is:

$$\frac{[H_3O^+] \text{ at equilibrium}}{\text{initial acid conc.}} \times 100\% = \frac{6 \times 10^{-4} \text{ M}}{7 \times 10^{-3} \text{ M}} \times 100\% = 9\%$$

Repeat the calculations for the saturated adipic acid solution.

$$\frac{1.44 \text{ g } C_5H_9O_2COOH}{100 \text{ g solution}} \times \frac{0.998 \text{ g solution}}{1 \text{ mL}} \times \frac{1000 \text{ mL}}{L} \times \frac{1 \text{ mol } C_5H_9O_2COOH}{146.1408 \text{ g } C_5H_9O_2COOH} = 9.9 \times 10^{-2} \text{ M}$$

Determine the concentration of H_3O^+ from pH = 2.7. $[H_3O^+] = 10^{-pH} = 10^{-3.2} = 2 \times 10^{-3}$

The percent ionization of the 0.1% adipic acid solution is: $\dfrac{2 \times 10^{-3} \text{ M}}{9.84 \times 10^{-2} \text{ M}} \times 100\% = 2\%$

Chapter 16: Additional Aqueous Equilibria

Introduction

Teaching for Conceptual Understanding

The applications of dynamic equilibria described Chapter 16 are important to many science, engineering and pre-professional health career majors, as well as many real-world experiences. For example, understanding buffers will be critical to biological science and pre-professional health career majors who are expected to learn about buffers in chemistry.

Though the problem solving in this chapter tends to focus on quantitative equilibrium. The key to success is figuring out which chemical equations to start with. You need to identify the chemical nature of the given materials and identify the most likely chemical reactions among them. Focus on what reactions could happen, and the relative value of their respective equilibrium constants. The largest K reaction should always be addressed first. Just because something is present initially, does not mean it must be a reactant in the written chemical reaction. Remember, reactions go both directions. Some of the given materials may actually be a product of a known reaction, not a reactant.

A common error that students make while doing solubility equilibrium calculation is forgetting the stoichiometric factor when determining the value of the molar solubility (s) at equilibrium. This happens because all the primary reactions studied in Chapter 15 (acid ionization, base ionization, and base hydrolysis) have 1:1 stoichiometry. Make sure you use the stoichiometric coefficients as multipliers for the changes in concentration as the reaction occurs.

Solutions to Blue-Numbered Questions
for Review and Thought for Chapter 16

Topical Questions

Buffer Solutions

14. *Answer:* **combination (c)**

    Strategy and Explanation: To determine the pH of a buffer, look up the pK_a (Table 16.1) or look up the K_a (Table 15.2) and calculate the pK_a ($pK_a = -\log K_a$). The pK_a closest to the desired pH is the best buffer, since close to equal quantities of the acid and base would be used, giving the solution approximately equal ability to neutralize added acid or added base.

    (a) The $CH_3COOH/NaCH_3COO$ buffer system has $pK_a = 4.74$.

    (b) The acid in $HCl/NaCl$ is HCl. It has K_a = very large. This is not a buffer.

    (c) The NH_3/NH_4Cl buffer system has $pK_a = 9.25$.

    The combination that would make the best pH 9 buffer system is (c) NH_3/NH_4Cl.

16. *Answer:* **(a) 2.1 (b) 7.21 (c) 12.46**

    Strategy and Explanation: To answer this Question quantitatively we need the value of pK_a, since that is equal to the pH in an equimolar buffer solution. In some cases, that value is not in the textbook. Without doing any calculations, we must estimate the pK_a from the K_a. We will look at the size of the K_a value and estimate its power of ten. We do this by determining which two powers of ten the number is between, then we can also use the values provided in Table 16.1 to give us advice about the fractional part of the pK_a.

(a) The acid of the pair is H_3PO_4. It has $K_a = 7.5 \times 10^{-3}$. Here the K_a is between 10^{-2} and 10^{-3}. In Table 16.1, the dihydrogen phosphate buffer K_a has a slightly smaller number multiplying its power of ten, 6.2, and its pK_a has a fractional component of .21, so we will estimate the $HNO_2/NaNO_2$ buffer system will have a pH of about 2.1.

(b) The NaH_2PO_4/Na_2HPO_4 buffer system has $pK_a = 7.21$, so its pH is 7.21.

(c) The acid of the pair is HPO_4^{2-}, with $K_a = 3.6 \times 10^{-13}$. Here the K_a is between 10^{-12} and 10^{-13}. In Table 16.1, the hypochlorous acid K_a has a similar number multiplying its power of ten, 3.5, and its pK_a has a fractional component of .46, so we will estimate the Na_2HPO_4/Na_3PO_4 buffer system will have a pH of about 12.46.

✓ *Reasonable Answer Check:* To check the answers, we will disobey the explicit instructions and do the calculation. In this case, you can also compare the estimates to the calculations you may have already done in Question 13.   (a) $pK_a = -\log(7.5 \times 10^{-3}) = 2.12$ (c) $pK_a = -\log(3.6 \times 10^{-13}) = 12.44$

18. *Answer:* **(a) lactic acid/lactate (b) acetic acid/acetate (c) hypochlorous acid/hypochlorite (d) hydrogen carbonate/carbonate; see explanations, below**

Strategy and Explanation: We will compare the pH to the values of pK_a in Table 16.1, since that is equal to the pH in an equimolar buffer solution. The pK_a closest to the desired pH is most suitable.

(a) pH = 3.45, needs a lactic acid/lactate buffer ($pK_a = 3.85$).

(b) pH = 5.48, needs an acetic acid/acetate buffer ($pK_a = 4.74$). Its pH is slightly closer to the desired pH than any of the other buffer systems in Table 16.1.

(c) pH = 8.32, needs a hypochlorous acid/hypochlorite buffer ($pK_a = 7.46$). Its pH is slightly closer to the desired pH than any of the other buffer systems.

(d) pH =10.15, needs a hydrogen carbonate/carbonate buffer ($pK_a = 10.32$).

21. *Answer:* **9.55; 9.51**

Strategy and Explanation: This Question uses methods learned in several previous chapters. First, calculate the initial concentration of the conjugate acid, NH_4^+ as described in Chapter 5.

$$(\text{conc. } NH_4^+) = \frac{0.125 \text{ mol } NH_4Cl}{500. \text{ mL}} \times \frac{1000 \text{ mL}}{1 \text{ L}} \times \frac{1 \text{ mol } NH_4^+}{1 \text{ mol } NH_4Cl} = 0.250 \text{ M } NH_4^+$$

Then, use the Henderson-Hasselbalch equation to find the pH in the equilibrium solution.

$$pH = pK_a + \log\left(\frac{[\text{conj. base}]}{[\text{conj. acid}]}\right)$$

In such calculations, we assume that the concentrations of conjugate acid and conjugate base are large enough to not be changed as the reaction equilibrium is established. This assumption is valid when the H_3O^+ concentration is small compared to the concentrations of conjugate acid and conjugate base.

Table 16.1 gives pK_a of ammonium/ammonia buffer as 9.25. The initial concentration of NH_3, the conjugate base, is 0.500 M.

$$pH = 9.25 + \log\left(\frac{0.500}{0.250}\right) = 9.55 \text{ before the HCl is added}$$

Assume that all the HCl gas bubbled through the solution is actually dissolved in the solution, find the initial concentration of H_3O^+ ions from the ionization of HCl, after it is dissolved but before it reacts:

$$(\text{conc. } H_3O^+) = \frac{0.0100 \text{ mol HCl}}{500. \text{ mL}} \times \frac{1000 \text{ mL}}{1 \text{ L}} \times \frac{1 \text{ mol } H_3O^+}{1 \text{ mol HCl}} = 0.0200 \text{ M } H_3O^+$$

The acid neutralizes the strongest base in the solution, NH_3. Write the product-favored neutralization equation:

$$NH_3(aq) + H_3O^+(aq) \longrightarrow NH_4^+(aq) + H_2O(\ell)$$

This reaction is product-favored, so we will first make as many products as possible. Using the method of limiting reactants, run the reaction towards products until one of the reactants runs out.

	$NH_3(aq)$	$H_3O^+(aq)$	$NH_4^+(aq)$
initial conc. (M)	0.500	0.0200	0.250
change as reaction occurs (M)	-0.0200	-0.0200	$+0.0200$
final conc. (M)	0.480	0	0.270

The solution is a still buffer solution, so, using the same technique, we can find the pH.

$$pH = 9.25 + \log\left(\frac{0.480}{0.270}\right) = 9.51 \text{ after the HCl is added}$$

✓ *Reasonable Answer Check:* The original buffer solution has a pH of 9.55. After adding some quantity of strong acid, it makes sense that the pH is slightly more acidic.

23. *Answer:* **Sample (b); see explanations below**

Strategy and Explanation: A 1-L solution of 0.20 M NaOH provides a strong base. The added solution must provide a source of a weak acid in sufficient quantity for some of it to completely neutralize the strong base and some of it to remain in the solution.

(a) Adding 0.10 mol CH_3COOH to the 1-L solution makes a 0.10 M CH_3COOH solution. The strong base neutralizes the acid in the solution, CH_3COOH.

$$CH_3COOH(aq) + OH^-(aq) \longrightarrow CH_3COO^-(aq) + H_2O(\ell)$$

This reaction is product-favored, so we will first make as many products as possible. Using the method of limiting reactants, run the reaction towards products until one of the reactants runs out.

	$CH_3COOH(aq)$	$OH^-(aq)$	$CH_3COO^-(aq)$
initial conc. (M)	0.10	0.20	0
change as reaction occurs (M)	-0.10	-0.10	$+0.10$
final conc. (M)	0.00	0.10	0.10

The solution produced is a not buffer solution. Too much base is present and all the CH_3COOH is neutralized.

(b) Adding 0.30 mol CH_3COOH to the 1-L solution makes a 0.30 M CH_3COOH solution. The strong base neutralizes the acid in the solution, as in (a). This reaction is product-favored, so we will first make as many products as possible. Using the method of limiting reactants, run the reaction towards products until one of the reactants runs out.

	$CH_3COOH(aq)$	$OH^-(aq)$	$CH_3COO^-(aq)$
initial conc. (M)	0.30	0.20	0
change as reaction occurs (M)	-0.20	-0.20	$+0.20$
final conc. (M)	0.10	0	0.20

The solution produced is a buffer solution, containing an acid-base conjugate pair.

(c) Adding a strong acid to a strong base will not produce a buffer. A buffer needs a weak acid-base conjugate pair in solution and this solution has neither.

(d) Adding a weak base to a strong base will not produce a buffer. A buffer needs a weak acid-base conjugate pair in solution and this solution has no significant acid concentration.

Therefore, of the four substances added to the base, only the addition of sample (b) produce buffer solutions.

25. *Answer:* **(a) ΔpH = 0.1 (b) ΔpH = 3.8 (c) ΔpH = 7.25**

Strategy and Explanation: Calculate the initial pH by finding the pK_a. Calculate initial OH^- concentration using the method from the solution to Question 42 in Chapter 5. Then adapt the method used in the answers to Question 23.

Equimolar buffer solutions have $pH = pK_a = -\log K_a$. So, $pH = -\log(1.8 \times 10^{-5}) = 4.74$

Calculate the concentration of hydroxide using the total volume of the solution after the addition, but before the reaction.

$$V = 0.100 \text{ L} \times \frac{1000 \text{ mL}}{1 \text{ L}} + 1.0 \text{ mL}$$

$$(\text{conc. } OH^-) = \frac{1.0 \text{ mL}}{101 \text{ mL}} \times \frac{1.0 \text{ mol } OH^-}{1 \text{ L}} = 0.0099 \text{ M } OH^-$$

The strong base neutralizes the acid in the solution, CH_3COOH.

$$CH_3COOH(aq) + OH^-(aq) \longrightarrow CH_3COO^-(aq) + H_2O(\ell)$$

This reaction is product-favored, so we will first make as many products as possible. Using the method of limiting reactants, run the reaction towards products until one of the reactants runs out.

(a)

	$CH_3COOH(aq)$	$OH^-(aq)$	$CH_3COO^-(aq)$
initial conc. (M)	0.10	0.0099	0.10
change as reaction occurs (M)	− 0.0099	− 0.0099	+ 0.0099
final conc. (M)	0.09	0	0.11

The solution is still a buffer solution, containing an acid-base conjugate pair, so we can find the pH using the Henderson-Hasselbalch equation. In Table 16.1, the pK_a is 4.74.

$$pH = 4.74 + \log\left(\frac{0.11}{0.09}\right) = 4.8$$

$$\Delta pH = 4.8 - 4.74 = 0.1$$

(b)

	$CH_3COOH(aq)$	$OH^-(aq)$	$CH_3COO^-(aq)$
initial conc. (M)	0.010	0.0099	0.010
change as reaction occurs (M)	− 0.0099	− 0.0099	+ 0.0099
final conc. (M)	0.000	0.0000	0.020

Within known significant figures (three decimal places), the solution is no longer a buffer solution. It contains only the conjugate base. We must find the pH using K_b of the base, as was done in Chapter 15, such as Question 38 in Chapter 15.

$$CH_3COO^-(aq) + H_2O(\ell) \rightleftharpoons CH_3COOH(aq) + OH^-(aq) \qquad K_b = \frac{[CH_3COOH][OH^-]}{[CH_3COO^-]}$$

As the reactants decompose, the concentrations of the products increase stoichiometrically, until they reach equilibrium concentrations.

	$CH_3COO^-(aq)$	$CH_3COOH(aq)$	$OH^-(aq)$
initial conc. (M)	0.020	0	0
change as reaction occurs (M)	− x	+ x	+ x
equilibrium conc. (M)	0.020 − x	x	x

At equilibrium $\qquad K_b = \dfrac{(x)(x)}{(0.020 - x)} = 5.6 \times 10^{-10}$

Assume x is very small and does not affect the difference.

$$5.6 \times 10^{-10} = \dfrac{x^2}{(0.020)}$$

$$x^2 = (5.6 \times 10^{-10})(0.20)$$

$$x = 3.3 \times 10^{-6} \text{ M} = [OH^-]$$

$$pOH = -\log[OH^-] = -\log(3.3 \times 10^{-6}) = 5.48$$

$$pH = 14.00 - pOH = 8.52$$

$$\Delta pH = 8.52 - 4.74 = 3.8$$

Notice: Even if we could ignore the limitation of the significant figures, we should not use the Henderson-Hassalbalch equation for such low acid concentrations.

(c) In this situation, the acetic acid is the limiting reactant:

	CH_3COOH(aq)	OH^-(aq)	CH_3COO^- (aq)
initial conc. (M)	0.0010	0.0099	0.0010
change as reaction occurs (M)	− 0.0010	− 0.0010	+ 0.0010
final conc. (M)	0.0000	0.0089	0.0020

The solution is no longer a buffer solution. This time, some OH^- ions are left over. The minor amount produced by the weak base reaction will not change this value, so:

$$pH = 14.00 + \log[OH^-] = 14.00 + \log(0.0089) = 11.95$$

$$\Delta pH = 11.95 - 4.74 = 7.25$$

27. *Answer:* **(a) 5.02 (b) 4.99 (c) 4.06**

Strategy and Explanation: Adapt the method described in the answers to Questions 21, 23, and 25.

Use the abbreviation HProp for the monoprotic propanoic acid. The salt sodium propanoate contains the Prop⁻ ion.

(a) Find the pH using Henderson-Hassalbalch. The K_a of propanoic acid is given: 1.4×10^{-5}. As described in Section 16.1, calculate the pK_a:

$$pK_a = -\log K_a = -\log(1.4 \times 10^{-5}) = 4.85$$

The concentration of the conjugate acid, HProp, is 0.20 M; the concentration of the conjugate base, Prop⁻, is 0.30 M.

$$pH = 4.85 + \log\left(\dfrac{0.30}{0.20}\right) = 4.85 + 0.18 = 5.02$$

(b) Calculate the concentration of hydrogen ion, using the total volume of the solution after the addition, but before the reaction.

$$V = 0.010 \text{ L} \times \dfrac{1000 \text{ mL}}{1 \text{ L}} + 1.0 \text{ mL} = 11 \text{ mL}$$

$$\text{(conc. } H_3O^+\text{)} = \dfrac{1.0 \text{ mL}}{11 \text{ mL}} \times \dfrac{1.0 \text{ mol } H_3O^+}{1 \text{ L}} = 0.0091 \text{ M } H_3O^+$$

The strong acid neutralizes the base in the solution, Prop⁻.

$$Prop^-\text{(aq)} + H_3O^+\text{(aq)} \longrightarrow HProp + H_2O(\ell)$$

This reaction is product-favored, so we will first make as many products as possible. Using the method of limiting reactants, run the reaction towards products until one of the reactants runs out.

	Prop⁻ (aq)	H₃O⁺(aq)	HProp(aq)
initial conc. (M)	0.30	0.0091	0.20
change as reaction occurs (M)	− 0.0091	− 0.0091	+ 0.0091
final conc. (M)	0.29	0	0.21

The solution is still a buffer solution, containing an acid-base conjugate pair, so determine the pH as in (a).

$$pH = 4.85 + \log\left(\frac{0.29}{0.21}\right) = 4.85 + 0.14 = 4.99$$

(c) Calculate the concentration of hydrogen ion, using the total volume of the solution after the addition, but before the reaction.

$$V = 0.010 \text{ L} \times \frac{1000 \text{ mL}}{1 \text{ L}} + 3.0 \text{ mL} = 13 \text{ mL}$$

$$(\text{conc. } H_3O^+) = \frac{3.0 \text{ mL}}{13 \text{ mL}} \times \frac{1.0 \text{ mol } H_3O^+}{1 \text{ L}} = 0.23 \text{ M } H_3O^+$$

The strong acid neutralizes the base in the solution, Prop⁻.

$$Prop^- (aq) + H_3O^+(aq) \longrightarrow HProp + H_2O(\ell)$$

This reaction is product-favored, so we will first make as many products as possible. Using the method of limiting reactants, run the reaction towards products until one of the reactants runs out.

	Prop⁻ (aq)	H₃O⁺(aq)	HProp(aq)
initial conc. (M)	0.30	0.23	0.20
change as reaction occurs (M)	− 0.23	− 0.23	+ 0.23
final conc. (M)	0.07	0	0.43

The solution is still a buffer solution, containing an acid-base conjugate pair, so we can find pH as in (a).

$$pH = 4.85 + \log\left(\frac{0.07}{0.43}\right) = 4.85 - 0.79 = 4.06$$

Titrations and Titration Curves

30. *Answer:* **Best choices: (a) bromthymol blue (b) phenolphthalein (c) methyl red (d) bromthymol blue; see explanations below**

Strategy and Explanation: The color change needs to be near the pH of the equivalence point.

(a) The strong base, NaOH, titrated with a strong acid, HClO₄, has a neutral equivalence point. It would be best to choose bromthymol blue, which is shown changing color in Figure 16.5 at a pH near 7. In practice, any of them would be suitable, because of the extreme change in the pH of the solution very close to the equivalence point.

(b) The weak acid, CH₃COOH, titrated with a strong base, KOH, has a basic equivalence point, due to the presence of the weak base CH₃COO⁻ in the solution. It would be best to choose phenolphthalein, which is shown changing color in Figure 16.5 at a pH near 9.

(c) The weak base, NH₃, titrated with a strong acid, HBr, has an acidic equivalence point, due to the presence of the weak acid NH₄⁺ in the solution. It would be best to choose methyl red, which is shown changing color in Figure 16.8 at a pH near 5.

(d) The strong base, KOH, titrated with a strong acid, HNO₃, has a neutral equivalence point. It would be best to choose bromthymol blue, which is shown changing color in Figure 16.5 at a pH near 7. In practice, any of them would be suitable, because of the extreme change in the pH of the solution very close to the equivalence point.

32. *Answer:* **0.0253 M HCl**

Strategy and Explanation: Use the method described in the solution to Question 52 in Chapter 5.

$$\frac{22.6 \text{ mL Ba(OH)}_2}{25.00 \text{ mL HCl}} \times \frac{1 \text{ L Ba(OH)}_2}{1000 \text{ mL Ba(OH)}_2} \times \frac{1000 \text{ mL HCl}}{1 \text{ L HCl}}$$

$$\times \frac{0.0140 \text{ mol Ba(OH)}_2}{1 \text{ L Ba(OH)}_2} \times \frac{2 \text{ mol OH}^-}{1 \text{ mol Ba(OH)}_2} \times \frac{1 \text{ mol HCl}}{1 \text{ mol OH}^-} = 0.0253 \frac{\text{mol HCl}}{\text{L HCl}} = 0.0253 \text{ M HCl}$$

34. *Answer:* **93.6%**

Strategy and Explanation: Use the method described in the answers to Questions 50-52 in Chapter 5.

$$24.4 \text{ mL NaOH} \times \frac{1 \text{ L}}{1000 \text{ mL}} \times \frac{0.110 \text{ mol NaOH}}{1 \text{ L NaOH}} \times \frac{1 \text{ mol C}_6\text{H}_8\text{O}_6}{1 \text{ mol NaOH}} \times \frac{176.1238 \text{ g C}_6\text{H}_8\text{O}_6}{1 \text{ mol C}_6\text{H}_8\text{O}_6} = 0.473 \text{ g C}_6\text{H}_8\text{O}_6$$

$$\frac{0.473 \text{ g C}_6\text{H}_8\text{O}_6}{0.505 \text{ g capsule}} \times 100\% = 93.6\%$$

36. *Answer:* **(a) 29.2 mL HCl (b) 600. mL HCl (c) 1.20 L HCl (d) 2.7 mL HCl**

Strategy and Explanation: Use the method described in the answers to Questions 50-52 in Chapter 5.

(a) $25.0 \text{ mL KOH} \times \dfrac{0.175 \text{ mol KOH}}{1 \text{ L KOH}} \times \dfrac{1 \text{ mol HCl}}{1 \text{ mol KOH}} \times \dfrac{1 \text{ L HCl}}{0.150 \text{ mol HCl}} = 29.2 \text{ mL HCl}$

(b) $15.0 \text{ mL NH}_3 \times \dfrac{6.00 \text{ mol NH}_3}{1 \text{ L NH}_3} \times \dfrac{1 \text{ mol HCl}}{1 \text{ mol NH}_3} \times \dfrac{1 \text{ L HCl}}{0.150 \text{ mol HCl}} = 600. \text{ mL HCl}$

(c) $15.0 \text{ mL C}_3\text{H}_7\text{NH}_2 \times \dfrac{0.712 \text{ g C}_3\text{H}_7\text{NH}_2}{1 \text{ mL C}_3\text{H}_7\text{NH}_2} \times \dfrac{1 \text{ mol C}_3\text{H}_7\text{NH}_2}{59.1099 \text{ g C}_3\text{H}_7\text{NH}_2}$

$$\times \frac{1 \text{ mol HCl}}{1 \text{ mol C}_3\text{H}_7\text{NH}_2} \times \frac{1 \text{ L HCl}}{0.150 \text{ mol HCl}} = 1.20 \text{ L HCl}$$

(d) $40.0 \text{ mL Ba(OH)}_2 \times \dfrac{0.0050 \text{ mol Ba(OH)}_2}{1 \text{ L Ba(OH)}_2} \times \dfrac{2 \text{ mol OH}^-}{1 \text{ mol Ba(OH)}_2}$

$$\times \frac{1 \text{ mol HCl}}{1 \text{ mol OH}^-} \times \frac{1 \text{ L HCl}}{0.150 \text{ mol HCl}} = 2.7 \text{ mL HCl}$$

39. *Answer:* **(a) 0.824 (b) 1.30 (c) 3.8 (d) 7.000 (e) 10.2 (f) 12.48; see titration curve below**

Strategy and Explanation: Adapt the method shown in Problem-Solving Example 16.6. For each of the points, determine the limiting reactant and use the concentration of the excess reactant to calculate the pH.total

$$\text{volume of acid in liters} = 50.00 \text{ mL} \times \frac{1 \text{ L}}{1000 \text{ mL}} = 0.05000 \text{ L}$$

$$\text{original mol H}_3\text{O}^+ \text{ added} = 0.05000 \text{ L} \times \frac{0.150 \text{ mol H}_3\text{O}^+}{1 \text{ L}} = 0.00750 \text{ mol H}_3\text{O}^+$$

(a) This solution contains only 0.150 M HCl, which ionizes to make 0.150 M H_3O^+, so

$$\text{pH} = \log[\text{H}_3\text{O}^+] = -\log(0.150) = 0.824$$

(b) After the addition of the titrant, use the equation given a the beginning of Problem-Solving Example 16.6 to determine the equilibrium concentration of H_3O^+

$$[\text{H}_3\text{O}^+] = \frac{\text{original moles acid} - \text{total moles base added}}{\text{volume of acid (L)} + \text{volume of base (L)}}$$

$$\text{volume of base in liters} = 25.00 \text{ mL} \times \frac{1 \text{ L}}{1000 \text{ mL}} = 0.02500 \text{ L}$$

$$\text{total mol OH}^- \text{ added} = 0.02500 \text{ L} \times \frac{0.150 \text{ mol OH}^-}{1 \text{ L}} = 0.00375 \text{ mol OH}^-$$

$$[H_3O^+] = \frac{0.00750 \text{ mol} - 0.00375 \text{ mol}}{0.05000 \text{ L} + 0.02500 \text{ L}} = 0.0500 \text{ M}$$

$$pH = -\log[H_3O^+] = -\log(0.050) = 1.30$$

(c) Use the equation given in (b), again.   volume of base (L) = $49.9 \text{ mL} \times \dfrac{1 \text{ L}}{1000 \text{ mL}} = 0.0490 \text{ L}$

$$\text{total mol OH}^- \text{ added} = 0.0499 \text{ L} \times \frac{0.150 \text{ mol OH}^-}{1 \text{ L}} = 0.007485 \text{ mol OH}^- \approx 0.00749 \text{ mol OH}^-$$

$$[H_3O^+] = \frac{0.00750 \text{ mol} - 0.007485 \text{ mol}}{0.05000 \text{ L} + 0.0499 \text{ L}} = \frac{0.000015 \text{ mol}}{0.0999 \text{ L}} = 0.00015 \text{ M} \approx 0.0002 \text{ M } (1 \text{ sig fig})$$

$$pH = -\log[H_3O^+] = -\log(0.00015) = 3.82 \approx 3.8 \ (1 \text{ decimal place})$$

(d) Calculate volume and total moles of base: volume of base (L) = $50.00 \text{ mL} \times \dfrac{1 \text{ L}}{1000 \text{ mL}} = 0.05000 \text{ L}$

$$\text{total mol H}_3O^+ \text{ added} = 0.05000 \text{ L} \times \frac{0.150 \text{ mol H}_3O^+}{1 \text{ L}} = 0.00750 \text{ mol H}_3O^+$$

Total moles of H_3O^+ added is equal to the original moles OH^-, so the solution is neutral, and the pH = 7.000.

(e) At this point, the moles of base begin to exceed the moles of acid, so we adapt the expression we used in (b) for a solution with excess base to determine the equilibrium concentration of OH^-:

$$[OH^-] = \frac{\text{total moles base added} - \text{original moles acid}}{\text{volume of acid (L)} + \text{volume of base (L)}}$$

$$\text{volume of base (L)} = 50.1 \text{ mL} \times \frac{1 \text{ L}}{1000 \text{ mL}} = 0.0501 \text{ L}$$

$$\text{total mol OH}^- \text{ added} = 0.0501 \text{ L} \times \frac{0.150 \text{ mol OH}^-}{1 \text{ L}} = 0.007515 \text{ mol OH}^- \approx 0.00752 \text{ mol OH}^-$$

$$[OH^-] = \frac{0.007515 \text{ mol} - 0.00750 \text{ mol}}{0.05000 \text{ L} + 0.0501 \text{ L}} = \frac{0.000015 \text{ mol}}{0.1001 \text{ L}} = 0.00015 \text{ M}$$

$$pH = -\log[H_3O^+] = -\log(0.000015) = 3.8$$

$$pH = 14.00 - pOH = 14.00 - 3.8 = 10.2$$

(f) Use the equation given in (e) again.

$$\text{volume of base (L)} = 75.00 \text{ mL} \times \frac{1 \text{ L}}{1000 \text{ mL}} = 0.0750 \text{ L}$$

$$\text{total mol OH}^- \text{ added} = 0.0750 \text{ L} \times \frac{0.150 \text{ mol H}_3O^+}{1 \text{ L}} = 0.0113 \text{ mol H}_3O^+$$

$$[OH^-] = \frac{0.0113 \text{ mol} - 0.00750 \text{ mol}}{0.05000 \text{ L} + 0.0750 \text{ L}} = \frac{0.00038 \text{ mol}}{0.1250 \text{ L}} = 0.030 \text{ M}$$

$$pH = -\log[H_3O^+] = -\log(0.030) = 1.52$$

$$pH = 14.00 - pOH = 14.00 - 1.52 = 12.48$$

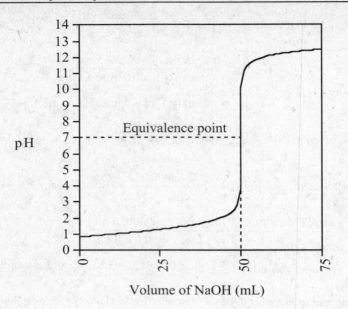

Volume of NaOH (mL)

Solubility Product

41. *Answer:* **see equations and expressions below**

Strategy and Explanation: Adapt the method developed in Problem-Solving Example 16.8.

(a) $FeCO_3(s) \rightleftharpoons Fe^{2+}(aq) + CO_3^{2-}(aq)$ $\qquad$ $K_{sp} = [Fe^{2+}][CO_3^{2-}]$

(b) $Ag_2SO_4(s) \rightleftharpoons 2\ Ag^+(aq) + SO_4^{2-}(aq)$ $\qquad$ $K_{sp} = [Ag^+]^2[SO_4^{2-}]$

(c) $Ca_3(PO_4)_2(s) \rightleftharpoons 3\ Ca^{2+}(aq) + 2\ PO_4^{3-}(aq)$ $\qquad$ $K_{sp} = [Ca^{2+}]^3[PO_4^{3-}]^2$

(d) $Mn(OH)_2(s) \rightleftharpoons Mn^{2+}(aq) + 2\ OH^-\ (aq)$ $\qquad$ $K_{sp} = [Mn^{2+}][OH^-]^2$

42. *Answer:* $\mathbf{K_{sp} = 2.2 \times 10^{-12}}$

Strategy and Explanation: Adapt the method developed in Problem-Solving Example 16.10.

Write the chemical equation and the equilibrium expression for the dissociation of the solute:

$$Ag_2CrO_4(s) \rightleftharpoons 2\ Ag^+(aq) + CrO_4^{2-}(aq) \qquad K_{sp} = [Ag^+]^2[CrO_4^{2-}]$$

At equilibrium, the moles of solid that dissolve per liter is:

$$\frac{2.7 \times 10^{-3}\ g\ Ag_2CrO_4}{100.\ mL} \times \frac{1\ mol\ Ag_2CrO_4}{331.7301\ g\ Ag_2CrO_4} \times \frac{1000\ mL}{1\ L} = 8.1 \times 10^{-5}\ M$$

The stoichiometry of the equation shows that the concentration of chromate ion is the same as the moles of solid that dissolve per liter, 8.1×10^{-5} M, and the fluoride ion concentration is twice that value, $2 \times (8.1 \times 10^{-5}$ M$) = 1.6 \times 10^{-4}$ M.

$$K_{sp} = (1.6 \times 10^{-4})^2(8.1 \times 10^{-5}) = 2.2 \times 10^{-12}$$

44. *Answer:* $\mathbf{K_{sp} = 1.7 \times 10^{-5}}$

Strategy and Explanation: Adapt the method developed in Problem-Solving Example 16.10.

Write the chemical equation and the equilibrium expression for the dissociation of the solute:

$$PbCl_2(s) \rightleftharpoons Pb^{2+} + 2\ Cl^-(aq) \qquad\qquad K_{sp} = [Pb^{2+}][Cl^-]^2$$

At equilibrium, the moles of solid that dissolve per liter is given as the solubility 1.62×10^{-2} M. The stoichiometry of the equation shows that the concentration of lead ion is the same as the moles of solid that dissolve per liter:

$$[Pb^{2+}] = 1.62 \times 10^{-2} \text{ M}$$

The chloride ion concentration is twice that value:

$$[Cl^-] = 2 \times (1.62 \times 10^{-2} \text{ M}) = 3.24 \times 10^{-2} \text{ M}$$

So,

$$K_{sp} = (1.62 \times 10^{-2})(3.24 \times 10^{-2})^2 = 1.70 \times 10^{-5}$$

Common Ion Effect

47. *Answer:* **3.1×10^{-5} mol/L**

Strategy and Explanation: Follow the method developed Problem-Solving Example 16.11.

The soluble Na_2SO_4 salt produces sulfate ions in the solution. Neglecting the reaction of sulfate as a base, (conc. SO_4^{2-}) = 0.010 mol/L.

	$SrSO_4(s) \rightleftharpoons$	$Sr^{2+}(aq)$	$+$	$SO_4^{2-}(aq)$
conc. initial (M)		0		0.010
change as reaction occurs (M)		+ S		+ S
equilibrium conc. (M)		S		0.010 + S

At equilibrium $K_{sp} = [Sr^{2+}][SO_4^{2-}] = (S)(0.010 + S) = 3.1 \times 10^{-7}$

Assuming S is small compared to 0.010, ignore its addition.

$$3.1 \times 10^{-7} = (S)(0.010)$$

$$S = 3.1 \times 10^{-5} \text{ mol/L}$$

The solubility of strontium sulfate in a 0.010 M solution of sodium sulfate is 3.1×10^{-5} mol/L.

49. *Answer:* **(a) 1×10^{-11} (b) 8.5×10^{-3} M or higher**

Strategy and Explanation: Follow the method developed in the answers to Questions 42 and 44.

(a) Write the chemical equation and the equilibrium expression for the dissociation of the solute:

$$Mg(OH)_2(s) \rightleftharpoons Mg^{2+}(aq) + 2 \text{ OH}^- (aq) \qquad K_{sp} = [Mg^{2+}][OH^-]^2$$

At equilibrium, the moles of solid that dissolve per liter are:

$$\frac{9 \text{ mg Mg(OH)}_2}{1 \text{ L}} \times \frac{1 \text{ g}}{1000 \text{ mg}} \times \frac{1 \text{ mol Mg(OH)}_2}{58.32 \text{ g Mg(OH)}_2} = 1.5 \times 10^{-4} \text{ M} \cong 2 \times 10^{-4} \text{ M} \text{ (1 sig. fig.)}$$

The stoichiometry of the equation shows that the concentration of magnesium ion is the same as the moles of solid that dissolve per liter, 1.5×10^{-4}, and the hydroxide ion concentration is two times that value, $2 \times (1.5 \times 10^{-4} \text{ M}) = 3.0 \times 10^{-4}$ M.

$$K_{sp} = (1.5 \times 10^{-4})(3.0 \times 10^{-4})^2 = 1.47 \times 10^{-11}$$

$$K_{sp} \cong 1 \times 10^{-11} \text{ (rounded to 1 sig. fig.)}$$

(b) First we determine the concentration of the iron(II) ion in a $1.0\mu g$ Fe^{2+} ion solution:

$$[Mg^{2+}] = \frac{5.0 \text{ μg Mg}^{2+}}{1 \text{ L}} \times \frac{10^{-6} \text{ g}}{1 \text{ μg}} \times \frac{1 \text{ mol Mg}^{2+}}{24.305 \text{ g Mg}^{2+}} = 2.1 \times 10^{-7} \text{ M}$$

Next, we use the calculated K_{sp} from (a) to calculate the hydroxide ion concentration in this solution:

$$K_{sp} = [Mg^{2+}][OH^-]^2$$

$$1.47 \times 10^{-11} = (2.1 \times 10^{-7})[OH^-]^2$$

$$[OH^-] = 8.5 \times 10^{-3} \text{ M}$$

An equilibrium $[OH^-]$ of 8.5×10^{-3} M or higher will keep the $[Mg^{2+}]$ at or below 1.0 µg/L.

Factors Affecting the Solubility of Sparingly Soluble Solutes

51. *Answer:* **4.5×10^{-9} M or lower**

Strategy and Explanation: Adapt the method developed in the solution to Question 47 and use the relationships provided before Question 24 in Chapter 15 to determine the hydroxide concentration from the pH.

Given the pH, we can determine the concentration of the aqueous ion, OH^-. The precipitation of sparingly soluble $Zn(OH)_2$ will occur above a certain concentration of Zn^{2+}.

$$[OH^-] = 10^{-pOH} = 10^{pH - 14.00} =$$

$$[OH^-] = 10^{10.00 - 14.00} = 1.0 \times 10^{-4} \text{ M}$$

$$K_{sp} = [Zn^{2+}][OH^-]^2 = 4.5 \times 10^{-17} \text{ (from Appendix H)}$$

$$4.5 \times 10^{-17} = [Zn^{2+}](1.0 \times 10^{-4})^2$$

$$[Zn^{2+}] = 4.5 \times 10^{-9} \text{ M}$$

An equilibrium $[Zn^{2+}]$ of 4.5×10^{-9} M or lower can exist in a solution with pH 10.00. Above that concentration, $Zn(OH)_2$ will precipitate.

53. *Answer:* **9.0**

Strategy and Explanation: Adapt the method developed in the solution to Question 51. We will use the K_{sp} derived for $Mg(OH)_2$ in part (a) of Question 49: $K_{sp} = 1 \times 10^{-11}$.

The acid added affects the concentration of the hydroxide. So, if all the solid is dissolved it must be acidic enough to hold all the magnesium ions without precipitation. So, we will calculate the resulting magnesium ion concentration:

$$[Mg^{2+}] = \frac{5.00 \text{ g } Mg(OH)_2}{1 \text{ L}} \times \frac{1 \text{ mol } Mg(OH)_2}{58.32 \text{ g } Mg(OH)_2} \times \frac{1 \text{ mol } Mg^{2+}}{1 \text{ mol } Mg(OH)_2} = 0.0857 \text{ M}$$

$$K_{sp} = [Mg^{2+}][OH^-]^2$$

$$1 \times 10^{-11} = (0.0857)[OH^-]^2$$

$$[OH^-] = 1 \times 10^{-5} \text{ M}$$

$$pOH = -\log[OH^-]$$

$$pOH = -\log(1 \times 10^{-5}) = 5.0$$

$$pH = 14.00 - pOH = 14.00 - 5.0 = 9.0$$

Enough acid must be added to drop the pH to 9.0, before all the solid will dissolve.

Complex Ion Formation

55. *Answer:* **see equations and expressions below**

Strategy and Explanation: The charge of the reactant metal ion is determined by subtracting the Lewis base's charge(s), if any, from the complex ion charge.

(a) $Ag^+ + 2\,CN^- \rightleftharpoons [Ag(CN)_2]^-$ $K_f = \dfrac{[[Ag(CN)_2]^-]}{[Ag^+][CN^-]^2}$

(b) $Cd^{2+} + 4\,NH_3 \rightleftharpoons [Cd(NH_3)_4]^{2+}$ $K_f = \dfrac{[[Cd(NH_3)_4]^{2+}]}{[Cd^{2+}][NH_3]^4}$

57. *Answer:* **7.8×10^{-3} mol or more**

Strategy and Explanation: Adapt the method described in the solution to Question 53, and Problem-Solving Example 16.12. The $Na_2S_2O_3$ salt provides a source of $S_2O_3^{2-}$, a Lewis base capable of forming a complex ion with the silver ion. If all the solid is dissolved it must be have enough $S_2O_3^{2-}$ to complex enough the silver ions to prevent the precipitation of AgBr in the resulting Br^- solution. First, we get the balanced equation for the reaction from page 601 and determine the value of its equilibrium constant using Tables 16.2 and 16.3:

$$AgBr(s) + 2\,S_2O_3^{2-}(aq) \rightleftharpoons [Ag(S_2O_3)_2]^{3-}(aq) + Br^-(aq) \qquad K = \frac{[[Ag(S_2O_3)_2]^{3-}][Br^-]}{[S_2O_3^{2-}]^2}$$

$$K = K_{sp} \times K_f = [Ag^+][Br^-] \times \frac{[[Ag(S_2O_3)_2]^{3-}]}{[Ag^+][S_2O_3^{2-}]^2} = \frac{[[Ag(S_2O_3)_2]^{3-}][Br^-]}{[S_2O_3^{2-}]^2}$$

$$K = K_{sp} \times K_f = (3.3 \times 10^{-13}) \times (2.0 \times 10^{13}) = 6.6$$

Now, calculate the $[Br^-]$ and $[[Ag(S_2O_3)_2]^{3-}]$, once all the AgBr has dissolved:

$$[Br^-] = \frac{0.020 \text{ mol AgBr}}{1.0 \text{ L}} \times \frac{1 \text{ mol Br}^-}{1 \text{ mol AgBr}} = 0.020 \text{ M},$$

Similarly, $[[Ag(S_2O_3)_2]^{3-}] = 0.020$ M

Now, we can calculate the necessary $[S_2O_3^{2-}]$: $6.6 = \dfrac{(0.020)(0.020)}{[S_2O_3^{2-}]^2}$

$$[S_2O_3^{2-}] = 7.8 \times 10^{-3} \text{ M}$$

One mole of $S_2O_3^{2-}$ is found in each mole of $Na_2S_2O_3$, so we must add 7.8×10^{-3} mol of $Na_2S_2O_3$ to this 1.0 L solution.

59. *Answer:* **see equations below**

Strategy and Explanation: The hydroxide anion of the solid is neutralized in excess acid, and the aqueous complex ion is formed in excess base.

(a) $Zn(OH)_2(s) + 2\,H_3O^+(aq) \longrightarrow Zn^{2+}(aq) + 4\,H_2O(\ell)$

          $Zn(OH)_2(s) + 2\,OH^-(aq) \longrightarrow [Zn(OH)_4]^{2-}(aq)$

(b) $Sb(OH)_3(s) + 3\,H_3O^+(aq) \longrightarrow Sb^{3+}(aq) + 6\,H_2O(\ell)$

          $Sb(OH)_3(s) + OH^-(aq) \longrightarrow [Sb(OH)_4]^-(aq)$

General Questions

61. *Answer:* **(a) H_2O, CH_3COO^-, Na^+, CH_3COOH, H_3O^+, OH^- (b) 4.95 (c) 5.05**
    (d) $CH_3COOH(aq) + H_2O(\ell) \rightleftharpoons H_3O^+(aq) + CH_3COO^-(aq)$

Strategy and Explanation:

(a) The solution contains an abundance of water, more sodium acetate than acetic acid, a small amount of H_3O^+ since the solution is an acidic buffer, and an even smaller amount of OH^-. Therefore, the ions and molecules in solution from the largest concentration to the smallest is: H_2O, CH_3COO^-, Na^+, CH_3COOH, H_3O^+, OH^-.

(b) Adapt the methods used in the answers to Question 25. The equation for the equilibrium and the equilibrium expression are:

$$CH_3COOH(aq) + H_2O(\ell) \rightleftharpoons H_3O^+(aq) + CH_3COO^-(aq) \qquad K_a = \frac{[H_3O^+][CH_3COO^-]}{[CH_3COOH]}$$

$$\frac{4.95 \text{ g NaCH}_3\text{COO}}{250. \text{ mL}} \times \frac{1 \text{ mol NaCH}_3\text{COO}}{82.0337 \text{ g NaCH}_3\text{COO}} \times \frac{1000 \text{ mL}}{1 \text{ L}} \times \frac{1 \text{ mol CH}_3\text{COO}^-}{1 \text{ mol NaCH}_3\text{COO}} = 0.241 \text{ M CH}_3\text{COO}^-$$

As the reactants decompose, the concentrations change stoichiometrically, until they reach equilibrium.

	$CH_3COOH(aq)$	$H_3O^+(aq)$	$CH_3COO^-(aq)$
initial conc. (M)	0.150	0	0.241
change as reaction occurs (M)	$-x$	$+x$	$+x$
equilibrium conc. (M)	$0.150 - x$	x	$0.241 + x$

At equilibrium $\qquad\qquad K_a = \dfrac{(x)(0.241 + x)}{(0.150 - x)} = 1.8 \times 10^{-5}$

Assume x is very small and does not affect the sum or the difference.

$$1.8 \times 10^{-5} = \frac{(x)(0.241)}{(0.150)}$$

$$x = 1.1 \times 10^{-5} \text{ M} = [H_3O^+]$$

$$pH = -\log[H_3O^+] = -\log(1.1 \times 10^{-5}) = 4.95$$

(c) The strong base neutralizes the acid in the solution, CH_3COOH.

$$CH_3COOH(aq) + OH^-(aq) \longrightarrow CH_3COO^-(aq) + H_2O(\ell)$$

$$(\text{conc. OH}^-) = \frac{80. \text{ mg NaOH}}{100. \text{ mL}} \times \frac{1000 \text{ mL}}{1 \text{ L}} \times \frac{1 \text{ g}}{1000 \text{ mg}} \times \frac{1 \text{ mol NaOH}}{40.00 \text{ g NaOH}} \times \frac{1 \text{ mol OH}^-}{1 \text{ mol NaOH}} = 0.020 \text{ M}$$

We take it as far to the products as possible, using the method of limiting reactants.

	$CH_3COOH(aq)$	$OH^-(aq)$	$CH_3COO^-(aq)$
initial conc. (M)	0.150	0.020	0.241
change as reaction occurs (M)	-0.020	-0.020	$+0.020$
conc. final (M)	0.130	0	0.261

The solution is still a buffer solution, containing an acid-base conjugate pair, so we can find the pH using K_a of the acid.

$$CH_3COOH(aq) + H_2O(\ell) \rightleftharpoons H_3O^+(aq) + CH_3COO^-(aq) \qquad\qquad K_a = \frac{[H_3O^+][CH_3COO^-]}{[CH_3COOH]}$$

As the reactants decompose, the concentrations change stoichiometrically, until they reach equilibrium.

	$CH_3COOH(aq)$	$H_3O^+(aq)$	$CH_3COO^-(aq)$
initial conc. (M)	0.130	0	0.261
change as reaction occurs (M)	$-x$	$+x$	$+x$
equilibrium conc. (M)	$0.130 - x$	x	$0.261 + x$

At equilibrium $\quad\quad\quad K_a = \dfrac{(x)(0.261+x)}{(0.130-x)} = 1.8 \times 10^{-5}$

Assume x is very small and does not affect the sum or the difference.

$$1.8 \times 10^{-5} = \dfrac{(x)(0.261)}{(0.130)} \quad\quad x = 9.0 \times 10^{-6}\ M = [H_3O^+]$$

$$pH = -\log[H_3O^+] = -\log(9.0 \times 10^{-6}) = 5.05$$

(d) $CH_3COOH(aq) + H_2O(\ell) \rightleftharpoons H_3O^+(aq) + CH_3COO^-(aq)$

62. *Answer:* **ratio = 1.6**

Strategy and Explanation: Adapt the methods used in the solution to Question 21.

$$pH = pK_a + \log\left(\dfrac{[o\text{-ethylbenzoate}]}{[o\text{-ethylbenzoic acid}]}\right)$$

At equilibrium, pH = 4.0 and pK_a = 3.79

$$pH - pK_a = \log\left(\dfrac{[o\text{-ethylbenzoate}]}{[o\text{-ethylbenzoic acid}]}\right)$$

$$4.0 - 3.79 = 0.2 = \log\left(\dfrac{[o\text{-ethylbenzoate}]}{[o\text{-ethylbenzoic acid}]}\right)$$

$$\dfrac{[o\text{-ethylbenzoate}]}{[o\text{-ethylbenzoic acid}]} = 10^{0.2} = 1.6 \cong 2 \text{ (must round to one sig. fig.)}$$

The [potassium o-ethylbenzoate] is approximately two times the [o-ethylbenzoic acid].

66. *Answer:* **3.5×10^{-6}**

Strategy and Explanation: Adapt the method used in Problem-Solving Example 16.7. We will use the equivalence point data to find the total moles of the weak acid, HA in the solution.

$$35.00\text{ mL NaOH} \times \dfrac{1\text{ L}}{1000\text{ L}} \times \dfrac{0.100\text{ mol NaOH}}{1\text{ L NaOH}} \times \dfrac{1\text{ mol HA}}{1\text{ mol NaOH}} = 0.00350\text{ mol HA}$$

Calculate the moles of hydroxide added, after 20.00 mL of titrant have been added.

$$\text{mol OH}^- = 20.00\text{ mL} \times \dfrac{1\text{ L}}{1000\text{ mL}} \times \dfrac{0.100\text{ mol OH}^-}{1\text{ L}} = 0.00200\text{ mol OH}^-$$

The strong base neutralizes the acid, HA, in the solution.

$$HA + OH^-(aq) \longrightarrow A^-(aq) + H_2O(\ell)$$

We take it as far to the products as possible, using the method of limiting reactants.

	Initial	After Reaction of HA with NaOH
mol A$^-$	0	0 + 0.00200 = 0.00200
mol HA	0.00350	0.00300 − 0.00200 = 0.00100

The equation for the equilibrium and the equilibrium expression are:

$$HA(aq) + H_2O(\ell) \rightleftharpoons H_3O^+(aq) + A^-(aq) \quad\quad K_a = \dfrac{[H_3O^+][A^-]}{[HA]}$$

We calculate the concentrations using the total volume after 20.0 mL of titrant has been added:

20.0 mL + 40.0 mL = 60.0 mL, which is 0.0600 L.

$$(\text{conc. A}^-) = \frac{0.00200 \text{ mol}}{0.0600 \text{ L}} = 0.0333 \text{ M} \qquad (\text{conc. HA}) = \frac{0.00100 \text{ mol}}{0.0600 \text{ L}} = 0.0167 \text{ M}$$

The solution is now a buffer solution, so use the Hendersen-Hasselbalch equation to calculate pK_a. The concentration of the conjugate acid, HA, is 0.0167 M; the concentration of the conjugate base, A^-, is 0.0333 M. The pH is 5.75.

$$pH = pK_a + \log\left(\frac{[\text{conj. base}]}{[\text{conj. acid}]}\right)$$

$$pK_a = 5.75 - \log\left(\frac{0.0333}{0.0167}\right) = 5.75 - 0.30 = 5.45$$

$$pK_a = -\log K_a, \text{ so } K_a = 10^{-pK_a} = 10^{-5.45} = 3.5 \times 10^{-6}$$

Applying Concepts

70. *Answer/Explanation:* The tiny amount of base (CH_3COO^-) present is insufficient to prevent the pH from changing dramatically if a strong acid is introduced into the solution.

72. *Answer:* **2.3×10^{-4}**

Strategy and Explanation: When exactly half of the acid in the original solution has been neutralized, that means that equal quantities of the acid and base are present in the solution. As described in the solution to Question 16, when equimolar quantities are present, $pH = pK_a$.

$$pH = 3.64 = pK_a$$

$$K_a = 10^{-pK_a} = 10^{-3.64} = 2.3 \times 10^{-4}$$

74. *Answer/Explanation:* Blood pH decreases because of an increase in H_2CO_3, which leads to acidosis, acidification of the blood.

75. *Answer/Explanation:* $Ca_5(PO_4)_3OH(s) \rightleftharpoons 5 Ca^{2+}(aq) + 3 PO_4^{3-}(aq) + OH^-(aq)$

(a) Apatite is a relatively insoluble compound. Drinking milk containing calcium ion increases the concentration of a product in the above reaction, and, according to LeChatelier's principle, drives the equilibrium toward the formation of apatite.

(b) Lactic acid will react with OH^-. The subsequent decrease in OH^- concentration will cause apatite to dissolve as equilibrium is re-established.

More Challenging Questions

76. *Answer:* **Sample A: $NaHCO_3$; Sample B: NaOH; Sample C: NaOH and Na_2CO_3 and/or $NaHCO_3$; Sample D: Na_2CO_3**

Strategy and Explanation: We must assume that any of these substances, if present, are present in "reasonable" concentrations.

Sample A: Phenolphthalein is colorless, suggesting the pH is below 8.3. The most acidic of the choices is $NaHCO_3$, and that is a likely guess for this sample's identity. (Notice: any of these compounds in solution, even NaOH, can have a pH below 8.3, if it is sufficiently dilute, that's why we have to assume reasonable concentrations.)

Sample B: We interpret the evidence to say that the methyl orange changed to its acidic color as soon as it was added. That means the solution became acidic as soon as the phenolphthalein end point was reached. That suggests the titration of a strong base with a very dramatic decline to acidic pH values after the equivalence point. In addition, at very low pH values (below pH = 3.01), bubbles would have been observed as the carbonate reacted to form $CO_2(g)$. All of these interpretations lead us to believe that this sample is probably NaOH.

Sample C: Presumably, both indicators changed color rapidly, suggesting that they were true endpoints, and not just a result of pH changes. Because two different acidic end points were reached with different indicators, this sample must be a mixture of excess strong base, NaOH, and one or more of the salts, either Na_2CO_3 or $NaHCO_3$. It is not possible to distinguish which salt is present, due to the immediate reaction of the strong base with $NaHCO_3$ to make the same product Na_2CO_3.

Sample D: This sample has a two endpoints the second at exactly twice the volume of the first, suggesting that two neutralizations occurred. That means the sample may have been pure Na_2CO_3, which undergoes two sequential neutralization reactions during the titration. Alternative scenarios, equally plausible, would be a mixture of equimolar quantities of NaOH and $NaHCO_3$ or a mixture of NaOH, Na_2CO_3, and $NaHCO_3$ with the following proportions:

$$(\text{conc. NaOH}) = (\text{conc. } NaHCO_3) = \frac{1}{2}(\text{conc. } Na_2CO_3)$$

Other proportions are also possible. Without some more data, or other limits on the quantities, no more definite answers are possible.

79. *Answer:* **3.22**

Strategy and Explanation: Distilled water with a pH of 5.6, has an H_3O^+ concentration.

$$[H_3O^+] = 10^{-pH} = 10^{-5.6} = 2.5 \times 10^{-6} \text{ M}$$

The acidic pH is a result of the following equlibria:

$$CO_2(aq) + H_2O(\ell) \rightleftharpoons H_2CO_3(aq)$$

$$H_2CO_3(aq) + H_2O(\ell) \rightleftharpoons H_3O^+(aq) + HCO_3^-(aq)$$

HCl is strong acid. It ionizes completely to produce H_3O^+. As a result, the equilibria reactions above shift the left in order to re-establish a new equilibrium state. That means the $H_3O^+(aq)$ concentration *resulting from the ionization of the weak acid* will be even less than that calculated, above.

Determine the $H_3O^+(aq)$ concentration resulting from the ionization of the HCl after one drop of concentrated HCl solution is added to the 1.0 L of water. Notice that the drop adds negligible volume to the 1.0 L solution.

$$\frac{1 \text{ drop}}{1.0 \text{ L}} \times \frac{1 \text{ mL}}{20 \text{ drop}} \times \frac{1 \text{ L}}{1000 \text{ mL}} \times \frac{12 \text{ mol HCl}}{1 \text{ L}} = 6.0 \times 10^{-4} \text{ M HCl}$$

It is clear that the addition of HCl produces so much H_3O^+ that the presence of aqueous CO_2 as an acid source is negligible. Ultimately, the pH depends only on the molarity of the HCl.

$$pH = -\log[H_3O^+] = -\log(6.0 \times 10^{-4}) = 3.22$$

81. *Answer:* **3.2 g glacial acetic acid**

Strategy and Explanation: Glacial acetic acid is pure liquid CH_3COOH. It contains no water.

Assuming that all 300 mL of 0.100 M sodium acetate is used to make the buffer, calculate the moles of CH_3COO^-.

$$300. \text{ mL} \times \frac{1 \text{ L}}{1000 \text{ mL}} \times \frac{0.100 \text{ mol } CH_3COONa}{1 \text{ L}} \times \frac{1 \text{ mol } CH_3COO^-}{1 \text{ mol } CH_3COONa} = 0.0300 \text{ mol}$$

The mole ratio between two solutes in a single solution is identical to their concentration ratio. Use this fact to adapt the Henderson-Hasselbalch equation (Section 16.1) to determine the moles of glacial acetic acid needed.

$$pH = pK_a + \log\left(\frac{[\text{conj. base}]}{[\text{conj. acid}]}\right)$$

The volume of the solution is constant for the acid and base solutions, so: $\dfrac{[\text{conj. base}]}{[\text{conj.acid}]} = \dfrac{\text{mol conj. base}}{\text{mol conj. acid}}$

Here
$$pH \doteq pK_a + \log\left(\frac{\text{mol CH}_3\text{COO}^-}{\text{mol CH}_3\text{COOH}}\right)$$

Solve the above equation for the moles of glacial acetic acid, x.

$$4.50 = 4.74 + \log\left(\frac{0.0300 \text{ mol}}{x}\right)$$

$$-0.24 = \log\left(\frac{0.0300 \text{ mol}}{x}\right)$$

$$\frac{0.0300 \text{ mol}}{x} = 10^{-0.24} = 0.57$$

$$x = \frac{0.0300 \text{ mol}}{0.57} = 0.053 \text{ mol CH}_3\text{COOH}$$

Use molar mass to determine the mass of glacial acetic acid:

$$0.053 \text{ mol CH}_3\text{COOH} \times \frac{60.0519 \text{ g}}{1 \text{ mol}} = 3.2 \text{ g CH}_3\text{COOH}$$

Notice: A small volume change does occur:

$$3.2 \text{ g} \times \frac{1 \text{ mL}}{1.05 \text{ g}} = 3.05 \text{ mL}$$

But the volume change affects both the numerator and the denominator of the concentration ratio. So, it is not necessary to use the given density of the glacial acetic acid.

Chapter 17: Thermodynamics: Directionality of Chemical Reactions

Introduction

The focus of Chapter 17 is the direction of chemical reactions. We use the terms product-favored reaction and reactant-favored reaction to relate to the position of equilibrium of reversible reactions. Sometimes these reactions do not favor either side of the reaction very much, so keep in mind that some equilibrium mixtures will contain sizeable amounts of both products and reactants.

Some simple errors to watch for when doing thermodynamic calculations are caused by units. The units of ΔH_f°, and ΔG_f° are kJ/mol while the units of S° are J/mol. Remember, when adding or subtracting terms, such as in the equation: $\Delta G = \Delta H - T\Delta S$, the units must be converted so that they are all the same before adding or subtracting.

Pre-professional health career majors, especially, must pay attention to the applications of Gibbs free energy to biological systems. Many biology courses will assume you have learned this material in depth in a chemistry course.

Solutions to Blue-Numbered Questions for Review and Thought for Chapter 17

Topical Questions

Reactant-Favored and Product-Favored Processes

10. *Answer:* **(a) 2 H$_2$O(ℓ) ⟶ 2 H$_2$(g) + O$_2$(g); reactant-favored (b) C$_8$H$_{18}$(ℓ) ⟶ C$_8$H$_{18}$(g); product-favored (c) C$_{12}$H$_{22}$O$_{11}$(s) ⟶ C$_{12}$H$_{22}$O$_{11}$(aq); product-favored**

Strategy and Explanation: Think about the process and whether you have observed it happening spontaneously. If it has, it will be classified as product-favored. If it has not, it will be classified as reactant-favored.

(a) Water decomposing into its elements. $2\ H_2O(\ell) \longrightarrow 2\ H_2(g) + O_2(g)$.

   The decomposition of water has not been observed to occur spontaneously at normal temperature and pressure, thus this reaction is classified as reactant-favored.

(b) Gasoline spills and evaporates, changing from liquid to gas phase: $C_8H_{18}(\ell) \longrightarrow C_8H_{18}(g)$

   The process has been observed to occur spontaneously at normal temperature and pressure, thus this reaction is classified as product-favored.

(c) Dissolving sugar at room temperature, changing it from solid to aqueous phase:

$$C_{12}H_{22}O_{11}(s) \longrightarrow C_{12}H_{22}O_{11}(aq)$$

   The process has been observed to occur spontaneously at normal temperature and pressure, thus this reaction is classified as product-favored.

Chemical Reactions and Dispersal of Energy

12. *Answer:* (a) $\frac{1}{2}$ (b) $\frac{1}{2}$ (c) **50 "heads" and 50 "tails"**

Strategy and Explanation: Determine the total number of possible outcomes and use that to build a fraction describing the probability of obtaining one outcome.

(a) "Heads" is one of two possible outcomes, so the probability is $\frac{1}{2}$.

(b) "Tails" is one of two possible outcomes, so the probability is $\frac{1}{2}$.

(c) If you flip a coin 100 times, you get 100 outcomes – each outcome has an equal probability, $\frac{1}{2}$, of being "heads" or "tails". The most likely number of "heads" or "tails" would be half of the total number of flips, or 50 "heads" and 50 "tails".

14. *Answer:* (a) $\frac{1}{2}$ **in A ;** $\frac{1}{2}$ **in B (b) 50 molecules in flask A and 50 molecules in flask B is most probably arrangement and has the highest entropy.**

Strategy and Explanation: Determine the total number of possible outcomes and use that to build a fraction describing the probability of obtaining one outcome. We will assume that the connecting tube between the two flasks is so short than there is no probability that the molecules will actually be found there.

(a) The probability of the molecule being in flask A is one of two possible outcomes, so the probability is $\frac{1}{2}$.

The probability of the molecule being in flask B is one of two possible outcomes, so the probability is $\frac{1}{2}$.

(b) If you put 100 molecules in the two-flask system, the most likely arrangement of the molecules in flask A and flask B would be half of the molecules in each, or 50 in flask A and 50 in flask B. This 50-50 arrangement has the highest entropy, because the molecules have a higher disorder than any other state that would crowd more than 50 atoms into one flask or the other.

Measuring Dispersal of Energy: Entropy

16. *Answer:* **(a) negative (b) positive (c) positive**

Strategy and Explanation: Use the qualitative guidelines for entropy changes described in Section 17.3.

(a) $H_2O(g) \longrightarrow H_2O(s)$

The solid product has lower entropy than the gas-phase reactant, so the entropy change is negative.

(b) $CO_2(aq) \longrightarrow CO_2(g)$

The gas-phase product has higher entropy than the aqueous reactant, so the entropy change is positive.

(c) $glass(s) \longrightarrow glass(\ell)$

The liquid product has higher entropy than the solid reactant, so the entropy change is positive.

18. *Answer:* **(a) Item 2 (b) Item 2 (c) Item 2**

Strategy and Explanation: Use the qualitative guidelines for entropy changes described in Section 17.3.

(a) Item 2 has higher entropy since it is identical to item 1 except that its temperature is higher. Molecules at higher temperature have higher entropy.

(b) Item 2, dissolved sugar, has higher entropy than item 1, solid sugar, because solute molecules are more random than those in a solid crystal.

(c) Item 2, the mixture of water and alcohol together has higher entropy than item 1, water and alcohol separate. Mixing makes the molecules more random.

20. *Answer:* **(a) NaCl (b) $P_4(g)$ (c) $NH_4NO_3(aq)$**

Strategy and Explanation: Use the methods described in Problem-Solving Example 17.2 and the qualitative guidelines for entropy changes described in Section 17.3.

(a) Comparing NaCl and CaO, we find that the biggest difference between these two ionic solids is the attractions due to the charges on the ions. According to Coulomb's law, the Ca^{2+} and O^{2-} ions have greater interaction than the Na^+ and Cl^- ions, so NaCl has a larger entropy/mol than CaO.

(b) P_4 molecules have more atoms than Cl_2 molecules, so P_4 has a larger entropy per mol than Cl_2.

(c) The solid NH_4NO_3 crystal is more ordered than the aqueous NH_4^+ and NO_3^- ions, so the aqueous NH_4NO_3 has a larger entropy per mol than the solid NH_4NO_3.

22. *Answer:* **(a) negative (b) negative (c) positive**

Strategy and Explanation: Use the qualitative guidelines for entropy changes described in Section 17.3.

(a) The reaction has more gas-phase reactants (3 mol) than gas-phase products (2 mol), so the entropy change will be negative.

(b) The reaction has more gas-phase reactants (3 mol) than gas-phase products (0 mol), so the entropy change will be negative.

(c) The reaction has fewer gas-phase reactants (2 mol) than gas-phase products (3 mol), so the entropy change will be positive.

Calculating Entropy Changes

24. *Answer:* **112 J $K^{-1}mol^{-1}$**

Strategy and Explanation: Rearrange Equation 17.5: $T = \dfrac{\Delta H^\circ}{\Delta S^\circ}$ (when $\Delta G = 0$) to solve for ΔS°.

Use the enthalpy of vaporization and the normal boiling point of ethanol:

$$\Delta S^\circ = \frac{\Delta H^\circ}{T} = \frac{\left(39.3 \ \frac{kJ}{mol}\right) \times \left(\frac{1000 \ J}{1 \ kJ}\right)}{(78.3 + 273.15)K} = 112 \ J \ K^{-1}mol^{-1}$$

26. *Answer:* **(a) 113.0 J/K (b) 38.17 kJ**

Strategy and Explanation:

(a) Adapt the method described in Problem-Solving Example 17.3. Write the balanced chemical equation and use the equation from the beginning of Section 17.4:

$$\Delta S^\circ = \sum \left[(\text{moles of product}) \times S^\circ(\text{product})\right] - \sum \left[(\text{moles of reactant}) \times S^\circ(\text{reactant})\right]$$

Vaporization equation: $CH_3OH(\ell) \longrightarrow CH_3OH(g)$

$$\Delta S^\circ = (1 \text{ mole}) \times S^\circ(CH_3OH(g)) - (1 \text{ mole}) \times S^\circ(CH_3OH(\ell))$$

$$\Delta S^\circ = (1 \text{ mole}) \times (239.81 \ J \ K^{-1}mol^{-1}) - (1 \text{ mole}) \times (126.8 \ J \ K^{-1}mol^{-1}) = 113.0 \ J \ K^{-1}$$

(b) Adapt the method described in the solution to Question 24.

$$\Delta H^\circ = T\Delta S^\circ = (64.6 + 273.15) \ K \times (113.0 \ J \ K^{-1}) \times \frac{1 \ kJ}{1000 \ J} = 38.17 \ kJ$$

28. *Answer:* **(a) −173.01 J/K; negative ΔS prediction confirmed (b) −326.69 J/K; negative ΔS prediction confirmed (c) 137.55 J/K; positive ΔS prediction confirmed**

Strategy and Explanation: Adapt the method described in the solution to Question 26(a).

(a) $\Delta S^\circ = (2 \text{ mol}) \times S^\circ(CO_2) - (2 \text{ mol}) \times S^\circ(CO) - (1 \text{ mol}) \times S^\circ(O_2)$

$= (2 \text{ mol}) \times (213.74 \ J \ K^{-1}mol^{-1}) - (2 \text{ mol}) \times (197.674 \ J \ K^{-1}mol^{-1})$

$- (1 \text{ mol}) \times (205.138 \ J \ K^{-1}mol^{-1}) = -173.01 \ J \ K^{-1}$

This confirms the negative ΔS prediction from Question 22.

(b) $\Delta S^\circ = (2 \text{ mol}) \times S^\circ(H_2O(\ell)) - (2 \text{ mol}) \times S^\circ(H_2) - (1 \text{ mol}) \times S^\circ(O_2)$

$$= (2 \text{ mol}) \times (69.91 \text{ J K}^{-1}\text{mol}^{-1}) - (2 \text{ mol}) \times (130.684 \text{ J K}^{-1}\text{mol}^{-1})$$
$$- (1 \text{ mol}) \times (205.138 \text{ J K}^{-1}\text{mol}^{-1}) = -326.69 \text{ J K}^{-1}$$

This confirms the negative ΔS prediction from Question 22.

(c) $\Delta S° = (3 \text{ mol}) \times S°(O_2) - (2 \text{ mol}) \times S°(O_3)$

$$= (3 \text{ mol}) \times (205.138 \text{ J K}^{-1}\text{mol}^{-1}) - (2 \text{ mol}) \times (238.93 \text{ J K}^{-1}\text{mol}^{-1}) = 137.55 \text{ J K}^{-1}$$

This confirms the positive ΔS prediction from Question 22.

Entropy and the Second Law of Thermodynamics

30. *Answer:* $\Delta S° = -247.7 \text{ J K}^{-1}$; **no, we need $\Delta H°$ also; the reaction is product-favored.**

Strategy and Explanation: Adapt the method described in the solution to Question 26(a).

$\Delta S° = (1 \text{ mol}) \times S°(C_2H_5OH) - (1 \text{ mol}) \times S°(C_2H_4) - (1 \text{ mol}) \times S°(H_2O(g))$

$$= (1 \text{ mol}) \times (160.7 \text{ J K}^{-1}\text{mol}^{-1}) - (1 \text{ mol}) \times (219.56 \text{ J K}^{-1}\text{mol}^{-1})$$
$$- (1 \text{ mol}) \times (188.825 \text{ J K}^{-1}\text{mol}^{-1}) = -247.7 \text{ J K}^{-1}$$

We cannot tell from the results of this calculation whether this reaction is product-favored. We need the $\Delta H°$, also. Use Equation 6.11 on page 211:

$$\Delta H° = \sum \left[(\text{moles of product}) \times \Delta H_f°(\text{product}) \right] - \sum \left[(\text{moles of reactant}) \times \Delta H_f°(\text{reactant}) \right]$$

$\Delta H° = (1 \text{ mol}) \times \Delta H_f°(C_2H_5OH) - (1 \text{ mol}) \times \Delta H_f°(C_2H_4) - (1 \text{ mol}) \times \Delta H_f°(H_2O(g))$

$$= (1 \text{ mol}) \times (-277.69 \text{ kJ/mol}) - (1 \text{ mol}) \times (52.26 \text{ kJ/mol}) - (1 \text{ mol}) \times (-241.818 \text{ kJ/mol}) = -88.13 \text{ kJ}$$

Since, both $\Delta H°$ and $\Delta S°$ are negative, the product-favored nature of the reaction depends on temperature.

$$T = \frac{(-88.13 \text{ kJ}) \times \left(\dfrac{1000 \text{ J}}{1 \text{ kJ}} \right)}{-247.7 \text{ J K}^{-1}} = 355.8 \text{ K}$$

When 355.8 K or lower, the reaction is product-favored. The temperature here is 25 °C + 273.15 = 298 K, so the reaction is product-favored.

32. *Answer:* **low temperatures**

Strategy and Explanation: Use the qualitative guidelines for entropy changes described in Section 17.3, then consult Table 17.2.

The reaction has more gas-phase reactants (1 mol) than gas-phase products (0 mol), so the entropy change will be negative. When $\Delta H°$ and $\Delta S°$ are both negative, the reaction is product-favored only at low temperatures. The exothermicity is sufficient to favor products, if the temperature is low enough to overcome the decrease in entropy.

34. *Answer:* **(a) $\Delta S°$ is positive; $\Delta H°$ is negative. (b) $\Delta H° = -184.28 \text{ kJ}$, which is negative as predicted; $\Delta S° = -7.7 \text{ J/K}$ which is not positive as predicted, but it is very small.**

Strategy and Explanation:

(a) The products of this reaction have a larger number of gas-phase products (1 mol) than gas-phase reactants (0 mol), so we predict the entropy change, $\Delta S°$, to be positive. Sodium reacts violently, suggesting that a great amount of heat is produced in the reaction. We will predict the enthalpy change, $\Delta H°$, is negative.

(b) Adapt the method described in the solution to Question 26(a).

$\Delta H° = (1 \text{ mol}) \times \Delta H_f°(NaOH) + (\frac{1}{2} \text{ mol}) \times \Delta H_f°(H_2) - (1 \text{ mol}) \times \Delta H_f°(Na) - (1 \text{ mol}) \times \Delta H_f°(H_2O(\ell))$

$$= (1 \text{ mol}) \times (-470.114 \text{ kJ/mol}) + (\tfrac{1}{2} \text{ mol}) \times (0 \text{ kJ/mol})$$
$$- (1 \text{ mol}) \times (0 \text{ kJ/mol}) - (1 \text{ mol}) \times (-285.83 \text{ kJ/mol}) = -184.28 \text{ kJ}$$

$$\Delta S° = (1 \text{ mol}) \times S°(NaOH) + (\tfrac{1}{2} \text{ mol}) \times S°(H_2) - (1 \text{ mol}) \times \Delta S°(Na) - (1 \text{ mol}) \times \Delta S°(H_2O(\ell))$$

$$= (1 \text{ mol}) \times (48.1 \text{ J K}^{-1}\text{mol}^{-1}) + (\tfrac{1}{2} \text{ mol}) \times (130.684 \text{ J K}^{-1}\text{mol}^{-1})$$

$$- (1 \text{ mol}) \times (51.21 \text{ J K}^{-1}\text{mol}^{-1}) - (1 \text{ mol}) \times (69.91 \text{ J K}^{-1}\text{mol}^{-1}) = -7.7 \text{ J K}^{-1}$$

The enthalpy change is negative, as predicted in (a); however, the entropy change is negative, not as predicted in (a). Looking at the values of S°, the aqueous solute has lower entropy than pure water, providing sufficient order to compensate for the higher disorder of the gas. The value of -7.7 J K^{-1} is pretty small.

36. *Answer:* **$\Delta S°$ is positive; $\Delta H°$ is positive. The reaction is reactant-favored. See explanation below.**

Strategy and Explanation: The equation for the chemical reaction is: $H_2O(\ell) \longrightarrow H_2(g) + \tfrac{1}{2} O_2(g)$

The reaction written here is the reverse of the highly exothermic reaction described in the first sentence of the Question, so this reaction is highly endothermic, with a positive enthalpy change ($\Delta H°$). With more gas products ($1\tfrac{1}{2}$ mol) than gas reactants (0 mol), the entropy change ($\Delta S°$) is positive. Because we do not see water spontaneously decomposing, we will conclude that the reaction is reactant-favored. The entropy increase is insufficient to drive this highly endothermic reaction to form products without assistance from the surroundings at the temperature of 25 °C.

38. *Answer:* **(a) -851.5 kJ; -37.52 J/K; product-favored at low temperatures (b) 66.36 kJ; -21.77 J/K; never product-favored**

Strategy and Explanation: Adapt the method described in the solution to Question 30 and use Table 17.2.

(a) $\Delta H° = (2 \text{ mol}) \times \Delta H_f°(Fe) + (1 \text{ mol}) \times \Delta H_f°(Al_2O_3) - (1 \text{ mol}) \times \Delta H_f°(Fe_2O_3) - (2 \text{ mol}) \times \Delta H_f°(Al)$

$$= (2 \text{ mol}) \times (0 \text{ kJ/mol}) + (1 \text{ mol}) \times (-1675.7 \text{ kJ/mol})$$
$$- (1 \text{ mol}) \times (-824.2 \text{ kJ/mol}) - (2 \text{ mol}) \times (0 \text{ kJ/mol}) = -851.5 \text{ kJ}$$

$\Delta S° = (2 \text{ mol}) \times S°(Fe) + (1 \text{ mol}) \times S°(Al_2O_3) - (1 \text{ mol}) \times S°(Fe_2O_3) - (2 \text{ mol}) \times S°(Al)$

$$= (2 \text{ mol}) \times (27.78 \text{ J K}^{-1}\text{mol}^{-1}) + (1 \text{ mol}) \times (50.92 \text{ J K}^{-1}\text{mol}^{-1})$$
$$- (1 \text{ mol}) \times (87.40 \text{ J K}^{-1}\text{mol}^{-1}) - (2 \text{ mol}) \times (28.3 \text{ J K}^{-1}\text{mol}^{-1}) = -37.52 \text{ J K}^{-1}$$

The reaction is exothermic, but its entropy change is negative, so it would stop being product-favored above a specific temperature. (Calculated from Equation 17.5, this temperature is 22,700 K.)

(b) $\Delta H° = (2 \text{ mol}) \times \Delta H_f°(NO_2) - (1 \text{ mol}) \times \Delta H_f°(N_2) - (2 \text{ mol}) \times \Delta H_f°(O_2)$
$$= (2 \text{ mol}) \times (33.18 \text{ kJ/mol}) - (1 \text{ mol}) \times (0 \text{ kJ/mol}) - (2 \text{ mol}) \times (0 \text{ kJ/mol}) = 66.36 \text{ kJ}$$

$\Delta S° = (2 \text{ mol}) \times S°(NO_2) - (1 \text{ mol}) \times S°(N_2) - (2 \text{ mol}) \times S°(O_2)$

$$= (2 \text{ mol}) \times (240.06 \text{ J K}^{-1}\text{mol}^{-1}) - (1 \text{ mol}) \times (191.61 \text{ J K}^{-1}\text{mol}^{-1})$$
$$- (2 \text{ mol}) \times (205.138 \text{ J K}^{-1}\text{mol}^{-1}) = -21.77 \text{ J K}^{-1}$$

The reaction is endothermic; the entropy change is negative, so it will never be product-favored.

Gibbs Free Energy

40. *Answer:* **(a) $\Delta S_{universe} = 4.92 \times 10^3$ J/K (b) $\Delta G°_{system} = -1.47 \times 10^3$ kJ and $\Delta G°_{system} = -T\Delta S_{universe}$ (c) Yes, since ethane is used as a fuel.**

Strategy and Explanation: Adapt the method described in Problem-Solving Example 17.4, using three equations:

Equation 17.2 with the Kelvin temperature: $\quad \Delta S_{universe} = -\dfrac{\Delta H°_{system}}{T} + \Delta S°_{system}$

$$T(K) = T(°C) + 273.15 = 25 °C + 273.15 = 298 \text{ K}$$

Equation 17.3 $\quad\quad\quad\quad\quad\quad\quad\quad \Delta G°_{system} = \Delta H°_{system} - T\Delta S°_{system}$

and Equation 17.4, as described in Problem-Solving Example 17.5.

$$\Delta G° = \sum\left[(\text{moles of product}) \times \Delta G_f°(\text{product})\right] - \sum\left[(\text{moles of reactant}) \times \Delta G_f°(\text{reactant})\right]$$

(a) $\Delta H°_{system} = (2 \text{ mol}) \times \Delta H_f°(CO_2) + (3 \text{ mol}) \times \Delta H_f°(H_2O(\ell)) - (1 \text{ mol}) \times \Delta H_f°(C_2H_6) - (\frac{7}{2} \text{ mol}) \times \Delta H_f°(O_2)$

$= (2 \text{ mol}) \times (-393.509 \text{ kJ/mol}) + (3 \text{ mol}) \times (-285.83 \text{ kJ/mol})$

$$- (1 \text{ mol}) \times (-84.68 \text{ kJ/mol}) - (\frac{7}{2} \text{ mol}) \times (0 \text{ kJ/mol}) = -1559.83 \text{ kJ}$$

$\Delta S°_{system} = (2 \text{ mol}) \times S°(CO_2) + (3 \text{ mol}) \times S°(H_2O(\ell)) - (1 \text{ mol}) \times S°(C_2H_6) - (\frac{7}{2} \text{ mol}) \times S°(O_2)$

$= (2 \text{ mol}) \times (213.74 \text{ J K}^{-1}\text{mol}^{-1}) + (3 \text{ mol}) \times (69.91 \text{ J K}^{-1}\text{mol}^{-1})$

$$- (1 \text{ mol}) \times (229.60 \text{ J K}^{-1}\text{mol}^{-1}) - (\frac{7}{2} \text{ mol}) \times (205.138 \text{ J K}^{-1}\text{mol}^{-1}) = -310.37 \text{ J K}^{-1}$$

$$\Delta S_{universe} = -\frac{(-1559.83 \text{ kJ})}{(298 \text{ K})} \times \frac{1000 \text{ J}}{1 \text{ kJ}} + (-310.37 \text{ J K}^{-1}) = 4.92 \times 10^3 \text{ J K}^{-1}$$

(b) $\Delta G°_{system} = \Delta H°_{system} - T\Delta S°_{system} = (-1559.83 \text{ kJ}) - (298 \text{ K})(-310.37 \text{ J K}^{-1}) \times \frac{1 \text{ kJ}}{1000 \text{ J}} = -1467.3 \text{ kJ}$

Independently, we can get $\Delta G°_{system}$ using Equation 17.4:

$\Delta G°_{system} = (2 \text{ mol}) \times \Delta G_f°(CO_2) + (3 \text{ mol}) \times \Delta G_f°(H_2O(\ell)) - (1 \text{ mol}) \times \Delta G_f°(C_2H_6) - (\frac{7}{2} \text{ mol}) \times \Delta G_f°(O_2)$

$= (2 \text{ mol}) \times (-394.359 \text{ kJ/mol}) + (3 \text{ mol}) \times (-237.129 \text{ kJ/mol})$

$$- (1 \text{ mol}) \times (-32.82 \text{ kJ/mol}) - (\frac{7}{2} \text{ mol}) \times (0 \text{ kJ/mol}) = -1467.28 \text{ kJ}$$

A negative $\Delta G°_{system}$ is consistent with a positive $\Delta S_{universe}$. In addition:

$$- T\Delta S_{universe} = (298 \text{ K})(4.92 \times 10^3 \text{ J K}^{-1}) \times \frac{1 \text{ kJ}}{1000 \text{ J}} = -1.47 \times 10^3 \text{ kJ} = \Delta G°_{system}.$$

(c) Yes. Ethane is used as a fuel; hence, we would expect its combustion reaction to be product–favored.

42. *Answer:* **First line: sign of ΔG_{system} = negative; last line: sign of ΔG_{system} = positive; see table below**

Strategy and Explanation: Adapt the method described in the solution to Question 38 using a generalized form of Equation 17.3:

$$\Delta G_{system} = \Delta H_{system} - T\Delta S_{system}$$

$$(\text{sign of } \Delta G_{system}) = (\text{sign of } \Delta H_{system}) - (\text{positive Kelvin temperature}) \times (\text{sign of } \Delta S°)$$

For the first line of the table: (sign of $\Delta G°$) = (negative) – (positive) × (positive) = negative

For the last line of the table: (sign of $\Delta G°$) = (positive) – (positive) × (negative) = positive

Sign of ΔH_{system}	Sign of ΔS_{system}	Product-favored?	Sign of ΔG_{system}
Negative (exothermic)	Positive	Yes	Negative
Negative (exothermic)	Negative	Yes at low T; no at high T	*(see Question 43)*
Positive (endothermic)	Positive	No at low T; yes at high T	*(see Question 43)*
Positive (endothermic)	Negative	No	Positive

44. *Answer/Explanation:* Adapt the method described in the solution to Question 40 using Equation 17.3:

$$\Delta G° = \Delta H° - T\Delta S°$$

Here: (sign of $\Delta H°$) = negative, and (sign of $\Delta S°$) = positive. The Kelvin temperature is always positive.

$$\Delta G° = (\text{negative } \Delta H°) - (\text{positive Kelvin temperature}) \times (\text{positive } \Delta S°)$$

Take the absolute values of the each term and then include the resulting signs explicitly:

$$\Delta G^\circ = -\,|\Delta H^\circ| - |T\Delta S^\circ| = -\,(|\Delta H^\circ| + |T\Delta S^\circ|)$$

$$(\text{sign of } \Delta G^\circ) = -\,(\text{positive}) = \text{negative}$$

So, $\Delta G^\circ < 0$ for all temperature values.

46. *Answer:* **ΔG° = 28.63 kJ, so reaction is reactant-favored.**

Strategy and Explanation: Adapt the method described in the solution to Question 40(b).

$$\Delta G^\circ = \Delta H^\circ - T\Delta S^\circ = (41.17 \text{ kJ}) - (298 \text{ K}) \times (42.08 \text{ J/K}) \times \frac{1 \text{ kJ}}{1000 \text{ J}} = 28.63 \text{ kJ}$$

The sign of ΔG° is positive, so the reaction is reactant-favored.

48. *Answer:* **(a) ΔG° = –141.05 kJ (b) ΔG° = 141.73 kJ (c) ΔG° = –959.43 kJ; reactions (a) and (c) are product-favored.**

Strategy and Explanation: Adapt the method described in the solution to Question 40(b).

(a) $\Delta G^\circ = (1 \text{ mol}) \times \Delta G_f^\circ(C_2H_4) - (1 \text{ mol}) \times \Delta G_f^\circ(C_2H_2) - (1 \text{ mol}) \times \Delta G_f^\circ(H_2)$

$= (1 \text{ mol}) \times (68.15 \text{ kJ/mol}) - (1 \text{ mol}) \times (209.20 \text{ kJ/mol}) - (1 \text{ mol}) \times (0 \text{ kJ/mol}) = -141.05 \text{ kJ}$

(b) $\Delta G^\circ = (2 \text{ mol}) \times \Delta G_f^\circ(SO_2) + (1 \text{ mol}) \times \Delta G_f^\circ(O_2) - (2 \text{ mol}) \times \Delta G_f^\circ(SO_3)$

$= (2 \text{ mol}) \times (-300.194 \text{ kJ/mol}) + (1 \text{ mol}) \times (0 \text{ kJ/mol}) - (2 \text{ mol}) \times (-371.06 \text{ kJ/mol}) = 141.73 \text{ kJ}$

(c) $\Delta G^\circ = (4 \text{ mol}) \times \Delta G_f^\circ(NO) + (6 \text{ mol}) \times \Delta G_f^\circ(H_2O(g)) - (4 \text{ mol}) \times \Delta G_f^\circ(NH_3) - (5 \text{ mol}) \times \Delta G_f^\circ(O_2)$

$= (4 \text{ mol}) \times (86.55 \text{ kJ/mol}) + (6 \text{ mol}) \times (-228.572 \text{ kJ/mol})$

$- (4 \text{ mol}) \times (-16.45 \text{ kJ/mol}) - (5 \text{ mol}) \times (0 \text{ kJ/mol}) = -959.43 \text{ kJ}$

50. *Answer:* **(a) 385.7 K (b) 835.1 K**

Strategy and Explanation: Adapt the method described in the solution to Question 26(a).

(a) $\Delta H^\circ = (1 \text{ mol}) \times \Delta H_f^\circ(CH_3OH) + (1 \text{ mol}) \times \Delta H_f^\circ(CO) - (2 \text{ mol}) \times \Delta H_f^\circ(H_2)$

$= (1 \text{ mol}) \times (-238.66 \text{ kJ/mol}) + (1 \text{ mol}) \times (-110.525 \text{ kJ/mol}) - (2 \text{ mol}) \times (0 \text{ kJ/mol}) = -128.14 \text{ kJ}$

$\Delta S^\circ = (1 \text{ mol}) \times S^\circ(CH_3OH) + (1 \text{ mol}) \times S^\circ(CO) - (2 \text{ mol}) \times S^\circ(H_2)$

$= (1 \text{ mol}) \times (126.8 \text{ J K}^{-1}\text{mol}^{-1}) + (1 \text{ mol}) \times (197.674 \text{ J K}^{-1}\text{mol}^{-1})$

$- (2 \text{ mol}) \times (130.684 \text{ J K}^{-1}\text{mol}^{-1}) = -332.2 \text{ J K}^{-1}$

$$T = \frac{\left(-128.14 \text{ kJ}\right) \times \left(\dfrac{1000 \text{ J}}{1 \text{ kJ}}\right)}{-332.2 \text{ J K}^{-1}} = 385.7 \text{ K}$$

(b) $\Delta H^\circ = (4 \text{ mol}) \times \Delta H_f^\circ(Fe) + (3 \text{ mol}) \times \Delta H_f^\circ(CO_2) - (2 \text{ mol}) \times \Delta H_f^\circ(Fe_2O_3) - (3 \text{ mol}) \times \Delta H_f^\circ(C)$

$= (4 \text{ mol}) \times (0 \text{ kJ/mol}) + (3 \text{ mol}) \times (-393.509 \text{ kJ/mol})$

$- (2 \text{ mol}) \times (-824.2 \text{ kJ/mol}) - (3 \text{ mol}) \times (0 \text{ kJ/mol}) = 467.9 \text{ kJ}$

$\Delta S^\circ = (4 \text{ mol}) \times S^\circ(Fe) + (3 \text{ mol}) \times S^\circ(CO_2) - (2 \text{ mol}) \times S^\circ(Fe_2O_3) - (3 \text{ mol}) \times S^\circ(C)$

$= (4 \text{ mol}) \times (27.78 \text{ J K}^{-1}\text{mol}^{-1}) + (3 \text{ mol}) \times (213.74 \text{ J K}^{-1}\text{mol}^{-1})$

$- (2 \text{ mol}) \times (87.40 \text{ J K}^{-1}\text{mol}^{-1}) - (3 \text{ mol}) \times (5.740 \text{ J K}^{-1}\text{mol}^{-1}) = 560.32 \text{ J K}^{-1}$

$$T = \frac{\left(467.9 \text{ kJ}\right) \times \left(\dfrac{1000 \text{ J}}{1 \text{ kJ}}\right)}{560.32 \text{ J K}^{-1}} = 835.1 \text{ K}$$

52. *Answer:* **(a) ΔH° = 178.32 kJ; ΔS° = 160.6 J K^{-1}; ΔG°= 130.5 kJ (b) reactant-favored (c) It is not product-favored at all temperatures. (d) 1110. K**

Strategy and Explanation: Adapt methods described in the solution to Questions 30 and 38.

(a) $\Delta H° = (1 \text{ mol}) \times \Delta H°_f(CaO) + (1 \text{ mol}) \times \Delta H°_f(CO_2) - (1 \text{ mol}) \times \Delta H°_f(CaCO_3)$

$= (1 \text{ mol}) \times (-635.09 \text{ kJ/mol}) + (1 \text{ mol}) \times (-393.509 \text{ kJ/mol})$
$- (1 \text{ mol}) \times (-1206.92 \text{ kJ/mol}) = 178.32 \text{ kJ}$

$\Delta S° = (1 \text{ mol}) \times S°(CaO) + (1 \text{ mol}) \times S°(CO_2) - (1 \text{ mol}) \times S°(CaCO_3)$

$= (1 \text{ mol}) \times (39.75 \text{ J K}^{-1}\text{mol}^{-1}) + (1 \text{ mol}) \times (213.74 \text{ J K}^{-1}\text{mol}^{-1})$
$- (1 \text{ mol}) \times (92.9 \text{ J K}^{-1}\text{mol}^{-1}) = 160.6 \text{ J K}^{-1}$

$\Delta G° = \Delta H° - T\Delta S° = (178.32 \text{ kJ}) - (298 \text{ K})(160.6 \text{ J K}^{-1}) \times \dfrac{1 \text{ kJ}}{1000 \text{ J}} = 130.5 \text{ kJ}$

(b) The change in Gibbs free energy is positive, so the reaction is reactant-favored.

(c) Both $\Delta H°$ and $\Delta S°$ are positive, so the reaction is only product-favored at high temperatures.

(d) $$T = \dfrac{(178.32 \text{ kJ}) \times \left(\dfrac{1000 \text{ J}}{1 \text{ kJ}}\right)}{160.6 \text{ J K}^{-1}} = 1110. \text{ K}$$

54. *Answer:* $\Delta G°_f(Ca(OH)_2) = -867.8 \text{ kJ/mol}$

Strategy and Explanation: Use Equation 17.4 to determine the value of $\Delta G°_f$ sought.

$\Delta G° = (1 \text{ mol}) \times \Delta G°_f(C_2H_2) + (1 \text{ mol}) \times \Delta G°_f(Ca(OH)_2) - (1 \text{ mol}) \times \Delta G°_f(CaC_2) - (2 \text{ mol}) \times \Delta G°_f(H_2O(\ell))$

$\Delta G°_f(Ca(OH)_2) = -\dfrac{\Delta G°}{(1 \text{ mol})} - \Delta G°_f(C_2H_2) + \Delta G°_f(CaC_2) + 2 \times \Delta G°_f(H_2O(\ell))$

$= (-119.282 \text{ kJ/mol}) - (209.20 \text{ kJ/mol}) - (-64.9 \text{ kJ/mol}) - 2 \times (-237.129 \text{ kJ/mol}) = -867.8 \text{ kJ/mol}$

Gibbs Free Energy Changes and Equilibrium Constants

56. *Answer:* (a) $K_p = 4 \times 10^{-34}$ (b) $K_p = 5 \times 10^{-31}$

Strategy and Explanation: Use the method shown in the solution to Question 40(b), then apply Equation 17.6 to calculate the equilibrium constant: $K°$, also called, K_p, for the gas phase reaction, here.

$$\Delta G° = -RT\ln K_p$$

(a) $\Delta G° = (1 \text{ mol}) \times \Delta G°_f(H_2) + (1 \text{ mol}) \times \Delta G°_f(Cl_2) - (2 \text{ mol}) \times \Delta G°_f(HCl)$

$= (1 \text{ mol}) \times (0 \text{ kJ/mol}) + (1 \text{ mol}) \times (0 \text{ kJ/mol}) - (2 \text{ mol}) \times (-95.299 \text{ kJ/mol}) = 190.598 \text{ kJ}$

$$K_p = e^{(-\Delta G°/RT)} = e^{-\left(\frac{190.598 \text{ kJ}}{(0.008314 \text{ kJ / mol·K}) \times (298 \text{ K})}\right)} = e^{-76.9} = 4 \times 10^{-34}$$

(b) $\Delta G° = (1 \text{ mol}) \times \Delta G°_f(NO) - (1 \text{ mol}) \times \Delta G°_f(N_2) - (2 \text{ mol}) \times \Delta G°_f(O_2)$

$= (2 \text{ mol}) \times (86.55 \text{ kJ/mol}) - (1 \text{ mol}) \times (0 \text{ kJ/mol}) - (1 \text{ mol}) \times (0 \text{ kJ/mol}) = 173.10 \text{ kJ}$

$$K_p = e^{(-\Delta G°/RT)} = e^{-\left(\frac{173.10 \text{ kJ}}{(0.008314 \text{ kJ / mol·K}) \times (298 \text{ K})}\right)} = e^{-69.9} = 5 \times 10^{-31}$$

58. *Answer:* (a) -100.97 kJ; product-favored (b) 5×10^{17}; since $\Delta G°$ is negative, K is greater than 1.

Strategy and Explanation: Use methods described in the solution to Question 56:

(a) $\Delta G° = (1 \text{ mol}) \times \Delta G°_f(H_3C-CH_3) - (1 \text{ mol}) \times \Delta G°_f(H_2C=CH_2) - (1 \text{ mol}) \times \Delta G°_f(H_2)$

$= (1 \text{ mol}) \times (-32.82 \text{ kJ/mol}) - (1 \text{ mol}) \times (-68.15 \text{ kJ/mol}) - (1 \text{ mol}) \times (0 \text{ kJ/mol}) = -100.97 \text{ kJ}$

The negative sign of $\Delta G°$ indicates that the reaction is product-favored.

(b) $K_p = e^{(-\Delta G^\circ / RT)} = e^{-\left(\dfrac{-100.97 \text{ kJ}}{(0.008314 \text{ kJ/mol·K}) \times (298 \text{ K})}\right)} = e^{+40.8} = 5 \times 10^{17}$

When ΔG° is positive, K is less than 1. When ΔG° is negative, K is greater than 1. The latter is the case.

62. *Answer:* **(a) –106 kJ/mol (b) 8.55 kJ/mol (c) –33.8 kJ/mol**

Strategy and Explanation: Adapt the strategy described in the solution to Question 56. In part (c), we will also need to use Equation 13.5 as in the solution to Question 19 in Chapter 13 to get K_p from K_c. Use the appropriate value of R in each equation. To convert K_p to K_c, keep in mind that standard pressure is 1 bar. 1 atm is 1.01325 bar, so calculate the correct value of R:

$$R = (0.08206 \text{ L·atm/mol·K}) \times \left(\frac{1 \text{ atm}}{1.01325 \text{ bar}}\right) = 0.083147 \text{ L·bar/mol·K}$$

(a) $\Delta G^\circ = -RT\ln K_p = -(0.008314 \text{ kJ/mol·K}) \times (298 \text{ K}) \times \ln(4.4 \times 10^{18}) = -106 \text{ kJ/mol}$

(b) $\Delta G^\circ = -RT\ln K_p = -(0.008314 \text{ kJ/mol·K}) \times (298 \text{ K}) \times \ln(3.17 \times 10^{-2}) = 8.55 \text{ kJ/mol}$

(c) Get K_p from K_c: $K_p = K_c(RT)^{\Delta n}$ In this equation use R = 0.083147 L·bar/mol·K

$$\Delta n = 2 \text{ mol product gas} - 4 \text{ mol reactant gas} = -2$$

$\Delta G^\circ = -RT\ln K_p = -RT\ln[K_c(RT)^{-2}] = -(0.008314 \text{ kJ/mol·K}) \times (298 \text{ K}) \times$

$$\ln[(0.083147 \text{ L·bar/mol·K} \times 298 \text{ K})^{-2} \times (3.5 \times 10^8)] = -33.8 \text{ kJ/mol}$$

Gibbs Free Energy and Maximum Work

64. *Answer:* **Reaction (a) can be used to do useful work; reactions (b) and (c) require work to be done.**

Strategy and Explanation: In Section 10.2 (page 344), we were told that for a gaseous substance the standard thermodynamic properties are given for a gas pressure of 1 bar. If we also assume that 25 °C is actually 25.00 °C, or 298.15 K, then we can use the standard thermodynamic Table in Appendix J. We will calculate the ΔG°, the Gibbs free energy change, which is a measure of maximum useful work. If it is negative, the reaction can be harnessed to do useful work. If it is positive, the reaction cannot be harnessed to do useful work.

(a) $\Delta G^\circ = (12 \text{ mol}) \times \Delta G_f^\circ(CO_2) + (6 \text{ mol}) \times \Delta G_f^\circ(H_2O(g))$

$$- (2 \text{ mol}) \times \Delta G_f^\circ(C_6H_6(\ell)) - (15 \text{ mol}) \times \Delta G_f^\circ(O_2)$$

$= (12 \text{ mol}) \times (-394.359 \text{ kJ/mol}) + (6 \text{ mol}) \times (-228.572 \text{ kJ/mol})$

$$- (2 \text{ mol}) \times (124.5 \text{ kJ/mol}) - (15 \text{ mol}) \times (0 \text{ kJ/mol}) = -6352.7 \text{ kJ}$$

The reaction can be harnessed to do useful work

(b) $\Delta G^\circ = (1 \text{ mol}) \times \Delta G_f^\circ(N_2) + (3 \text{ mol}) \times \Delta G_f^\circ(F_2) - (2 \text{ mol}) \times \Delta G_f^\circ(NF_3)$

$= (1 \text{ mol}) \times (0 \text{ kJ/mol}) + (3 \text{ mol}) \times (0 \text{ kJ/mol}) - (2 \text{ mol}) \times (-83.2 \text{ kJ/mol}) = 166.4 \text{ kJ}$

The reaction requires work to be done.

(c) $\Delta G^\circ = (1 \text{ mol}) \times \Delta G_f^\circ(Ti) + (3 \text{ mol}) \times \Delta G_f^\circ(O_2) - (2 \text{ mol}) \times \Delta G_f^\circ(TiO_2)$

$= (1 \text{ mol}) \times (0 \text{ kJ/mol}) + (3 \text{ mol}) \times (0 \text{ kJ/mol}) - (1 \text{ mol}) \times (-884.5 \text{ kJ/mol}) = 884.5 \text{ kJ}$

The reaction requires work to be done.

66. *Answer:* **Reaction (b) needs 5.068 g; reactions (c) needs 26.94 g**

Strategy and Explanation: Adapt the method described at the end of the solution to Question 52 in Chapter 6. Find the Gibbs free energy of the graphite reaction and use that as a thermochemical conversion factor to supply the Gibbs free energy of the reactions that require work be done in Question 64(b) and (c).

$$C(graphite) + O_2(g) \longrightarrow CO_2(g)$$

This equation is identical to the equation describing the formation reaction for CO_2, so $\Delta G° = -394.359$ kJ/mol

The calculation for Question 64(b) looks like this:

$$166.4 \text{ kJ endergonic reaction} \times \frac{1 \text{ mol C}}{394.359 \text{ kJ provided}} \times \frac{12.011 \text{ g C}}{1 \text{ mol C}} = 5.068 \text{ g C}$$

The rest of the results are described in the table below.

Endergonic Reaction Reference	Calculated $\Delta G°$ (X) (kJ)	Mass of graphite needed (g)
Question 64(b)	166.4 kJ	5.068
Question 64(c)	884.5 kJ	26.94

68. *Answer:* **(a) 2 C(s) + 2 Cl$_2$(g) + TiO$_2$(s) $\longrightarrow$ TiCl$_4$(g) + 2 CO(g) (b) $\Delta H° = -44.6$ kJ; $\Delta S° = 242.7$ J K^{-1}; $\Delta G° = -116.5$ kJ (c) product-favored (d) more reactant-favored**

Strategy and Explanation: Balance the equation, then use the method described in the solution to Questions 34 and 40.

(a) $\qquad$ 2 C(s) + 2 Cl$_2$(g) + TiO$_2$(s) $\longrightarrow$ TiCl$_4$(g) + 2 CO(g) $\qquad$ Check: 2 C, 4 Cl, 1 Ti, 2 O

(b) $\Delta X° = (1 \text{ mol}) \times X°(\text{TiCl}_4) + (2 \text{ mol}) \times X°(\text{CO}) - (2 \text{ mol}) \times X°(\text{C})$

$$- (2 \text{ mol}) \times X°(\text{Cl}_2) - (1 \text{ mol}) \times X°(\text{TiO}_2) \qquad (X = \Delta H_f, S, \text{ or } \Delta G_f)$$

$\Delta H° = (1 \text{ mol}) \times (-763.2 \text{ kJ/mol}) + (2 \text{ mol}) \times (-110.525 \text{ kJ/mol})$

$$- (2 \text{ mol}) \times (0 \text{ kJ/mol}) - (2 \text{ mol}) \times (0 \text{ kJ/mol})$$
$$- (1 \text{ mol}) \times (-939.7 \text{ kJ/mol}) = -44.6 \text{ kJ}$$

$\Delta S° = (1 \text{ mol}) \times (354.9 \text{ J K}^{-1}\text{mol}^{-1}) + (2 \text{ mol}) \times (197.674 \text{ J K}^{-1}\text{mol}^{-1})$

$$- (2 \text{ mol}) \times (5.740 \text{ J K}^{-1}\text{mol}^{-1}) - (2 \text{ mol}) \times (223.066 \text{ K}^{-1}\text{mol}^{-1})$$
$$- (1 \text{ mol}) \times (49.92 \text{ J K}^{-1}\text{mol}^{-1}) = 242.7 \text{ J K}^{-1}$$

$\Delta G° = (1 \text{ mol}) \times (-726.7 \text{ kJ/mol}) + (2 \text{ mol}) \times (-137.168 \text{ kJ/mol})$

$$- (2 \text{ mol}) \times (0 \text{ kJ/mol}) - (2 \text{ mol}) \times (0 \text{ kJ/mol})$$
$$- (1 \text{ mol}) \times (-884.5 \text{ kJ/mol}) = -116.5 \text{ kJ}$$

(c) $\Delta G°$ is negative, so the reaction is product-favored.

(d) As the reaction temperature increases, heat energy is added to the system. The reaction is exothermic, evolving heat, so the reaction becomes more reactant-favored,

70. *Answer:* **CuO, Ag$_2$O, HgO, and PbO**

Strategy and Explanation: Adapt the method used in the solution to Questions 30 and 56. (X, here, stands for ΔH_f or S)

Kelvin temperature = $800 + 273.15 = 1.1 \times 10^3$ K

(a) $\Delta X° = (1 \text{ mol}) \times X°(\text{Cu}) + (1 \text{ mol}) \times X°(\text{CO}) - (1 \text{ mol}) \times X°(\text{CuO}) - (1 \text{ mol}) \times X°(\text{C})$

$\Delta H° = (1 \text{ mol}) \times (0 \text{ kJ/mol}) + (1 \text{ mol}) \times (-110.525 \text{ kJ/mol})$

$$- (1 \text{ mol}) \times (-157.3 \text{ kJ/mol}) - (1 \text{ mol}) \times (0 \text{ kJ/mol}) = 46.8 \text{ kJ}$$

$\Delta S° = (1 \text{ mol}) \times (33.15 \text{ J K}^{-1}\text{mol}^{-1}) + (1 \text{ mol}) \times (197.674 \text{ J K}^{-1}\text{mol}^{-1})$

$$- (1 \text{ mol}) \times (42.63 \text{ J K}^{-1}\text{mol}^{-1}) - (1 \text{ mol}) \times (5.740 \text{ J K}^{-1}\text{mol}^{-1}) = 182.45 \text{ J K}^{-1}$$

$\Delta G°_{800} = \Delta H° - T\Delta S° = (46.8 \text{ kJ}) - (1.1 \times 10^3 \text{ K}) \times (182.45 \text{ J K}^{-1}) \times \dfrac{1 \text{ kJ}}{1000 \text{ J}}$

$$= 46.8 \text{ kJ} - 2.0 \times 10^2 \text{ kJ} = -1.5 \times 10^2 \text{ kJ}$$

(b) $\Delta X°$ = (2 mol) × X°(Ag) + (1 mol) × X°(CO) – (1 mol) × X°(Ag$_2$O) – (1 mol) × X°(C)

$\Delta H°$ = (2 mol) × (0 kJ/mol) + (1 mol) × (–110.525 kJ/mol)

$$– (1 \text{ mol}) × (–31.05 \text{ kJ/mol}) – (1 \text{ mol}) × (0 \text{ kJ/mol}) = –79.47 \text{ kJ}$$

$\Delta S°$ = (2 mol) × (42.55 J K^{-1}mol^{-1}) + (1 mol) × (197.674 J K^{-1}mol^{-1})

$$– (1 \text{ mol}) × (121.3 \text{ J K}^{-1}\text{mol}^{-1}) – (1 \text{ mol}) × (5.740 \text{ J K}^{-1}\text{mol}^{-1}) = 155.7 \text{ J K}^{-1}$$

This is an exothermic reaction ($\Delta H°$ = negative) with a positive entropy change, so it will be spontaneous at any temperature, including 800. °C.

(c) $\Delta X°$ = (1 mol) × X°(Hg) + (1 mol) × X°(CO) – (1 mol) × X°(HgO) – (1 mol) × X°(C)   (X = ΔH_f or S)

$\Delta H°$ = (1 mol) × (0 kJ/mol) + (1 mol) × (–110.525 kJ/mol)

$$– (1 \text{ mol}) × (–90.83 \text{ kJ/mol}) – (1 \text{ mol}) × (0 \text{ kJ/mol}) = –19.70 \text{ kJ}$$

$\Delta S°$ = (1 mol) × (29.87 J K^{-1}mol^{-1}) + (1 mol) × (197.674 J K^{-1}mol^{-1})

$$– (1 \text{ mol}) × (70.29 \text{ J K}^{-1}\text{mol}^{-1}) – (1 \text{ mol}) × (5.740 \text{ J K}^{-1}\text{mol}^{-1}) = 151.51 \text{ J K}^{-1}$$

This is an exothermic reaction ($\Delta H°$ = negative) with a positive entropy change, so it will be spontaneous at any temperature, including 800. °C.

(d) $\Delta X°$ = (1 mol) × X°(Mg) + (1 mol) × X°(CO) – (1 mol) × X°(MgO) – (1 mol) × X°(C)

$\Delta H°$ = (1 mol) × (0 kJ/mol) + (1 mol) × (–110.525 kJ/mol)

$$– (1 \text{ mol}) × (–601.70 \text{ kJ/mol}) – (1 \text{ mol}) × (0 \text{ kJ/mol}) = 491.18 \text{ kJ}$$

$\Delta S°$ = (1 mol) × (32.68 J K^{-1}mol^{-1}) + (1 mol) × (197.674 J K^{-1}mol^{-1})

$$– (1 \text{ mol}) × (26.94 \text{ J K}^{-1}\text{mol}^{-1}) – (1 \text{ mol}) × (5.740 \text{ J K}^{-1}\text{mol}^{-1}) = 197.67 \text{ J K}^{-1}$$

$$\Delta G°_{800} = \Delta H° - T\Delta S° = (491.18 \text{ kJ}) - (1.1 × 10^3 \text{ K}) × (197.67 \text{ J K}^{-1}) × \frac{1 \text{ kJ}}{1000 \text{ J}}$$

$$= 491.18 \text{ kJ} - 2.1 × 10^2 \text{ kJ} = 2.8 × 10^2 \text{ kJ}$$

(e) $\Delta X°$ = (1 mol) × X°(Pb) + (1 mol) × X°(CO) – (1 mol) × X°(PbO) – (1 mol) × X°(C)

$\Delta H°$ = (1 mol) × (0 kJ/mol) + (1 mol) × (–110.525 kJ/mol)

$$– (1 \text{ mol}) × (–217.32 \text{ kJ/mol}) – (1 \text{ mol}) × (0 \text{ kJ/mol}) = 106.80 \text{ kJ}$$

$\Delta S°$ = (1 mol) × (64.81 J K^{-1}mol^{-1}) + (1 mol) × (197.674 J K^{-1}mol^{-1})

$$– (1 \text{ mol}) × (68.7 \text{ J K}^{-1}\text{mol}^{-1}) – (1 \text{ mol}) × (5.740 \text{ J K}^{-1}\text{mol}^{-1}) = 188.0 \text{ J K}^{-1}$$

$$\Delta G°_{800} = \Delta H° - T\Delta S° = (106.80 \text{ kJ}) - (1.1 × 10^3 \text{ K}) × (188.0 \text{ J K}^{-1}) × \frac{1 \text{ kJ}}{1000 \text{ J}}$$

$$= 106.80 \text{ kJ} - 2.0 × 10^2 \text{ kJ} = -1 × 10^2 \text{ kJ}$$

The coupled reactions that have negative $\Delta G°$ can be used to produce the respective metals, so CuO, Ag$_2$O, HgO, and PbO can be used to obtain Cu, Ag, Hg, and Pb, respectively, by this method at 800 °C.

Conservation of Gibbs Free Energy

72. *Answer/Explanation:* Food we eat provides us with a supply of Gibbs free energy. Coal, petroleum, and natural gas are the most common fuel sources used to supply Gibbs free energy by combustion. We also use solar and nuclear energy, as well as the kinetic energy of wind and water.

Thermodynamic and Kinetic Stability

74. *Answer:* **(a) $\Delta G°$ = –86.5 kJ so the reaction is product-favored (b) $\Delta G°$ = –873.1 kJ (c) No (d) Yes**

Strategy and Explanation: Balance the equation, when not provided, and adapt the method in Question 40(b).

(a) $\Delta G°$ = (1 mol) × $\Delta G°_f$(CH$_3$COOH) – (1 mol) × $\Delta G°_f$(CH$_3$OH) – (1 mol) × $\Delta G°_f$(CO)

$$= (1 \text{ mol}) × (–389.9 \text{ kJ/mol}) – (1 \text{ mol}) × (–166.27 \text{ kJ/mol}) – (1 \text{ mol}) × (–137.168 \text{ kJ/mol}) = –86.5 \text{ kJ}$$

Because $\Delta G°$ is negative, the reaction is product-favored.

(b) $CH_3COOH(\ell) + 2\,O_2(g) \longrightarrow 2\,CO_2(g) + 2\,H_2O(\ell)$

$\Delta G° = (2\text{ mol}) \times \Delta G_f°(CO_2) + (2\text{ mol}) \times \Delta G_f°(H_2O(\ell)) - (1\text{ mol}) \times \Delta G_f°(CH_3COOH) - (2\text{ mol}) \times \Delta G_f°(O_2)$

$= (2\text{ mol}) \times (-394.359\text{ kJ/mol}) + (2\text{ mol}) \times (-237.129\text{ kJ/mol})$
$- (1\text{ mol}) \times (-389.9\text{ kJ/mol}) - (2\text{ mol}) \times (0\text{ kJ/mol}) = -873.1\text{ kJ}$

(c) Based on the answer to (b), the products of oxidation are more stable, so acetic acid is not thermodynamically stable.

(d) Acetic acid can be kept both in liquid form and in solution form if stored properly. In the presence of air, it does not explode, so we will classify it as kinetically stable.

General Questions

76. *Answer:* **(a) Reaction 2 (b) Reactions 1 & 5 (c) Reaction 2 (d) Reactions 2 & 3 (e) None**

Strategy and Explanation: Adapt the method described in the solution to Question 62(c) and concepts described in Sections 13.2 and 13.5.

(a) Because of the relationship: $K_p = K_c(RT)^{\Delta n}$, all we need to do is check Δn for each reaction. K_p is larger than K_c when Δn is positive, as long as RT >1(RT <1 only below T $= 12$ K!)

Reaction	Δn	$K_p > K_c$
1	2 mol product gas – 2 mol reactant gas = 0	No
2	1 mol product gas – 0 mol reactant gas = 1	Yes
3	2 mol product gas – 2 mol reactant gas = 0	No
4	1 mol product gas – 3 mol reactant gas = –2	No
5	2 mol product gas – 2 mol reactant gas = 0	No

Only reaction 2 has K_p larger than K_c.

(b) Now we actually need the value of K_p, because $K_p = K° > 1$ for a gas-phase reaction that is product-favored.

$$(RT) = (0.08206\text{ L atm K}^{-1}\text{mol}^{-1})(298\text{ K}) = 24.5$$

$$K_p = K_c(RT)^{\Delta n}$$

Reaction	$(RT)^{\Delta n}$	K_c	K_p	Product-favored
1	1	3.6×10^{20}	3.6×10^{20}	Yes
2	$(24.5)^1$	1.24×10^{-5}	3.03×10^{-4}	No
3	1	9.5×10^{-13}	9.5×10^{-13}	No
4	$(24.5)^{-2}$	3.76	6.29×10^{-3}	No
5	1	2×10^9	2×10^9	Yes

Only reactions 1 and 5 are product-favored.

(c) Every gas contributes a concentration to the K_c expression, as described in Section 13.2. Only one reaction given has just one gas-phase component: Reaction 2.

(d) The product concentrations get higher when the value of K_c increases. In Section 13.6, we learned that the K_c increases with an increase in temperature when the reaction is endothermic. So, reactions 2 and 3 will have an increase in the product concentrations when the temperature increases.

(e) Only two reactions (1 and 2) have a product that is H_2O. The difference between $\Delta G_f°(H_2O(\ell))$ and $\Delta G_f°(H_2O(g))$ is 8.557 kJ/mol, so the switch would make the reaction $\Delta G°$ less negative by 8.557 kJ. If these reactions have positive $\Delta G°$ between 8.557 kJ and zero, then the sign would switch. If these reactions have negative $\Delta G°$ values with gas-phase water, then the $\Delta G°$ will still be negative with liquid water.

Reaction	K_p	$\Delta G° = - RT\ln K°$	$\Delta G°$ sign affected by (ℓ) to (g)?
1	3.6×10^{20}	-117	No
2	3.03×10^{-4}	20.1	No

None of these has a $\Delta G°$ whose sign is affected by a switch from $H_2O(g)$ to $H_2O(\ell)$.

78. *Answer:* **(a) 141.73 kJ (b) no (c) yes (d) $K_p = 1 \times 10^4$ (d) $K_c = 7 \times 10^1$**

Strategy and Explanation: $2\ SO_3(g) \longrightarrow 2\ SO_2(g) + O_2(g)$

(a) $\Delta G° = (2\ mol) \times \Delta G_f°(SO_2) + (1\ mol) \times \Delta G_f°(O_2) - (2\ mol) \times \Delta G_f°(SO_3)$

   $= (2\ mol) \times (-300.194\ kJ/mol) + (1\ mol) \times (0\ kJ/mol) - (2\ mol) \times (-371.06\ kJ/mol) = 141.73\ kJ$

(b) The positive Gibbs free energy change indicates that the reaction is not product-favored at 25 °C.

(c) $\Delta H° = (2\ mol) \times \Delta H_f°(SO_2) + (1\ mol) \times \Delta H_f°(O_2) - (2\ mol) \times \Delta H_f°(SO_3)$

   $= (2\ mol) \times (-296.830\ kJ/mol) + (1\ mol) \times (0\ kJ/mol) - (2\ mol) \times (-395.72\ kJ/mol) = 197.78\ kJ$

   $\Delta S° = (2\ mol) \times S°(SO_2) + (1\ mol) \times S°(O_2) - (2\ mol) \times S°(SO_3)$

   $= (2\ mol) \times (248.22\ J\ K^{-1}mol^{-1}) + (1\ mol) \times (205.138\ J\ K^{-1}mol^{-1})$
   $$- (2\ mol) \times (256.76\ J\ K^{-1}mol^{-1}) = 188.06\ J\ K^{-1}$$

   With both $\Delta H°$ and $\Delta S°$ positive, the reaction will be product-favored at some high temperature, one higher than 25 °C.

(d) $T = 1500\ °C + 273.15 = 1773\ K = 1.8 \times 10^3\ K$ *(round to hundreds place)*

   $\Delta G_{1773}° = \Delta H° - T\Delta S° = (197.78\ kJ) - (1.8 \times 10^3\ K) \times (188.06\ J\ K^{-1}) \times \dfrac{1\ kJ}{1000\ J} = -1.4 \times 10^2\ kJ$

   $$K_{1773}° = e^{(-\Delta G°/RT)} = e^{-\left(\frac{-1.4\times10^2\ kJ}{(0.008314\ kJ/mol\cdot K)\times(1.8\times10^3\ K)}\right)} = e^{+9.2} = 1 \times 10^4$$

(e) $K_c = K_p(RT)^{-\Delta n}$ $\Delta n = 3\ mol - 2\ mol = 1\ mol$

   $K_c = (1 \times 10^4) \times [(0.08206\ L\ atm\ K^{-1}mol^{-1}) \times (1.8 \times 10^3\ K)]^{-1} = 7 \times 10^1$

80. *Answer:* **(a) $\Delta G° = 31.8$ kJ (b) $K_p = P_{Hg(g)}$ (c) $K° = 2.7 \times 10^{-6}$ (d) 2.7×10^{-6} atm (e) 450 K**

Strategy and Explanation: Use the method described in the solution to Question 56. Then derive the equation an equation relating $K°$ and T.

(a) $\Delta G° = (1\ mol) \times \Delta G_f°(Hg(g)) - (1\ mol) \times \Delta G_f°(Hg(\ell))$

   $= (1\ mol) \times (31.8\ kJ/mol) - (1\ mol) \times (0\ kJ/mol) = 31.8\ kJ$

(b) $K° = K_p = P_{Hg(\ell)}$

(c) $K° = e^{(-\Delta G°/RT)} = e^{-\left(\frac{31.8\ kJ/mol}{(0.008314\ kJ/mol\cdot K)\times(298\ K)}\right)} = e^{-12.8} = 10^{-5.56} = 2.7 \times 10^{-6}$

(d) $K° = K_p = P_{Hg(\ell)} = 2.7 \times 10^{-6}$ atm

(e) $K_1° = 2.7 \times 10^{-6}$ atm at $T_1 = 298$ K, $K_2° = P_{Hg(\ell)} = 10$ mm Hg at T_2

$$\Delta H° = (1 \text{ mol}) \times \Delta H°_f(Hg(g)) - (1 \text{ mol}) \times \Delta H°_f(Hg(\ell))$$

$$= (1 \text{ mol}) \times (61.4 \text{ kJ/mol}) - (1 \text{ mol}) \times (0 \text{ kJ/mol}) = 61.4 \text{ kJ}$$

To derive and equation relating K° to T, use Equation 17.6 and Equation 17.3:

$$\Delta G° = -RT\ln K° = \Delta H° - T\Delta S°$$

Solve for $\Delta S°$:
$$\Delta S° = \frac{\Delta H°}{T} - R\ln K°$$

Only K° changes when T changes, so write this equation with two different T and K° pairs:

$$\Delta S° = \frac{\Delta H°}{T_1} - R\ln K°_1 \qquad \Delta S° = \frac{\Delta H°}{T_2} - R\ln K°_2$$

Because $\Delta S°$ is the same value in both equations, the two equations can be combined:

$$\frac{\Delta H°}{T_1} - R\ln K°_1 = \frac{\Delta H°}{T_2} - R\ln K°_2$$

Once we rearrange the equation into a more compact form, we get a functional form exactly like that derived in Problem-Solving Example 12.9.

$$\ln\left(\frac{K°_1}{K°_2}\right) = \frac{\Delta H°}{R}\left(\frac{1}{T_2} - \frac{1}{T_1}\right)$$

Plug in the values of $K°_1$, $K°_2$, and T_1:

$$\ln\left(\frac{2.7 \times 10^{-6} \text{ atm}}{10 \text{ mmHg} \times \dfrac{1 \text{ atm}}{760 \text{ mmHg}}}\right) = \frac{61.4 \text{kJ} / \text{mol}}{0.008314 \text{kJ} / \text{mol} \cdot \text{K}}\left(\frac{1}{T_2} - \frac{1}{298 \text{ K}}\right)$$

Solve for T_2:
$$T_2 = 450 \text{ K}$$

Applying Concepts

82. *Answer/Explanation:* If you don't know the Humpty Dumpty poem, it should be possible to look it up at http://en.wikipedia.org/wiki/Humpty_Dumpty. Humpty Dumpty is a fictional character who was also an egg. Scrambled is a very disordered state for an egg. The second law of thermodynamics says that the more disordered state is the more probable state. Putting the delicate tissues and fluids back where they were before the scrambling occurred would take a great deal of energy. Humpty Dumpty fell off a wall. A very probable result of that fall is for the egg to become scrambled. The story goes on to tell that all the energy of the king's horses and men was not sufficient to put Humpty together again.

83. *Answer/Explanation:* It is possible to define conditions for which the entropy of a substance has its lowest possible value, namely zero at T = 0 K, so absolute entropies can be determined. It is not possible to define conditions for a specific minimum value for internal energy, enthalpy, or Gibbs free energy of a substance, so relative quantities must be used.

85. *Answer/Explanation:* NaCl, in an orderly crystal structure, and pure water, with only O–H hydrogen bonding interactions in the liquid state, are far more ordered than the dispersed hydrated sodium and chloride ions interacting with the water molecules.

87. *Answer:* **(a) false (b) false (c) true (d) true (e) true**

Strategy and Explanation: Use Equation 17.3 ($\Delta G° = \Delta H° - T\Delta S°$) and Equation 17.6 ($\Delta G° = -RT\ln K°$) to relate equilibrium constant to thermodynamic values. When K = 1.0, then lnK = ln(1.0) = 0.0, so $\Delta G° = 0.0$ and $\Delta H° - T\Delta S° = 0.0$

(a) It is false to say that a chemical reaction with K = 1.0 has $\Delta H° = 0$, unless $\Delta S° = 0$, also (this is unlikely to be true, except in the trivial cases where the reaction has the same reactants as products).

(b) It is false to say that a chemical reaction with K = 1.0 has $\Delta S° = 0$, unless $\Delta H° = 0$, also (this is unlikely to be true, except in the trivial cases where the reaction has the same reactants as products).

(c) It is true to say that a chemical reaction with K = 1.0 has $\Delta G° = 0$.

(d) It is true to say that $\Delta H°$ and $\Delta S°$ have equal sign, since $\Delta H°$ must have the same sign as $T\Delta S°$ and T is always positive.

(e) It is true to say that $\Delta H°/T = \Delta S°$ at the temperature T. The relationship is obtained when we solve this equation: $\Delta H° - T\Delta S° = 0$ for $\Delta S°$.

89. *Answer/Explanation:* $\Delta G < 0$ means products are favored; however, the equilibrium state will always have some reactants present, too. To get all the reactants to go away requires the removal of the products from the reactants, so that the reaction continues forward.

More Challenging Questions

91. *Answer:* **(a) $\Delta S° = 58.78$ J/K (b) $\Delta S° = -53.29$ J/K (c) $\Delta S° = -173.93$ J/K; Adding more H atoms decreases the ΔS (i.e., makes it more negative).**

Strategy and Explanation: Adapt the method described in the solution to Question 26. Formation is defined as producing one mol of a substance from standard state elements.

(a) 2 C(graphite) + H_2(g) $\longrightarrow$ C_2H_2(g)

$\Delta S° = (1 \text{ mol}) \times S°(C_2H_2) - (2 \text{ mol}) \times S°(C) - (1 \text{ mol}) \times S°(H_2)$

$= (1 \text{ mol}) \times (200.94 \text{ J K}^{-1}\text{mol}^{-1}) - (2 \text{ mol}) \times (5.740 \text{ J K}^{-1}\text{mol}^{-1})$
$- (1 \text{ mol}) \times (130.684 \text{ J K}^{-1}\text{mol}^{-1}) = 58.78 \text{ J K}^{-1}$

(b) 2 C(graphite) + 2 H_2(g) $\longrightarrow$ C_2H_4(g)

$\Delta S° = (1 \text{ mol}) \times S°(C_2H_4) - (2 \text{ mol}) \times S°(C) - (2 \text{ mol}) \times S°(H_2)$

$= (1 \text{ mol}) \times (219.56 \text{ J K}^{-1}\text{mol}^{-1}) - (2 \text{ mol}) \times (5.740 \text{ J K}^{-1}\text{mol}^{-1})$
$- (2 \text{ mol}) \times (130.684 \text{ J K}^{-1}\text{mol}^{-1}) = -53.29 \text{ J K}^{-1}$

(c) 2 C(graphite) + 3 H_2(g) $\longrightarrow$ C_2H_6(g)

$\Delta S° = (1 \text{ mol}) \times S°(C_2H_6) - (2 \text{ mol}) \times S°(C) - (3 \text{ mol}) \times S°(H_2)$

$= (1 \text{ mol}) \times (229.60 \text{ J K}^{-1}\text{mol}^{-1}) - (2 \text{ mol}) \times (5.740 \text{ J K}^{-1}\text{mol}^{-1})$
$- (3 \text{ mol}) \times (130.684 \text{ J K}^{-1}\text{mol}^{-1}) = -173.93 \text{ J K}^{-1}$

Adding more H atoms decreases the ΔS (i.e., makes it more negative).

93. *Answer:* **product-favored; see explanation below**

Strategy and Explanation: Adapt the method described in the solution to Questions 26 and 24.

$\Delta H° = (1 \text{ mol}) \times \Delta H°_f(C_8H_{18}) - (1 \text{ mol}) \times \Delta H°_f(C_8H_{16}) - (1 \text{ mol}) \times \Delta H°_f(H_2)$

$= (1 \text{ mol}) \times (-208.45 \text{ kJ/mol}) - (1 \text{ mol}) \times (-82.93 \text{ kJ/mol}) - (1 \text{ mol}) \times (0 \text{ kJ/mol}) = -125.52 \text{ kJ}$

We do not have $S°(C_8H_{16})$, but we can assume that it is very close to that of $S°(C_8H_{18})$. The flexibility and freedom of motion of the atoms in these large eight-carbon-chain molecules is probably about the same except near the one double bond in 1-octene. In support of this presumption, we compare $S°(C_2H_6)$ and $S°(C_2H_4)$, which are both larger than 200 J K^{-1}mol^{-1} and differ by only about 10 J K^{-1}mol^{-1}.

$\Delta S° = (1 \text{ mol}) \times S°(C_8H_{18}) - (1 \text{ mol}) \times S°(C_8H_{16}) - (1 \text{ mol}) \times S°(H_2)$

$\cong - (1 \text{ mol}) \times S°(H_2) = - (1 \text{ mol}) \times (130.684 \text{ J K}^{-1}\text{mol}^{-1}) = -130.684 \text{ J K}^{-1}$

We can determine the temperature at which this reaction becomes reactant-favored.

$$T \cong \frac{(-125.52 \text{ kJ}) \times \left(\dfrac{1000 \text{ J}}{1 \text{ kJ}}\right)}{-130.684 \text{ J K}^{-1}} = 960.48 \text{ K}$$

The given temperature of 25 °C (298K) is much lower, so the reaction is confirmed to be product-favored.

(We could also have estimated the $\Delta G°$ for this reaction and shown that it is negative.)

95. *Answer:* **(a) $K_p = 1.5 \times 10^7$ (b) product-favored (c) $K_c = 3.7 \times 10^8$**

Strategy and Explanation: Adapt methods described in the solution to Questions 56 and 62.

(a) $\Delta G° = (2 \text{ mol}) \times \Delta G_f°(NOCl) + (2 \text{ mol}) \times \Delta G_f°(NO) - (1 \text{ mol}) \times \Delta G_f°(Cl_2)$

$= (2 \text{ mol}) \times (66.08 \text{ kJ/mol}) + (2 \text{ mol}) \times (86.55 \text{ kJ/mol}) - (1 \text{ mol}) \times (0 \text{ kJ/mol}) = -40.94 \text{ kJ}$

$K_p = e^{(-\Delta G°/RT)} = e^{-\left(\frac{-40.94 \text{ kJ}}{(0.008314 \text{ kJ/mol·K}) \times (298 \text{ K})}\right)} = e^{16.5} = 10^{7.16} = 1.5 \times 10^7$

(b) The negative Gibbs free energy change indicates that the reaction is product-favored.

(c) Get K_c from K_p, using $K_p = K_c(RT)^{\Delta n}$

$\Delta n = 2 \text{ mol product gas} - 3 \text{ mol reactant gas} = -1$

$K_c = K_p(RT)^{-\Delta n} = (1.5 \times 10^7) \times (0.08206 \text{ L·atm/mol·K} \times 298 \text{ K})^{-(-1)} = 3.7 \times 10^8$

97. *Answer:* **(a) See equations below (b) C_2H_6 reaction: $\Delta H° = 101.02$ kJ; $\Delta S° = -72.71$ J/K; $\Delta G° = 122.72$ kJ; C_3H_8 reaction: $\Delta H° = 76.5$ kJ; $\Delta S° = -86.6$ J/K; $\Delta G° = 102.08$ kJ; CH_3OH reaction: $\Delta H° = 96.44$ kJ; $\Delta S° = -305.2$ J/K; $\Delta G° = 187.39$ kJ; None are feasible.**

Strategy and Explanation: Balance the equations, then use the method described in Question 40.

(a) $7 \text{ C(s)} + 6 \text{ H}_2\text{O(g)} \longrightarrow 2 \text{ C}_2\text{H}_6\text{(g)} + 3 \text{ CO}_2\text{(g)}$

$5 \text{ C(s)} + 4 \text{ H}_2\text{O(g)} \longrightarrow \text{C}_3\text{H}_8\text{(g)} + 2 \text{ CO}_2\text{(g)}$

$3 \text{ C(s)} + 4 \text{ H}_2\text{O(g)} \longrightarrow 2 \text{ CH}_3\text{OH}(\ell) + \text{CO}_2\text{(g)}$

(b) C_2H_6 reaction:

$\Delta X° = (2 \text{ mol}) \times X°(C_2H_6) + (3 \text{ mol}) \times X°(CO_2)$

$- (7 \text{ mol}) \times X°(C) - (6 \text{ mol}) \times X°(H_2O(g))$ $(X = \Delta H_f, S, \text{ or } \Delta G_f)$

$\Delta H° = (2 \text{ mol}) \times (-84.68 \text{ kJ/mol}) + (3 \text{ mol}) \times (-393.509 \text{ kJ/mol})$
$- (7 \text{ mol}) \times (0 \text{ kJ/mol}) - (6 \text{ mol}) \times (-241.818 \text{ kJ/mol}) = 101.02 \text{ kJ}$

$\Delta S° = (2 \text{ mol}) \times (229.60 \text{ J K}^{-1}\text{mol}^{-1}) + (3 \text{ mol}) \times (213.74 \text{ J K}^{-1}\text{mol}^{-1})$
$- (7 \text{ mol}) \times (5.740 \text{ J K}^{-1}\text{mol}^{-1}) - (6 \text{ mol}) \times (188.825 \text{ J K}^{-1}\text{mol}^{-1}) = -72.71 \text{ J K}^{-1}$

$\Delta G° = (2 \text{ mol}) \times (-32.82 \text{ kJ/mol}) + (3 \text{ mol}) \times (-394.359 \text{ kJ/mol})$
$- (7 \text{ mol}) \times (0 \text{ kJ/mol}) - (6 \text{ mol}) \times (-228.572 \text{ kJ/mol}) = 122.72 \text{ kJ}$

C_3H_8 reaction:

$\Delta X° = (1 \text{ mol}) \times X°(C_3H_8) + (2 \text{ mol}) \times X°(CO_2)$

$- (5 \text{ mol}) \times X°(C) - (4 \text{ mol}) \times X°(H_2O(g))$ $(X = \Delta H_f, S, \text{ or } \Delta G_f)$

$\Delta H° = (1 \text{ mol}) \times (-103.8 \text{ kJ/mol}) + (2 \text{ mol}) \times (-393.509 \text{ kJ/mol})$
$- (5 \text{ mol}) \times (0 \text{ kJ/mol}) - (4 \text{ mol}) \times (-241.818 \text{ kJ/mol}) = 76.5 \text{ kJ}$

$\Delta S° = (1 \text{ mol}) \times (269.9 \text{ J K}^{-1}\text{mol}^{-1}) + (2 \text{ mol}) \times (213.74 \text{ J K}^{-1}\text{mol}^{-1})$
$- (5 \text{ mol}) \times (5.740 \text{ J K}^{-1}\text{mol}^{-1}) - (4 \text{ mol}) \times (188.825 \text{ J K}^{-1}\text{mol}^{-1}) = -86.6 \text{ J K}^{-1}$

$\Delta G° = (1 \text{ mol}) \times (-23.49 \text{ kJ/mol}) + (2 \text{ mol}) \times (-394.359 \text{ kJ/mol})$
$- (5 \text{ mol}) \times (0 \text{ kJ/mol}) - (4 \text{ mol}) \times (-228.572 \text{ kJ/mol}) = 102.08 \text{ kJ}$

CH$_3$OH reaction

$$\Delta X° = (2 \text{ mol}) \times X°(CH_3OH) + (1 \text{ mol}) \times X°(CO_2)$$
$$- (3 \text{ mol}) \times X°(C) - (4 \text{ mol}) \times X°(H_2O(g)) \qquad (X = \Delta H_f, S, \text{ or } \Delta G_f)$$

$$\Delta H° = (2 \text{ mol}) \times (-238.66 \text{ kJ/mol}) + (1 \text{ mol}) \times (-393.509 \text{ kJ/mol})$$
$$- (3 \text{ mol}) \times (0 \text{ kJ/mol}) - (4 \text{ mol}) \times (-241.818 \text{ kJ/mol}) = 96.44 \text{ kJ}$$

$$\Delta S° = (2 \text{ mol}) \times (126.8 \text{ J K}^{-1}\text{mol}^{-1}) + (1 \text{ mol}) \times (213.74 \text{ J K}^{-1}\text{mol}^{-1})$$
$$- (3 \text{ mol}) \times (5.740 \text{ J K}^{-1}\text{mol}^{-1}) - (4 \text{ mol}) \times (188.825 \text{ J K}^{-1}\text{mol}^{-1}) = -305.2 \text{ J K}^{-1}$$

$$\Delta G° = (2 \text{ mol}) \times (-166.27 \text{ kJ/mol}) + (1 \text{ mol}) \times (-394.359 \text{ kJ/mol})$$
$$- (3 \text{ mol}) \times (0 \text{ kJ/mol}) - (4 \text{ mol}) \times (-228.572 \text{ kJ/mol}) = 187.39 \text{ kJ}$$

None of these is feasible. $\Delta G°$ is positive. In addition, $\Delta H°$ is positive and $\Delta S°$ is negative, suggesting that there is no temperature at which the products would be favored.

99. *Answer:* **(i) Reaction (b)   (ii) Reactions (a) & (c)   (iii) None of them   (iv) None of them**

Strategy and Explanation: Determine the qualitative changes in $\Delta H°$ and $\Delta S°$, then refer to Table 17.2.

(a) Using the hint at the end of the Question, we look up bond enthalpies in Section 6.7 page 202. We find that forming a bond is exothermic ($\Delta H°$ = negative). Two mol of gas-phase reactants form one mol of gas-phase products, so the entropy decreases ($\Delta S°$ = negative). That means this reaction is (ii) product-favored at low temperatures but not at high temperatures.

(b) A combustion reaction for a hydrocarbon is exothermic ($\Delta H°$ = negative). Nine mol of gas-phase reactants form eleven mol of gas-phase products, so the entropy increases ($\Delta S°$ = positive). That means this reaction is (i) always product-favored.

(c) Going farther with the hint at the end of the Question, we can look up actual bond energies in Chapter 8 (Table 8.2). We conclude that forming very strong P–F bonds (490 kJ/mol) produces more energy than is used breaking the weak P–P (209 kJ/mol) and F–F (158 kJ/mol) bonds. Hence, we will predict that the reaction is exothermic ($\Delta H°$ = negative). Eleven mol of gas-phase reactants form four mol of gas-phase products, so the entropy decreases ($\Delta S°$ = negative). That means this reaction is (ii) product-favored at low temperatures but not at high temperatures.

Chapter 18: Electrochemistry and Its Applications

Introduction

It may be useful to review the principles of oxidation and redox reactions from Chapter 5. The balancing of redox reactions and the concept of half reactions will be used again when discussing electrochemical cells. Quantitative and qualitative questions in this chapter rely on you being able to balance redox equations.

Electrochemical cells may also be referred to as galvanic cells, voltaic cells, and batteries. The language varies in part due to the historical nomenclature, but sometimes identifies what the cell is being used for.

Understanding how to use the information in Table 18.1, Standard Reduction Potentials in Aqueous Solution, is key to predicting electrochemical processes.

A good mnemonic to use to remember the type of reaction occurring at an electrode is this: both anode and oxidation begin with a vowel (a and o) and both cathode and reduction begin with a consonant (c and r).

As technological advances are made in battery technology, more and more energy will be supplied in this form.

Solutions to Questions for Review and Thought for Chapter 18

Topical Questions

Using Half-Reactions to Understand Redox Reactions

8. *Answer:* **(a)** $Zn(s) \longrightarrow Zn^{2+} + 2\ e^-$ **(b)** $2\ H_3O^+ + 2\ e^- \longrightarrow 2\ H_2O + H_2(g)$ **(c)** $Sn^{4+} + 2\ e^- \longrightarrow Sn^{2+}$
(d) $Cl_2(g) + 2\ e^- \longrightarrow 2\ Cl^-$ **(e)** $6\ H_2O + SO_2 \longrightarrow SO_4^{2-} + 4\ H_3O^+ + 2\ e^-$

Strategy and Explanation: Follow the methods described in Problem-Solving Example 18.3.

(a) Balance Zn atoms, then balance charge with electrons:

$$Zn(s) \longrightarrow Zn^{2+}(aq) + 2\ e^- \qquad \text{(Check: 1 Zn, zero net charge)}$$

(b) Balance H atoms, then balance O atoms with H_2O, then balance charge with electrons:

$$2\ H_3O^+(aq) + 2\ e^- \longrightarrow 2\ H_2O(\ell) + H_2(g) \qquad \text{(Check: 6 H, 2 O, zero net charge)}$$

(c) Balance Sn atoms, then balance charge with electrons:

$$Sn^{4+}(aq) + 2\ e^- \longrightarrow Sn^{2+}(aq) \qquad \text{(Check: 1 Sn, +2 net charge)}$$

(d) Balance Cl atoms, then balance charge with electrons:

$$Cl_2(g) + 2\ e^- \longrightarrow 2\ Cl^-(aq) \qquad \text{(Check: 2 Cl, –2 net charge)}$$

(e) Balance S atoms, then balance O atoms with H_2O, then balance H atoms with H^+, then balance charge with electrons:

$$2\ H_2O(\ell) + SO_2(g) \longrightarrow SO_4^{2-}(aq) + 4\ H^+(aq) + 2\ e^-$$

Then add four water molecules to each side to convert the H^+ ions into H_3O^+ ions:

$$6\ H_2O(\ell) + SO_2(g) \longrightarrow SO_4^{2-}(aq) + 4\ H_3O^+(aq) + 2\ e^- \quad \text{(Check: 12 H, 8 O, 1 S, zero net charge)}$$

10. *Answer:* **4 Zn(s) + 7 OH⁻(aq) + NO₃⁻(aq) + 6 H₂O(ℓ) ⟶ 4 Zn(OH)₄²⁻(aq) + NH₃(aq)**

Strategy and Explanation: Follow methods described in the solution to Question 8 and Problem-Solving Example 18.4.

The chemicals for oxidation all go in one half-reaction and the chemicals for reduction all go in the other half-reaction. Balance the half-reactions, neutralize the acid, then combine the half-reactions.

Put Zn and $Zn(OH)_4^{2-}$ in one half-reaction. Put NO_3^- and NH_3 in the other. Balance as in Question 8.

Note that hydroxide ion is combining with the zinc ion, and we eventually want the reaction in basic solution, so it's simpler to just add hydroxide, than to balance in acid.

$$Zn(s) + 4\ OH^-(aq) \longrightarrow Zn(OH)_4^{2-} + 2\ e^- \qquad \text{(Check: 1 Zn, 4 O, 4 H, –4 net charge)}$$

Balanced the second half-reaction in acid then neutralized the acid.

$$NO_3^-(aq) + 9\ H^+(aq) + 8\ e^- \longrightarrow NH_3(aq) + 3\ H_2O(\ell) \qquad \text{(Check: 1 N, 9 H, 3 O, zero net charge)}$$

$$NO_3^-(aq) + 9\ H^+(aq) + 9\ OH^- + 8\ e^- \longrightarrow NH_3(aq) + 3\ H_2O(\ell) + 9\ OH^-$$

$$NO_3^-(aq) + 9\ H_2O(\ell) + 8\ e^- \longrightarrow NH_3(aq) + 3\ H_2O(\ell) + 9\ OH^-$$

$$NO_3^-(aq) + 6\ H_2O(\ell) + 8\ e^- \longrightarrow NH_3(aq) + 9\ OH^- \qquad \text{(Check: 1 N, 6 O, 12 H, –9 net charge)}$$

Multiply the first half-reaction by 4 so that the electrons balance, then add the half-reactions.

$$4\ Zn(s) + 16\ OH^-(aq) \longrightarrow 4\ Zn(OH)_4^{2-} + 8\ e^-$$
$$\underline{NO_3^-(aq) + 6\ H_2O(\ell) + 8\ e^- \longrightarrow NH_3(aq) + 9\ OH^-}$$

$$4\ Zn(s) + 7\ OH^-(aq) + NO_3^-(aq) + 6\ H_2O(\ell) \longrightarrow 4\ Zn(OH)_4^{2-} + NH_3(aq)$$

(Check: 4 Zn, 16 O, 19 H, 1 N, –8 net charge)

12. *Answer:* **(a) 3 CO + O₃ ⟶ 3 CO₂; oxidizing agent: O₃; reducing agent: CO (b) H₂ + Cl₂ ⟶ 2 HCl; oxidizing agent: Cl₂; reducing agent: H₂ (c) 2 H₃O⁺ + H₂O₂ + Ti²⁺ ⟶ 4 H₂O + Ti⁴⁺; oxidizing agent: H₂O₂; reducing agent: Ti²⁺ (d) 2 MnO₄⁻ + 6 Cl⁻ + 8 H₃O⁺ ⟶ 2 MnO₂ + 3 Cl₂ + 12 H₂O; oxidizing agent: MnO₄⁻; reducing agent: Cl⁻ (e) 4 FeS₂ + 11 O₂ ⟶ 2 Fe₂O₃ + 8 SO₂; oxidizing agent: O₂; reducing agent: FeS₂ (f) O₃ + NO ⟶ O₂ + NO₂; oxidizing agent: O₃; reducing agent: NO (g) Zn(Hg) + HgO ⟶ ZnO + 2 Hg; oxidizing agent: HgO; reducing agent: Zn(Hg)**

Strategy and Explanation: Follow the methods described in the solution to Questions 8 and 10 and Problem-Solving Examples 18.3 and 18.4. After the half-reactions are separated and balanced, equalize the electrons with appropriate multipliers and add the two half-reactions. The conversion of the H^+ ions into H_3O^+ ions can wait until the redox reaction is balanced.

(a) Put CO and CO_2 in one half-reaction. Put O_3 in the other. Balance each, then add.

$$3\ H_2O(\ell) + 3\ CO(g) \longrightarrow 3\ CO_2(g) + 6\ H^+(aq) + 6\ e^-$$
$$\underline{6\ e^- + 6\ H^+(aq) + O_3(g) \longrightarrow 3\ H_2O(\ell)}$$
$$3\ CO(g) + O_3(g) \longrightarrow 3\ CO_2(g)$$

O_3 is the oxidizing agent. CO is the reducing agent.

(b) This reaction is trivial to balance the old-fashioned way: $H_2(g) + Cl_2(g) \longrightarrow 2\ HCl(g)$

Cl_2 is the oxidizing agent. H_2 is the reducing agent.

(c) Put H_2O_2 in one half-reaction. Put Ti^{2+} and Ti^{4+} in the other. Balance each, then add.

$$2\ e^- + 2\ H^+(aq) + H_2O_2(aq) \longrightarrow 2\ H_2O(\ell)$$
$$\underline{Ti^{2+}(aq) \longrightarrow Ti^{4+}(aq) + 2\ e^-}$$
$$2\ H^+(aq) + H_2O_2(aq) + Ti^{2+}(aq) \longrightarrow 2\ H_2O(\ell) + Ti^{4+}(aq)$$

Then add two water molecules to each side to convert the H^+ ions into H_3O^+ ions:

$$2\ H_3O^+(aq) + H_2O_2(aq) + Ti^{2+}(aq) \longrightarrow 4\ H_2O(\ell) + Ti^{4+}(aq) \quad (\text{Check: 8 H, 4 O, 1 Ti, +4 net charge})$$

H_2O_2 is the oxidizing agent. Ti^{2+} is the reducing agent.

(d) Put MnO_4^- and MnO_2 in one half-reaction. Put Cl^- and Cl_2 in the other.

$$3\ e^- + MnO_4^-(aq) + 4\ H^+(aq) \longrightarrow MnO_2(s) + 2\ H_2O(\ell)$$

$$2\ Cl^-(aq) \longrightarrow Cl_2(g) + 2\ e^-$$

Multiply each half-reaction by a constant to get the same number of electrons:

$$2 \times [3\ e^- + MnO_4^-(aq) + 4\ H^+(aq) \longrightarrow MnO_2(s) + 2\ H_2O(\ell)]$$

$$3 \times [2\ Cl^-(aq) \longrightarrow Cl_2(g) + 2\ e^-]$$

Now add them:

$$6\ e^- + 2\ MnO_4^-(aq) + 8\ H^+(aq) \longrightarrow 2\ MnO_2(s) + 4\ H_2O(\ell)$$

$$6\ Cl^-(aq) \longrightarrow 3\ Cl_2(g) + 6\ e^-$$

_____

$$2\ MnO_4^-(aq) + 6\ Cl^-(aq) + 8\ H^+(aq) \longrightarrow 2\ MnO_2(s) + 3\ Cl_2(g) + 4\ H_2O(\ell)$$

Then add two water molecules to each side to convert the H^+ ions into H_3O^+ ions:

$$2\ MnO_4^-(aq) + 6\ Cl^-(aq) + 8\ H_3O^+(aq) \longrightarrow 2\ MnO_2(s) + 3\ Cl_2(g) + 12\ H_2O(\ell)$$

$$(\text{Check: 2 Mn, 16 O, 24 H, zero net charge})$$

MnO_4^- is the oxidizing agent. Cl^- is the reducing agent.

(e) Put FeS_2, Fe_2O_3 and SO_2 in one half-reaction. Put O_2 in the other.

$$11\ H_2O(\ell) + 2\ FeS_2(s) \longrightarrow Fe_2O_3(s) + 4\ SO_2(g) + 22\ H^+(aq) + 22\ e^-$$

$$4\ e^- + 4\ H^+(aq) + O_2(g) \longrightarrow 2\ H_2O(\ell)$$

Multiply each half-reaction by a constant to get the same number of electrons:

$$2 \times [11\ H_2O(\ell) + 2\ FeS_2(s) \longrightarrow Fe_2O_3(s) + 4\ SO_2(g) + 22\ H^+(aq) + 22\ e^-]$$

$$11 \times [4\ e^- + 4\ H^+(aq) + O_2(g) \longrightarrow 2\ H_2O(\ell)]$$

Now add them:

$$22\ H_2O(\ell) + 4\ FeS_2(s) \longrightarrow 2\ Fe_2O_3(s) + 8\ SO_2(g) + 44\ H^+(aq) + 44\ e^-]$$

$$44\ e^- + 44\ H^+(aq) + 11\ O_2(g) \longrightarrow 22\ H_2O(\ell)$$

_____

$$4\ FeS_2(s) + 11\ O_2(g) \longrightarrow 2\ Fe_2O_3(s) + 8\ SO_2(g)$$

$$(\text{Check: 4 Fe, 8 S, 22 O, zero net charge})$$

O_2 is the oxidizing agent. FeS_2 is the reducing agent.

(f) This reaction is already balanced: $O_3(g) + NO(g) \longrightarrow O_2(g) + NO_2(g)$

O_3 is the oxidizing agent. NO is the reducing agent.

(g) This reaction is already balanced: $Zn(Hg)(\text{amalgam}) + HgO(s) \longrightarrow ZnO(s) + 2\ Hg(\ell)$

HgO is the oxidizing agent. $Zn(Hg)$ is the reducing agent.

Electrochemical Cells

14. *Answer/Explanation:* The generation of electricity occurs when electrons are transmitted through a wire from the metal to the cation. Here, the transfer of electrons would occur directly from the metal to the cation and the electrons would not flow through any wire.

16. *Answer:* **(a) Zn + Pb^{2+} ⟶ Zn^{2+} + Pb (b) Anode reaction: Zn(s) ⟶ Zn^{2+} + 2 e$^-$ Anode: Zn(s)**
Cathode reaction: 2 e$^-$ + Pb^{2+} ⟶ Pb(s)   Cathode: Pb(s) (c) see diagram below

Strategy and Explanation: (a) Zn(s) + Pb^{2+}(aq) ⟶ Zn^{2+}(aq) + Pb(s)

(b) Oxidation of zinc atoms occurs at the anode, which is metallic zinc:

$$Zn(s) ⟶ Zn^{2+}(aq) + 2 e^-$$

The reduction of lead(II) ions occurs at the cathode, which is metallic lead:

$$2 e^- + Pb^{2+}(aq) ⟶ Pb(s)$$

(c)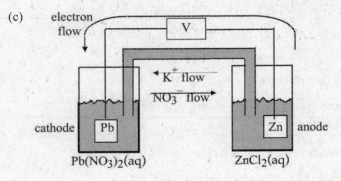

Electrochemical Cells and Voltage

18. *Answer:* **7500 C**

Strategy and Explanation: This is a standard conversion factor problem: divide the time by the voltage and use the definition of "watt" given in the Question to show that 25 J are transferred per second. 1 J = 1C × 1V

$$\frac{1.0 \text{ hr}}{12 \text{ V}} \times \frac{60 \text{ min}}{1 \text{ hr}} \times \frac{60 \text{ s}}{1 \text{ min}} \times \frac{25 \text{ J}}{1 \text{ s}} \times \frac{1 \text{ C} \times 1 \text{ V}}{1 \text{ J}} = 7500 \text{ C}$$

19. *Answer:* **(a) Cu ⟶ Cu^{2+} + 2 e$^-$; Ag$^+$ + e$^-$ ⟶ Ag (b) Oxidation: Cu half-reaction; Reduction: Ag half-reaction; Anode compartment: Cu half-reaction; Cathode compartment: Ag half-reaction**

Strategy and Explanation:

(a) Copper is converted to copper(II) ion and silver ion is converted to silver metal.

$$Cu(s) ⟶ Cu^{2+}(aq) + 2 e^-$$

$$Ag^+(aq) + e^- ⟶ Ag(s)$$

(b) The copper half-reaction is oxidation (since Cu is losing electrons) and it occurs in the anode compartment. The silver half-reaction is reduction (since Ag$^+$ is gaining electrons) and it occurs in the cathode compartment.

Using Standard Reduction Potentials

21. *Answer:* **Strongest reducing agent: Li Weakest oxidizing agent: Li$^+$ Strongest oxidizing agent: F$_2$**
Weakest reducing agent: F$^-$

Strategy and Explanation: On a chart where the standard reduction potentials are listed from most positive to most negative (Table 18.1), the reactant of the first reaction is the strongest oxidizing agent, because it has the largest reduction potential. The reactant of the last reaction is the weakest oxidizing agent, because that reaction has the smallest reduction potential. The product of the first reaction is the weakest reducing agent, because that reaction, in reverse, has the smallest oxidation potential. The product of the last reaction is the strongest reducing agent, because that reaction, in reverse, has the largest oxidation potential. The Li is the strongest reducing agent and Li$^+$ is the weakest oxidizing agent. F$_2$ is the strongest oxidizing agent and F$^-$ is the weakest reducing agent.

23. *Answer:* **(d) < (c) < (a) < (b)**

Strategy and Explanation: Look up the reduction potential for the half-reactions that have the given species as reactants. The largest positive reduction potential is the best at reducing and represents the best oxidizing agent.

$$H_2O\ (-0.8277\ V) < PbSO_4\ (-0.356\ V) < O_2\ (+1.229\ V) < H_2O_2\ (+1.77\ V)$$

25. *Answer:* **(a) 2.91 V (b) –0.028 V (c) 0.65 V (d) 1.16 V; (a), (c) and (d) are product-favored.**

Strategy and Explanation: Use the method described in Problem-Solving Example 18.7.

(a) $Mg(s) \longrightarrow Mg^{2+}(aq) + 2\ e^-$ oxidation half-reaction $E°_{anode} = -2.37\ V$

   $I_2(s) + 2\ e^- \longrightarrow 2\ I^-(aq)$ reduction half-reaction $E°_{cathode} = +0.535\ V$

   ───

   $I_2(s) + Mg(s) \longrightarrow Mg^{2+}(aq) + 2\ I^-(aq)$ $E°_{cell} = (+0.535\ V) - (-2.37\ V) = 2.91\ V$

   This reaction is product-favored.

(b) $Ag(s) \longrightarrow Ag^+(aq) + e^-$ oxidation half-reaction $E°_{anode} = +0.7994\ V$

   $Fe^{3+}(aq) + e^- \longrightarrow Fe^{2+}(aq)$ reduction half-reaction $E°_{cathode} = +0.771\ V$

   ───

   $Ag(s) + Fe^{3+}(aq) \longrightarrow Fe^{2+}(aq) + Ag^+(aq)$ $E°_{cell} = (+0.771\ V) - (+0.7994\ V) = -0.028\ V$

   This reaction is not product-favored.

(c) $Sn^{2+}(aq) \longrightarrow Sn^{4+}(aq) + 2\ e^-$ oxidation half-reaction $E°_{anode} = +0.15\ V$

   $2\ Ag^+(aq) + 2\ e^- \longrightarrow 2\ Ag(s)$ reduction half-reaction $E°_{cathode} = +0.7994\ V$

   ───

   $Sn^{2+}(aq) + 2\ Ag^+(aq) \longrightarrow Sn^{4+}(aq) + 2\ Ag(s$ $E°_{cell} = (+0.7994\ V) - (+0.15\ V) = 0.65\ V$

   This reaction is product-favored.

(d) $2\ Zn(s) \longrightarrow 2\ Zn^{2+}(aq) + 4\ e^-$ oxidation half-reaction $E°_{anode} = -0.763\ V$

   $O_2(s) + 2\ H_2O(\ell) + 4\ e^- \longrightarrow 4\ OH^-(aq)$ reduction half-reaction $E°_{cathode} = +0.40\ V$

   ───

   $2\ Zn(s) + O_2(s) + 2\ H_2O(\ell) \longrightarrow Zn^{2+}(aq) + 4\ OH^-(aq)$ $E°_{cell} = (+0.40\ V) - (-0.763\ V) = 1.16\ V$

   This reaction is product-favored.

27. *Answer:* **(a) Al³⁺ (b) Ce⁴⁺ (c) Al (d) Ce³⁺ (e) Yes (f) No (g) cerium(IV) ion, silver ion, and mercury(I) ion (h) Hg, Sn, Ni, and Al**

Strategy and Explanation:

(a) The reactant of the last reaction, Al^{3+}, is the weakest oxidizing agent.

(b) The reactant of the first reaction, Ce^{4+}, is the strongest oxidizing agent.

(c) The product of the last reaction, Al, is the strongest reducing agent.

(d) The product of the first reaction, Ce^{3+}, is the weakest reducing agent

(e) $Sn(aq) \longrightarrow Sn^{2+}(aq) + 2\ e^-$ oxidation half-reaction $E°_{anode} = -0.14\ V$

   $2\ Ag^+(aq) + 2\ e^- \longrightarrow 2\ Ag(s)$ reduction half-reaction $E°_{cathode} = +0.7994\ V$

   ───

   $Sn^{2+}(aq) + 2\ Ag^+(aq) \longrightarrow Sn^{4+}(aq) + 2\ Ag(s)$ $E°_{cell} = (+0.7994\ V) - (-0.14\ V) = 0.94\ V$

   Yes, Sn(s) will reduce Ag^+ to Ag(s).

(f) $2\ Hg(\ell) \longrightarrow Hg_2^{2+}(aq) + 2\ e^-$ oxidation half-reaction $E°_{anode} = +0.79\ V$

   $Sn^{2+}(aq) + 2\ e^- \longrightarrow Sn(s)$ reduction half-reaction $E°_{cathode} = -0.14\ V$

   ───

   $2\ Hg(\ell) + Sn^{2+}(aq) \longrightarrow Hg_2^{2+}(aq) + Sn(s)$ $E°_{cell} = (-0.14\ V) - (+0.79\ V) = -0.93\ V$

   No, Hg(ℓ) will not reduce Sn^{2+} to Sn(s).

(g) Sn(s) can reduce any ion whose reduction potential is more positive than -0.14 V. In the table provided, Ce^{4+}, Ag^+, and Hg_2^{2+} can reduce Sn(s). Their names are cerium ion, silver ion, and mercury(I) ion.

(h) $Ag^+(aq)$ can be oxidized by any metal whose reduction potential is smaller than 0.7994 V. In the table provided, Hg, Sn, Ni, and Al can oxidize $Ag^+(aq)$.

$E°$ and Gibbs Free Energy

29. *Answer:* **(a) 1.55 V (b) –1196 kJ, 1.55 V**

Strategy and Explanation:

(a) Adapt the method described in Problem-Solving Example 18.8: $\Delta G° = -nFE°_{cell}$

Two O atoms are going from Ox. # = 0 to Ox. # = –2. So, n = 4 mol.

$$E°_{cell} = \frac{-\Delta G°}{nF} = \frac{-(-598 \text{ kJ}) \times \left(\frac{1000 \text{ J}}{1 \text{ kJ}}\right) \times \left(\frac{1 \text{ C} \times 1 \text{ V}}{1 \text{ J}}\right)}{(4 \text{ mol})(96485 \text{ C/mol})} = 1.55 \text{ V}$$

(b) When the reaction is written with all the coefficients doubled:

$$\Delta G°_{double} = 2 \times \Delta G°_{original} = 2 \times (-598 \text{ kJ}) = -1196 \text{ kJ}$$

$E°_{cell}$ is independent of scale, so $E°_{cell} = 1.55$ V.

31. *Answer:* **–409 kJ**

Strategy and Explanation: Adapt the method described in the solution to Question 29 and Problem-Solving Example 18.8.

$$Zn(s) + Cl_2(g) \longrightarrow Zn^{2+}(aq) + 2 \text{ Cl}^-(aq)$$

Zn is going from Ox. # = 0 to Ox. # = +2. So, n = 2 mol.

$$\Delta G° = -(2 \text{ mol}) \times (96485 \text{ C/mol}) \times (2.12 \text{ V}) \times \frac{1 \text{ J}}{1 \text{ C} \times 1 \text{ V}} \times \frac{1 \text{ kJ}}{1000} = -409 \text{ kJ}$$

33. *Answer:* **$K° = 1 \times 10^{-18}$, $\Delta G° = 102$ kJ**

Strategy and Explanation: Adapt the method described in the solution to Question 29 and Problem-Solving Example 18.9.

$$2 \text{ Br}^-(aq) \longrightarrow Br_2(\ell) + 2 \text{ e}^- \qquad E°_{anode} = +1.066 \text{ V}$$

$$I_2(s) + 2 \text{ e}^- \longrightarrow 2 \text{ I}^-(aq) \qquad E°_{cathode} = +0.535 \text{ V}$$

$$I_2(s) + 2 \text{ Br}^-(aq) \longrightarrow 2 \text{ I}^-(aq) + Br_2(\ell) \quad E°_{cell} = (+0.535 \text{ V}) - (+1.066 \text{ V}) = -0.531 \text{ V}$$

$$\log K° = \frac{nE°_{cell}}{0.0592 \text{ V}} = \frac{(2 \text{ mol}) \times (-0.531 \text{ V})}{0.0592 \text{ V}} = -17.9$$

$$K° = 1 \times 10^{-18}$$

$$\Delta G° = -(2 \text{ mol}) \times (96485 \text{ C/mol}) \times (-0.531 \text{ V}) \times \frac{1 \text{ J}}{1 \text{ C} \times 1 \text{ V}} \times \frac{1 \text{ kJ}}{1000} = 102 \text{ kJ}$$

35. *Answer:* **$K° = 4 \times 10^1$**

Strategy and Explanation: Adapt the method described in the solution to Question 29 and Problem-Solving Example 18.9.

$$\log K° = \frac{nE°_{cell}}{0.0592 \text{ V}}$$

$E°_{cell}$ is given to be 0.046 V and the number of moles of electrons passed to take 2+ ions to neutral is 2.

$$\log K^\circ = \frac{nE^\circ_{cell}}{0.0592 \text{ V}} = \frac{(2 \text{ mol}) \times (0.046 \text{ V})}{0.0592 \text{ V}} = 1.5$$

$$K^\circ = 10^{1.5} = 36 \cong 4 \times 10^1 \quad \textit{(round to 1 sig fig)}$$

Effect of Concentration on Cell Potential

37. *Answer:* **(a)** $E^\circ_{cell} = 1.20$ **V (b)** $E_{cell} = 1.16$ **V (c) (conc Ag⁺) = 3 M**

Strategy and Explanation: Adapt the method described in the solution to Question 25 and Problem-Solving Example 18.7, then use the Nernst equation at T = 298 K, by adapting the method described in the Problem-Solving Example 18.10.

$$E_{cell} = E^\circ_{cell} - \frac{0.0592 \text{ V}}{n} \log Q$$

(a) $Cd(s) \longrightarrow Cd^{2+}(aq) + 2 \text{ e}^- \qquad\qquad E^\circ_{anode} = -0.403 \text{ V}$

$\underline{2 \text{ Ag}^+(aq) + 2 \text{ e}^- \longrightarrow 2 \text{ Ag}(s) \qquad\qquad E^\circ_{cathode} = +0.7994 \text{ V}}$

$Cd(s) + 2 \text{ Ag}^+(aq) \longrightarrow Cd^{2+}(aq) + 2 \text{ Ag}(s) \quad E^\circ_{cell} = (+0.7994 \text{ V}) - (-0.403 \text{ V}) = 1.202 \text{ V}$

For this reaction, $\qquad Q = \dfrac{(\text{conc Cd}^{2+})}{(\text{conc Ag}^+)^2}$ and $n = 2$

(b) $E_{cell} = E^\circ_{cell} - \dfrac{0.0592 \text{ V}}{2} \log\left(\dfrac{(\text{conc Cd}^{2+})}{(\text{conc Ag}^+)^2}\right) = 1.20 \text{ V} - \dfrac{0.0592 \text{ V}}{2} \log\left(\dfrac{2.0 \text{ M}}{(0.25 \text{ M})^2}\right)$

$\qquad = 1.202 \text{ V} - \dfrac{0.0592 \text{ V}}{2} \log(32) = 1.202 \text{ V} - \dfrac{0.0592 \text{ V}}{2} \times 1.5 = 1.202 \text{ V} - 0.04 \text{ V}$

$$E_{cell} = 1.16 \text{ V}$$

(c) $\log\left(\dfrac{(\text{conc Cd}^{2+})}{(\text{conc Ag}^+)^2}\right) = \dfrac{n}{0.0592 \text{ V}}(E^\circ_{cell} - E_{cell}) = \dfrac{2}{0.0592 \text{ V}} \times (1.20 \text{ V} - 1.25 \text{ V})$

$$= \frac{2}{0.0592 \text{ V}} \times (-0.05 \text{ V}) = -1.69 \cong -2$$

(subtraction of voltages leaves only 1 sig.fig, so the answer must be round to 1 sig.fig.)

$$\log\left(\frac{0.100 \text{ M}}{(\text{conc Ag}^+)^2}\right) = -2$$

$$10^{-2} = \frac{0.100}{(\text{conc Ag}^+)^2}$$

$$(\text{conc Ag}^+)^2 = 0.100(10^{+2})$$

$$(\text{conc Ag}^+) = 3 \text{ M}$$

40. *Answer:* $\dfrac{\text{conc Pb}^{2+}}{\text{conc Sn}^{2+}} = 0.3$

Strategy and Explanation: Adapt the method described in the solution to Question 37.

$Pb(s) \longrightarrow Pb^{2+} + 2e^- \quad E^\circ_{anode} = -0.126 \text{ V}$

$\underline{+ \quad Sn^{2+} + 2e^- \longrightarrow Sn(s) \qquad E^\circ_{cathode} = -0.14 \text{ V}}$

$Pb(s) + Sn^{2+} \longrightarrow Pb^{2+} + Sn(s)$

$E^\circ_{cell} = -0.014 \text{ V} = -0.01 \text{ V}$ *(must be rounded to two decimal places)* $\qquad\qquad n = 2$

$$E_{cell} = E°_{cell} - \frac{0.0592 \text{ V}}{n} \log\left(\frac{\text{conc Pb}^{2+}}{\text{conc Sn}^{2+}}\right)$$

$$0 \text{ V} = -0.014 \text{ V} - \frac{0.0592 \text{ V}}{2} \log\left(\frac{\text{conc Pb}^{2+}}{\text{conc Sn}^{2+}}\right)$$

$$-0.47 = \log\left(\frac{\text{conc Pb}^{2+}}{\text{conc Sn}^{2+}}\right)$$

$$0.34 = \frac{\text{conc Pb}^{2+}}{\text{conc Sn}^{2+}} \ (= 0.3 \text{ with proper sig figs})$$

Common Batteries

42. *Answer:* **(a) Ni^{2+} + Cd $\longrightarrow$ Ni + Cd^{2+} (b) oxidized: Cd; reduced: Ni^{2+}; oxidizing agent: Ni^{2+}; reducing agent: Cd (c) anode: Cd; cathode: Ni (d) $E°_{cell}$ = 0.15 V (e) from the Cd electrode to the Ni electrode (f) toward the anode compartment**

Strategy and Explanation: Apply the methods described in the answers to Questions 16, 19, and 27.

(a) Nickel ion reacts with cadmium metal to form Nickel metal and cadmium ion.

$$Ni^{2+}(aq) + Cd(s) \longrightarrow Ni(s) + Cd^{2+}(aq)$$

(b)

Substance oxidized	Substance reduced	Oxidizing agent	Reducing agent
Cd	Ni^{2+}	Ni^{2+}	Cd

(c) Metals in the half-reaction will serve as the electrodes. Metallic Cd is the anode and metallic Ni is the cathode.

(d) $Cd(s) \longrightarrow Cd^{2+}(aq) + 2 e^-$ $\qquad E°_{anode} = -0.403$ V

$Ni^{2+}(aq) + 2 e^- \longrightarrow Ni(s)$ $\qquad E°_{cathode} = -0.25$ V

$Cd(s) + Ni^{2+}(aq) \longrightarrow Cd^{2+}(aq) + Ni(s)$ $\quad E°_{cell} = (-0.25 \text{ V}) - (-0.403 \text{ V}) = 0.15$ V

(e) The half-reactions above show that electrons flow spontaneously from the Cd electrode to the Ni electrode.

(f) The NO_3^- anions in the salt bridge flow toward the anode compartment to replenish the negative charges to neutralize the Cd^{2+} ions being formed.

Fuel Cells

44. *Answer/Explanation:* A fuel cell has a continuous supply of reactants, and will be useable for as long as the reactants are supplied. A battery contains all the reactants of the reaction. Once the reactants are gone, the battery is no longer useable.

46. *Answer:* **(a) anode: N_2H_4 oxidation; cathode: O_2 reduction (b) $N_2H_4(g) + O_2(g) \longrightarrow N_2(g) + 2H_2O(\ell)$ (c) 7.5 g N_2H_4 (d) 7.5 g O_2**

Strategy and Explanation: Adapt the methods described in the answers to Question 37, Question 33 in Chapter 5, and Problem-Solving Example 18.12.

(a) The N_2H_4 oxidation occurs at the anode. The O_2 reduction occurs at the cathode.

(b) Adding the two half-reactions gives: $N_2H_4(g) + O_2(g) \longrightarrow N_2(g) + 2H_2O(\ell)$

(c) $50.0 \text{ hr} \times 0.50 \text{ A} \times \dfrac{3600 \text{ s}}{1 \text{ hr}} \times \dfrac{1 \text{ C}}{1 \text{ A} \cdot 1 \text{ s}} \times \dfrac{1 \text{ mol e}^-}{96485 \text{ C}} \times \dfrac{1 \text{ mol N}_2\text{H}_4}{4 \text{ mol e}^-} \times \dfrac{32.05 \text{ g N}_2\text{H}_4}{1 \text{ mol N}_2\text{H}_4} = 7.5 \text{ g N}_2\text{H}_4$

(d) $7.5 \text{ g N}_2\text{H}_4 \times \dfrac{1 \text{ mol N}_2\text{H}_4}{32.05 \text{ g N}_2\text{H}_4} \times \dfrac{1 \text{ mol O}_2}{1 \text{ mol N}_2\text{H}_4} \times \dfrac{32.00 \text{ g O}_2}{1 \text{ mol O}_2} = 7.5 \text{ g O}_2$

Electrolysis—Reactant-Favored Redox Reactions

47. *Answer:* **anode O_2; cathode H_2; 2 moles of H_2 produced per mole O_2**

Strategy and Explanation: Electrolysis of water in sulfuric acid involves the oxidation of water to form oxygen gas and the reduction of the strong acid to form hydrogen gas.

Anode - oxidation: $6 H_2O(\ell) \longrightarrow O_2(g) + 4 H_3O^+(aq) + 4 e^-$

Cathode - reduction: $4 H_3O^+(aq) + 4 e^- \longrightarrow 2 H_2(g) + 4 H_2O(\ell)$

$$6 H_2O(\ell) + 4 H_3O^+(aq) \longrightarrow O_2(g) + 2 H_2(g) + 4 H_3O^+(aq) + 4 H_2O(\ell)$$

$$2 H_2O(\ell) + 4 H_3O^+(aq) \longrightarrow O_2(g) + 2 H_2(g) + 4 H_3O^+(aq)$$

Gaseous O_2 is produced is at the anode, gaseous H_2 is produced at the cathode, and 2 moles of $H_2(g)$ are produced for each mole of $O_2(g)$.

48. *Answer:* **Au^{3+}, Hg^{2+}, Ag^+, Hg_2^{2+}, Fe^{3+}, Cu^{2+}, Sn^{4+}, Sn^{2+}, Ni^{2+}, Cd^{2+}, Fe^{2+}, Zn^{2+}**

Strategy and Explanation: All the metals with a reduction potential more positive than the half-reaction with water as a reactant ($- 0.8277$ V) can be electrolyzed from their aqueous ions to the corresponding metals. The metals in Table 18.1 that qualify are listed above.

50. *Answer:* **$H_2(g)$, $Br_2(\ell)$, and 2 $OH^-(aq)$ are formed; H_2O, Na^+, OH^- (and small amounts of dissolved Br_2 and H_3O^+) are in the solution; H_2 is formed at the cathode. Br_2 is formed at the anode.**

Strategy and Explanation: Electrolysis of NaBr involves the oxidation of the anion Br^- to Br_2 and the reduction of the water.

Anode - oxidation: $2 Br^-(aq) \longrightarrow Br_2(\ell) + 2 e^-$

Cathode - reduction: $2 H_2O(\ell) + 2 e^- \longrightarrow H_2(g) + 2 OH^-(aq)$

$$2 H_2O(\ell) + 2 Br^-(\ell) \longrightarrow H_2(g) + Br_2(\ell) + 2 OH^-(aq)$$

H_2 and Br_2 are produced in a basic solution. After the reaction is complete, Br^- is used up and the solution contains Na^+, OH^-, a small amount of dissolved Br_2 (though it has low solubility in water), and a very small amount of H_3O^+. H_2 is formed in the reduction reaction at the cathode. Br_2 is formed in the oxidation reaction at the anode.

Counting Electrons

52. *Answer:* **0.16 g Ag**

Strategy and Explanation: Follow the method described in the answers to Question 46 and Problem-Solving Example 18.12 on page 679-80:

$$Ag^+(aq) + e^- \longrightarrow Ag(s)$$

$$155 \text{ min} \times 0.015 \text{ A} \times \frac{60 \text{ s}}{1 \text{ min}} \times \frac{1 \text{ C}}{1 \text{ A} \cdot 1 \text{ s}} \times \frac{1 \text{ mol e}^-}{96485 \text{ C}} \times \frac{1 \text{ mol Ag}}{1 \text{ mol e}^-} \times \frac{107.9 \text{ g Ag}}{1 \text{ mol Ag}} = 0.16 \text{ g Ag}$$

53. *Answer:* **5.93 g Cu**

Strategy and Explanation: Follow the method described in the solution to Question 52.

$$Cu^{2+}(aq) + 2 e^- \longrightarrow Cu(s)$$

$$2.00 \text{ hr} \times 2.50 \text{ A} \times \frac{3600 \text{ s}}{1 \text{ hr}} \times \frac{1 \text{ C}}{1 \text{ A} \cdot 1 \text{ s}} \times \frac{1 \text{ mol e}^-}{96485 \text{ C}} \times \frac{1 \text{ mol Cu}}{2 \text{ mol e}^-} \times \frac{63.55 \text{ g Cu}}{1 \text{ mol Cu}} = 5.93 \text{ g Cu}$$

56. *Answer:* **1.9×10^2 g Pb**

Strategy and Explanation: Follow the method described in the solution to Question 52. Equations for chemical reactions are found in Section 18.8.

$$50. \text{ hr} \times 1.0 \text{ A} \times \frac{3600 \text{ s}}{1 \text{ hr}} \times \frac{1 \text{ C}}{1 \text{ A} \cdot 1 \text{ s}} \times \frac{1 \text{ mol e}^-}{96485 \text{ C}} \times \frac{1 \text{ mol Pb}}{2 \text{ mol e}^-} \times \frac{207.2 \text{ g Pb}}{1 \text{ mol Pb}} = 1.9 \times 10^2 \text{ g Pb}$$

58. *Answer:* **The lithium battery uses 0.043 g Li, while the lead battery uses 0.64 g Pb.**

Strategy and Explanation: Follow the method described in the answers to Question 52.

$$10. \ \text{min} \times 1.0 \ \text{A} \times \frac{60 \ \text{s}}{1 \ \text{min}} \times \frac{1 \ \text{C}}{1 \ \text{A} \cdot 1 \ \text{s}} \times \frac{1 \ \text{mol e}^-}{96485 \ \text{C}} = 6.2 \times 10^{-3} \ \text{mol e}^-$$

$$6.2 \times 10^{-3} \ \text{mol e}^- \times \frac{1 \ \text{mol Li}}{1 \ \text{mol e}^-} \times \frac{6.941 \ \text{g Li}}{1 \ \text{mol Li}} = 0.043 \ \text{g Li}$$

$$6.2 \times 10^{-3} \ \text{mol e}^- \times \frac{1 \ \text{mol Pb}}{2 \ \text{mol e}^-} \times \frac{207.2 \ \text{g Pb}}{1 \ \text{mol Pb}} = 0.64 \ \text{g Pb}$$

The lithium battery consumes only 0.043 g Li, while the lead-acid storage battery consumes 0.64 g Pb.

60. *Answer:* **6.85 min**

Strategy and Explanation: Adapt the method described in the solution to Question 52.

$$\frac{0.500 \ \text{g Ni}}{4.00 \ \text{A}} \times \frac{1 \ \text{mol Ni}}{58.6934 \ \text{g Ni}} \times \frac{2 \ \text{mol e}^-}{1 \ \text{mol Ni}} \times \frac{96485 \ \text{C}}{1 \ \text{mol e}^-} \times \frac{1 \text{A} \cdot 1 \text{s}}{1 \text{C}} \times \frac{1 \ \text{min}}{60 \ \text{s}} = 6.85 \ \text{min}$$

Corrosion—Product-Favored Redox Reactions

63. *Answer/Explanation:* As described in Section 18.12, one requirement for corrosion is an electrolyte in contact with both the anode and the cathode. The presence of sodium and chloride ions in salt water increases the electrolytic capacity of the solution.

65. *Answer/Explanation:* Chromium is highly resistant to corrosion and protects the more active iron metal in the steel from oxidizing.

General Questions

68. *Answer:* **0.00689 g Cu; 0.00195 g Al**

Strategy and Explanation: Adapt the method described in the solution to Question 52.

$$\text{Ag}^+(\text{aq}) + \text{e}^- \longrightarrow \text{Ag(s)}, \qquad 0.0234 \ \text{g Ag} \times \frac{1 \ \text{mol Ag}}{107.9 \ \text{g Ag}} \times \frac{1 \ \text{mol e}^-}{1 \ \text{mol Ag}} = 2.17 \times 10^{-4} \ \text{mol e}^-$$

$$\text{Cu}^{2+}(\text{aq}) + 2 \ \text{e}^- \longrightarrow \text{Cu(s)}, \qquad 2.17 \times 10^{-4} \ \text{mol e}^- \times \frac{1 \ \text{mol Cu}}{2 \ \text{mol e}^-} \times \frac{63.55 \ \text{g Cu}}{1 \ \text{mol Cu}} = 0.00689 \ \text{g Cu}$$

$$\text{Al}^{3+}(\text{aq}) + 3 \ \text{e}^- \longrightarrow \text{Al(s)}, \qquad 2.17 \times 10^{-4} \ \text{mol e}^- \times \frac{1 \ \text{mol Al}}{3 \ \text{mol e}^-} \times \frac{26.98 \ \text{g Al}}{1 \ \text{mol Al}} = 0.00195 \ \text{g Al}$$

Applying Concepts

70. *Answer:* **B < D < A < C**

Strategy and Explanation: The strongest reducing agent is the most reactive metal. From the information in (b), we find that metal C reacts with all the other metals' ions, so it will be last on the list. From the information in (a), we find that metals A and C are more reactive than the other metals, so A will precede C in the list. From the information in (c), we find that metal D reacts with the ions of metal B, so B will be first on the list.

72. *Answer:* **(a) oxidized: B(s); reduced: A^{2+} (b) oxidizing agent: A^{2+}; reducing agent: B(s) (c) anode: B(s); cathode: A(s) (d) see equations below (e) A(s) (f) from B(s) to A(s) (g) towards the A^{2+} solution**

Strategy and Explanation: The solution labeled "A^{2+}" is getting lighter and the solution labeled "B^{2+}" is getting darker. If we assume that means A^{2+} is getting less concentrated and that B^{2+} is getting more concentrated we can make the following conclusions:

(a) B(s) is being oxidized to B^{2+} and A^{2+} is being reduced to A(s).

(b) A^{2+} is the oxidizing agent since it is being reduced, and B(s) is the reducing agent since it is being oxidized.

(c) B(s) is the anode, the solid at the site of oxidation, and A(s) is the cathode, the solid at the site of reduction.

(d) $A^{2+} + 2\,e^- \longrightarrow A(s)$ and $B(s) \longrightarrow B^{2+} + 2\,e^-$

(e) The A(s) metal gains mass.

(f) Electrons flow from the B(s) electrode to the A(s) electrode.

(g) K^+ ions in the salt bridge will migrate towards the A^{2+} solution to replace the cations that plated out as A(s).

More Challenging Questions

76. *Answer:* **(a) 9.5×10^6 g HF (b) 1.7×10^3 kWh**

Strategy and Explanation: Perform a standard stoichiometry problem (Chapter 4) and then adapt the method described in the answers to Question 69(d) and at the end of Section 18.11, page 680–681.

(a) $9.0 \text{ metric tons} \times \dfrac{1000 \text{ kg}}{1 \text{ metric ton}} \times \dfrac{1000 \text{ g}}{1 \text{ kg}} \times \dfrac{1 \text{ mol F}_2}{38.00 \text{ g F}_2} \times \dfrac{2 \text{ mol HF}}{1 \text{ mol F}_2} \times \dfrac{20.01 \text{ g HF}}{1 \text{ mol HF}} = 9.5 \times 10^6 \text{ g HF}$

(b) $24.\,\text{hr} \times (6.0 \times 10^3 \text{ A}) \times \dfrac{3600 \text{ s}}{1 \text{ hr}} \times \dfrac{1 \text{ C}}{1 \text{ A} \cdot 1 \text{ s}} \times 12 \text{ V} \times \dfrac{1 \text{ J}}{1 \text{ C} \cdot 1 \text{ V}} \times \dfrac{1 \text{ kWh}}{3.60 \times 10^6 \text{ J}} = 1.7 \times 10^3 \text{ kWh}$

78. *Answer:* $Cl_2 + 2\,Br^- \longrightarrow Br_2 + 2\,Cl^-$

Strategy and Explanation: $Br_2 + 2\,e^- \longrightarrow 2\,Br^-$ +1.066 V

$Cl_2 + 2\,e^- \longrightarrow 2\,Cl^-$ +1.36 V

Reverse the first half-reaction to make a spontaneous reaction (with a positive cell potential).

$2\,Br^- \longrightarrow Br_2 + 2\,e^-$ $E^\circ_{anode} = +1.066$ V

$+\quad Cl_2 + 2\,e^- \longrightarrow 2\,Cl^-$ $E^\circ_{cathode} = +1.358$ V

$Cl_2 + 2\,Br^- \longrightarrow Br_2 + 2\,Cl^-$ $E^\circ_{cell} = +0.292$ V

80. *Answer:* **4+**

Strategy and Explanation: Adapt methods described in the solution to Question 52 and in Section 18.11.

$3.00\,\text{hr} \times 2.00 \text{ A} \times \dfrac{1 C}{1 A \cdot 1 s} \times \dfrac{3600 \text{ s}}{1 \text{ hr}} \times \dfrac{1 \text{ mol e}^-}{96485 \text{ C}} = 0.224 \text{ mol e}^-$

$10.9 \text{ g Pt} \times \dfrac{1 \text{ mol Pt}}{195.08 \text{ g Pt}} = 0.0559 \text{ mol Pt}$

$\dfrac{0.224 \text{ mole e}^-}{0.0559 \text{ mol Pt}} = 4.01$

The platinum ion is a 4+ ion, Pt^{4+}.

82. *Answer:* **75 s**

Strategy and Explanation: Use the methods described in the solution to Question 60 and in Section 18.11.

$V = 1200 \text{ mm}^2 \times 1.0\,\mu\text{m} \times \left(\dfrac{1 \text{ m}}{1000 \text{ mm}} \times \dfrac{100 \text{ cm}}{1 \text{ m}}\right)^2 \times \left(\dfrac{10^{-6} \text{ m}}{1\,\mu\text{m}} \times \dfrac{100 \text{ cm}}{1 \text{ m}}\right) = 0.0012 \text{ cm}^3$

$0.0012 \text{ cm}^3 \times \dfrac{10.5 \text{ g}}{1 \text{ cm}^3} \times \dfrac{1 \text{ mol}}{107.8682 \text{ g}} = 1.2 \times 10^{-4} \text{ mol Ag}$

$$1.2 \times 10^{-4} \text{ mol Ag} \times \frac{1 \text{ mol Ag}^+}{1 \text{ mol Ag}} \times \frac{1 \text{ mol e}^-}{1 \text{ mol Ag}^+} \times \frac{96485 \text{ C}}{1 \text{ mol e}^-} = 11 \text{ C}$$

$$\frac{11 \text{ C}}{150.0 \text{ mA}} \times \frac{1000 \text{ mA}}{1 \text{ A}} \times \frac{1\text{A} \cdot 1\text{s}}{1 \text{ C}} = 75 \text{ s}$$

83. *Answer:* **conc Cu^{2+} = 5 × 10^{-6} M**

Strategy and Explanation: Use the method shown in the solution to Question 37 involving the Nernst equation.

$$Cu(s) \longrightarrow Cu^{2+} + 2e^- \qquad\qquad E°_{anode} = 0.337 \text{ V}$$

$$+ \quad 2(Ag^+ + e^- \longrightarrow Ag(s)) \qquad\qquad E°_{cathode} = 0.7994 \text{ V}$$

$$Cu(s) + 2 \text{ Ag}^+ \longrightarrow Cu^{2+} + 2 \text{ Ag(s)} \qquad E°_{cell} = 0.462 \text{ V} \qquad n = 2$$

$$E_{cell} = E°_{cell} - \frac{0.0592 \text{ V}}{n} \log\left(\frac{\text{conc Cu}^{2+}}{(\text{conc Ag}^+)^2}\right)$$

$$0.62 \text{ V} = 0.462 \text{ V} - \frac{0.0592 \text{ V}}{2} \log\left(\frac{\text{conc Cu}^{2+}}{(1.00 \text{ M})^2}\right)$$

$$\log(\text{conc Cu}^{2+}) = -5.3$$

$$\text{conc Cu}^{2+} = 10^{-5.3} = 5 \times 10^{-6} \text{ M} \text{ *(round to 1 sig fig)*}$$

Chapter 19: Nuclear Chemistry

Introduction

Nuclear chemistry is usually associated with nuclear energy; however, nuclear chemistry has both energy- and non-energy-related applications, including as applications in medicine and archeology.

Radioactive decay particles can be predicted, sometimes. A good device for predicting products or types of nuclear decay is found on page 697. Radiation emission associated with nuclear reactions and the effects of radiation are a critical topic for discussing the usefulness of nuclear power.

All radioactive decays follow first-order kinetics. So, it may be useful reviewing first-order reactions in Chapter 12.

Solutions Blue-Numbered to Questions for Review and Thought for Chapter 19

Topical Questions

Nuclear Reactions

11. *Answer:* **(a) alpha emission (b) beta emission (c) electron capture or positron emission (d) beta emission**

    Strategy and Explanation: The mass numbers and atomic numbers must balance. Use that to determine the mass number and the atomic number of the decay particle. Use the periodic table and the atomic number to get the symbol of an element. Use the identity of the decay particle to identify what transformation that occurs.

    (a) $^{230}_{90}\text{Th} \longrightarrow {}^{226}_{88}\text{Ra} + {}^{4}_{2}\text{He}$ $230 - 226 = 4,\ 90 - 88 = 2$

    Thorium-230 decays by alpha emission to form radium-226.

    (b) $^{137}_{55}\text{Cs} \longrightarrow {}^{137}_{56}\text{Ba} + {}^{0}_{-1}\text{e}$ $137 - 137 = 0,\ 55 - 56 = -1$

    Cesium-137 decays by beta emission to form barium-137.

    (c) $^{38}_{19}\text{K} + {}^{0}_{-1}\text{e} \longrightarrow {}^{38}_{18}\text{Ar}$ $38 - 38 = 0,\ 19 + x = 18,\ x = -1$

    $^{38}_{19}\text{K} \longrightarrow {}^{38}_{18}\text{Ar} + {}^{0}_{+1}\text{e}$ $38 - 38 = 0,\ 19 - 18 = 1$

    Potassium-38 decays by electron capture or positron emission to form argon-38.

    (d) $^{97}_{40}\text{Zr} \longrightarrow {}^{97}_{41}\text{Nb} + {}^{0}_{-1}\text{e}$ $97 - 97 = 0,\ 41 - 40 = -1$

    Zirconium-97 decays by beta emission to form niobiom-137.

    ✓ *Reasonable Answer Check:* Each reaction matches a known radioactive decay process, atomic numbers and mass numbers are conserved.

13. *Answer:* **(a)** $^{238}_{92}\text{U}$ **(b)** $^{32}_{15}\text{P}$ **(c)** $^{10}_{5}\text{B}$ **(d)** $^{0}_{-1}\text{e}$ **(e)** $^{15}_{7}\text{N}$

    Strategy and Explanation: The mass numbers and atomic numbers must balance. Use that to determine the mass number and the atomic number of the missing entry. Use the periodic table and the atomic number to get the symbol of an element.

    (a) $^{242}_{94}\text{Pu} \longrightarrow {}^{4}_{2}\text{He} + {}^{238}_{92}\text{U}$ $242 - 4 = 238,\ 94 - 2 = 92,\ \text{Element 92 is U.}$

    (b) $^{32}_{15}\text{P} \longrightarrow {}^{32}_{16}\text{S} + {}^{0}_{-1}\text{e}$ $32 + 0 = 32,\ 16 + (-1) = 15,\ \text{Element 15 is P.}$

(c) $^{252}_{98}\text{Cf} + {}^{10}_{5}\text{B} \longrightarrow 3\,{}^{1}_{0}\text{n} + {}^{259}_{103}\text{Lr}$ $3 \times (1) - 259 - 252 = 10$

$3 \times (0) - 103 - 98 = 5$, Element 5 is B.

(d) $^{55}_{26}\text{Fe} + {}^{0}_{-1}\text{e} \longrightarrow {}^{55}_{25}\text{Mn}$ $55 - 55 = 0$, $25 - 26 = -1$, electron captured ${}^{0}_{-1}\text{e}$.

(e) $^{15}_{8}\text{O} \longrightarrow {}^{15}_{7}\text{N} + {}^{0}_{+1}\text{e}$ $15 - 0 = 15$, $8 - 1 = 7$, Element 7 is N.

✓ *Reasonable Answer Check:* Atomic numbers and mass numbers are conserved.

15. *Answer:* (a) $^{28}_{12}\text{Mg} \longrightarrow {}^{28}_{13}\text{Al} + {}^{0}_{-1}\text{e}$ (b) $^{238}_{92}\text{U} + {}^{12}_{6}\text{C} \longrightarrow 4\,{}^{1}_{0}\text{n} + {}^{246}_{98}\text{Cf}$

(c) $^{2}_{1}\text{H} + {}^{3}_{2}\text{He} \longrightarrow {}^{4}_{2}\text{He} + {}^{1}_{1}\text{H}$ (d) $^{38}_{19}\text{K} \longrightarrow {}^{38}_{18}\text{Ar} + {}^{0}_{+1}\text{e}$ (e) $^{175}_{78}\text{Pt} \longrightarrow {}^{4}_{2}\text{He} + {}^{171}_{76}\text{Os}$

Strategy and Explanation: Interpret the statement by identifying the nuclear symbol(s) for the given reactant and/or product isotope(s) and identifying the details of the radioactive decay process. Then follow the balancing method described in the solution to Question 11.

(a) Magnesium-28 is $^{28}_{12}\text{Mg}$ and β emission is the production of ${}^{0}_{-1}\text{e}$.

$$^{28}_{12}\text{Mg} \longrightarrow {}^{28}_{13}\text{Al} + {}^{0}_{-1}\text{e}$$

(b) Uranium-238 is $^{238}_{92}\text{U}$, carbon-12 is $^{12}_{6}\text{C}$, and the neutron symbol is ${}^{1}_{0}\text{n}$.

$$^{238}_{92}\text{U} + {}^{12}_{6}\text{C} \longrightarrow 4\,{}^{1}_{0}\text{n} + {}^{246}_{98}\text{Cf}$$

(c) Hydrogen-2 is $^{2}_{1}\text{H}$, helium-3 is $^{3}_{2}\text{He}$, and helium-4 is $^{4}_{2}\text{He}$.

$$^{2}_{1}\text{H} + {}^{3}_{2}\text{He} \longrightarrow {}^{4}_{2}\text{He} + {}^{1}_{1}\text{H}$$

(d) Argon-38 is $^{38}_{18}\text{Ar}$ and positron emission is the production of ${}^{0}_{+1}\text{e}$.

$$^{38}_{19}\text{K} \longrightarrow {}^{38}_{18}\text{Ar} + {}^{0}_{+1}\text{e}$$

(e) Platinum-175 is $^{175}_{78}\text{Pt}$, and osmium-171 is $^{171}_{76}\text{Os}$.

$$^{175}_{78}\text{Pt} \longrightarrow {}^{4}_{2}\text{He} + {}^{171}_{76}\text{Os}$$

✓ *Reasonable Answer Check:* Atomic numbers and mass numbers are conserved.

17. *Answer:* $^{231}_{90}\text{Th}$, $^{231}_{91}\text{Pa}$, $^{227}_{89}\text{Ac}$, $^{227}_{90}\text{Th}$, and $^{223}_{88}\text{Ra}$

Strategy and Explanation: The first five steps of the decay series are: α, β, α, β, α. Therefore, start with uranium-235 undergoing an α decay reaction. Then take the radioisotope produced and make it the reactant of the second β decay reaction. Repeat this process for the remaining three steps undergoing α, then β, then α.

$$^{235}_{92}\text{U} \longrightarrow {}^{4}_{2}\text{He} + {}^{231}_{90}\text{Th}$$

$$^{231}_{90}\text{Th} \longrightarrow {}^{0}_{-1}\text{e} + {}^{231}_{91}\text{Pa}$$

$$^{231}_{91}\text{Pa} \longrightarrow {}^{4}_{2}\text{He} + {}^{227}_{89}\text{Ac}$$

$$^{227}_{89}\text{Ac} \longrightarrow {}^{0}_{-1}\text{e} + {}^{227}_{90}\text{Th}$$

$$^{227}_{90}\text{Th} \longrightarrow {}^{4}_{2}\text{He} + {}^{223}_{88}\text{Ra}$$

So, the radioisotopes produced in the first five steps are: $^{231}_{90}\text{Th}$, $^{231}_{91}\text{Pa}$, $^{227}_{89}\text{Ac}$, $^{227}_{90}\text{Th}$, and $^{223}_{88}\text{Ra}$.

✓ *Reasonable Answer Check:* Atomic numbers and mass numbers are conserved.

19. *Answer:* $^{222}_{86}\text{Rn} \longrightarrow {}^{4}_{2}\text{He} + {}^{218}_{84}\text{Po}$, $\quad {}^{218}_{84}\text{Po} \longrightarrow {}^{4}_{2}\text{He} + {}^{214}_{82}\text{Pb}$, $\quad {}^{214}_{82}\text{Pb} \longrightarrow {}^{0}_{-1}\text{e} + {}^{214}_{83}\text{Bi}$,

$^{214}_{83}\text{Bi} \longrightarrow {}^{0}_{-1}\text{e} + {}^{214}_{84}\text{Po}$, $\quad {}^{214}_{84}\text{Po} \longrightarrow {}^{4}_{2}\text{He} + {}^{210}_{82}\text{Pb}$, $\quad {}^{210}_{82}\text{Pb} \longrightarrow {}^{0}_{-1}\text{e} + {}^{210}_{83}\text{Bi}$,

$^{210}_{83}\text{Bi} \longrightarrow {}^{0}_{-1}\text{e} + {}^{210}_{84}\text{Po}$, $\quad {}^{210}_{84}\text{Po} \longrightarrow {}^{4}_{2}\text{He} + {}^{206}_{82}\text{Pb}$

Strategy and Explanation: The decay series of radon-222 has eight steps: α, α, β, β, α, β, β, α. Start with radon-222 undergoing an α decay reaction. Then take the radioisotope produced and make it the reactant of the second α decay reaction. Repeat this process for the remaining six steps.

$$^{222}_{86}\text{Rn} \longrightarrow {}^{4}_{2}\text{He} + {}^{218}_{84}\text{Po}$$

$$^{218}_{84}\text{Po} \longrightarrow {}^{4}_{2}\text{He} + {}^{214}_{82}\text{Pb}$$

$$^{214}_{82}\text{Pb} \longrightarrow {}^{0}_{-1}\text{e} + {}^{214}_{83}\text{Bi}$$

$$^{214}_{83}\text{Bi} \longrightarrow {}^{0}_{-1}\text{e} + {}^{214}_{84}\text{Po}$$

$$^{214}_{84}\text{Po} \longrightarrow {}^{4}_{2}\text{He} + {}^{210}_{82}\text{Pb}$$

$$^{210}_{82}\text{Pb} \longrightarrow {}^{0}_{-1}\text{e} + {}^{210}_{83}\text{Bi}$$

$$^{210}_{83}\text{Bi} \longrightarrow {}^{0}_{-1}\text{e} + {}^{210}_{84}\text{Po}$$

$$^{210}_{84}\text{Po} \longrightarrow {}^{4}_{2}\text{He} + {}^{206}_{82}\text{Pb} \qquad \text{Lead-206 is stable.}$$

✓ *Reasonable Answer Check:* Atomic numbers and mass numbers are conserved.

Nuclear Stability

20. *Answer:* (a) $^{19}_{10}\text{Ne} \longrightarrow {}^{19}_{9}\text{F} + {}^{0}_{+1}\text{e}$ (b) $^{230}_{90}\text{Th} \longrightarrow {}^{0}_{-1}\text{e} + {}^{230}_{91}\text{Pa}$ (c) $^{82}_{35}\text{Br} \longrightarrow {}^{0}_{-1}\text{e} + {}^{82}_{36}\text{Kr}$

(d) $^{212}_{84}\text{Po} \longrightarrow {}^{4}_{2}\text{He} + {}^{208}_{82}\text{Pb}$

Strategy and Explanation: Identify which type of radioactive decay is most likely for the isotope, by identifying the N/Z ratio and seeing where it falls on Figure 19.2.

(a) The neon-19 isotope has 9 neutrons and 10 protons, giving an N/Z ratio of 0.90. That means it has too few neutrons and undergoes positron emission or electron capture. Neon is relatively small, so we will predict that it undergoes positron emission:

$$^{19}_{10}\text{Ne} \longrightarrow {}^{19}_{9}\text{F} + {}^{0}_{+1}\text{e}$$

(b) The thorium-230 isotope has 140 neutrons and 90 protons, giving an N/Z ratio of 1.56. Stable isotopes in this range of the graph have an N/Z ratio of about 1.52. That means it has too many neutrons, so it undergoes β emission:

$$^{230}_{90}\text{Th} \longrightarrow {}^{0}_{-1}\text{e} + {}^{230}_{91}\text{Pa}$$

(c) The bromine-82 isotope has 47 neutrons and 35 protons, giving an N/Z ratio of 1.34. Stable isotopes in this range of the graph have an N/Z ratio of about 1.20. That means it has too many neutrons, so it undergoes β emission:

$$^{82}_{35}\text{Br} \longrightarrow {}^{0}_{-1}\text{e} + {}^{82}_{36}\text{Kr}$$

(d) The lead-212 isotope has 128 neutrons and 84 protons, more than the threshold limit of 126 neutrons and 83 protons above which all isotopes undergo α decay:

$$^{212}_{84}\text{Po} \longrightarrow {}^{4}_{2}\text{He} + {}^{208}_{82}\text{Pb}$$

✓ *Reasonable Answer Check:* Each reaction is a well-known radioactive decay process. Atomic numbers and mass numbers are conserved.

22. *Answer:* **6.252×10^8 kJ/mol nucleons ^{10}B; 6.688×10^8 kJ/ mol nucleon ^{11}B; ^{11}B is more stable than ^{10}B**

Strategy and Explanation: Use the method described in Section 19.3. We want to compare the binding energy per nucleon. Calculate the mass defect (Δm) when the nucleus is formed from the protons and neutrons. Use

Einstein's equation: $E = (\Delta m)c^2$ to calculate the total energy generated, then divide that number by the number of nucleons to determine the binding energy per nucleon.

$\Delta m = 10.01294$ g ^{10}B $- [(5\text{ mol }{}_{1}^{1}\text{H} \times 1.00783$ g/mol ${}_{1}^{1}\text{H}) + (5\text{ mol }{}_{0}^{1}\text{n} \times 1.00867$ g/mol ${}_{0}^{1}\text{n})] = -0.06956$ g

The nuclear binding energy $= E_b = (\Delta m)c^2$

$E_b = (\Delta m)c^2 = \left(-0.06956 \text{ g} \times \dfrac{1 \text{ kg}}{1000 \text{ g}}\right) \times (2.99792 \times 10^8 \text{ m/s})^2 \times \dfrac{1 \text{ J}}{1 \text{ kg m}^2 \text{ s}^{-2}} \times \dfrac{1 \text{ kJ}}{1000 \text{ J}} = -6.252 \times 10^9 \text{ kJ}$

As described in Section 19.3 on page 701, "Binding energy per nucleon" is really E_b per mol nucleons.

One mol of boron-10 nuclei has 5 mol of protons and 5 mol of neutrons, or a total of 10 mol of nucleons:

$$E_b \text{ per mol nucleon} = \dfrac{6.252 \times 10^9 \text{ kJ}}{10 \text{ mol nucleons}} = 6.252 \times 10^8 \text{ kJ/mol nucleons}$$

$\Delta m = 11.00931$ g ^{11}B $- [(5\text{ mol }{}_{1}^{1}\text{H} \times 1.00783$ g/mol ${}_{1}^{1}\text{H}) + (6\text{ mol }{}_{0}^{1}\text{n} \times 1.00867$ g/mol ${}_{0}^{1}\text{n})] = -0.08186$ g

$E_b = (\Delta m)c^2 = \left(-0.08186 \text{ g} \times \dfrac{1 \text{ kg}}{1000 \text{ g}}\right) \times (2.99792 \times 10^8 \text{ m/s})^2 \times \dfrac{1 \text{ J}}{1 \text{ kg m}^2 \text{ s}^{-2}} \times \dfrac{1 \text{ kJ}}{1000 \text{ J}} = -7.357 \times 10^9 \text{ kJ}$

One mol of boron-11 nuclei has 5 mol of protons and 6 mol of neutrons, or a total of 11 mol of nucleons:

$$E_b \text{ per nucleon} = \dfrac{7.357 \times 10^9 \text{ kJ}}{11 \text{ mol nucleons}} = 6.688 \times 10^8 \text{ kJ/ mol nucleon}$$

Boron-11 has a larger E_b per mol nucleon than boron-10, so ^{11}B is more stable than ^{10}B.

✓ *Reasonable Answer Check:* The average atomic mass of boron (10.811) is closer to 11 than 10, consistent with the calculation that boron-11 is more stable than boron-10.

24. *Answer:* **-1.7394×10^{11} kJ/mol ^{238}U; 7.3086×10^8 kJ/mol nucleon**

Strategy and Explanation: Use the method described in Question 22. One mol of uranium-238 nuclei has 92 mol of protons and 146 mol of neutrons, or a total of 238 mol of nucleons:

$\Delta m = 238.0508$ g ^{238}U $- [(92\text{ mol }{}_{1}^{1}\text{H} \times 1.00783$ g/mol ${}_{1}^{1}\text{H}) + (146\text{ mol }{}_{0}^{1}\text{n} \times 1.00867$ g/mol ${}_{0}^{1}\text{n})] = -1.9354$ g

$E_b = (\Delta m)c^2 = \left(-1.9354 \text{ g} \times \dfrac{1 \text{ kg}}{1000 \text{ g}}\right) \times (2.99792 \times 10^8 \text{ m/s})^2 \times \dfrac{1 \text{ J}}{1 \text{ kg m}^2 \text{ s}^{-2}} \times \dfrac{1 \text{ kJ}}{1000 \text{ J}} = -1.7394 \times 10^{11} \text{ kJ}$

"Binding energy per nucleon" (as described in Section 19.3) is E_b per mol nucleon.

$$E_b \text{ per mol nucleon} = \dfrac{1.7394 \times 10^{11} \text{ kJ}}{238 \text{ mol nucleons}} = 7.3086 \times 10^8 \text{ kJ/mol nucleon}$$

Rates of Disintegration Reactions

27. *Answer:* **5 mg**

Strategy and Explanation: Because of the special circumstances where we are given an amount of time that is a multiple of the half-life, we can use the method shown in Problem-Solving Example 19.4:

$$t = 1 \text{ d} \times \left(\dfrac{24 \text{ h}}{1 \text{ d}}\right) + 6 \text{ h} = 30. \text{ h} \quad \textit{(Assume in this context that 1 d is exact.)}$$

$$30. \ h \times \frac{1 \ \text{half - life}}{15 \ \text{h}} = 2.0 \ \text{half-lives}$$

30. hours is 2.0 times the half-life of 15 hours, so the sample mass is reduced by half twice:

$$\frac{1}{2} \times \frac{1}{2} \times 20 \ \text{mg} = 5 \ \text{mg}.$$

It is only legitimate to use the quick method described above when the elapsed time is a whole-number multiple of the half-life.

If the elapsed time was not a whole-number multiple of the half-life, it would be necessary to adapt the methods described in other Problem-Solving Examples in Section 19.4 and use these equations:

$$A = kN \qquad \ln\left(\frac{N}{N_0}\right) = -kt \qquad \ln\left(\frac{A}{A_0}\right) = -kt \qquad t_{1/2} = \frac{\ln 2}{k}$$

(Notice: ln2 is used in the last equation instead of 0.693 for two reasons: it is faster to type into a calculator – requiring only two buttons be pushed instead of four, and it is more precise than its 3 sig. fig. approximation.)

Use the last equation to determine the value of k:

$$k = \frac{\ln 2}{t_{1/2}} = \frac{\ln 2}{15 \ \text{h}} = 4.6 \times 10^{-3} \ \text{h}^{-1}$$

The mass (m) of a sample of a pure isotopic substance is directly proportional to the number of atoms (N) because its atomic mass is a constant value and the number of atoms in a mole is also a constant. Therefore, the atom ratio is the same as the mass ratio:

$$\frac{N}{N_0} = \frac{m}{m_0}$$

Now, use the second equation to determine the new mass:

$$\ln\left(\frac{N}{N_0}\right) = \ln\left(\frac{m}{m_0}\right) = -kt = -(4.6 \times 10^{-3} \ \text{h}^{-1}) \times (30. \ \text{h}) = -1.4$$

$$m = m_0 e^{-1.4} = (20 \ \text{mg}) \times e^{-1.4} = (20 \ \text{mg}) \times (0.25) = 5 \ \text{mg}$$

✓ *Reasonable Answer Check:* The quick method and the actual calculation produce the same answer, which is a small fraction of the initial mass.

29. *Answer:* **(a)** $^{131}_{53}\text{I} \longrightarrow \ ^{0}_{-1}\text{e} + \ ^{131}_{54}\text{Xe}$ **(b) 1.56 mg**

Strategy and Explanation: Adapt the methods described in the answers to Questions 15 and 27.

(a) The reaction's equation is: $\qquad ^{131}_{53}\text{I} \longrightarrow \ ^{0}_{-1}\text{e} + \ ^{131}_{54}\text{Xe} \qquad\qquad 131 = 0 + 131; \ 53 = -1 + 54.$

(b) $\qquad\qquad\qquad 32.2 \ \text{d} \times \dfrac{1 \ \text{half - life}}{8.05 \ \text{d}} = 4.00 \ \text{half-lives}$

32.2 days is four half-lives, and the sample mass is reduced by half four times:

$$\frac{1}{2} \times \frac{1}{2} \times \frac{1}{2} \times \frac{1}{2} \times 25.0 \ \text{mg} = 1.56 \ \text{mg}$$

If the elapsed time was not a whole-number multiple of the half-life, it would be necessary to do the actual calculations:

$$k = \frac{\ln 2}{t_{1/2}} = \frac{\ln 2}{8.05 \ \text{d}} = 8.61 \times 10^{-2} \ \text{d}^{-1}$$

The mass (m) of a sample of a compound is directly proportional to the number of radioactive atoms (N) because there is a fixed percentage of that atom in the compound, so the ratios are interchangeable, as described in the solution to Question 27:

$$\ln\left(\frac{m}{m_0}\right) = -kt = -(8.61 \times 10^{-2} \ \text{d}^{-1}) \times (32.2 \ \text{d}) = -2.77$$

$$m = m_0 e^{-1.4} = (25.0 \ \text{mg}) \times e^{-2.77} = (25.0 \ \text{mg}) \times (0.063) = 1.6 \ \text{mg}$$

✓ *Reasonable Answer Check:* The quick method and the actual calculation produce the same answer with what is known about the significant figures. The mass remaining is a small fraction of the initial mass.

31. *Answer:* **4.0 y**

Strategy and Explanation: Adapt the methods described in the solution to Question 27.

We can adapt the quick method by applying the definition of half-life. Initially, the radioactivity measured is defined as 100%. After one half-life, the radioactivity will drop by half to 50.0% of the original. After the second half-life, the radioactivity will drop by half again to 25.0% of the original. After the third half-life, the radioactivity will drop by half again to 12.5% of the original (i.e., $\frac{1}{2} \times \frac{1}{2} \times \frac{1}{2} \times 100\% = 12.5\%$). We are told that it takes 12 years to get to 12.5% of the original radioactivity; therefore, 12 years must represent three half-lives:

$$t_{1/2} = \frac{1}{3} \times (12 \text{ y}) = 4.0 \text{ y}$$

If the percentage given did not correspond exactly to one of the subsequent reductions of 100% by half, it would be necessary to do the actual calculations: $A_0 = 100.0\%$ and $A = 12.5\%$

$$\ln\left(\frac{A}{A_0}\right) = -kt \qquad \ln\left(\frac{12.5\%}{100.0\%}\right) = -2.079 = -k \times (12 \text{ y}) \qquad k = 0.17 \text{ y}^{-1}$$

$$t_{1/2} = \frac{\ln 2}{k} = \frac{\ln 2}{0.17 \text{ y}^{-1}} = 4.0 \text{ y}$$

✓ *Reasonable Answer Check:* The quick method and the actual calculation produce the same answer. The half life must be less than the elapsed time, since the remaining radioactivity is less than half of the original radioactivity.

33. *Answer:* **34.8 d**

Strategy and Explanation: Adapt the methods described in the answers to Questions 27, 29 and 31.

The percentage decrease given (5.0%) does not correspond exactly to one of the sequential reductions of 100% by half, so we can't use the quick method. Set up the time calculation using the equation given in Question 27.

In the solution to Question 29, we calculated $k = 8.61 \times 10^{-2} \text{ d}^{-1}$. $A_0 = 100.0\%$ and $A = 5.0\%$

$$\ln\left(\frac{A}{A_0}\right) = -kt \qquad \ln\left(\frac{5.0\%}{100.0\%}\right) = -3.00 = -(8.61 \times 10^{-2} \text{ d}^{-1})t \qquad t = 34.8 \text{ d}$$

✓ *Reasonable Answer Check:* The elapsed time (34.8 d) is much longer than the half-life (8.05 d) and is consistent with the small remaining percent (5.0%).

35. *Answer:* **2.58×10^3 y**

Strategy and Explanation: Adapt the methods described in the solution to Question 33 and Problem-Solving Example 19.5.

$$k = \frac{\ln 2}{5.73 \times 10^3 \text{ y}} = 1.21 \times 10^{-4} \text{ y}^{-1}$$

$A_0 = 15.3 \text{ d min}^{-1}\text{g}^{-1}$ and $A = 11.2 \text{ d min}^{-1}\text{g}^{-1}$

$$\ln\left(\frac{11.2 \text{ d min}^{-1} \text{g}^{-1}}{15.3 \text{ d min}^{-1} \text{g}^{-1}}\right) = -0.312 = -(1.21 \times 10^{-4} \text{ y}^{-1})t \qquad t = 2.58 \times 10^3 \text{ y}$$

✓ *Reasonable Answer Check:* The elapsed time is shorter than the half-life and is consistent with the small reduction in the measured activity.

37. *Answer:* **1.6×10^5 y**

Strategy and Explanation: Adapt the methods described in the solution to Question 33.

The percentage decrease (10.0%) does not correspond exactly to one of the sequential reductions of 100% by half, so we can't use the quick method. Set up the calculations for k and then t: $N_0 = 100.\%$ and $N = 1\%$.

$$k = \frac{\ln 2}{2.4 \times 10^4 \ y} = 2.9 \times 10^{-5} \ y^{-1}$$

$$\ln\left(\frac{1\%}{100\%}\right) = -4.605 = -(2.9 \times 10^{-5} \ y^{-1})t \qquad\qquad t = 1.6 \times 10^5 \ y$$

✓ *Reasonable Answer Check:* The elapsed time is much longer than the half-life and is consistent with the small remaining percent.

Nuclear Fission and Fusion

39. *Answer/Explanation:* Three components represent the fundamental parts of a nuclear fission reactor. Cadmium rods are used as a neutron absorber to control the rate of the fission reaction. Uranium rods are the source of fuel, since uranium is a reactant in the nuclear equation. Water is used for cooling by removing excess heat energy. It is also used in the form of steam in the steam/water cycle to produce the turning torque for the generator.

41. *Answer:* **(a) $_{54}^{140}Xe$ (b) $_{41}^{104}Nb$ (c) $_{36}^{92}Kr$**

Strategy and Explanation: Use the method described in Question 13.

(a)
$$_{92}^{235}U + {}_0^1 n \longrightarrow {}_y^x ? + {}_{38}^{93}Sr + 3\ {}_0^1 n$$

$$235 + 1 = x + 93 + 3\times(1) = 236; \ \ 92 + 0 = y + 38 + 3\times(0)$$

$$x = 140, \text{ and } y = 54, \text{ so the isotope is: } {}_{54}^{140}Xe$$

(b)
$$_{92}^{235}U + {}_0^1 n \longrightarrow {}_y^x ? + {}_{51}^{132}Sb + 3\ {}_0^1 n$$

$$235 + 1 = x + 132 + 3\times(1) = 236; \ 92 + 0 = y + 51 + 3\times(0)$$

$$x = 101, \text{ and } y = 41, \text{ so the isotope is: } {}_{41}^{101}Nb$$

(c)
$$_{92}^{235}U + {}_0^1 n \longrightarrow {}_y^x ? + {}_{56}^{141}Ba + 3\ {}_0^1 n$$

$$235 + 1 = x + 141 + 3\times(1) = 236; \ 92 + 0 = y + 56 + 3\times(0)$$

$$x = 92, \text{ and } y = 36, \text{ so the isotope is: } {}_{36}^{92}Kr$$

43. *Answer:* **6.9×10^3 barrels**

Strategy and Explanation: This is a typical conversion factor Question.

$$1.0 \text{ lb } {}^{235}U \times \frac{453.6 \text{ g}}{1 \text{ lb}} \times \frac{1 \text{ mol } {}^{235}U}{235 \text{ g U}} \times \frac{2.1 \times 10^{10} \text{ kJ}}{1 \text{ mol } {}^{235}U} \times \frac{1 \text{ barrel oil}}{5.9 \times 10^6 \text{ kJ}} = 6.9 \times 10^3 \text{ tons of coal}$$

Effects of Nuclear Radiation

45. *Answer/Explanation:* The unit "rad" is the measure of the amount of radiation absorbed. The unit "rem" includes a quality factor that better describes the biological impact of a radiation dose. The unit rem would be more appropriate when talking about the effects of an atomic bomb on humans. The unit gray (Gy) is 100 rad.

47. *Answer/Explanation:* Since most elements have some proportion of unstable isotopes that decay and we are composed of these elements (e.g., ^{14}C), our bodies emit radiation particles.

General Questions

49. *Answer:* (a) $_{-1}^0 e$ (b) $_2^4 He$ (c) $_{35}^{87}Br$ (d) $_{84}^{216}Po$ (e) $_{31}^{68}Ga$

Strategy and Explanation: Follow the method described in the answers to Questions 13 and 15.

(a) $\ _{83}^{214}Bi \longrightarrow {}_{-1}^0 e + {}_{84}^{214}Po \qquad 214 = \underline{0} + 214; \ \ 83 = \underline{-1} + 84$

(b) $4 \, {}^{1}_{1}H \longrightarrow {}^{4}_{2}He + 2 \, {}^{0}_{+1}e$ $4 \times (1) = \underline{4} + 2 \times (0); \quad 4 \times (1) = \underline{2} + 2 \times (1)$

(c) $\, {}^{249}_{99}Es + {}^{1}_{0}n \longrightarrow 2 \, {}^{1}_{0}n + {}^{87}_{35}Br + {}^{161}_{64}Gd$ $249 + 1 = 2 \times (1) + \underline{87} + 161; \quad 99 + 0 = 2 \times (0) + \underline{35} + 64$

(d) $\, {}^{220}_{86}Rn \longrightarrow {}^{216}_{84}Po + {}^{4}_{2}He$ $220 = \underline{216} + 4; \quad 86 = \underline{84} + 2$

(e) $\, {}^{68}_{32}Ge + {}^{0}_{-1}e \longrightarrow {}^{68}_{31}Ga$ $68 + 0 = \underline{68}; \quad 32 + (-1) = \underline{31}$

51. *Answer:* **2 mg**

Strategy and Explanation: Use the methods described in the solution to Question 27.

$$t = (1 \text{ h}) \times \left(\frac{60 \text{ min}}{1 \text{ h}} \right) = 60 \text{ min}$$

The elapsed time (60 min) is six times the half-life (10 min), so the quantity remaining will be reduced by half six times:

$$\frac{1}{2} \times \frac{1}{2} \times \frac{1}{2} \times \frac{1}{2} \times \frac{1}{2} \times \frac{1}{2} \times 96 \text{ mg} = 2 \text{ mg}$$

If the time elapsed were not a whole-number multiple of the half-life, it would be necessary to do the calculations:

$$k = \frac{\ln 2}{10 \text{ min}} = 7 \times 10^{-2} \text{ min}^{-1} \text{ (one sig fig)}$$

$$\ln \left(\frac{m}{m_0} \right) = -kt = -(7 \times 10^{-2} \text{ min}^{-1}) \times (60 \text{ min}) = -4 \text{ (one sig fig)}$$

$$m = m_0 e^{-4} = (96 \text{ mg}) \times e^{-4} = (96 \text{ mg}) \times (0.02) \cong 2 \text{ mg} \text{ (one sig fig)}$$

53. *Answer:* **3.6×10^9 y**

Strategy and Explanation: Adapt the methods described in the solution to Question 35.

Because the quantity ratio is not a multiple of $\frac{1}{2}$, we cannot use the quick method. Therefore, we will set up the calculations for k and t:

$$\frac{N}{N_0} = 0.951, \quad \ln \left(\frac{N}{N_0} \right) = -kt$$

$$k = \frac{\ln 2}{4.9 \times 10^{10} \text{ y}} = 1.4 \times 10^{-11} \text{ y}^{-1}$$

$$\ln(0.951) = -0.0502 = -(1.4 \times 10^{-11} \text{ y}^{-1})t$$

$$t = 3.6 \times 10^9 \text{ y}$$

Applying Concepts

57. *Answer/Explanation:* The ^{20}Ne isotope is stable. The ^{17}Ne isotope is likely to decay by positron emission, to increase the ratio of neutrons to protons. The ^{23}Ne isotope is likely to decay by beta emission to decrease the ratio of neutrons to protons. For more details, refer to the discussion in Section 19.3 and examine Figure 19.2.

59. *Answer/Explanation:* A nuclear reaction occurred, making products. Therefore, some of the lost mass is found in the decay particles, if the decay is alpha or beta decay, and almost all the rest is found in the element produced by the reaction.

More Challenging Questions

61. *Answer/Explanation:* All radioactive decays are first order because, to occur, there needs to be one and only one reactant. The reaction involves only the nucleus of an atom.

63. *Answer:* 2.8×10^4 kg

Strategy and Explanation: Use the methods described in Section 19.3. Calculate the amount of energy that must be produced in an hour to provide 7.0×10^{14} kJ/s, then use $E = (\Delta m)c^2$ to determine the mass of solar material.

$$1 \text{ hour} \times \frac{3600 \text{ s}}{1 \text{ hr}} \times \frac{7.0 \times 10^{14} \text{ kJ}}{1 \text{ s}} = 2.5 \times 10^{18} \text{ kJ}$$

$$\Delta m = \frac{E}{c^2} = \frac{2.5 \times 10^{14} \text{ kJ} \times \dfrac{1000 \text{ J}}{1 \text{ kJ}}}{(2.99792 \times 10^8 \text{ m/s})^2 \times \dfrac{1 \text{ J}}{1 \text{ kg m}^2 \text{ s}^{-2}}} = 2.8 \times 10^4 \text{ kg of solar material}$$

65. *Answer:* **75.1 y**

Strategy and Explanation: Adapt the methods described in the solution to Question 33.

$$N_0 = 100\%, N = 1.45\%$$

The ratio of percentage activity is not a multiple of $\dfrac{1}{2}$, so we can't use the quick method. Therefore, set up the calculation for k and t.

$$k = \frac{\ln 2}{12.3 \text{ y}} = 0..0564 \text{ y}^{-1}$$

$$\ln\left(\frac{1.45 \%}{100 \%}\right) = -4.234 = -(0..0564 \text{ y}^{-1})t$$

$$t = 75.1 \text{ y}$$

67. *Answer:* **3.92×10^3 y**

Strategy and Explanation: Use methods described in Chapter 3 to determine the mass of carbon, then calculate the disintegrations per minute per gram (d min^{-1} g^{-1}) of carbon, and use the method described in the solution to Question 35.

$$1.14 \text{ g CaCO}_3 \times \frac{1 \text{ mole CaCO}_3}{100.087 \text{ g CaCO}_3} \times \frac{1 \text{ mole C}}{1 \text{ mole CaCO}_3} \times \frac{12.0107 \text{ g C}}{1 \text{ mole C}} = 0.137 \text{ g C}$$

$$k = \frac{\ln 2}{5730 \text{ y}} = 1.21 \times 10^{-4} \text{ y}^{-1}$$

$$A = \frac{\dfrac{2.17 \times 10^{-2} \text{ dis}}{\text{s}} \times \dfrac{60 \text{ s}}{1 \text{ min}}}{0.137 \text{ g C}} = 9.52 \text{ d min}^{-1} \text{ g}^{-1} \quad \text{and} \quad A_0 = 15.3 \text{ d min}^{-1} \text{ g}^{-1}$$

$$\ln\left(\frac{9.52 \text{ d min}^{-1} \text{ g}^{-1}}{15.3 \text{ d min}^{-1} \text{ g}^{-1}}\right) = -0.474 = -(1.21 \times 10^{-4} \text{ y}^{-1})t$$

$$t = 3.92 \times 10^3 \text{ y}$$

Appendix A: Problem Solving and Mathematical Operations

Appendix Contents:

Introduction

This appendix is designed to give you some help with basic concepts of problem solving and mathematics. The methods described in this appendix are very basic, but also very important. If you are struggling with mastering these things, it might be worthwhile to see about finding a tutor to help you solidify your basic skills.

The skills described in the first several sections will be important immediately. The subject of logarithms is used extensively in Chapter 12. Quadratic equations will be used in some calculations in Chapters 13, 15, and 16. Graphing will be very important in Chapters 10 and 12.

Solutions to Blue-Numbered Questions in Appendix A

Numbers, Units, and Quantities

4. *Answer/Explanation:* If the units for the answer to a calculation do not make sense, then we must conclude that the solution to the problem contains one or more errors. Faced with a situation like this, we must review the plan and its execution to find out where errors were made. Check for incomplete unit conversions (e.g., cm^3 to L, g to kg, mm Hg to atm, etc.) and check to see that the numerators and denominators of unit factors are placed such that the unwanted units can cancel (e.g., g/mL or mL/g).

6. *Answer:* **(c); see explanation below.**

Strategy and Explanation: We can immediately eliminate choice (d) since it has no units at all and the quantity measured must be reported with associated units. We can eliminate (a) and (b), since the units of this measured quantity will not be feet or meters. The answer provided in (c) looks like a suitable way to record the measurement of 8 inches and 6.1 sixteenths, because $6.1 \times 0.0625 = 0.38$, the decimal part of the reported quantity. The answer in (e) may also be used to describe the length of the pencil, but it should not be used to describe the physical quantity determined from this measurement, because the measuring device is calibrated in inches and fractions of inches, not in feet. The answer to (e) is the result of a conversion calculation, not the observed measurement.

8. *Answer:* **(a) units should not be squared; 9.95 g (b) the conversion factor is upside-down; 1.07 mL (c) the conversion factor needs to be cubed and the units of the answer should be cm^3; 9.26×10^6 cm**

Strategy and Explanation: Given flawed calculations, determine what is wrong and show the correct calculation and result for each.

Review the calculation and determine what is wrong. Fix what is wrong and redo the calculation to get the answer.

(a) The sum of two quantities with the same units will also have those units. The calculation shown indicates that two masses added together have units of mass squared. That is incorrect. The correct calculation looks like this: 7.86 g – 5.63 g = 2.23 g

(b) The conversion shown looks like it was intended to calculate the mass of a sample from the volume. If that is the case, then the conversion factor used is upside-down and the resulting units should be grams. The correct calculation looks like this:

$$5.23 \text{ g} \times \frac{1.00 \text{ mL}}{4.87 \text{ g}} = 1.07 \text{ mL}$$

(c) The conversion shown looks like it was intended to convert cubic centimeters to cubic meters. If that is the case, then the conversion factor needs to be cubed and the final units need to be in cubic centimeters. The correct calculation looks like this:

$$3.57 \text{ cm}^3 \times \left(\frac{1 \text{ m}}{100 \text{ cm}}\right)^3 = 3.57 \times 10^{-6} \text{ m}^3$$

✓ *Reasonable Answer Check:* The units cancel properly and the size of the result of each calculation makes sense. In (a), when mass is removed, the resulting mass will be smaller and in units of mass. In (b), when the density is greater than 1, the volume quantity will have a smaller numerical value than the mass quantity. In (c), the number of cubic meters will be much smaller than the number of cubic centimeters, since a meter is much longer than a centimeter.

Precision, Accuracy, and Significant Figures

10. *Answer:* **(a) 95.9 ±0.59 in (b) three (c) It is accurate.**

Strategy and Explanation: Given several measured quantities for the length of a pole, determine what should be reported as the length, determine the number of significant figures to be reported, and, given the actual length, determine whether the result is accurate.

(a) Add up the five individual measurements and divide by five to get the average:

$$\text{Average length of the pole} = \frac{95.31 \text{ in} + 96.44 \text{ in} + 96.02 \text{ in} + 95.78 \text{ in} + 95.94 \text{ in}}{5} = \frac{479.49 \text{ in}}{5} = 95.898 \text{ in}$$

The scatter includes measurements lower than the average by as much as (95.898 – 95.31 =) 0.59 inches and higher than the average by as much as (96.44 – 95.898 =) 0.54 inches. So, the pole is reported to be 95.9 ± 0.59 inches long.

(b) The result should be reported with three significant figures, since the uncertainty is found in the first decimal place.

(c) 95.90 ± 0.59 inches is 7.99 ± 0.049 feet. The pole's actual height is 8 feet. The correct height is within the range of the uncertainty, so the result is accurate, though not very precise.

12. *Answer:* **(a) Four (b) Two (c) Two (d) Four**

Strategy and Explanation: Given several measured quantities, determine the number of significant figures.

Use rules given in Sections 2.4 and A.3, summarized here: All non-zeros are significant. Zeros that precede (sit to the left of) non-zeros are never significant (e.g., 0.003). Zeros trapped between non-zeros are always significant (e.g., 3.003). Zeros that follow (sit to the right of non-zeros may be significant or may not be significant. They are significant if a decimal point is explicitly given (e.g., 3300.) and not significant, if a decimal point is not specified (e.g., 3300).

(a) 3.274 has **four** significant figures (The 3, 2, 7, and 4 digits are each significant.)

(b) 0.0034 L has **two** significant figures (The 3 and the 4 digits are each significant. The zeros are all before the first non-zero-digit, 3, and therefore they are not significant.)

(c) 43,000 m has **two** significant figures (The 4 and the 3 digits are significant. The zeros after the 3 are not significant because the number does not have an explicit decimal point specified.)

(d) 6200. mL has **four** significant figures (The 6 and 2 digits are each significant. The zeros after the 2 are also significant because the number has an explicit decimal point specified.)

✓ *Reasonable Answer Check:* The significant figures rules have been properly applied.

14. *Answer:* **(a) 43.32 (b) 43.32 (c) 43.32 (d) 43.32 (e) 43.32 (f) 43.32**

Strategy and Explanation: Given several measured quantities, round them to four significant figures. Use rules for rounding given in Sections 2.4 and A.3, as summarized here. If the last digit is below 5, then rounding does not change the digit before it. If the last digit is above a five, the digit before it is made one larger. If the last digit is exactly five, round the digit before it to an even number, up if odd and down if even.

(a) 43.3250 has six significant figures. To round it to four significant figures, we need to examine the fifth significant figure, the 5. Since the digit we're rounding is a 5, we look at the number before the 5, which is 2. Since 2 is even, we will round down and leave the 2 unchanged: **43.32**.

(b) 43.3165 has six significant figures. To round it to four significant figures, we need to look at the fifth significant figures, the 6. The removal of the fifth digit 6 rounds up the 1 next to it to a 2. The result is **43.32**.

(c)   43.3237 has six significant figures. To round it to four significant figures, we need to look at the fifth significant figures, the 3.   The removal of 3 does not change the 2 next to it. The result is **43.32**.

(d)   43.32499 has seven significant figures. To round it to four significant figures, we need to look at the fifth significant figures, the 4.   The removal of 4 doesn't change the 2 next to it. The result is **43.32**.

(e)   43.3150 has six significant figures. To round it to four significant figures, we need to look at the fifth significant figures, the 5. Since the digit we're rounding is a 5, we look at the number before the 5, which is 1. Since 1 is odd, we will round up and change the 1 to a 2: **43.32**.

(f)   43.32501 has seven significant figures. To round it to four significant figures, we need to examine the fifth significant figure, the 5. Since the digit we're rounding is a 5, we look at the number before the 5, which is 2. Since 2 is even, we will round down and leave the 2 unchanged: **43.32**.

✓ *Reasonable Answer Check:* Each answer has four significant figures.

16. *Answer:* **(a) 13.7 (b) 0.247 (c) 12.0**

Strategy and Explanation: Given some numbers combined using calculations, determine the result with proper significant figures.

Perform the mathematical steps according to order of operations, applying the proper significant figures (addition and subtraction retains the least number of decimal places in the result; multiplication and division retain the least number of significant figures in the result). *Notice: if operations are combined that use different rules, it is important to stop and determine the intermediate result any time the rule switches.*

(a)
$$\frac{4.47}{0.3260}$$

The ratio uses the division rule. The numerator has three significant figures and the denominator has four significant figures, so the answer will have three significant figures. Therefore, we get **13.7**.

(b)
$$\frac{4.03 + 3.325}{29.75}$$

The numerator uses the addition rule. The first number has two decimal places (the 4 and the 7 are both decimal places -- digit that follow the decimal point to the right) and the second number has three decimal places (the 3, the 2, and the 5 are all decimal places), so the result of the addition has two decimal places.

$$\frac{7.355}{29.75} \cong \frac{7.36}{29.75}$$

The ratio uses the division rule. The numerator has three significant figures and the denominator has four significant figures, so the answer will have three significant figures. Therefore, we get **0.247**.

(c)
$$\frac{8.234}{5.673 - 4.987}$$

The denominator uses the subtraction rule. The first number has three decimal places (the 6, the 7, and the 3 are all decimal places) and the second number has three decimal places (the 9, the 8, and the 7 are all decimal places), so the result of the subtraction has three decimal places.

$$\frac{8.234}{0.686}$$

The ratio uses the division rule. The numerator has four significant figures and the denominator has three significant figures, so the answer will have three significant figures. Therefore, we get **12.0**.

✓ *Reasonable Answer Check:* The proper significant figures rules were used. The size and units of the answers are appropriate.

Exponential or Scientific Notation

18. *Answer:* **(a) 7.6003 × 10⁴ (b) 3.7 × 10⁻⁴ (c) 3.4 × 10⁴**

Strategy and Explanation: Adjust the appearance of the number to scientific notation (i.e., write it as a number between 1 and 9.999…, that is multiplied by ten to a whole-number power).

(a)   76,003 is $7.6003 \times 10,000$ or 7.6003×10^4

(b)   0.00037 is 3.7×0.0001 or 3.7×10^{-4}

(c)   34,000 is $3.4 \times 10,000$ or 3.4×10^4

20. *Answer:* **(a)** $\mathbf{2.415 \times 10^{-3}}$ **(b)** $\mathbf{2.70 \times 10^8}$ **(c)** **3.236** **(d)** **116 cm^3**

Strategy and Explanation: Given some numbers combined using calculations, determine the result with proper significant figures.

Perform the mathematical steps according to order of operations, applying the proper significant figures (addition and subtraction retains the least number of decimal places in the result, or rounds the result to the position of greatest uncertainty; multiplication and division retain the least number of significant figures in the result).

(a)
$$\frac{0.7346}{304.2}$$

The ratio uses the division rule. The numerator has four significant figures and the denominator has four significant figures, so the answer will have four significant figures. Therefore, we get $\mathbf{2.415 \times 10^{-3}}$.

(b)
$$\frac{(3.45 \times 10^{-3})(1.83 \times 10^{12})}{23.4}$$

The calculation uses the division and multiplication rules. The numbers in the numerator have three significant figures and the denominator has three significant figures, so the answer will have three significant figures. Therefore, we get $\mathbf{2.70 \times 10^8}$.

(c)
$$3.240 - 4.33 \times 10^{-3}$$

The calculation uses the subtraction rule. The first number has three decimal places (the 2, the 4, and the 0 are all decimal places) and the second number, 4.33×10^{-3} is also 0.00433, so it has five decimal places (the zeros after the decimal point, the 4, the 3, and the 3 are all decimal places), so the result of the subtraction has three decimal places.

$$3.240 - 0.00433 = \mathbf{3.236}$$

(d)
$$(4.87 cm)^3 = 4.87 cm \times 4.87 cm \times 4.87 cm$$

The calculation uses the multiplication rule. The numbers have three significant figures, so the answer will have three significant figures. Therefore, we get **116 cm^3**.

✓ *Reasonable Answer Check:* The proper significant figures rules were used. The size and units of the answers are appropriate.

Logarithms

22. *Answer:* **(a)** **−0.1351** **(b)** **3.541** **(c)** **23.7797** **(d)** **54.7549** **(e)** **−7.455**

Strategy and Explanation: Given some numbers and calculations using logarithms, determine the answers with appropriate significant figures.

Use the information in Section A.6 describing the operations of logarithms and their significant figures. When taking the log of a number, the mantissa (the digits to the right of the decimal point) should have as many significant figures as the numbers whose log was found. So, the number of significant figures in the number we start with gives the number of decimal places in the result.

(a)   The log of 4 significant figures, gives an answer with 4 decimal places: $\log(0.7327) = -0.1351$

(b)   The ln of 3 significant figures, gives an answer with 3 decimal places: $\ln(34.5) = 3.541$

(c)   The log of 4 significant figures, gives an answer with 4 decimal places: $\log(6.022 \times 10^{23}) = 23.7797$

(d)   The ln of 4 significant figures, gives an answer with 4 decimal places: $\ln(6.022 \times 10^{23}) = 54.7549$

(e)   The ratio of 3 significant figures gives a result with 3 significant figures. The log of 3 significant figures, gives an answer with 3 decimal places:

$$\log\left(\frac{8.34\times10^{-5}}{2.38\times10^3}\right)=\log(3.50\times10^{-8})=-7.455$$

✓ *Reasonable Answer Check:* For log calculations, the power of ten is approximately reflected in the ordinate (the number to the right of the decimal point), as in (e), where we see –7.xx when the number was a little larger than 10^{-8}. For ln calculations, the power of ten will be approximately reflected as the ordinate times 2.303. The answer in (d) should be 2.303 times the answer in (c). The ratio is 2.3034 times.

24. *Answer:* (a) **5.404** (b) **9×10^{14}** (c) **110.** (d) **1.000320** (e) **3.75**

Strategy and Explanation: Given some numbers and calculations using antilogarithms, determine the answers with appropriate significant figures.

Use the information in Section A.6 describing the operations of antilogarithms and their significant figures. When taking the antilog of a number, the number of decimal places in the number whose antilog was taken should be the same as the number of significant figures in the result of the antilogarithm operation.

(a) The antilog of 4 decimal places, gives an answer with 4 significant figures: antilog(0.7327) = 5.404

(b) The antiln of 1 decimal place, gives an answer with 1 significant figure: $\ln(34.5) = 9\times10^{14}$

(c) The 10^x of 3 decimal places, gives an answer with 3 significant figures: $10^{2.043} = 110.$

(d) The number 3.20×10^{-4} is also 0.000320. Therefore, it has 6 decimal places. The e^x of 6 decimal place, gives an answer with 6 significant figure: : $e^{0.000320} = 1.000320$

(e) The ratio of 4 significant figures gives a result with 4 significant figures. The exp of 3 decimal places, gives an answer with 3 significant figures:

$$\exp\left(\frac{4.333}{3.275}\right)=\exp(1.323)=3.75$$

✓ *Reasonable Answer Check:* Antilogarithms of small numbers (i.e., numbers near zero) give answers near one. Antilog on larger numbers, give answers with large powers of ten.

Quadratic Equations

26. *Answer:* (a) **0.480 and –1.80** (b) **3.23 and 1.34**

Strategy and Explanation: Given quadratic equations, find the roots.

Set up the quadratic equation in the form of: $ax^2 + bx + c = 0$, then plug into the quadratic equation:

$$x = \frac{-b\pm\sqrt{b^2-4ac}}{2a}$$

The two roots are generated when we use the "+" or the "–" of the "±" in this equation.

(a) $3.27x^2 + 4.32x - 2.83 = 0$, so, a = 3.27, b = 4.32, and c = –2.83:

$$x = \frac{-4.32\pm\sqrt{4.32^2-4(3.27)(-2.83)}}{2(3.27)}$$

$$x = \frac{-4.32+7.46}{6.54} = \frac{3.14}{6.54} = 0.480$$

$$x = \frac{-4.32-7.46}{6.54} = \frac{-11.78}{6.54} = -1.80$$

(b) Rearrange: $x^2 + 4.32 = 4.57x$

$x^2 - 4.57x + 4.32 = = 0$, so, $a = 1$, $b = -4.57$, and $c = 4.32$:

$$x = \frac{-(-4.57) \pm \sqrt{(-4.57)^2 - 4(1)(4.32)}}{2(1)}$$

$$x = \frac{4.57 + 1.90}{2} = \frac{6.47}{2} = 3.23$$

$$x = \frac{4.57 - 1.90}{2} = \frac{2.67}{2} = 1.34$$

✓ *Reasonable Answer Check:* Each root should solve the equation:

(a) $3.27(0.480)^2 + 4.32(0.480) - 2.83 = 0.753 + 2.07 - 2.83 = 0.00$

$3.27(-1.80)^2 + 4.32(-1.80) - 2.83 = 10.6 - 7.78 - 2.83 = 0.0$

(b) $(3.23)^2 + 4.32 = 10.4 + 4.32 = 14.8$ matches $4.57(3.23) = 14.8$

$(1.34)^2 + 4.32 = 1..80 + 4.32 = 6.12$ matches $4.57(1.34) = 6.12$

Graphing

28. *Answer:* **see graph below; yes**

Strategy and Explanation: Given the masses of several volumes of a sample, draw the graph and determine if mass is directly proportional to volume.

Plot the graph and see if the data fits a linear equation.

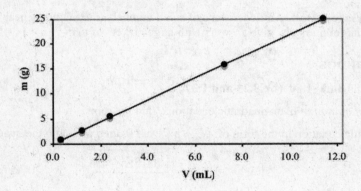

The graph is linear so the mass is directly proportional to volume.

✓ *Reasonable Answer Check:* Mass and volume are extensive properties that are related to each other through a substances fixed density, so it makes sense that the graph is linear, since we expect the mass is directly proportional to volume.

Appendix B: Units, Equivalences, and Conversion Factors

Introduction

This appendix reviews the international system of units, which are the standard units of science. It can serve as a supplement to the problems solving that is introduced in the first chapter and used throughout the entire book.

Conversion of units is something that you will be required to do in nearly every chapter. If you have difficulties understanding conversions, you may want to consider finding a tutor.

Solutions to Blue-Numbered Questions in Appendix B

Units of the International System

1. *Answer/Explanation:* Determine which SI units and prefix would be used to measure a mass, a volume, and a thickness. We use the Tables B.1-B.3.

    (a) Mass would use SI unit of **kilogram**. The mass of the book could be measured in kilograms, so no prefix would be needed.

    (b) Volume would use SI unit of **cubic meters**, m^3. The volume of a glass of water could be measured in cubic centimeters, cm^3 which also milliliters, mL. $1000000\ cm^3$ is a m^3; therefore, the measurement using SI units would need the prefix **nano-**.

    (c) Thickness is a unit of length and the SI unit is **meter**, m. The thickness of this page could be measured in **milli**meters, mm.

3. *Answer/Explanation:* SI base (fundamental) units are set by convention; derived units are derived from the SI base fundamental units.

Conversion of Units for Physical Quantities

5. *Answer:* **(a) 4.75×10^{-10} m (b) 5.6×10^7 kg (c) 4.28×10^{-6} A**

    Strategy and Explanation: Given several measured quantities, express it in SI base units.

    (a) pico- is 10^{-12}, so 475 pm is 475×10^{-12} m or 4.75×10^{-10} m, the SI base unit of length.

    (b) giga- is 10^9, so 56 Gg is 56×10^9 g or 5.6×10^{10} g, which is 5.6×10^7 kg, the SI base unit of mass.

    (c) micro- is 10^{-6}, so 4.28 µA is 4.28×10^{-6} A, the SI base unit of electric charge.

7. *Answer:* **(a) 8.7×10^{-18} m^2 (b) 2.73×10^{-17} J (c) 2.73×10^{-5} N**

    Strategy and Explanation: Given several measured quantities, express it in SI base units.

    (a) nano- is 10^{-9}, so 8.7 nm^2 is $8.7 \times (10^{-9})^2$ m^2 or 8.7×10^{-18} m^2, area using the SI base unit of length.

    (b) atto- is 10^{-18}, so 27.3 aJ is 27.3×10^{-18} J or 2.73×10^{-17} J, the SI base unit of energy.

    (c) micro- is 10^{-6}, so 27.3 µN is 27.3×10^{-6} N or 2.73×10^{-5} N, the SI derived unit for force.

9. *Answer:* **(a) 2.20 pounds (b) 2.22×10^3 kg (c) 60 in^3 (d) 1×10^5 pascal, 1 bar**

    Strategy and Explanation: Express quantities in units given with scientific notation.

    (a) Table B.4 gives: 1 pound = 0.45359 kg, so: $1.00\ kg \times \dfrac{1\ pound}{0.45359\ kg} = 2.20\ pounds$

    (b) Table B.4 gives: 1 short ton = 907.2 kg, so: $2.45\ ton \times \dfrac{907.2\ kg}{1\ ton} = 2.22 \times 10^3\ kg$

(c) Table B.4 gives: $1 \text{ L} = 1000 \text{ cm}^3$ and $1 \text{ inch} = 2.54 \text{ cm}$ so: $1 \text{ L} \times \dfrac{1000 \text{ cm}^3}{1 \text{ L}} \times \left(\dfrac{1 \text{ in}}{2.54 \text{ cm}}\right)^3 = 6 \times 10^1 \text{ in}^3$

(d) Table B.4 gives: $1 \text{ atm} = 1.01325 \times 10^5$ pascals and $1 \text{ bar} = 10^5$ pascals, so:

$$1 \text{ atm} \times \dfrac{1.01325 \times 10^5 \text{ pascals}}{1 \text{ atm}} = 1 \times 10^5 \text{ pascals} \qquad \text{and} \qquad 1 \times 10^5 \text{ pascals} \times \dfrac{1 \text{ bar}}{10^5 \text{ pascals}} = 1 \text{ bar}$$

11. *Answer:* **(a) 99° F (b) -- 10.5 °F (c) -- 40.0 °F**

Strategy and Explanation: Use temperature conversion given in Table B.4: $t_F = \dfrac{9}{5} t_C + 32$

(a) $t_F = \dfrac{9}{5}(37\ ^\circ\text{C}) + 32 = 99\ ^\circ\text{F}$ (b) $t_F = \dfrac{9}{5}(-23.6\ ^\circ\text{C}) + 32 = -10.5\ ^\circ\text{F}$ (c) $t_F = \dfrac{9}{5}(-40.0\ ^\circ\text{C}) + 32 = -40.0\ ^\circ\text{F}$

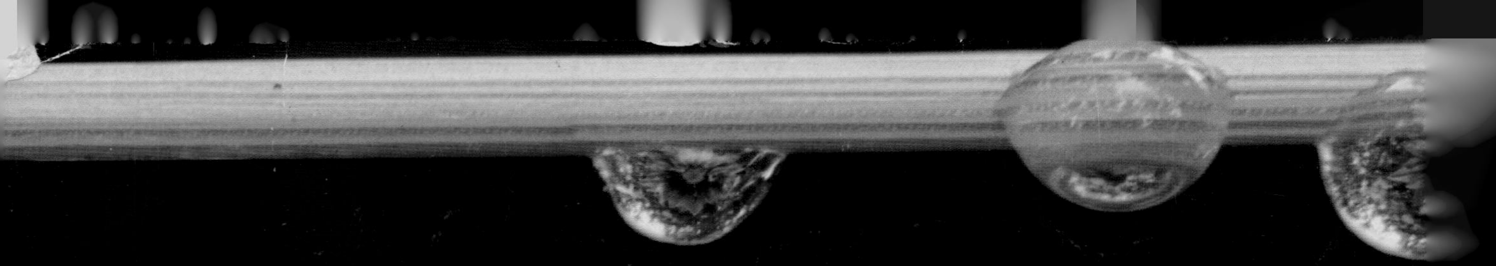

www.cengage.com/chemistry

ISBN-13: 978-0-495-39158-6
ISBN-10: 0-495-39158-1

BROOKS/COLE
CENGAGE Learning

To learn more about Brooks/Cole, visit **www.cengage.com/brookscole**

Purchase any of our products at your local college store or at our preferred
online store **www.ichapters.com**